MW01119032

TEACHING MATHEMATICS

second edition

TEACHING MATHEMATICS
METHODS AND CONTENT

FREDRICKA K. REISMAN University of Georgia

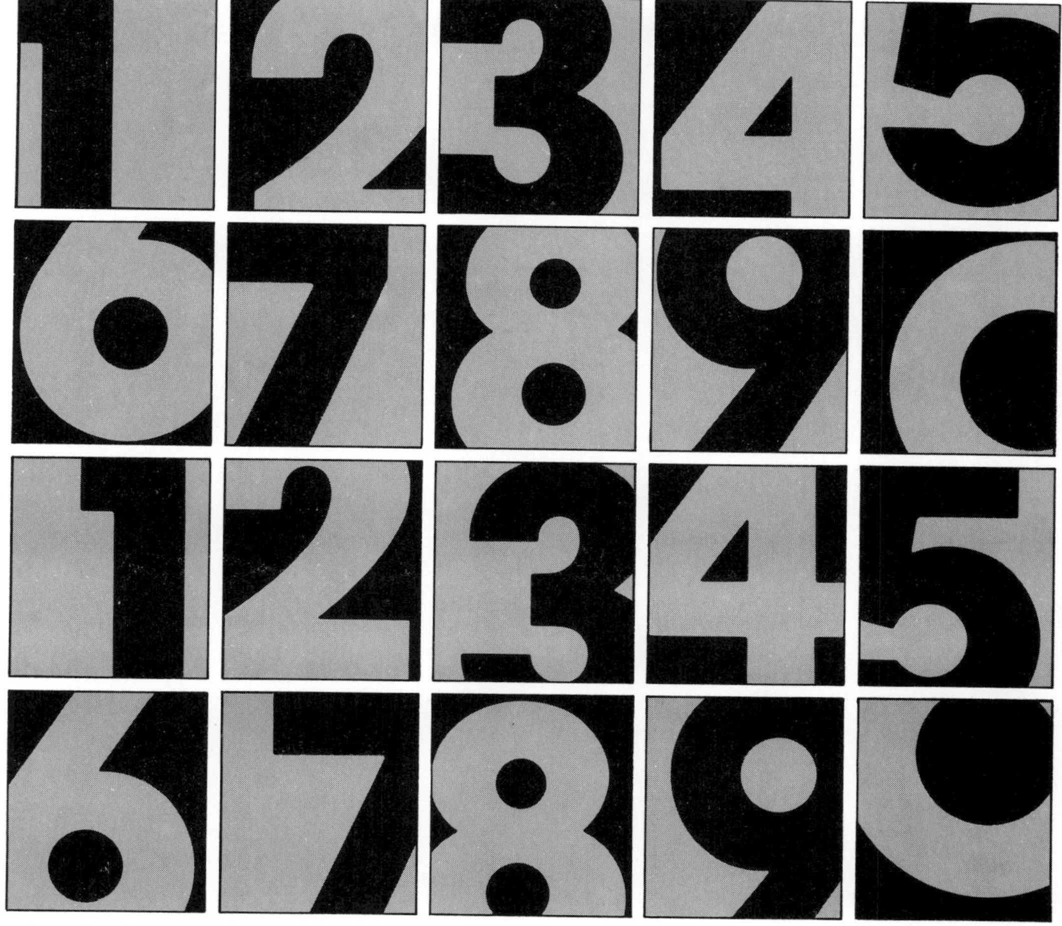

HOUGHTON MIFFLIN COMPANY BOSTON
DALLAS GENEVA, ILLINOIS HOPEWELL, NEW JERSEY PALO ALTO LONDON

To Lisa, Rodger, and Dad

B. Othanel Smith, *Advisory Editor*

Printed in the U.S.A.

Library of Congress Catalog Card Number: 80-50976

ISBN: 0-395-30706-6

Contents

Editor's Foreword

The first edition of this book emphasized diagnostic instruction. This approach has been retained in the present edition. It is treated so as to enable teachers to understand and gain a feeling for diagnosis by themselves doing the sorts of problems their students will be expected to work; to note their own mistakes, how they were made, and what to do to correct them. A new and significant addition is a list of student difficulties for the teacher to watch for as each new set of operations is taken up. These will prove invaluable as the teacher assesses each student's progress.

Much research has been done on the teaching of arithmetic since the first edition of this book appeared. Most of this research supports direct teaching where students are kept on task and objectives are clearly stated and understood by students; where the teacher engages the student in exercises and checks to see that they are properly done, and where the teacher identifies errors students make and then helps them make corrections. This new approach has been integrated with diagnostic and heuristic modes of instruction and thus synthesized into a humane and effective approach to mathematics education.

The historical materials contained in the first edition have been eliminated in the interest of brevity. While set theory is not as prominent in this new edition, its essential elements have been retained and woven into various mathematical processes so as to help the student understand and perform the operations themselves.

The importance of mathematics in the elementary curriculum is crucial to the further education of the student at all levels of subsequent study. This fact underlies the call today for more thorough preparation in the fundamentals of arithmetic. The reader will find in this new edition a renewed emphasis on the need for teachers to understand the mathematics they teach, to use effective techniques and procedures of instruction, and to maintain an abiding interest in their students.

B. Othanel Smith

Preface

This new edition of TEACHING MATHEMATICS reflects a more direct presentation of the fundamentals of instruction than was true of the earlier DIAGNOSTIC TEACHING OF ELEMENTARY SCHOOL MATHEMATICS. Various tasks are analyzed for their main goals, for by understanding tasks in this way, teachers can more readily become aware of underlying reasons for students errors.

Instructional issues, based on the research literature, are considered throughout the book. These include consideration of the rates of development of children in cognitive, affective, and visual-motor areas, and mistakes made by children because of gaps in their mathematical knowledge. Mathematical content or information related to factors that affect mathematics learning is also given in each chapter.

Fundamentals of mathematics instruction are presented along with theory in the 80 mini-lessons, each of which begins with a broad goal, key vocabulary, and objectives for understanding the goal. "Possible Trouble Spots" also appear in each mini-lesson as a diagnostic aid to understanding student errors. The section of each mini-lesson entitled "Additional Considerations" serves three main functions. These include all or some of the following:

1. Summary statements, made to draw together the "Possible Trouble Spots" with the activities
2. Diagnostic tasks
3. Text material as related to the mini-lesson goal or activities

The boxed diagrams at the end of the mini-lessons refer to the goals in the lessons. The top box gives the main goal; the two boxes below state the instructional objectives that might be taught next. These objectives and those given in the numbered order of the scope and sequence chart in Appendix B illustrate the flexibility in the text available to teachers in preparing mathematics lessons.

The scope-and-sequence chart in Appendix B is designed to serve as a road map through a basal mathematics text book, with its emphasis on the school year, or grade. This focus on grades from kindergarten to eighth (sequence) is viewed in terms of mathematical strands (scope), which include integers, rationals, decimals, and geometry. The numbering system of the chart identifies a sequential development of strands at a particular grade and allows for easy use of both a basal text and mathematics activities according to students' educational needs. The chart indicates the grade, the level within the grade, and the number of the objective to be accomplished by both grade and level. This numbering system allows for differential instruction, meaning that groups of students within a classroom may move at different rates through the objectives as numbered. This system is particularly useful for teachers whose students need either more practice or exposure to broader and more complex mathematics. The task analysis nature of this numbering system also accommodates the special educational needs of slower learning students who may be mainstreamed into a regular classroom. For these students, the teacher may select objectives that are of most use as self-help skills, such as shopping for food and clothing, telling time, and simple budgeting. On the other hand, for gifted students, objectives in the geometry or graphs and functions strands would probably be utilized to a greater degree.

Chapter 1 sets the tone of the book as a reconciliation of the strengths of the modern mathematics era with those of back-to-basics. Generic influences on learning mathematics are proposed as an alternative to cognitive stages so that learning characteristics can be diagnosed. Chapters 2 through 7 deal with addition and multiplication of whole numbers; Chapters 8 through 10 focus on integers and the subtraction operation; Chapters 11 and 12 discuss division. Prime and composite whole numbers are presented in Chapter 13. Chapter 14 introduces rational numbers and sets the stage for renaming fractions in Chapter 15 and for fraction computations in Chapter 16. Decimals are taught in Chapters 17 and 18. Chapter 19 addresses rounding whole numbers, estimation, simplifying expressions, and exponents. Geometries are presented in Chapters 20, 21, and 22. Chapter 23 deals with mathematical applications, including percent, probability, statistics, and graphs. Solving word problems is the subject matter of Chapter 24. Appendix A comprises the Answers to Exercises followed by the Scope and Sequence Chart in Appendix B.

The author wishes to thank the following reviewers for their help in developing this second edition: Charles Lamb of the University of Texas at Austin, Randell Drum of Corpus Christi State University, Preston Feden of LaSalle College, Tiny Goodman of Central Missouri State University, Marilyn Gutmann of Wayne State University, David O'Neil of Georgia State University, and George Willson of North Texas State University.

Thanks are also due to members of the editorial staff: Charles Heinle, Erek Smith, Lorraine Wolf, Iris Rothstein, Bernice Rappoport; and my secretarial staff: Gale Dorough, Wanda Callaham and Paulette Bone.

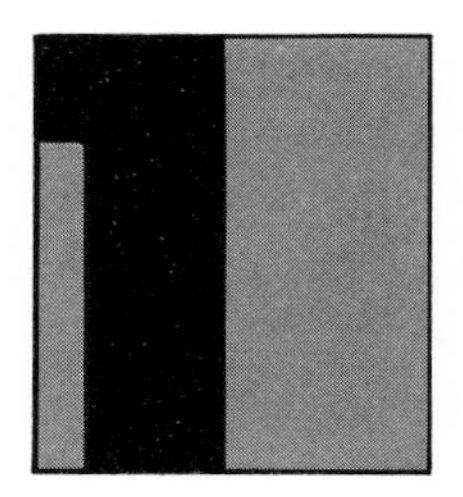

Reconciling Modern Math, Back-to-Basics, and Teaching Mathematics Today

Back-to-basics should not mean back-to-rote. It is important that today's mathematics teachers interpret "back-to-basics" in a broader sense than drilling on computations. For example, 10 basic skill areas were cited by the National Council for Teachers of Mathematics.[1]

Today's basic skill areas

1. Problem solving
2. Applying mathematics in everyday situations
3. Alertness to the reasonableness of results
4. Estimation and approximation

---

1. *An Agenda for Action: Recommendations for School Mathematics of the 1980s* (Reston, Va.: National Council for Teachers of Mathematics, 1980), p. 6.

5. Appropriate computational skills
6. Geometry
7. Measurement
8. Reading; interpreting; and constructing tables, charts, and graphs
9. Using mathematics to predict
10. Computer literacy

Today's instructional activities

These 10 basic skill areas can be attained with activities such as the following:[2]

1. Locating and processing quantitative information
2. Collecting, organizing, presenting, and interpreting data
3. Drawing inferences and predicting from data
4. Estimating measures and using appropriate tools for measuring
5. Mentally estimating results of calculations
6. Calculating with numbers rounded to one or two digits
7. Using technological aids to calculate
8. Using ratio and proportion to deal with rate problems in general and with percent problems in particular
9. Using imagery, maps, sketches, and diagrams as aids to visualizing and conceptualizing a problem
10. Using concrete representations and puzzles that aid in improving the perception of spatial relationships

There will most probably be a movement away from engaging students in tasks that can be done faster and with consistent accuracy by calculators, such as paper-pencil calculations of more than two digits. Also, there will be a de-emphasis on mastering highly specialized vocabulary that is not useful in daily living (for example, some of the language and symbols dealing with sets). There will be a de-emphasis on converting from one system of measures to another (for example, English to metric, just as when learning French or Spanish, students are encouraged not to translate but rather to think in the language being learned).

The results of the second National Assessment of Educational Progress (NAEP) pointed out that students can compute but that they do not necessarily know when to use various computations to solve mathematics problems. In fact, competence in problem solving was found to be at a very low point.

So where do we go from here? What is a classroom teacher to do?

1. *Do you plan your curriculum on Piagetian stage theory?* Many educators base selection of mathematics topics and methods of teaching on whether a student is at the concrete operational or the formal

2. *An Agenda for Action,* p. 7.

operational stage of cognitive development. This has been useful to some degree, but there are many students who do not fit into Piaget's stage theory.

2. *Do you teach page by page from your basal textbook?* This is unsatisfactory for several reasons:

a. There are usually too few practice activities to allow students to consolidate their learning before moving on to the next topic.
b. The sequence of topics is not usually arranged developmentally and gaps in learning result. These gaps are the cause of trouble later in the students' mathematics experiences.
c. There is often no control for reading level difficulties. For example, many third grade basal series books are written at too difficult a reading level for most of the third graders in the class.
d. Moving through a text page by page does not allow for different rates of learning.
e. There is a tendency to emphasize paper-pencil tasks and to have too little activity at the concrete, manipulative level.
f. Word problems are stilted and artificial and therefore inhibit the development of problem-solving skills. In fact, problem solving in this sense is very narrow and offers no bridge to solving real world problems.

3. *Do you form three or four (or more) mathematics groups and teach one group at a time?* The latest research shows that grouping loses its effectiveness when the majority of students are off task as the teacher works with one group.

4. *Do you employ an individualized mathematics program where students are independently working their way through a series of booklets?* This approach has questionable effectiveness. Think of the social and communication experiences that are lost to students who have to interact mainly with workbooks instead of people.

5. *Do you teach the entire class at a time and ignore individual students' needs?* This instructional classroom organization, discarded in earlier times, is being reviewed to increase students' time on task.

6. *Do you emphasize a guided discovery method of instruction?* Students can become so frustrated with this approach that they say, "I wish my instructor would tell me what it is I am supposed to discover, and I'll discover it!"

7. *Are you completely didactic?* Do you think creative problem-solving is only for "gifteds" who have their heads in the clouds? Have you ever engaged in cognitive monitoring?

How *do* you teach mathematics? Can you *teach* it, or can you only arrange the instructional situation so that the student can *learn?* There are whole bodies of literature that support both sides of these questions. Those who advocate the teacher *teaching* are "behaviorist," whereas those who believe the teacher structures learning situations to allow the student to construct knowledge are "cognitive" theorists.

Discussed briefly in the following sections are Piagetian stage theory and generic factors that influence learning mathematics, cognitive monitoring, and classroom organization for instruction.

GENERIC FACTORS AS AN ALTERNATIVE TO COGNITIVE STAGES

Piaget, a leading proponent of developmental phases, identified four stages of intellectual growth: sensorimotor, preoperational, concrete operational, and formal operational. In the earliest stage, *sensorimotor* (birth to 2 years of age), the infant progresses from reflexive, involuntary movements to voluntary organized actions. Reflexes such as sucking, eye movements, and palmer reflex become purposeful for eating, looking, and grasping.

During the *preoperational* stage (ages 2 to 7), the child attends to only one attribute of a situation at a time and cannot perform multiple discriminations. During this stage, children believe there is more clay in a ball of clay if it is rolled into a rod shape. They believe the number of objects increases when the objects are spread out in space so they are farther apart. If water or beads are poured from a large flat container into a tall, thin bottle, they state there are more in the tall vessel.

In the *concrete operational* stage (usually ages 7 or 8 to 11), the child is aware that perceptual changes in arrangements of objects or in the configuration of an object do not affect the quantitative aspects of the situation. In this stage, children take into account the fact that if nothing has been added or taken away from a quantity, then no change in amount has occurred. Thus, the ball of clay, the water, or the beads may change in appearance or spatial arrangement, but the amount remains the same.

At the *formal operational* stage (ages 11 to 14), the child is able to formulate hypotheses; to make inferences and interpretations in a hypothetical, deductive sense; and to engage in propositional thinking, such as "either a or b exists, but not both a and b." At this stage, students can begin to deal with combinations, permutations, probabilities, and correlations. Thinking takes on an ordered, systematic nature.

These stages are sequential but do not have fixed limits or boundaries. The underlying assumption is that children at a particular level of mental development can do certain things and not other things.

Implications for applying Piagetian theory to teaching have been strongly supported by Furth and Wachs,[3] questioned by Smedslund,[4] and vigorously opposed by Engelman.[5]

Assessing students' levels of cognitive development for the purpose of observing the presence or absence of cognitive requisites was a popular pastime of the sixties and early seventies. Implications for instruction stemmed from three basic notions of Piagetian theory:

1. Intellectual development occurs through a series of stages that must always occur in the same order.
2. Stages are defined by clusters of mental operations (seriating, conserving, classifying, hypothesizing, inferring) that underlie intellectual behavior.
3. Movement through the stages is accomplished by equilibration, the process of development that describes the interactions between experience (assimilation) and growing cognitive structures (accommodation).

There is a large amount of overlap among children who perform traditional Piagetian tasks before or after a given specified age. As a result, an alternative scheme for assessing cognitive development was presented by Reisman and Kauffman.[6] They identified generic factors that influence learning and used these to describe the developmental progress of students in terms of strengths and weaknesses related to learning mathematics. Generic factors of a cognitive nature may involve the following:

1. Learning at rapid rates
2. Attending to salient aspects of situations
3. Conserving (e.g., relates transformation and accompanying invariant. This is the definition of the term *conservation* as it is used in the mini-lessons in this book labeled "Possible Trouble Spots")
4. Retaining information with minimal repetition
5. Understanding complex materials
6. Constructing relationships, concepts, and generalizations

---

3. H. G. Furth and H. Wachs, *Thinking Goes to School* (New York: Oxford University Press, 1974).

4. J. Smedslund, "Symposium: Practical and Theoretical Issues in Piagetian Psychology III—Piaget's Psychology in Practice," *British Journal of Educational Psychology* 47 (1977): 1–6.

5. S. Engelman, "Does the Piagetian Approach Imply Instruction?" in *Measurement and Piaget,* ed. D. R. Green, M. P. Ford, and G. B. Flamer (New York: McGraw-Hill, 1971) and "Teaching Formal Operations to Pre-School Advantaged and Disadvantaged Children," *Ontario Journal of Educational Research* 9 (1967): 3.

6. Fredricka K. Reisman and S. H. Kauffman, *Teaching Mathematics to Children with Special Needs* (Columbus, Ohio: Charles E. Merrill Publishing Co., 1980).

Chart 1.1. Profile of a Child with a Learning Disability—Average Rate of Cognitive Development

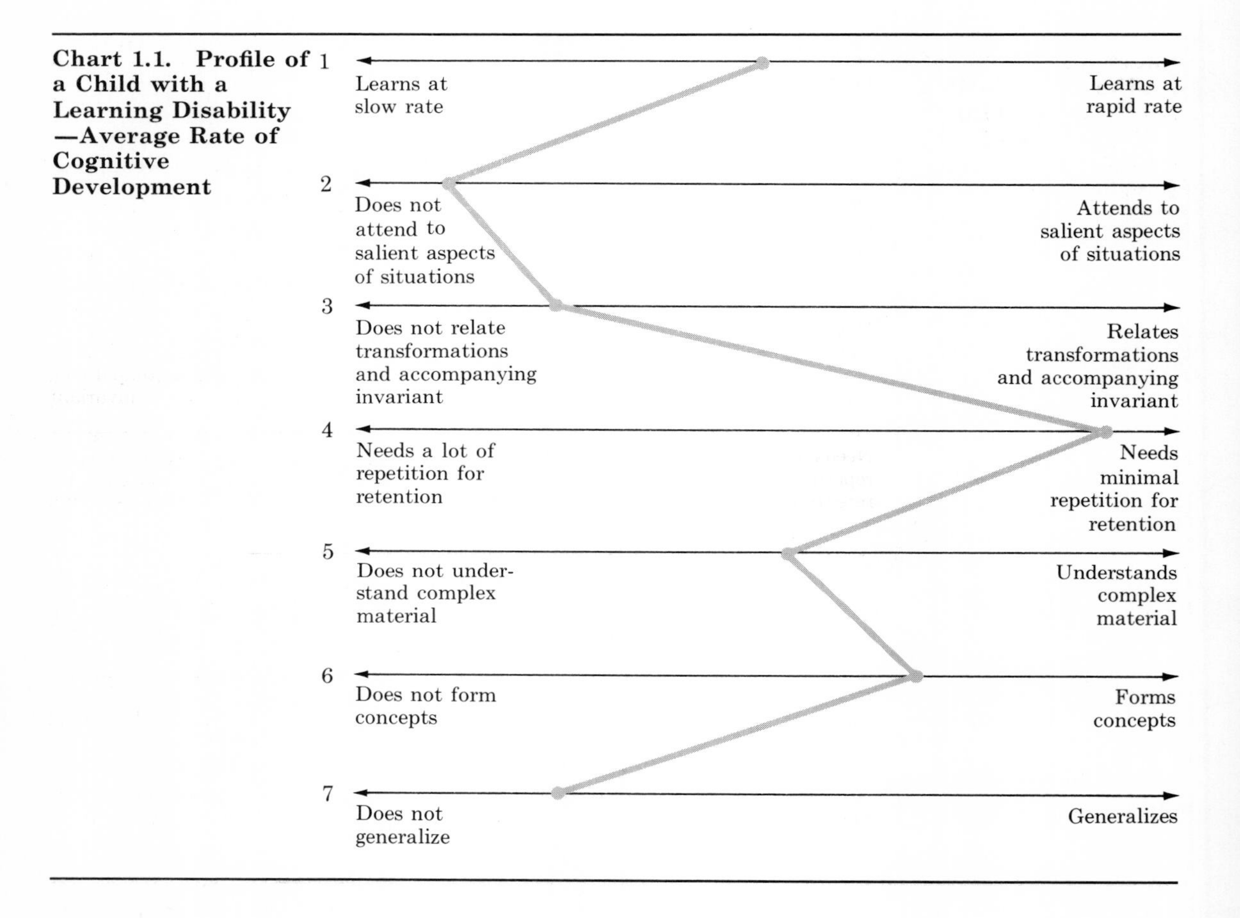

Instead of focusing on a categorical stage description, Reisman and Kauffman suggest a "function model" that utilizes generic factors as aids for grouping students for instruction and for selecting instructional procedures.[7]

Charts 1.1, 1.2, and 1.3 present profiles of three types of students graphed in terms of generic factors that influence the learning of mathematics.

7. Reisman and Kauffman, *Teaching Mathematics to Children with Special Needs.*

Chart 1.2. Profile of a Gifted Child with a Learning Disability—Rapid Rate of Cognitive Development

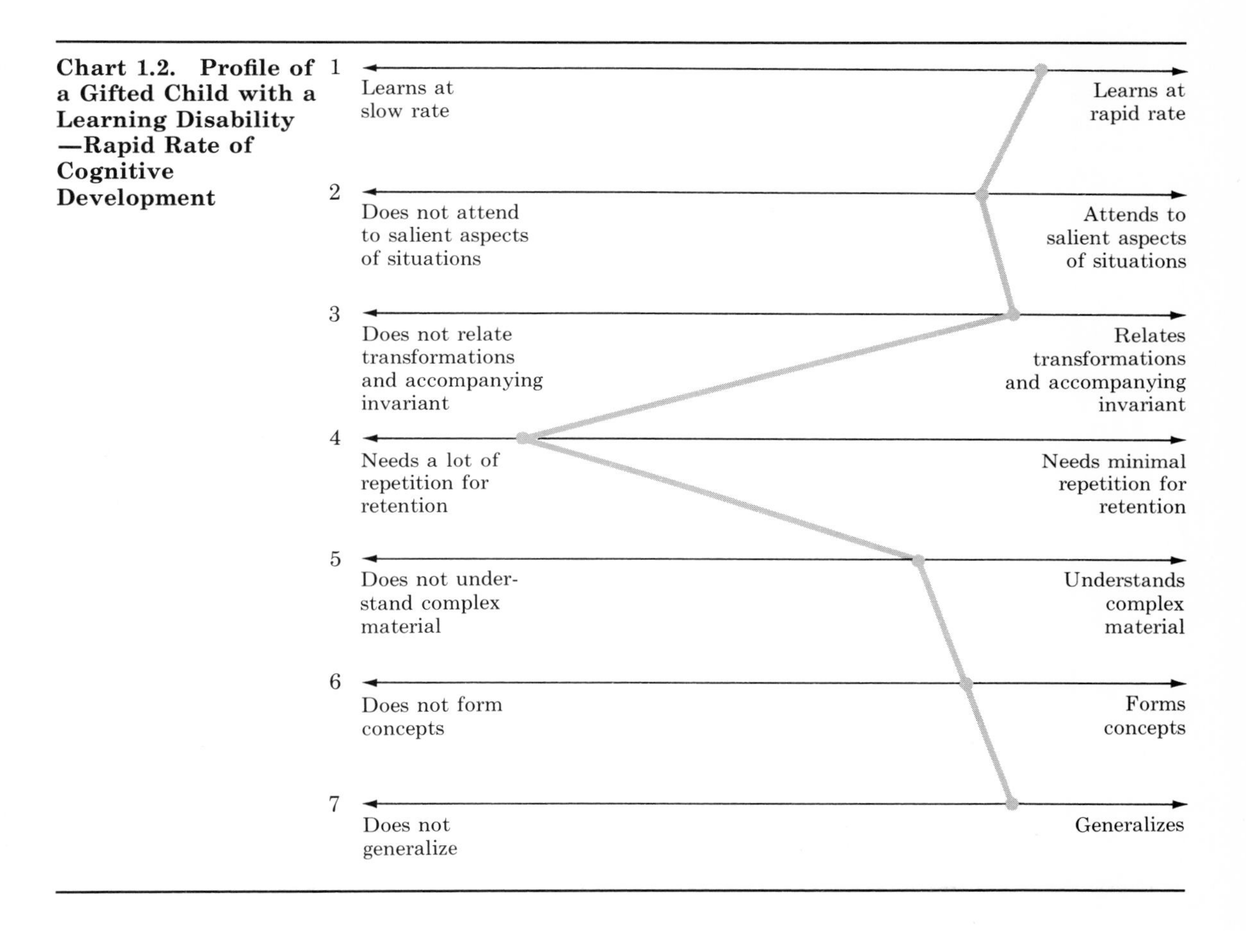

Interpretation of these profiles may be both inter- and intra-individual in nature. *Inter-individual* comparisons compare a child with others within the same age range and may be used for purposes of forming instructional groups. *Intra-individual* analyses compare children's strengths and weaknesses within themselves. For example, some students may memorize rapidly with minimal drill activities but need cues to help them notice the significant conditions of a situation (see Chart 1.1). Other students may learn in an opposite manner, thus calling for a different instructional approach (see Chart 1.2). The first

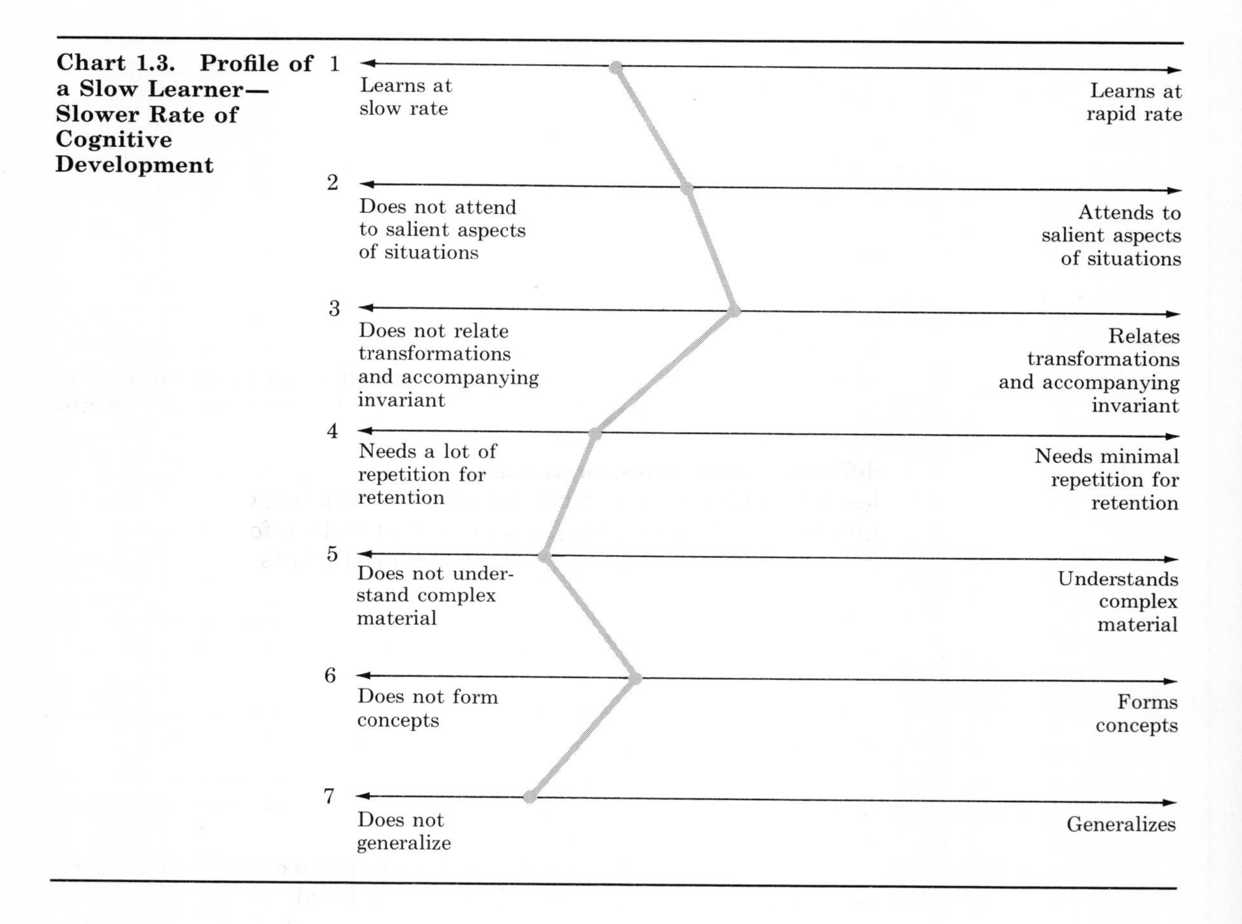

Chart 1.3. Profile of a Slow Learner—Slower Rate of Cognitive Development

group of students needs less practice, but the teacher has to provide cues to help them notice crucial aspects of situations.[8] The second group of students would need to be taught strategies to enhance their memory skills. A third group of students may have both poor memory skills and not know how to attend to crucial points. Some of these students may be described as slow learners or retarded (see Chart 1.3) if their rate of learning is slower than age peers.

---

8. See Reisman and Kauffman, *Teaching Mathematics to Children with Special Needs* for instructional strategies for special needs.

The Piagetian stages have been used diagnostically to make broad interpretations, suggesting, for example, that 8-year-olds who do not correctly perform a certain task may not yet be concrete operational in terms of their cognitive development. However, reasons for failing a Piagetian task are unclear, and alternative hypotheses (for example, memory failure, not understanding the task) are being generated to explain the overlap in age-related performance on the various tasks developed by Piaget.

It is suggested that assessing developmental progress on charts allows teachers to cluster groups of students. More students may then be kept to the task longer as larger instructional groups are created. Methods of instruction can also be selected so as to accommodate generic influences. For example, cues given to students who do not pick out salient aspects may include highlighting sequences by using different colors, sizes, or textures; by grouping numbers, as when learning a telephone number, license plate number, or social security number; by directly telling students what to look for, including providing advanced organizers; or by teaching students to engage in cognitive monitoring (talking to oneself).

There are other advantages of using generic factors as assessment indicators and as guides for mathematics instruction. These include:

1. Enabling teachers to observe student performance across tasks, since generic influences are not task-specific
2. Identifying and therefore easily assessing generic influences, so that the teacher can identify specific strengths and weaknesses that influence learning mathematics
3. Forming intra-individual profiles in order to observe the interaction among generic factors within an individual
4. Grouping students with similar learning strengths and weaknesses for some tasks; at other times, groups may be formed by mixing students who are strong in a particular skill with those who are weak in that same skill.
5. Enabling teachers to base hypotheses and inferences from observations over time and across tasks in regard to student levels of learning and styles of learning
6. Planning instruction based on learning strengths and weaknesses, so that instructional procedures and materials, as well as criteria for curriculum selection, can be directly related to generic influences on learning mathematics. (For example, if complexity of material interferes with learning, the teacher may have to simplify; if retention is poor, the teacher may have to provide more drill; if relationships are not noticed, the teacher may have to point them out

directly; if style of performance is unplanned, using trial and error, the teacher may encourage cognitive monitoring to aid in developing better problem-solving skills.)

COGNITIVE MONITORING

Students are often unaware of the cognitive processing demands of a learning task. They may also be unaware that they do not understand something. Markman found that "third through sixth graders do not spontaneously carry out those processes that they are capable of carrying out" in tasks involved in assessing their own comprehension failures.[9] Markman also found that children in the first through third grades who were asked to evaluate instructions that contained glaring omissions seemed satisfied with the instructions. It was not until the third grade that children noticed the gaps.[10]

When students do become aware of comprehension failure, they begin to monitor their thinking (for example, "Do I need to add or subtract?" "What is missing?" "Oh, I've had this same kind of problem wrong before." "I have to pay attention to the important stuff and not let my mind wander." "I need to draw this in order to see what's missing.")

Cognitive monitoring is used in some of the mini-lessons in this text as a cueing technique. This strategy of talking to oneself helps a person to keep on task, aids retention, and is useful for developing self-awareness of strengths and weaknesses.

CLASSROOM ORGANIZATION FOR INSTRUCTION

Individualized instruction has been described by Kulik and Kulik as "teaching adapted to the background and aptitude of individual learners. Most systems of individualized instruction are mastery oriented and self-paced, and they rely heavily on instructional materials."[11] Evertson, Anderson, Anderson, and Brophy define indi-

---

9. Ellen M. Markman, "Realizing that You Don't Understand: Elementary School Children's Awareness of Inconsistencies," *Child Development* 50 (September 1979): 643–55.

10. Ellen M. Markman, "Realizing that You Don't Understand: A Preliminary Investigation," *Child Development* 48 (1977): 986–92.

11. J. A. Kulik and C. C. Kulik, "College Teaching," in *Research on Teaching: Concepts, Findings, and Implications,* ed. Penelope L. Peterson and Herbert J. Walberg (Berkeley, Calif.: McCutchan Publishing Co., 1979), p. 78.

vidualized instruction as "extensive use of self-paced materials, individualized contracting, allowing some student choice about what is studied, and infrequent use of whole-class instruction."[12]

Research in mathematics achievement and attitudes toward mathematics have consistently favored more emphasis on large group instruction. In the Evertson, Anderson, Anderson, and Brophy study, the authors conclude:

> In mathematics classes, the results form a consistent picture of the practices of "good" teachers (using both achievement and student attitudes as criteria). The more effective teachers were active, well organized, and strongly academically oriented. They tended to emphasize whole-class instruction, but with some time also devoted to seat work. They managed their class efficiently, and tended to "nip trouble in the bud," stopping a disturbance before it could seriously disrupt the class. They asked many questions during class discussions. Most were "lower order" product questions *calling for short answers* (italics mine), but "higher order" process questions *calling for explanations from the students* (italics mine) were also fairly common.[13]

Good and Grouws reported that effective fourth grade teachers taught the whole class more, went over homework for a shorter amount of time, asked few questions that involved explanations on the part of the student, maintained a more relaxed classroom atmosphere, and were clear in their presentations.[14] Evertson, Emmer, and Brophy, in a study of seventh and eighth grade mathematics teachers found the same pattern; that is, effective teachers "used substantially more of the time to present content . . . and somewhat less time for individual seatwork."[15] Previously Dewar reported that grouping for mathematics instruction within a sixth grade class benefited both high- and low-achieving groups more than did whole-class instruction,[16] and Davis and Tracy summarized studies on intraclass grouping concluding that no one grouping plan appeared best.[17] The most

---

12. C. M. Evertson, C. W. Anderson, L. M. Anderson, and J. E. Brophy, "Relationships Between Classroom Behaviors and Student Outcomes in Junior High Mathematics and English Classes," *American Educational Research Journal* 17 (Spring 1980): 55.

13. Evertson, Anderson, Anderson, and Brophy, "Relationships Between Classroom Behaviors and Student Outcomes," p. 58.

14. T. L. Good and D. Grouws, "Teaching Effects: A Process-Product Study in Fourth Grade Mathematics Classes," *Journal of Teacher Education* 28 (1977): 49–54.

15. C. M. Evertson, E. T. Emmer, and J. E. Brophy, "Predictors of Effective Teaching in Junior High Mathematics Classrooms," *Journal for Research in Mathematics Education* 11 (May 1980): 167–79.

16. J. A. Dewar, "Grouping for Arithmetic Instruction in the Sixth Grade," *Elementary School Journal* 63 (1963): 266–69.

17. O. L. Davis, Jr., and N. H. Tracy, "Arithmetic Achievement and Instructional Grouping," *The Arithmetic Teacher* 10 (1963): 12–17.

recent research, however, favors group instruction over working with one or two students at a time.[18]

Rosenshine summarized studies on grouping students for learning:

> Students spend more time off-task in transition when they are working alone, whereas the use of large-group settings allows for more adult supervision. Although many educators prefer that teachers work with one or two children at a time, the reality is that when teachers were working with only one or two children they were unable to provide supervision for the remaining children, who, as a result, had less academically engaged time. . . . The message is clear: What is not taught and attended to in academic areas is not learned.[19]

The diagnostic teaching cycle described by Reisman is preventive rather than remedial (see Figure 1.1).[20] This model may be generalized to the total classroom. The teacher can administer a mathematics screening assessment to the entire class or to a large group of students in order to identify their strengths and weaknesses across a broad range of topics. Those students with similar academic needs can then be grouped for instruction. Members of such learning groups will change as student weaknesses are overcome.

Figure 1.1. The Diagnostic Teaching Cycle Model

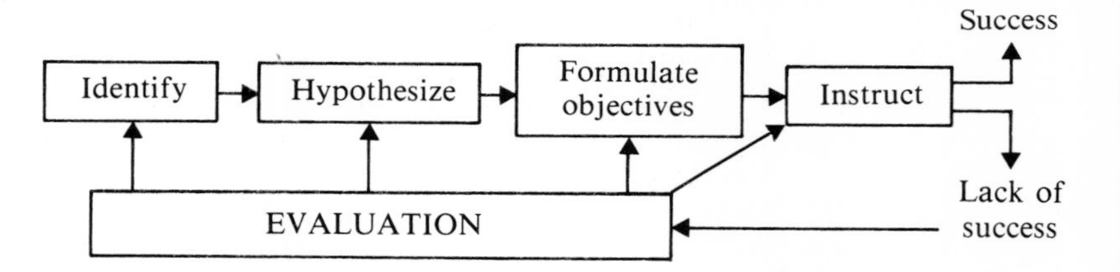

Source: Adapted from Fredricka K. Reisman, *A Guide to the Diagnostic Teaching of Arithmetic* (Columbus, Ohio: Charles E. Merrill, 1972), p. 5. Reprinted with permission.

---

18. J. A. Stallings and D. H. Kaskowitz, *Follow Through Classroom Observation Evaluation, 1972–73* (Menlo Park, Calif.: Stanford Research Institute, 1974); R. S. Soar, *Follow Through Classroom Process Measurement and Pupil Growth (1969–71); Final Report* (Gainesville, Fla.: College of Education, University of Florida, 1973); C. W. Fisher, N. N. Filby, and R. S. Marliave, *Descriptions of Distributions of ALT Within and Across Classes During the B-C Period,* Technical Note IV-lb (San Francisco: Far West Laboratory for Educational Research and Development, 1977); J. S. Kounin and P. V. Gump, "Signal Systems of Lesson Settings and the Task-Related Behavior of Preschool Children," *Journal of Educational Psychology* 66 (1974): 554–62; J. B. Carroll, "A Model of School Learning," *Teacher's College Record* 64 (1963): 723–32; B. S. Bloom, *Human Characteristics and School Learning* (New York: McGraw-Hill, 1976); A. Harnischfeger and D. Wiley, "The Teaching-Learning Process in Elementary Schools: A Synoptic View," *Curriculum Inquiry* 6 (1976): 6–43; and D. C. Berliner et al, *Proposal for Phase III of the Beginning Teaching Evaluation Study* (San Francisco: Far West Laboratory for Educational Research and Development, 1976).

19. S. V. Rosenshine, "Content, Time, and Direct Instruction," in *Research on Teaching: Concepts, Findings, and Implications,* ed. P. L. Peterson and H. J. Walberg (Berkeley, Calif.: McCutchan Publishing Co., 1979), pp. 42–43.

20. Fredricka K. Reisman, *Diagnostic Teaching of Elementary School Mathematics: Methods and Content* (Chicago: Rand McNally College Publishing Co., 1977) and *A Guide to the Diagnostic Teaching of Arithmetic,* 2nd ed. (Columbus, Ohio: Charles E. Merrill Publishing Co., 1978).

The teacher should be a good observer and a good detective who identifies mathematics skills that need to be developed in students. Inferences that serve as corrective procedures should follow from the observations. Next, instructional goals for a group of students or, if appropriate, for the entire class may be developed. These goals are then translated into mathematics activities and, finally, an evaluation of the lesson is done.

Allowing students to take some of their mathematics tests as a team has been effective in reducing test anxiety. College students report that they learn quite well as they study for a test when they do not want to let their partners down.

In summary, stage theory, individualized instruction, and discovery learning are terms that have permeated education courses for the last two decades. However, these educational foundations are being scrutinized as we enter the decade of the eighties. The applicability of Piaget's stage theory to educational practice continues to be questionable; evidence has been provided for advantages of large group instruction that allows for more time on task for greater numbers of students; and direct teaching in contrast to discovery methods is again recognized as an effective instructional procedure. Ways in which these changes relate to mathematics instruction were discussed briefly in this chapter. Recommendations for mathematics curricula and instruction by the board of directors of the National Council of Teachers of Mathematics were presented. The reader who wishes more in-depth treatments of these issues should refer to the Suggested Readings at the end of the chapter.

SUGGESTED READINGS

Cognitive Monitoring

Flavell, J. "The Development of Metacommunication." Paper given at the Symposium on Language and Cognition, 21st International Congress of Psychology, Paris, July 1976.

Flavell, J., and Wellman, H. M. "Metamemory." In *Memory in Cognitive Development,* edited by R. V. Kail and J. W. Hagen. Hillsdale, N.J.: Erlbaum, 1969.

Markman, Ellen M. "Realizing that You Don't Understand: Elementary School Children's Awareness of Inconsistencies." *Child Development* 50 (September 1979): 643–55.

Markman, Ellen M. "Realizing that You Don't Understand: A Preliminary Investigation." *Child Development* 48 (1977): 986–92.

Mead, G. H. *Mind, Self, and Society.* Chicago: University of Chicago Press, 1934.

Reisman, Fredricka K., and Kauffman, Samuel H. *Teaching Mathematics to Children with Special Needs.* Columbus, Ohio: Charles E. Merrill, 1980. P. 197.

Organization for Instruction

Berliner, D. C. et al. *Proposal for Phase III of the Beginning Teaching Evaluation Study.* San Francisco: Far West Laboratory for Educational Research and Development, 1976.

Bloom, B. S. *Human Characteristics and School Learning.* New York: McGraw-Hill, 1976.

Carroll, J. B. "A Model of School Learning." *Teacher's College Record* 64 (1963): 723–32.

Davis, O. L., Jr., and Tracy, N. H. "Arithmetic Achievement and Instructional Grouping." *The Arithmetic Teacher* 10 (1963): 12–17.

Dewar, J. A. "Grouping for Arithmetic Instruction in the Sixth Grade." *Elementary School Journal* 63 (1963): 266–69.

Evertson, C. M.; Anderson, C. W.; Anderson, L. M.; and Brophy, J. E. "Relationships Between Classroom Behaviors and Student Outcomes in Junior High Mathematics and English Classes." *American Educational Research Journal* 17 (Spring 1980): 43–60.

Evertson, C. M.; Emmer, E. T.; and Brophy, J. E. "Predictors of Effective Teaching in Junior High Mathematics Classrooms." *Journal for Research in Mathematics Education* 11 (May 1980): 167–78.

Fisher, C. W.; Filby, N. N.; and Marliave, R. S. *Descriptions of Distributions of ALT Within and Across Classes During the B-C Period.* Technical Note IV-lb. San Francisco: Far West Laboratory for Educational Research and Development, 1977.

Good, T. L., and Grouws, D. "Teaching Effects: A Process-Product Study in Fourth Grade Mathematics Classes." *Journal of Teacher Education* 28 (1977): 49–54.

Harnischfeger, A., and Wiley, D. "The Teaching-Learning Process in Elementary Schools: A Synoptic View." *Curriculum Inquiry* 6 (1976): 6–43.

Kounin, J. S., and Gump, P. V. "Signal Systems of Lesson Settings and the Task-Related Behavior of Preschool Children." *Journal of Educational Psychology* 66 (1974): 554–62.

Kulik, J. A., and Kulik, C. C. "College Teaching." In *Research on Teaching: Concepts, Findings, and Implications,* edited by Penelope L. Peterson and Herbert J. Walberg. Berkeley, Calif.: McCutchan Publishing Co., 1979.

Peterson, P. L. "Direct Instruction Reconsidered." In *Research on Teaching: Concepts, Findings, and Implications,* edited by P. L. Peterson and H. J. Walberg. Berkeley, Calif.: McCutchan Publishing Co., 1979.

Rosenshine, S. V. "Content, Time, and Direct Instruction." In *Research on Teaching: Concepts, Findings, and Implications,* edited by P. L. Peterson and J. J. Walberg. Berkeley, Calif.: McCutchan Publishing Co., 1979.

Soar, R. S. *Follow Through Classroom Process Measurement and Pupil Growth (1969–71): Final Report.* Gainesville, Fla.: College of Education, University of Florida, 1973.

Stallings, J. A., and Kaskowitz, D. H. *Follow Through Classroom Observation Evaluation, 1972–73.* Menlo Park, Calif.: Stanford Research Institute, 1974.

Diagnostic Teaching

Reisman, F. K. *A Guide to the Diagnostic Teaching of Arithmetic.* 2nd ed. Columbus, Ohio: Charles E. Merrill Publishing Co., 1978.

Reisman, F. K. *Diagnostic Teaching of Elementary School Mathematics: Methods and Content.* Chicago: Rand McNally College Publishing Co., 1977.

Reisman, F. K. and Kauffman, S. H. *Teaching Mathematics to Children with Special Needs.* Columbus, Ohio: Charles E. Merrill Publishing Co., 1980.

Stage Theory

Bower, R. G. R. *Development in Infancy.* San Francisco: Freeman, 1974.

Brainerd, C. J. "Neo-Piagetian Training Experiments Revisited: Is There Any Support for the Cognitive-Developmental Stage Hypothesis?" *Cognition* 2 (1973): 349–70.

Brown, G., and Desforges, C. "Piagetian Psychology and Education: Time for Revision." *British Journal of Educational Psychology* 47 (1977): 7–17.

Bryant, P. *Perception and Understanding in Young Children.* London: Methuen, 1974.

Engelman, S. "Does the Piagetian Approach Imply Instruction?" In *Measurement and Piaget,* edited by D. R. Green, M. P. Ford, and G. B. Flamer. New York: McGraw-Hill, 1971.

Engelman, S. "Teaching Formal Operations to Pre-School Advantaged and Disadvantaged Children." *Ontario Journal of Educational Research* 9 (1967): 3.

Flavell, J. H. "Thought in the Young Child." In *Monograph of the Society of Research in Child Development,* edited by W. Kessen and C. Kuhlman, 27 (1962): 65–82.

Furth, H. G., and Wachs, H. *Thinking Goes to School.* New York: Oxford University Press, 1974.

Gelman, R. and Tucker, M. F. "Further Investigations of the Young Child's Conception of Number." *Child Development* 46 (1975): 167–75.

Price-Williams, D. R.; Gordon, W.; and Ramirez, R. M. "Skill and Conservation: A Study of Pottery-Making Children." Developmental Psychology 1 (1969): 769.

Reisman, F. K., and Torrance, E. P. "Alternative Procedures for Assessing Intellectual Strengths of Young Children." *Psychological Reports* 46 (1980): 227–30.

Smedslund, J. "Symposium: Practical and Theoretical Issues in Piagetian Psychology III—Piaget's Psychology in Practice." *British Journal of Educational Psychology* 47 (1977): 1–6.

Wohlwill, J. F. "Piaget's System as a Source of Empirical Research." In *Logical Thinking in Children,* edited by I. E. Sigel and F. H. Hooper. New York: Holt, Rinehart and Winston 1968.

Whole Numbers and the Concept of Set

Number ideas and relationships emerged from a human desire to put structure on the environment. For example, the caveman may have wanted to know if he or his neighbor had a greater *herd* of sheep and the cave woman if she had a large enough *collection* of shells for her dowry. The size of a *flock* of birds was of interest for hunting as was the number of fish in a *school* of fish. Today, a teacher is anxious to know how large a *class* he or she is assigned; parents are affected financially by the size of their *family;* truck drivers are concerned with the cost of a *tank* of gas, and we are all concerned with the *amount* of money it takes to live in times of inflation.

The term *set* could have been substituted for all of the italicized words above. During the 1960s and 1970s, set theory was an important part of the elementary school mathematics curriculum. The trend in the 1980s is away from a formal treatment of set notation as a result of public pressure to focus on more basic skills of reading and writing numbers and computations on numbers. However, there are some basic notions about set that help one to understand number, and these ideas will still be used in mathematics instruction.

SET IDEAS RELATED TO NUMBER

A *set* may be described as a collection of objects or ideas. The listing or description of elements of the collection must leave no doubt as to membership in the particular set. A *number* tells how large the set is.

> *Note:* An issue has been made in the recent past about distinguishing between number and numeral to the point of sometimes blocking a pupil's expression. (For simplicity, the word *number* has been used more often than *numeral* throughout this book.) A child should eventually be able to distinguish between an idea (number) and a symbol (numeral) for a quantitative idea. If the learner needs to compare sizes of groups, he or she soon learns to use oral or written symbols—word names or numerals. For example, when children wish to discuss how many gumdrops they have eaten, they need words or numerals to express the number of objects consumed. Oral number names evolved first, followed by written numbers, or numerals, and written word names followed these.

Cardinal Number Property

A *cardinal number,* then, answers the question, How many elements? For example, $\{a, b, \square\}$ shows 3 elements and indicates that this set has a cardinal number of 3. This may be written

$$n(A) = 3$$

Finite sets
Infinite sets

Sets having a countable number of elements are said to be *finite.* A set that is not countable is called *infinite.* For example, the whole numbers go on and on; they are infinite in number. An infinite set may be represented by three dots to indicate a particular sequence has been established as in

$$\{1, 2, 3, \ldots\}$$

Empty sets

The three dots, called "ellipsis points," indicate that there are more elements than those listed but that enough were listed so that a pattern is apparent.

If a set has no elements it is said to be *empty.* The empty set is finite. Notation for stating that A is an empty set (sometimes called the "null set") is either

$$A = \{\ \}$$

or

$$A = \varnothing$$

Elements of sets paired in one-to-one correspondence

The cardinal number property of an empty set is zero. This may be written

$$n(A) = \varnothing$$

Elements of $\{a, b, c\}$ and $\{1, 2, 3\}$ can be paired in a *one-to-one correspondence* and have the same cardinal number:

$$\begin{matrix} a & b & c \\ \updownarrow & \updownarrow & \updownarrow \\ 1 & 2 & 3 \end{matrix}$$

Each element of the first set is paired to just one element of the second set, and to each element of the second set is paired just one element of the first set. The cardinal number property of a set is as real as such properties as color, size, shape, taste, length, and width. Number is that property of a set of objects that the set has in common with every other set with which it can be placed in a one-to-one correspondence.

If $A = \{1, 2, 3, 4, 5\}$ and $B = \{\square, \triangle, x, o, \Leftrightarrow\}$, then $n(A) = n(B) = 5$.

Notation for the number of elements in a set

But does $n(A) = n(B)$ imply that $A = B$? Are the elements of A and B the same? No. Then $A \neq B$ (set A is not equal to set B), and the notation $n(A) = n(B)$ does not imply equal sets; they are equivalent sets.

Naming Sets

Sets are named with capital letters; lower-case letters denote elements of sets. Sets may be described in three ways:

1. *Tabulation, or roster method:* Listing the members, or elements: If a set, A, has as its elements 1, 2, 3, we would write

$$A = \{1, 2, 3\}$$

2. *Descriptive method:* Describing the members of a set with words: If

$$A = \{1, 2, 3\},$$

we may write

$$A = \{\text{the first three counting numbers}\}$$

3. *Rule, or set-builder notation:*

$$\begin{aligned} A &= \{x \,/\, x \in C;\ 0 < x < 4;\ C \text{ is a counting number}\} \\ &= \{1, 2, 3\} \end{aligned}$$

(A is the set of all x's such that x is a counting number and x is greater than zero and less than 4.)

$$B = \{x \mid x + 3 = 9;\ x \text{ is a counting number}\}$$

(B is the set of all x's such that x plus 3 is 9 and x is a counting number. This is the same set as $\{6\}$).

This method is usually relegated to use in high school or with gifted mathematics pupils in the upper elementary grades.

To show in mathematical language that "x is an element of set A," we write

$$x \in A$$

using the Greek letter $\in$ for epsilon to mean "is a member of." To write that "y is *not* an element of set A," we use the symbol $\notin$; thus,

$$y \notin A$$

Set notation uses braces and commas when listing the elements. The order of listing the elements does not matter:

$$\{\text{Billie, Betsy, Mary}\} = \{\text{Betsy, Mary, Billie}\}$$
$$= \{\text{Mary, Billie, Betsy}\}$$

and so on. Sets may be composed of tangible and intangible elements.

In this chapter, teaching sets is assimilated into lessons on number. The use of sets of objects helps to provide students with activities at the concrete level, where manipulation of objects is a foundation for building number ideas. For the reader who wishes information on sets of numbers that comprise the Real Number System, see "For Further Study" at the end of this chapter.

Mini-lesson 1.1.5 Indicate the Cardinal Number Property of Sets

Vocabulary: object, collection set, member, element, bunch, group, describe, list, thing, alike, same shapes (round, square, big, little), colors (red, green, blue, yellow), match, not the same, different, equivalent, nonequivalent, correspondence, relationship, zero, one, two, three, four, five, six, seven, eight, nine, more, less, most, least, greater, less than, fewer, fewest, greatest.

Possible Trouble Spots: does not understand how to sort objects according to physical appearance, function, or other attribute; does not attend to salient aspects of situations; does not seem aware of the one-to-one relationship; does not perform at age-appropriate levels in motor coordination, visual functioning, or perceptual functioning; counts rotely but does not enumerate.

Requisite Objectives:[1]
Group objects by physical attributes and by function

1. Those requisite objectives that are italicized comprise the Scope and Sequence Chart in the Appendix.

Select underlying attribute of a set

Sets 1.1.1: Denote elements of a set

Sets 1.1.2: State the attributes common to two or more sets

Match sets in a one-to-one relationship

Sets 1.1.3: Recognize equivalent and nonequivalent sets

Use succession relationship

Place objects and numbers in sequence

Count by rote

Match number name to objects in one-to-one relationship

Sets 1.1.4: Compare nonequivalent sets

Activity 1
Have students match sets according to a given attribute.

A. To show that they can group objects by physical attributes, have students match shapes to holes on a form board.

B. Ask students to identify sets in the room by physical appearance (e.g., red things, square, etc.). Sort objects by physical attributes.

C. Then provide activities that involve grouping objects according to their functions. For example, ask how the following are alike: chair and stool, apple and banana, car and train. Ask students to identify sets in the room by their functions (e.g., things you write with).

Activity 2
In order to form concepts, it is necessary that students can select underlying attributes of groups of objects. Ask all children wearing clothing of a particular color (red, for example) to come forward. Gather them together and place a rope on the floor to form a boundary around them. Have some children draw a picture of this set. Others may describe the attribute that made them members of the set. Then have some children list the names of the children in the set.

Activity 3
To denote elements of a set, show a variety of groups of objects and ask students to describe and list what is the same about objects within a set. For example, show pictures of furniture, toys, or fruit,

with only one object per picture. Ask the students to group the pictures of furniture by how it is used; group the toys by what they do, e.g., bounce, make noise; and group the fruit by taste, e.g., bananas are slimy; lemons, oranges, and grapefruits are juicy; peaches and apricots are fuzzy.

Activity 4
In order to construct simple generalizations, students should engage in activities that require stating the attributes common to two or more sets.

Procedure A: Place 3 natural-colored wood blocks next to one another on a table. Ask the child to tell you everything that is the same about the 3 blocks in the set. The child may respond that they are the same size, same color, same shape, or that all of them are wood, all have corners, or all are blocks.

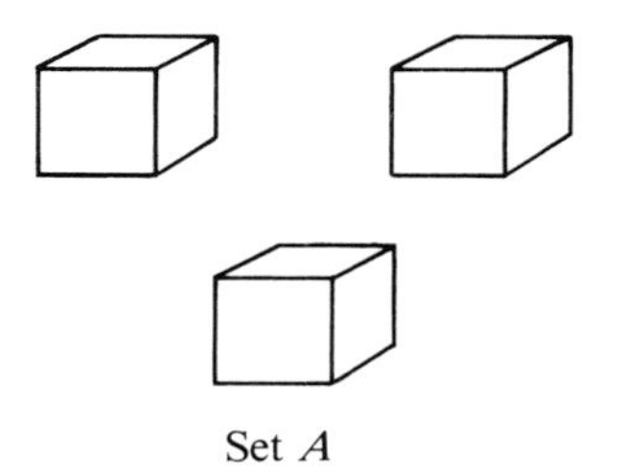
Set *A*

Procedure B: Place a set of 3 red blocks on the table but slightly apart from the first set. Ask the child to look at the *two* sets and tell what is the *same about both sets.* Be sure he understands that you want him to compare sets and *not* single elements within a set, or even single elements across sets. This activity requires the child to tell what is the same about set *A and* set *B.* Perhaps placing yarn around each set will help. The child may state that both sets have blocks, have wooden things, have 3 blocks, or have blocks that are the same size and shape.

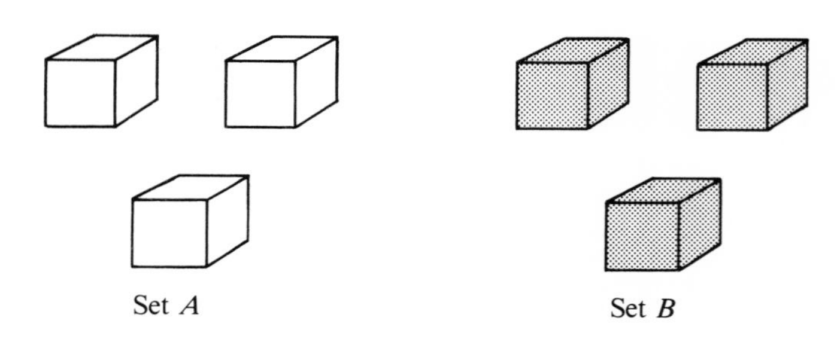
Set *A* Set *B*

Procedure C: Add a third set, *C*, made up of a red block, a natural-colored wood block, and a tongue depressor. Ask the child to tell what is the *same about all three sets*. He may respond that they all have wood things or they all have 3 things.

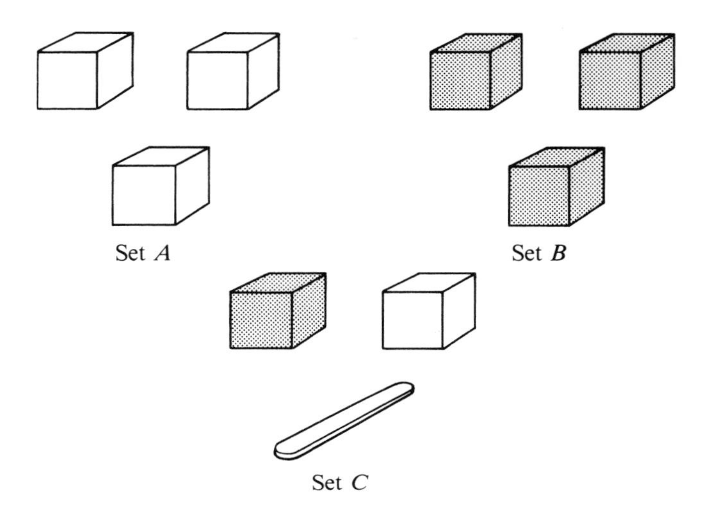

Procedure D: Finally, add a fourth set, *D*, having a blue block, a tongue depressor, and a marble. Ask the child to tell what is the *same about all four sets*. He may say that they all have the same number or they all have 3 things.

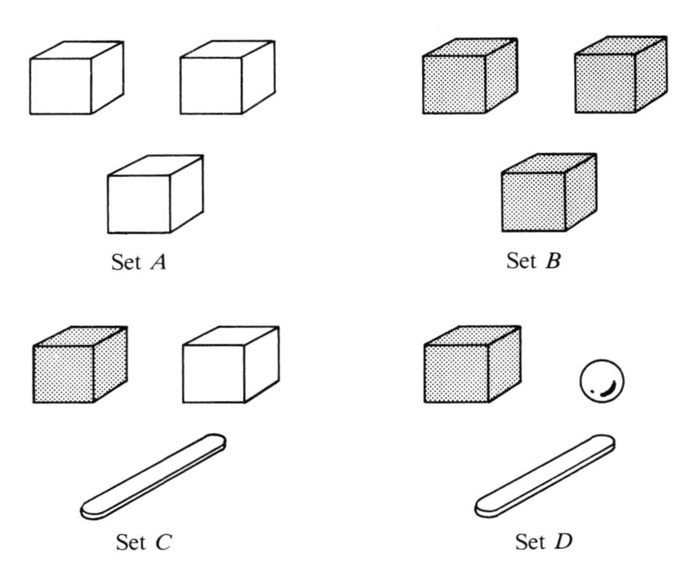

Activity 5
A basic relationship that underlies much of elementary school mathematics is the ability to match sets in a one-to-one correspondence—for example, matching cups to saucers, toy cowboys to their horses, cookies to children, and number names to objects counted. This one-for-one matching underlies equivalent sets. Two sets are *equivalent* if the elements of one set can be matched in a one-to-one correspondence with the elements of the other, thereby having the same cardinal number property. Therefore, if set $A = \{1, 2, 3\}$ and $B = \{x, y, z\}$, A and B are equivalent sets, written $A \sim B$. (Note, however, that equivalent sets are not necessarily equal sets. For two sets to be equal, the elements of the sets *must be the same*. If $A = \{\text{Mary, Laura, Betsy}\}$ and $B = \{\text{Mary, Laura, Betsy}\}$, then $A = B$. The same children are members of both sets.

Activity 6
To understand equivalent sets, the student must generalize one-to-one correspondence between two sets—to three or more sets. Also involved in understanding equivalence among three or more sets is the *transitive property*. This property states that if sets A and B are in one-to-one correspondence, and if sets A and C are also, then it follows that sets B and C are in a one-to-one relationship.

To show that students can recognize equivalent and nonequivalent sets, an ability that is a requisite of understanding cardinality of sets, provide the following experiences.

A. To recognize equivalent sets, show pictures of three sets (4 horses, 4 cowboys, 3 cowboys). Ask a child to point to the set of cowboys who would need all of the horses so each cowboy has his own horse to ride.

B. To recognize nonequivalent sets, give the child dot cards and ask her to hand you the card whose number of dots differs from the others.

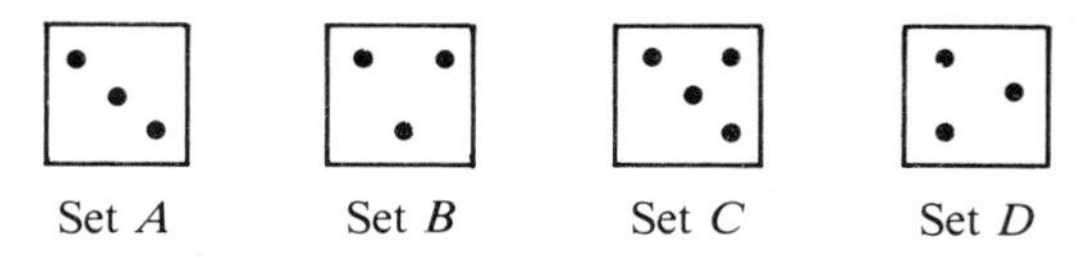

"Card *C* has a different number of dots."

C. To discriminate among equivalent and nonequivalent sets, give the child a set of dominoes. Ask her to hand you all of the dominoes whose dots match one-for-one. Note that at this point the child may not be able to state the number property of a set.

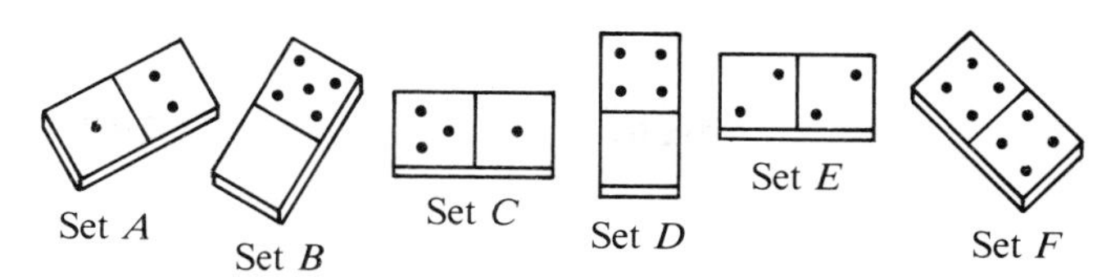

"Sets C, D, and E have four dots on them."

Activity 7
In order to decide whether or not various sets have the same number property, called *cardinality* or *cardinal number property,* have the students compare nonequivalent sets. Have them select sets both greater in number and sets less than other sets.

A. Ask students to select sets by putting an X on the set with more objects.

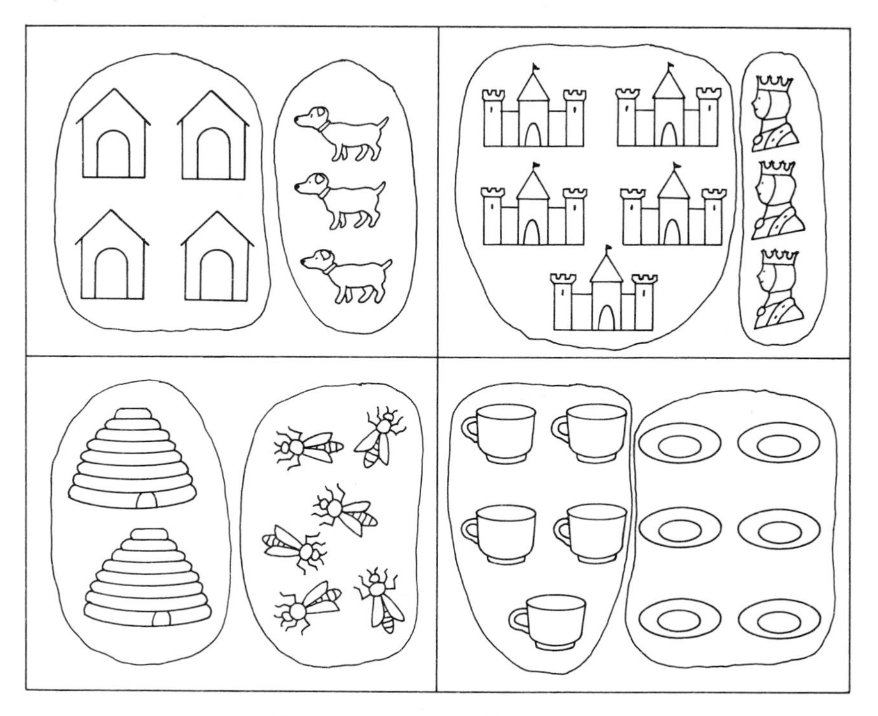

B. Then show children artwork similar to that in Activity 1A and ask them to select sets greater in number by putting an X on the set with fewer objects.

C. Ask students to draw a set of X's to create a set less than a given set. For example, show a set of 5 balls and ask students to draw a set of X's fewer in number than the balls.

D. Ask children to select a set that is greatest in number. For example, give students three or more sets of pennies, with from 3 to 6 pennies in each set. Ask them to point to the set with the greatest number. Use other objects for the same procedure. Next have students create sets that are fewer or greater than given sets.

Activity 8
Students usually enter school able to count by rote. Some may need practice, such as singing number songs (e.g., *Ten Little Indians* or *This Old Man*). Choral counting is also a method of drill that youngsters enjoy.

Activity 9
Matching number names to objects in a one-to-one relationship is called *enumeration.* Some children can say the names in order but when counting do not assign one number to one object. The following experiences help students develop this skill.

A. Give each child sets of from 1 to 9 objects and ask the students to put a finger on a different object in a given set as he says each number.

B. Provide recognition activities for relating the digits 0 to 9 to their oral names. Use picture and digit recognition activities. Using different sets of pictured objects, ask students to name the number property of sets zero to 9 and write the number for each.

Additional considerations: Students need a great deal of practice in recognizing and comparing equivalent and nonequivalent sets at both concrete and picture levels. Whether these are group or individual activities, it is important that students discuss these activities as well as participate in them.

Students who have difficulty in using number names in order should participate in activities involving succession and sequencing. The succession relationship merely involves one event following another. Children may find sequencing more complex because they must recognize a pattern in order to further the particular sequence. Examples of such activities follow.

A. As students walk, emphasize the succession in placing one foot after the other.

B. Give students a pile of beads to string. Point out how the beads can be put on the string one at a time, i.e., there is no pattern that is required.

C. Sequencing is an extension of succession. In these activities, students must recognize an underlying simple pattern or criterion that establishes the sequence. Have students repeat simple tunes, clap their hands in various rhythms, and reproduce simple designs. Copy simple bead pattern on string as shown below. Ask students to complete each pattern.

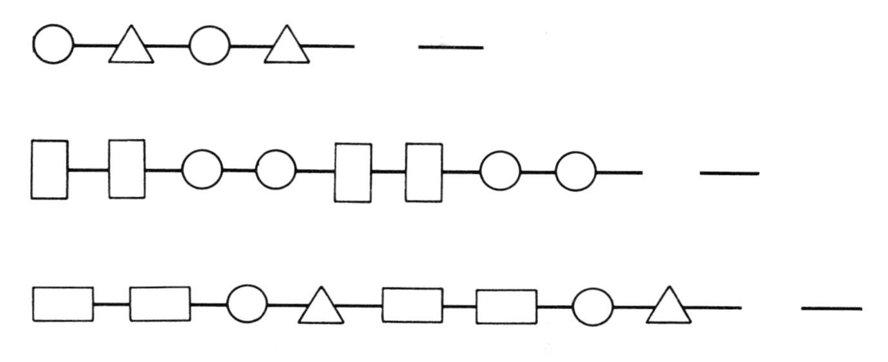

Note: Mini-lesson 1.1.5 is a prerequisite for Mini-lessons 1.1.6 and 1.1.8.

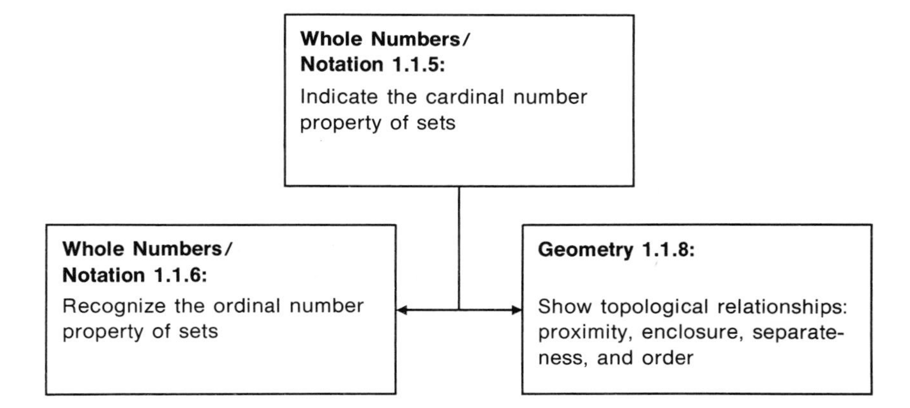

PHYSICAL AND SENSORY FUNCTIONING

Every teacher needs to realize that the effectiveness of the instructional methods depends on the child's physical and sensory functioning. Therefore, it is important that the teacher keep the individual child in mind when implementing the mini-lessons in the text. The following information is presented to help a classroom teacher identify those children who may need to have the mini-lessons tailored to their needs or who may have to be referred to a resource teacher or professional specialist in vision, hearing, or movement education.

Motor Coordination

A regular classroom teacher should be aware of the fact that imbalance and poor coordination in the young child interferes with development of spatial and temporal relationships, both of which are very much involved in developing mathematics ideas, especially those involving spatial relationships. The teacher should observe whether or not the child displays difficulty in holding a pencil firmly or in placing objects in a designated position. Is the child's gait smooth or spastic while walking, climbing, or going up stairs? Does he put both feet on a stair and then proceed to the next or does he use one foot per stair?

Perceptual Functioning

It is fine to use a set approach for teaching fundamental mathematics, but the teacher must insure that this technique is appropriate for each of the students. For example, children with learning disabilities who have perceptual discrimination problems involving inability to differentiate among various shapes, colors, objects, or symbols may be hindered in their learning by use of sets. Following are some diagnostic procedures for assessing perceptual association performance and suggestions for helping a young child learn to discriminate.

Competence in matching dot configurations is related to matching sets in one-to-one correspondence and in identifying equivalent sets. Both of these are basic to constructing the cardinal number property of sets. Present a set of 7 cards bearing configurations of from 1 to 7 dots on a table before the child:

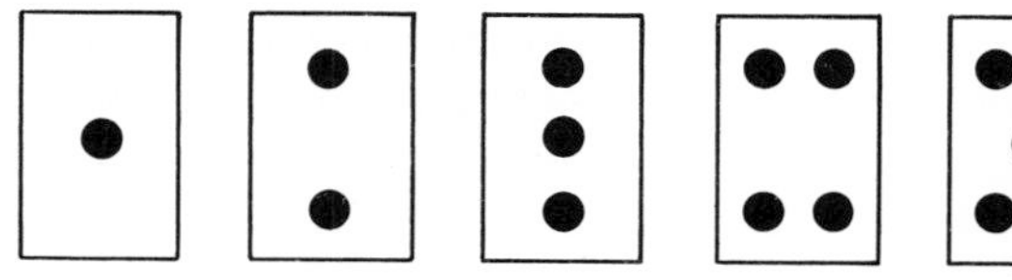

From a second set of identical cards, show one at a time in a vertical position and ask the child to point to the card in the set on the table that looks exactly the same as the one you are holding. Present your cards in the following order for the child to match: 2 dots, 1, 7, 4, 3, 6, and 5 dots. For the student with poor motor coordination, it is often best that the teacher demonstrate the use of concrete instructional materials used to develop whole number ideas.

Visual Functioning

Impairment in visual acuity, visual pursuit, and depth perception may restrict mathematics learning. *Visual acuity* refers to keenness of vision and is different from visual discrimination. Someone may be able to see an object perfectly (acuity) but be unable to differentiate or discriminate it from another form. *Visual pursuit* is the ability to track a spot of light from a flashlight or an object in space with one's eyes alone, without moving the head or body. *Depth perception,* also called "binocularity," involves effective use of visual position-in-space cues regarding objects around the person. Many young children who have been identified as "learning disabled" demonstrate binocular impairment. For example, they may misjudge the distances between objects.

For children with impaired or immature visual functioning who learn tactilely, it may help to glue sets of objects to cardboard or paste sets of flannel forms on boards to maintain the stability of the various sets. Also appropriate are sets having textural attributes, e.g., sets of sandpaper forms, velvet shapes, wooden shapes, and glass shapes.

Competence in using two sense systems simultaneously is also important to school learning. For example, ability to coordinate touch and vision facilitates learning attributes of objects that relate to these two senses. Auditory skills may be combined with touch to develop sequencing skill, as in copying patterns of hand clapping, drum beats, and so on.

This activity is a good diagnostic assessment to identify those children who find it difficult to copy from a vertical arrangement on the blackboard to a horizontal arrangement on their paper. For these children, use the overhead projector as a teaching aid. By placing boardwork on completed transparencies, these students can work from the horizontally positioned transparency.

Difficulty in matching dot configurations may be due to inadequate attention to details, failure to recognize the significance of certain details, inattention to the salient aspects of a situation, or inability to concentrate on and respond to what the child may see as a complex task.

COUNTING AND WRITING NUMBERS

Counting, according to Webster, is "the naming of numbers in regular order." When we count, we are using the words assigned to number values. Counting that involves assigning number names to objects in a one-to-one correspondence is *rational counting,* or *enumeration.* We need to know the agreed-upon order of saying the number names (which depends upon their number values), since the last number named tells us how many elements are in a set of things. This last number named tells the *cardinality,* or the "how-muchness," of a set.

The mere reciting of the number names in the correct order but without citing a one-to-one correspondence is *rote counting.* Rote counting is related to an ordinal idea of number, in which the order of the number names is sometimes designated by "first," "second," "third," and so on. Ordinal numbers answer the question, "Where in the sequence?"

Mini-lesson 1.2.2 Count and Write Numbers from 0 to 9 in Sequence

Vocabulary: order, first, second, third, fourth, fifth, sixth, seventh, eighth, ninth, 0, 1, 2, 3, 4, 5, 6, 7, 8, 9

Possible Trouble Spots: does not place sets in order from smallest to largest or vice versa; does not properly form the digits 0 to 9; does not correctly count and/or write numbers in sequence from 0 to 9

Requisite Objectives:
Sets 1.1.4: Compare nonequivalent sets

Whole Numbers/Notation 1.1.5: Indicate the cardinal number property of sets

Whole Numbers/Notation 1.1.6: Recognize the ordinal number property of sets

Whole Numbers/Notation 1.1.7: Write the digits 0 to 9

Activity 1
Use blocks to represent linear differences among sets of numbers, as shown in the following activity.

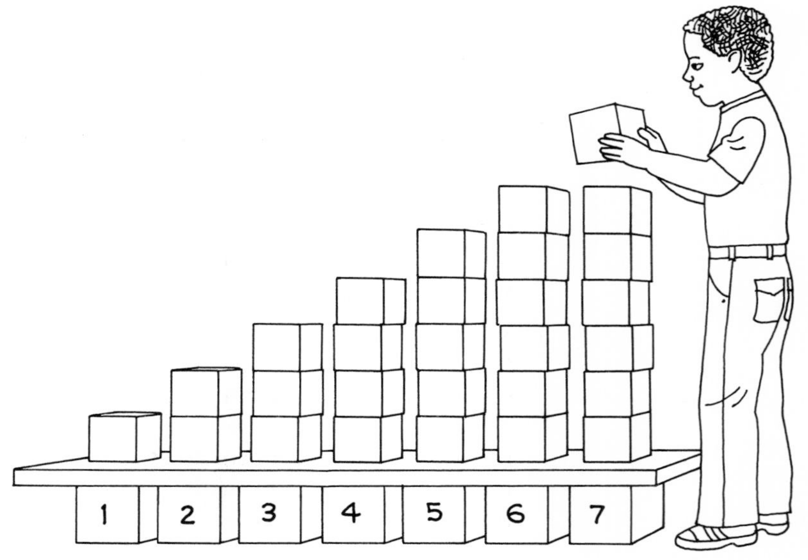

Activity 2
Identify how many objects are in a given set. Point out the relationship between the cardinality of a set and the name of the last object

to be enumerated. Activities to assess this skill are suggested as follows.

A. Ask the student to put an *X* on the ninth object and a *O* on the fourth object.

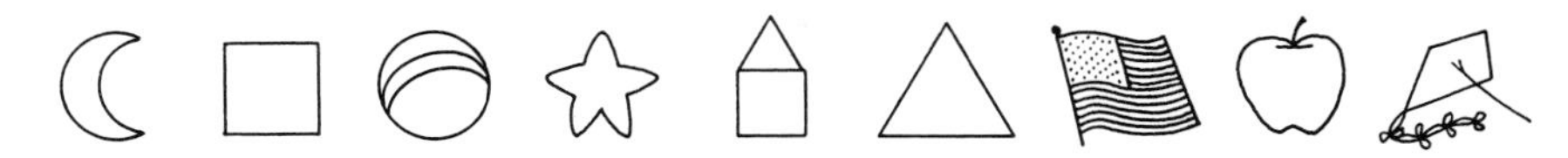

B. Ask the students to write the number of objects in each set and then to put an *X* on the set that ends with the ninth object.

Activity 3
Have students rearrange sets of 4 to 9 objects in linear sequence. Have them point to the first and last objects in each of the linear sequences they create. Tell them the ordinal names of the last object in the various sequences. After providing many practice activities of this type, instruct students to assign each object in a linear sequence its ordinal name. This lends itself to group cooperation as some may remember some ordinal names and other children can fill in names not remembered. Students may also arrange in sequence flashcards with ordinal names written on them.

Activity 4
Children need guidance in the direction of movement when writing digits. You may hold the child's hand as you and the child form the digits. Tracing ditto sheet numerals with directional arrows may be helpful but only as a drill technique after you are sure the child writes in the desired direction. Ask children to write the digits while listening to music.[2]

2. The author still remembers the fun she had when teaching a remedial handwriting class during her lunch break. I had children (grades 4 to 6) whose handwriting was illegible write the numerals to the blasting of the Beatles's records. After a few weeks of voluntary attendance, handwriting improved.

The following are directions for forming the digits:

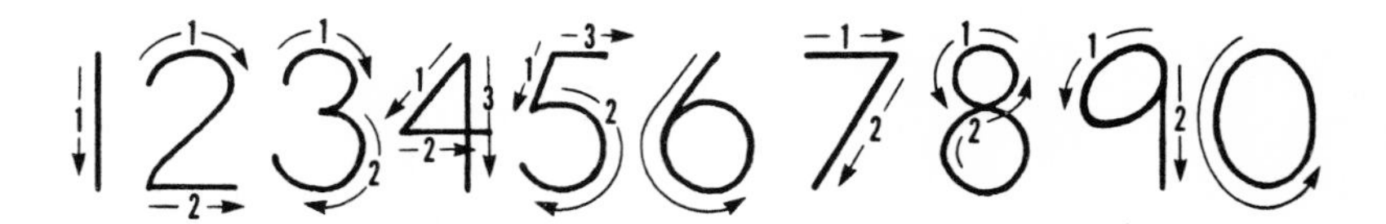

Activity 5
Have students write the digits 0 to 9 from dictation.

Activity 6
Place objects in a linear sequence. Ask students to enumerate these objects aloud and then write the counted sequence.

Additional Considerations: In order for children to understand that number symbols stand for number values when participating in these activities, they should record on paper what they are noticing in regard to cardinality and ordinality of sets of objects. This practice serves as a bridge from concrete and picture activities to those involving symbols. The following are examples of practice exercises to help children consolidate their skills and remember what they learned.

A. Show students sets having different numbers of objects. Ask them to draw a line from the digit 2 to each set having 2 objects.

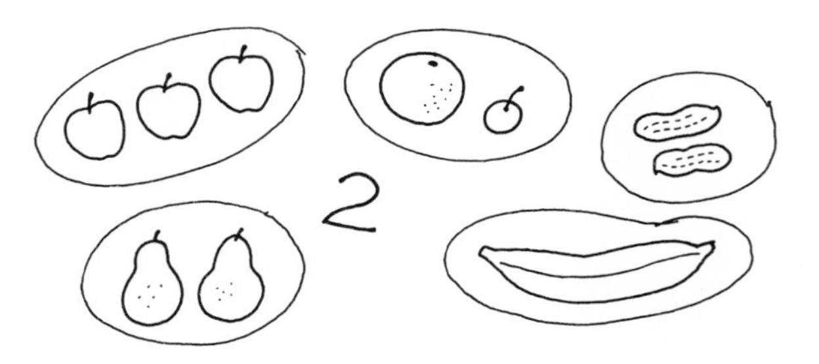

B. Show the students pictures of blocks with different numbers of dots on them. Ask them to write the number that tells the number of dots on the block.

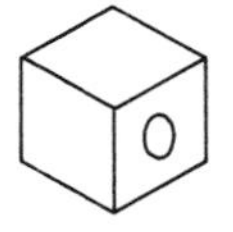

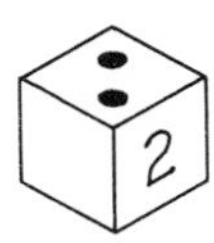

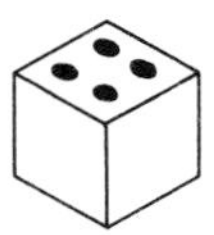
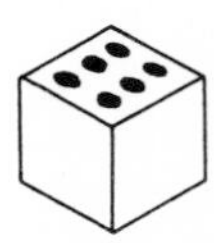
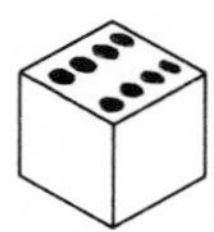

Note: Mini-lesson 1.1.2 is a prerequisite for Mini-lessons 1.2.3 and 1.3.3.

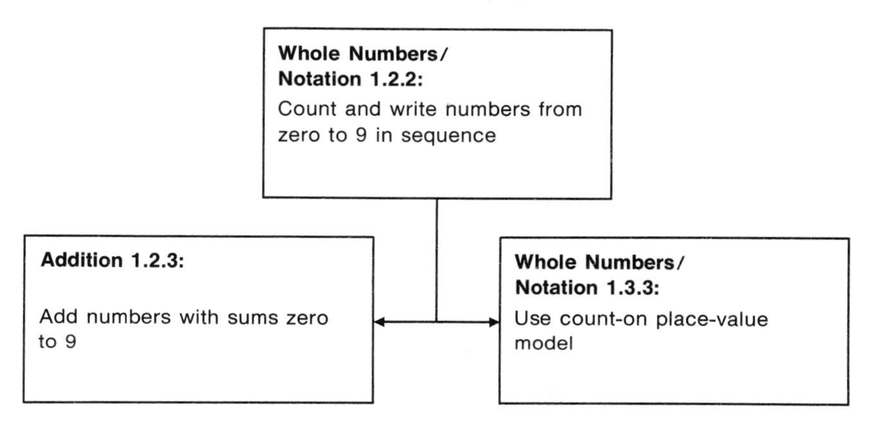

EXERCISES

Match the following instructional goals with the corresponding activities. A goal can be chosen more than once or not at all; some goals make no sense as stated. Choose the best goal.

Instructional goals

1. Select specified set(s).
2. Match specified sets.
3. Count specified set(s).
4. Write specified set(s).
5. Divide specified set(s).
6. Select equivalent set(s).
7. Match equivalent sets.
8. Count equivalent sets.
9. Write equivalent sets.
10. Divide equivalent sets.
11. Select objects in a set.
12. Match objects in a set.
13. Count objects in a set.
14. Write objects in a set.
15. Divide objects in a set.
16. Select numerals showing cardinality of a set.
17. Match numerals showing cardinality of a set.
18. Count numerals showing cardinality of a set.
19. Write numerals showing cardinality of a set.
20. Divide numerals showing cardinality of a set.
21. Select numerals.
22. Match numerals.
23. Count numerals.
24. Write numerals.

25. Divide numerals.
26. Select equation.
27. Match equation.
28. Count equation.
29. Write equation.
30. Divide equation.

Activities

The child is asked to:

a. Point to objects of a specified color
b. Make a set of blocks matching a set placed before him or her.
c. Circle pictures of equivalent sets

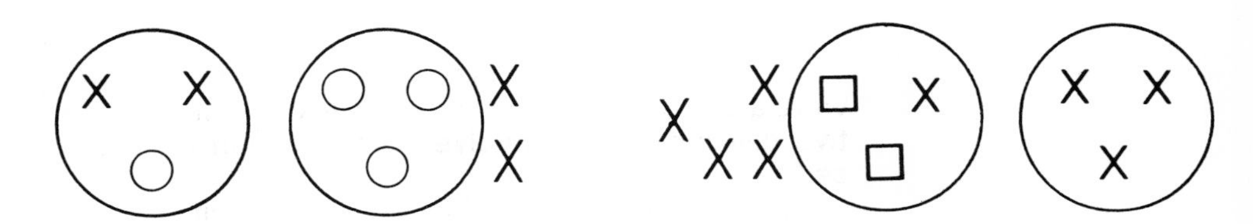

d. Enumerate sets when shown different sets of objects
e. Match a set of objects by selecting objects from a group of mixed items
f. Write the two-digit numeral represented on an abacus
g. Circle the correct numeral when it is spoken aloud:

25 53 (35) 30 "Thirty-five"

h. Color all of the squares in the picture below:

i. Match two sets having the same number of objects
j. Draw lines to connect like objects in sets
k. Circle all of the yellow objects in a picture
l. Place the appropriate numerals before different sets of objects
m. Write two-digit numerals from picture representation

tens	ones	
XX X	XX	32

n. Write the correct numeral when it is spoken aloud

o. Write the numeral that represents each set

```
XXX      XX
XXX      XX
XX       X
 8        5
```

p. Tell how many objects are in different sets of objects

q. Write addition and multiplication equations from pictorial representation

```
X X X        3 + 3 = 6
X X X        2 × 3 = 6
```

r. Divide a set of objects into a specified number of sets and write a corresponding division equation (This is an example of partitive division, which involves the question of "how many in each set?")

s. Divide a group of objects into a specified number of sets with some extra

```
(X X X)  X        8 ÷ 3 = 2 rem. 2
(X X X)  X
```

t. Write three- and four-digit numerals from the following picture representations

```
XX   XX    XXXXX
XX   X      XXXX   XX
 4    3       9     2  = 4392
```

u. Count groups of tens on an abacus

SUGGESTED READINGS

Boyd, Henry. "Zero the Troublemaker." *The Arithmetic Teacher* 16 (May 1969): 365–67.

Denmark, Tom, and Kepner, Henry S., Jr. "Basic Skills in Mathematics: A Survey." *Journal for Research in Mathematics Education* 11 (March 1980): 104–123.

Yates, Daniel S. "Magic Triangles and a Teacher's Discovery." *Mathematics Teacher* 69 (May 1976): 351–54.

FOR FURTHER STUDY

The union of sets of numbers called *natural counting* numbers, *whole* numbers, *integers, rationals,* and *irrationals* comprise the set of *real numbers.* The sets of numbers in the Real Number System are presented in Chart 2.1. A rationale for this sequence is based on the closure property. The closure property states that for a given set of numbers and a given operation, the number resulting from the operation will be a member of the same set of numbers as the original numbers. Therefore, if there exists in a set any two elements that yield an answer that is *not* in the set, then the system is not closed for that operation. The closure axiom for a number system implies that the system strives for closure.

Chart 2.1. Sets of Numbers in the Real Number System

Set of *natural counting numbers:* $\{1, 2, 3, 4, \ldots\}$*
Set of *whole numbers:* $\{0, 1, 2, 3, 4, \ldots\}$
Set of *integers:* $\{\ldots -4, -3, -2, -1, 0, 1, 2, 3, 4, \ldots\}$

Set of *rationals:*
$$\left\{\begin{matrix} \ldots \frac{-4}{1}, \frac{-3}{1}, \frac{-2}{1}, \frac{-1}{1}, \frac{0}{1}, \frac{1}{1}, \frac{2}{1}, \frac{3}{1}, \frac{4}{1}, \ldots \\ \ldots \frac{-4}{2}, \frac{-3}{2}, \frac{-2}{2}, \frac{-1}{2}, \frac{0}{2}, \frac{1}{2}, \frac{2}{2}, \frac{3}{2}, \frac{4}{2}, \ldots \\ \ldots \frac{-4}{3}, \frac{-3}{3}, \frac{-2}{3}, \frac{-1}{3}, \frac{0}{3}, \frac{1}{3}, \frac{2}{3}, \frac{3}{3}, \frac{4}{3}, \ldots \\ \vdots \end{matrix}\right\}^{\dagger}$$

Set of *irrationals:* The set of numbers that may be expressed as an infinite nonrepeating decimal‡
Set of *complex numbers:* Numbers of the form $a + bi$, where a and b are real numbers and $i = \sqrt{-1}$ such that $i^2 = -1$§

*Some authors are now including zero in the set of counting numbers to answer such questions as "How many three-dollar bills do you have in your pocket?"

†For any two integers a and b ($b \neq 0$), there exists a unique number q such that $q \times b = a$. The number q is called a "rational number" (see Chapter 14).

‡See Chapter 17 for explanation of infinite nonrepeating decimal.

§The set of real numbers may be considered a proper subset of the set of *complex numbers:* Numbers of the form $a + bi$, where a and b are real numbers and $i = \sqrt{-1}$ such that $i^2 = -1$.

Teaching Place Value

The fact that many children arrive at kindergarten already able to count up through the teens is both a blessing and a problem. The number names *ten, eleven, twelve,* and possibly *thirteen* are thought of in the same way as are the numbers from one to nine. Students count in a linear sequence where "nine" triggers "ten"; "ten" is the stimulus for "eleven"; "eleven" triggers "twelve"; and so on. At this point, there is probably no awareness of "I have ten units; that's one group of ten and no extra."

THE PLACE-VALUE SYSTEM

Encipherment

Encipherment, the assigning of a separate symbol to a number, was a key step in the maturing of place-value notation. This concept was apparent in the Egyptian hieratic script found on papyrus rolls and in the Indian Brahmi numerals for 10 (∝), 20 (ꝏ), 30 (↗), 40 (X), 50 (J), 60 (⊣), 70 (ꭓ), 80 (ↀ), and 90 (⊕).

Eventually, encipherment was restricted to units values only. This allowed for the development of our present-day place-value notation, in which the names of the ranks are not written but implied by position. Enciphering allowed each number up to the quantity of the base to be represented by a single symbol (which we call a *digit*). This concept along with using a symbol to represent zero led to our set of the 10 digits: 0, 1, 2, 3, 4, 5, 6, 7, 8, 9.

Characteristics of a place-value system

The Hindu-Arabic system has (a) a finite set of symbols (the 10 digits 0 to 9), (b) a base of ten, and (c) rules for combining these symbols. Some of these rules are unique to place-value systems, and some have remained from less-mature systems of numeration.

Underlying principle of a place-value system

The *place-value system* involves this principle: *Each digit in a numeral represents the product of the number it names* (*face value*) *and the value assigned to its position in the numeral* (*place value*). *The entire numeral named is the sum of the products of face value times place value.*

Function of base

The *model group,* or *base,* has an important function in determining the values of the positions in a place-value numeration system. (Any counting number greater than 1 can be used as a base for a place-value system.) The base becomes a *factor* (a term in multiplication referring to numbers multiplied to obtain a product), and the value of each position becomes a *power* of the base (obtained by repeated multiplication of the base). Thus, in our numeration system, in which the base is ten, moving a digit one place to the left multiplies its face value by a power of ten. Place value in our system, then, is a power of the base, with values successively increasing by 10 from right to left. Place value refers to the number assigned to a position.

Use of decimal dot

Place-value systems are also used to write numerals for numbers that are not whole numbers. For this purpose, the *decimal dot* is used, and place values less than 1 are assigned positions to the right of the position having a value of 1 (base to the zero power, b^0). Each digit, then, does two jobs:

Function of digit in a numeral

1. Its position in a numeral names its place value.
2. Its arbitrary number assignment, referred to as its *face value* (what the digit looks like no matter where in a numeral it may be found), tells how many of a particular value are indicated.

Then what does the base-ten numeral 364.02 mean?

$$\begin{aligned}364.02 &= 3 \text{ hundreds} + 6 \text{ tens} + 4 \text{ units} + 0 \text{ tenths} + 2 \text{ hundredths}\\ &= (3 \times 100) + (6 \times 10) + (4 \times 1) + (0 \times 1/10) + (2 \times 1/100)\end{aligned}$$

The expanded form of 364.02 shows that a numeral having more than one digit is a sum. What makes up this sum? The sum consists of products resulting from the multiplication of face value times place value.

So we find in our system that rather than having a new symbol to designate that we have reached a power of the base, the position immediately to the left does this job. Digits are necessary to specify how many of a particular power are involved. For example, if in base ten, the value for position b^2 is hundreds, the number of hundreds can range from zero (0) to nine (9). As soon as another hundred is added to 9 hundreds, we have 10 hundreds, which makes a new positional grouping, or power to the base—b^3, or thousands. We use the next larger power, since we have no single digit for 10 in base ten.

Exponential notation

The term *base* has a general use beyond that of a model group in a numeration system. In algebra, the number that is used as a repeating factor is called the "base" (this number may not be zero). A special numeral called an *exponent* is used to designate the power to which the base is raised. An exponent is a superscript numeral, a numeral written to the top right of a base numeral. Thus, $10 \times 10 \times 10$ in exponential notation would be 10^3. The 10 is the base; the exponent 3 indicates the power to which the base is raised. If the base is represented by b, and the exponent by n, then b^n is read "base to the nth power." Chart 3.1 illustrates the process of generating place values (using Chart 3.1 as a model, the reader may find it helpful to show successive powers when the base is equal to three, five, seven, and eleven. This exercise will help prepare the reader for a better understanding of the section dealing with exponents [see Chapter 19]).

Chart 3.1. Generating Place Values

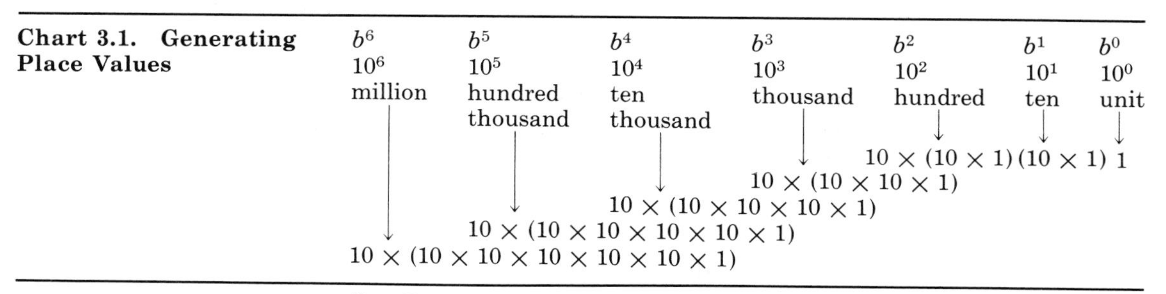

Note: For an explanation of b^0, see Chapter 19.

TASK ANALYSIS OF PLACE VALUE

A task analysis reveals that two very different relationships are involved in understanding place value—a count-on-by-one sequence and an exchange model.

The *count-on-by-one model* emphasizes the fact that when recording the counting sequence, a change in thinking from units to tens occurs after the count of 9. The bundling of sticks model, on the other hand, involves a change in thinking after the count of 10. We count 10 sticks and then bundle them to represent 1 ten.

In the beginning stages of learning, children do not think "I have ten units; that's one group of ten and no units." (In fact, this is a division relationship.) Until they construct some basic multiplication ideas, they should not be expected to understand a positional notation system because the values of the positions are products, i.e., $10 = 10 \times 1$, $100 = (10 \times 10 \times 1)$, $1{,}000 = 10 \times (10 \times 10 \times 1)$, and so on. In the numeral 7*3*6, the italicized digit has a face value of 3, a place value of 10, and a total value of 30, obtained by multiplying the digit's place value by its face value: $3 \times 10 = 30$.

Mini-lesson 1.3.4 Count and Write Numbers Through 19 in Sequence

Vocabulary: counting board, units, ones, tens, place value, face value, space value, ten, eleven, twelve, thirteen, fourteen, fifteen, sixteen, seventeen, eighteen, nineteen

Possible Trouble Spots: does not copy patterns correctly, showing lack of awareness of sequencing; poor memory skills; makes errors in configuration of digits, i.e., reversals, illegible

Requisite Objectives:
Whole Numbers/Notation 1.3.3: Use count-on place value model

Whole Numbers/Notation 1.2.2: Count and write numbers from 0 to 9 in sequence

Construct pattern for counting beyond 13

Activity 1
The count-on place value model described below is prerequisite to the bundling-by-tens model as it involves a simple count-on-by-one sequence. The suggested activities should be spaced over one to three weeks to allow enough practice time for students to understand how the value of positions in our notational system are developed.

A. Give each student his own counting board and a counting chip or token.[1] An empty board represents a value of zero; each space has the value of the space just below plus 1, thus representing an add-1 sequence.

---

1. Fredricka K. Reisman and S. H. Kauffman, *Teaching Mathematics to Children with Special Needs* (Columbus, Ohio: Charles E. Merrill Publishing Co., 1980).

9
8
7
6
5
4
3
2
1

In beginning work with a counting board, write the space values on the board. Tell the students that a rule of the game involves putting only 1 chip on a board at a time. Have students practice counting board values from zero to 9 by jumping a chip onto the board to count 1, moving the chip up 1 space to show 2, and so on up to 9. Thus, a chip on a board is analagous to the face value of a digit. When only one board is used, it represents units or ones place value.

B. Provide a second counting board and chip. Position the boards as follows for continuing the counting sequence beyond 9.

9	
8	
7	
6	
5	
4	
3	
2	
1	
	9
	8
	7
	6
	5
	4
	3
	2
	1

The board to the upper left represents the tens place value. Remind students that as the chip is moved up a space, the board value increases by 1. Present the problem of what the count is if the chip is moved up 1 move beyond 9; it is now just above the units board. If necessary tell the students the name of the next number, *10*. However, most children will probably be able to count up through 10, 11, or 12 by rote. Move the chip to the left onto the bottom space on the board to the upper left, as shown below.

Since only a horizontal move and no upper move occurred, there was no increase in value. Thus, the value of the bottom space on the left-hand board is 10.

For some students, it may be better to omit writing the digits on the tens board. This will emphasize the horizontal move to the left at the count of 10 rather than the idea of "1 ten."

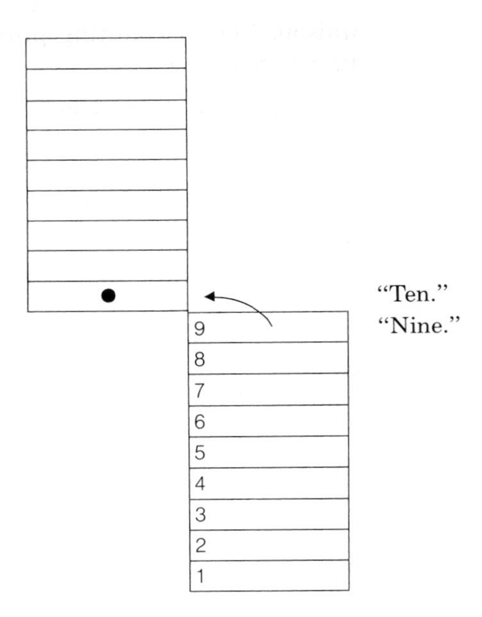

Place the second chip next to the 1-space on the units board and instruct a child to show the next number in the counting sequence. Be sure that the student does not attempt to show 11 by moving the chip on the 10-space to the 20-space. Have the child continue counting in sequence on the counting boards.

C. After several days of practice, lower the tens board to a position adjacent to the units. Have students show counting sequences beyond 9 and record the numerals represented on the boards as shown.

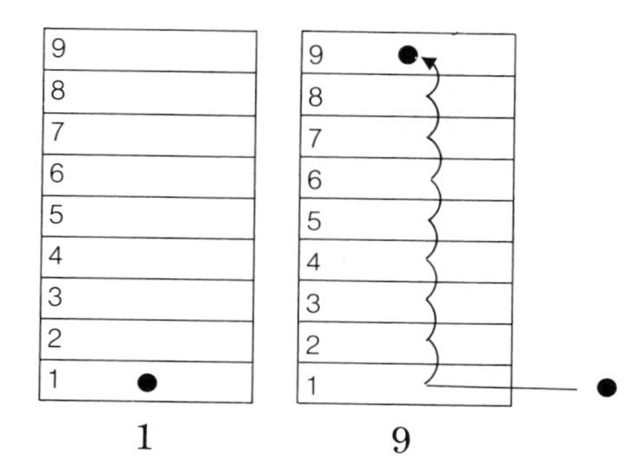

Activity 2
Have students participate in verbal counting activities.

A. Provide experiences that allow students to count up to 19 in unison. Choral counting provides practice in saying the "teens" in proper sequence.

B. Have students count out 19 counters (chips, strips, pennies, etc.). Then, commence counting from various numbers and ask students to continue the count to 19: 9, 10, 11, . . . 19. This task may be done as both group and individual activities.

Activity 3
Apply the count-on sequence model to show numbers through 19 by using the counting boards as follows. Construct a model of the counting boards by pasting strips of paper (each one numbered 1 to 9) to cardboard. Then cut a see-through window in another piece of cardboard and insert strips as shown below.

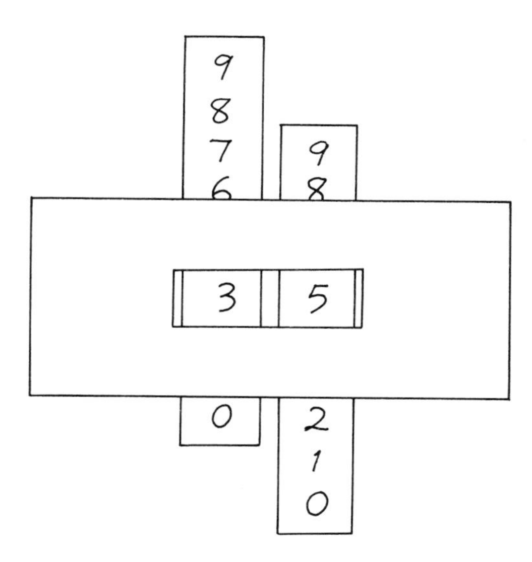

A. Indicate the number 1 by showing 0 on the left board and 1 on the right board that represents units. Increase the units board values 1 space at a time. When 9 shows, pull out the units board and insert it again to show 0 while simultaneously moving the tens board down to 1, thus showing 10. Continue pulling the units board down through 19.

B. Dictate numbers and direct students to continue the counting sequence orally and in writing; e.g., say, "Two, three, four, . . . continue." Shown below is a paper-pencil activity.

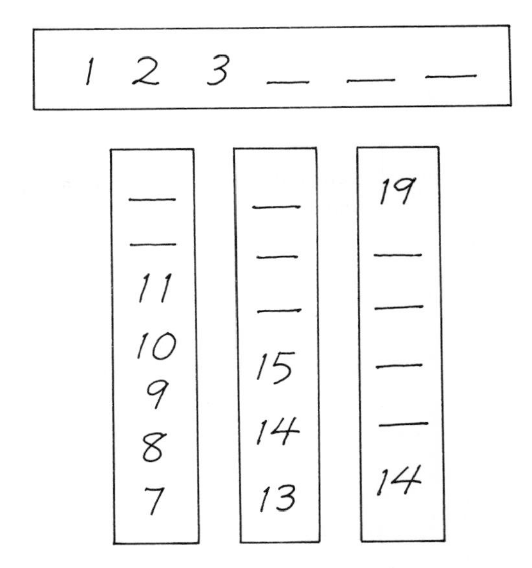

Additional considerations: The jump onto the counting board to show the movement from a value of zero to a value of 1 is crucial. The importance of counting moves rather than spaces underlies obtaining a correct count on the number line and on a clockface as well. When students do not realize that it is the moves that are counted, they end up with one too many in their count. The empty board to represent zero points up a discrepancy between a physical representation of the number zero and a graphic representation. We show zero value with nothing at the concrete level, but at the symbolic level we use something, namely, the digit 0. Another crucial movement, the jump off the top of the board to show the number 10, is a direct physical representation of the fact that we run out of 1-digit symbols at the count of the number 10. The horizontal move to the bottom space on the board that represents the next higher place value is a concrete representation of the way we reuse the digits to represent numbers greater than 9.

The activities for this mini-lesson lend well to whole class participation. Practice in drilling the sequence of numbers helps those who have poor memory skills.

Note: Mini-lesson 1.3.4 is a prerequisite for Mini-lessons 1.3.7 and 1.3.9.

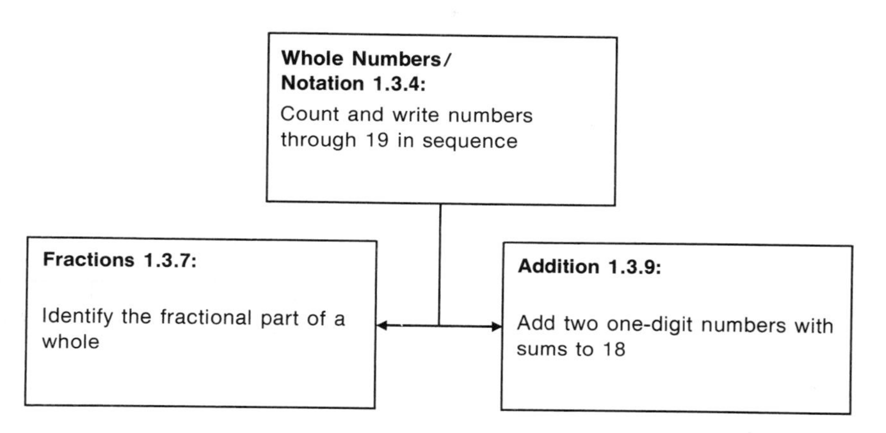

Mini-lesson 1.3.5 Show Many-to-One and One-to-Many Representations of the Same Number

Vocabulary: many-to-one, two-for-one, five-for-one

Possible Trouble Spots: does not know one-to-one relationship; cannot handle money in artificial situations

Requisite Objective: Match sets in one-to-one correspondence

Activity 1
Students may role play life of cavemen and have 1 pebble stand for so many sheep; a knot in a rope or a notch on a stick stand for a group of objects.

Activity 2
Have students match 5 pennies to a case nickel; 5 nickels to a case quarter; 2 nickels to a case dime; 10 pennies to a dime; nickel to 5 pennies, etc. A case nickel, dime, or quarter refers to the single coin of a value, i.e., 5 cents, 10 cents, 25 cents respectively.

Additional considerations: Students who perform one-to-one exchanges may not necessarily know how to perform the many-to-one exchange. This process must be taught directly for many students. Proponents of the back-to-basics movement suggest movement away from discovery learning and toward a greater emphasis on direct teaching followed by ample practice. It is especially important that students learn the topic of this mini-lesson, since the many-to-one exchange underlies two major mathematics topics: place value and multiplication.

Mini-lesson 1.3.6 Count By 2s, 5s, and 10s

Vocabulary: twenty, thirty, forty, fifty, sixty, seventy, eighty, ninety

Possible Trouble Spots: does not sequence according to simple patterns; poor ability to generalize; poor memory skills

Requisite Objectives:
Whole Number/Notation 1.3.3: Use count-on place value model

Place objects in sequence according to a given pattern

Recognize duality

Sets 1.3.1: Separate sets into subsets

Activity 1
Using the counting boards described in Mini-lesson 1.3.4, have students count in sequence through 19. Then work with the total class to extend the bulletin board table shown below through 90 by 10s.

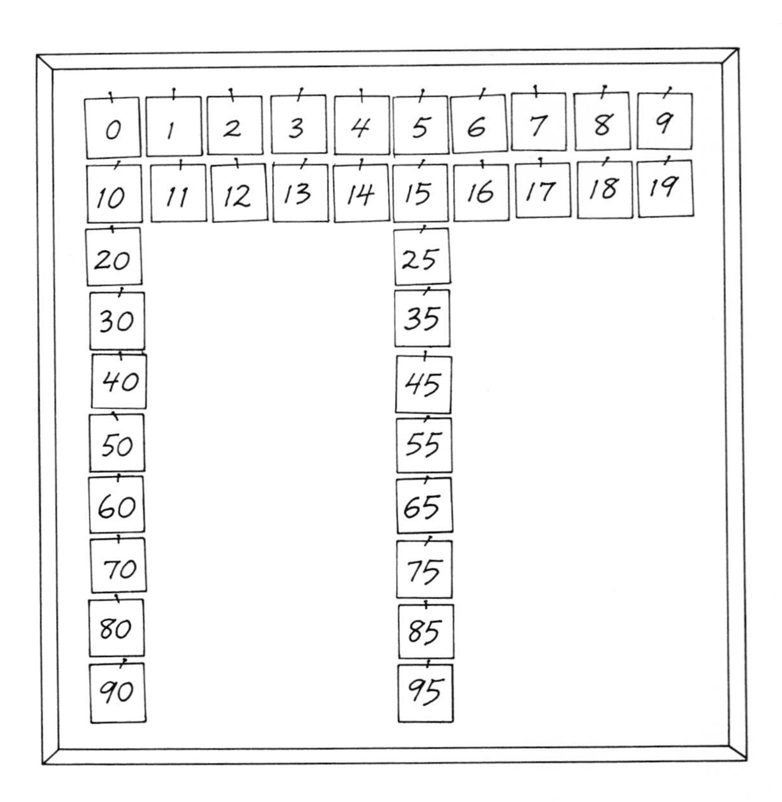

Activity 2
Next, initiate the 5s pattern by cueing as follows: "ten, fifteen, twenty, twenty-five, what comes next?" Complete the 5s column on the table and allow students to use this bulletin board as a check on related seat work.

Activity 3
As a preparation for counting by 2s, students must be aware of duality. Have them point out duality in nature, their bodies, and geometric shapes.

A. Have students separate sets of 4, 6, 8, and 10 objects into 2 equivalent sets. Then ask them to make subsets with 2 objects in each from sets of 4, 6, 8, and 10.

B. Ask the students to count a set of 20 objects by 2s.

Activity 4
Given a set of 20 pebbles, have students make sets of 2s and count by 2s; do the same with sets of 5s and then 10s. Use sets of 2, 5, and 10 objects at both concrete and picture representations (e.g., blocks, dominoes, dot cards) for counting by 2s, 5s, and 10s. This may be a group activity.

Activity 5
Have students complete grids as shown.

5	15	25			55				95
10	20			50				90	

2	4		8		12			18	20

5	10	15			30	35	

Additional considerations: The skill of counting by 2s is especially necessary if students are to understand the forward move on a counting board when 2 checkers on the 1-space are exchanged for a single checker placed on the 2-space.

Note: Mini-lesson 1.3.6 is a prerequisite for Mini-lessons 1.3.7 and 1.3.16.

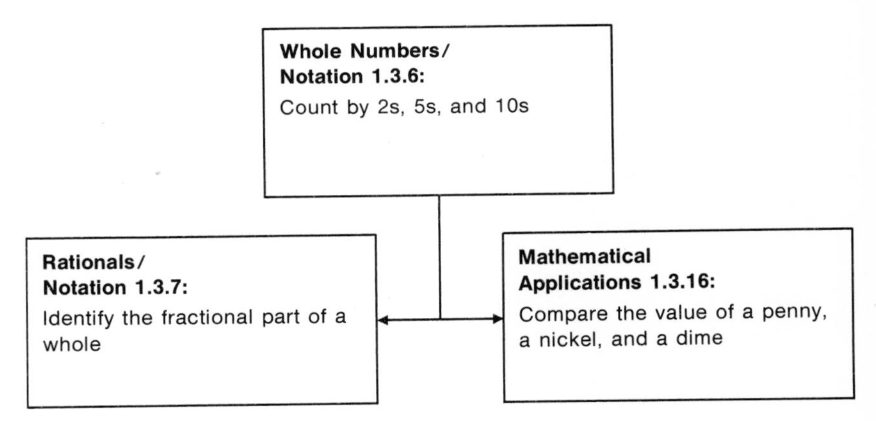

Mini-lesson 2.2.1 Use Exchange Model for Place Value

Vocabulary: 2-for-1 exchange, many-to-one, place value, ones, units, tens, digit, numeral, odd, even, same as

Possible Trouble Spots: does not realize that a number may be expressed using a variety of names; is not a conserver of number; does not realize that place value is related to multiplication; does not understand two-for-one or many-to-one exchanges

Requisite Objectives:
Make two-for-one exchanges

Multiplication 1.2.5: Become aware of beginning multiplication ideas

Whole Numbers/Notation 1.3.3: Use count-on place value model

Whole Numbers/Notation 1.3.6: Count by 2s

Addition 1.3.9: Add two 1-digit numbers with sums to 18

Recognize conservation of number

Whole Numbers/Notation 2.1.2: Apply many-to-one generalization to exchange model of place value

Whole Numbers/Notation 2.1.4: Recognize odd and even numbers

Whole Numbers/Notation 2.1.3: Extend number notation to 99

Multiplication 2.1.8: Show multiplication situations related to many-to-one relationships

Graphs and Functions 2.1.12: Select situations showing the reflexive relation

Activity 1
Overload spaces on the units counting board by allowing more than 1 chip on a space (see facing page). Then move up on the board by

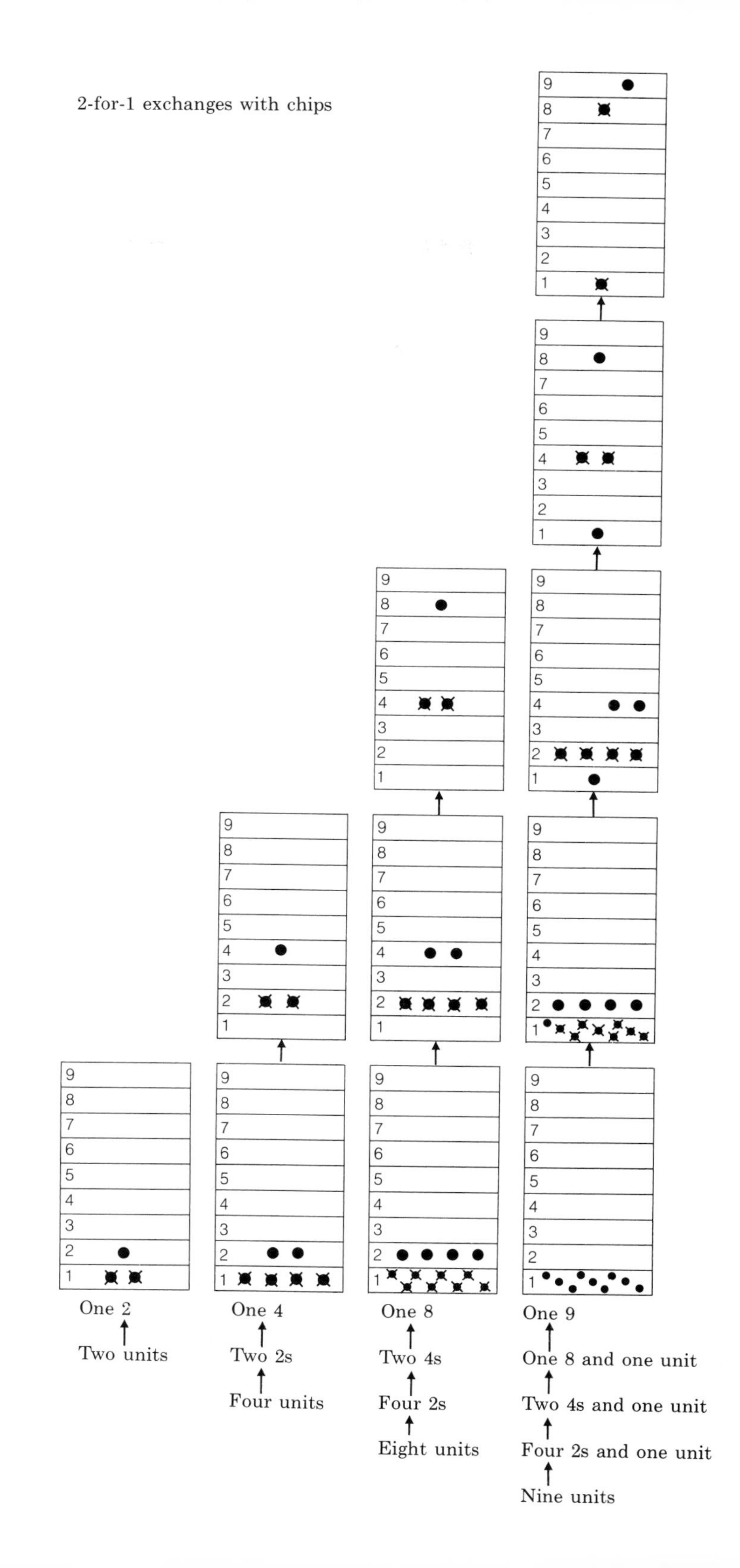
2-for-1 exchanges with chips
One 2
Two units
One 4
Two 2s
Four units
One 8
Two 4s
Four 2s
Eight units
One 9
One 8 and one unit
Two 4s and one unit
Four 2s and one unit
Nine units

2-for-1 exchanges. Have students show 2-for-1 exchanges on the counting board while recording number exchanges. This is preparation for expanded notation, e.g., "four 2s and one unit."

Record movements on the counting boards as follows on the chalkboard.

two units ⟶ 2
two 2s ⟶ 4
two 4s ⟶ 8

Activity 2
To enable students to practice beginning multiplication relationships, extend the 2-for-1 exchanges to many-for-one exchanges on the counting boards as shown below.

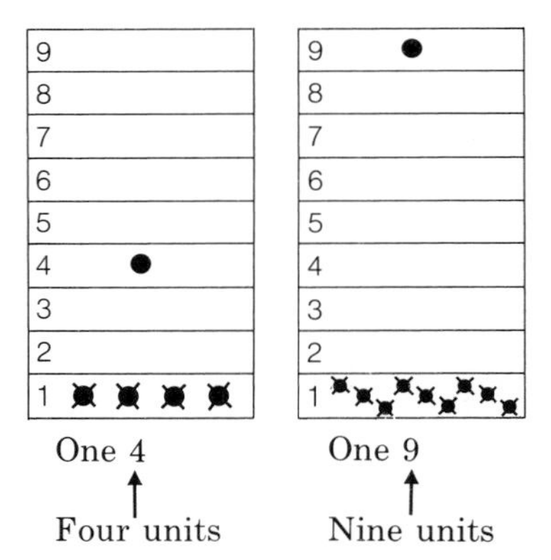

Activity 3
Have a group of 8 students form the following: 2 teams of 4; 4 teams of 1; 8 teams of 1; and 1 team of 8.

Observe students' language as they describe the situations, e.g., "two fours."

Activity 4
To enable students to apply the many-to-one generalization to the exchange model of place value, extend use of the units counting board to show a variety of many-to-one exchanges that illustrate different physical representations for the same number using only the units counting board. Set the rule that the final representation of a number should use the fewest number of chips.

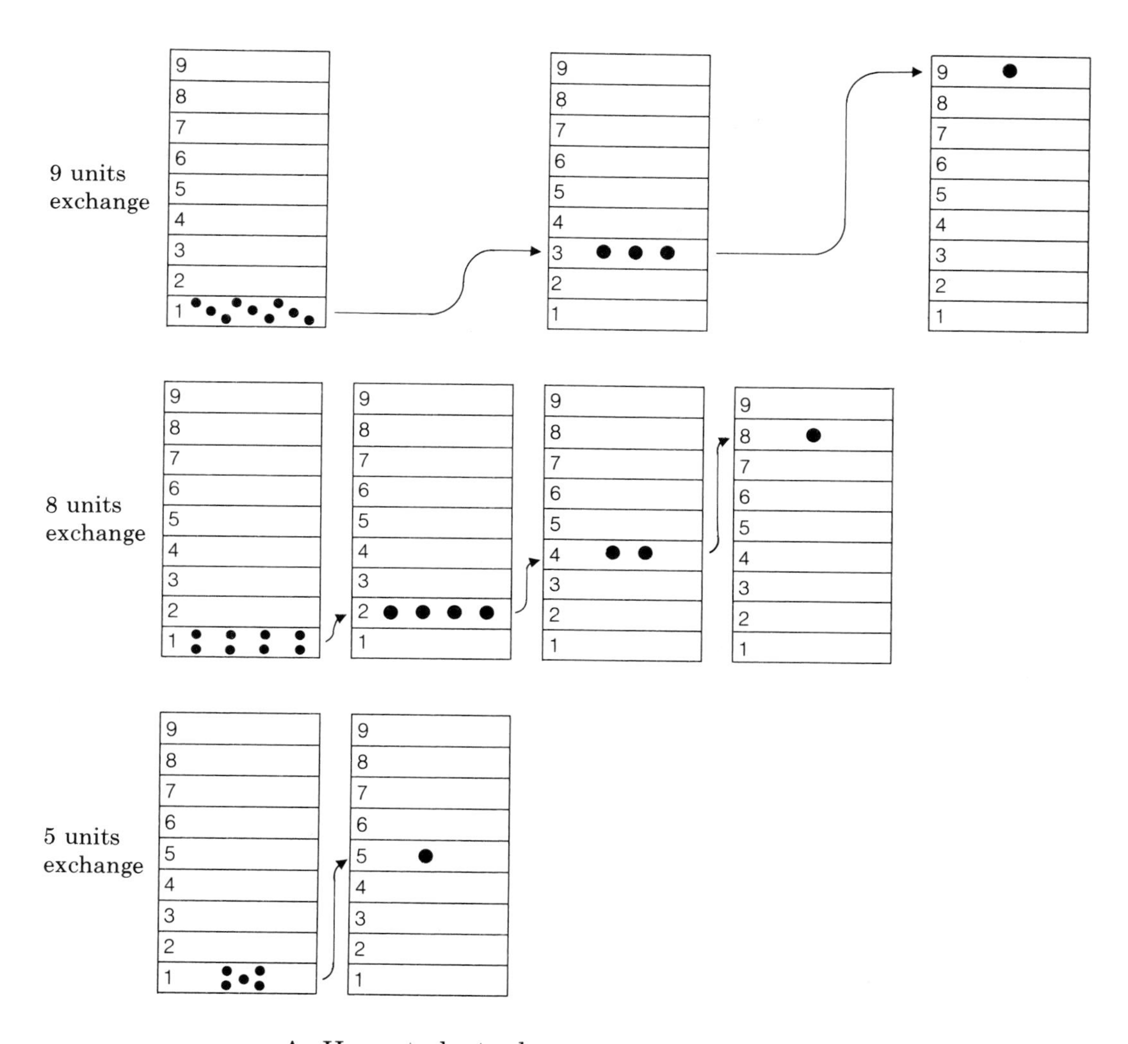

A. Have students show:

three units ⟶ one 3
three 3s ⟶ one 9
four units ⟶ one 4
four 2s ⟶ one 8
five units ⟶ one 5
six units ⟶ one 6
⋮
nine units ⟶ one 9

B. Next, overload the units board with 10 or more tokens. Provide the tens board as shown. Have students continue the many-to-one exchanges that represent the same number, through 19. Allow students to make individual exchanges with the goal of eventually using only 2 chips to represent two-digit numbers.

Shown on the counting boards are 10 units as 1 ten and 15 units as 1 ten, 5 units. Students may practice other many-for-one exchanges, such as 13 units → 1 ten, 3 units, and 19 units → 1 ten, 9 units. Note that the words *units* and *ones* are used synonymously by some teachers.

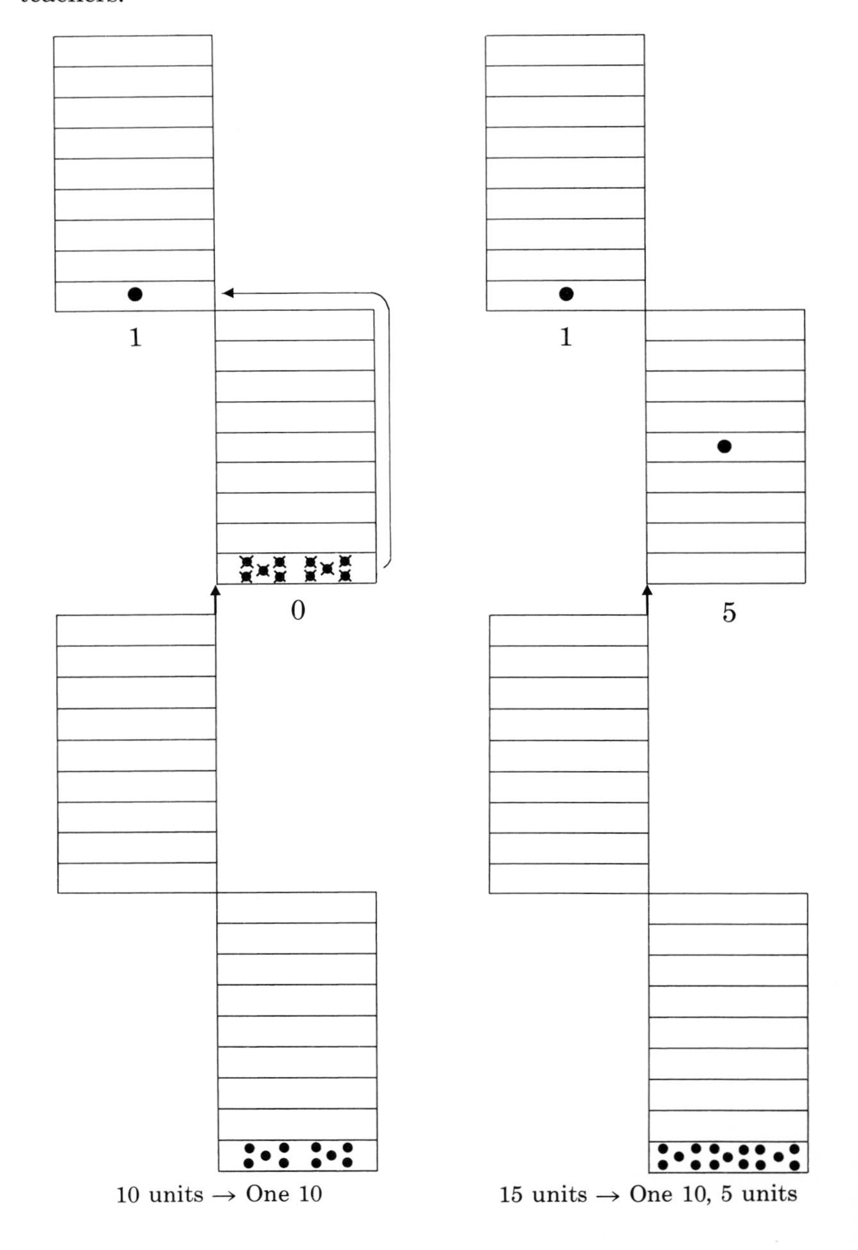

10 units → One 10

15 units → One 10, 5 units

Activity 5
Extend number notation to 99 by having students use units and tens counting boards to keep counting tallies up through 99. Start with both boards empty to represent zero; count to 10; continue, and, at the count of 20, students should make a 2-for-1 exchange from the 10-space to the 20-space as shown.

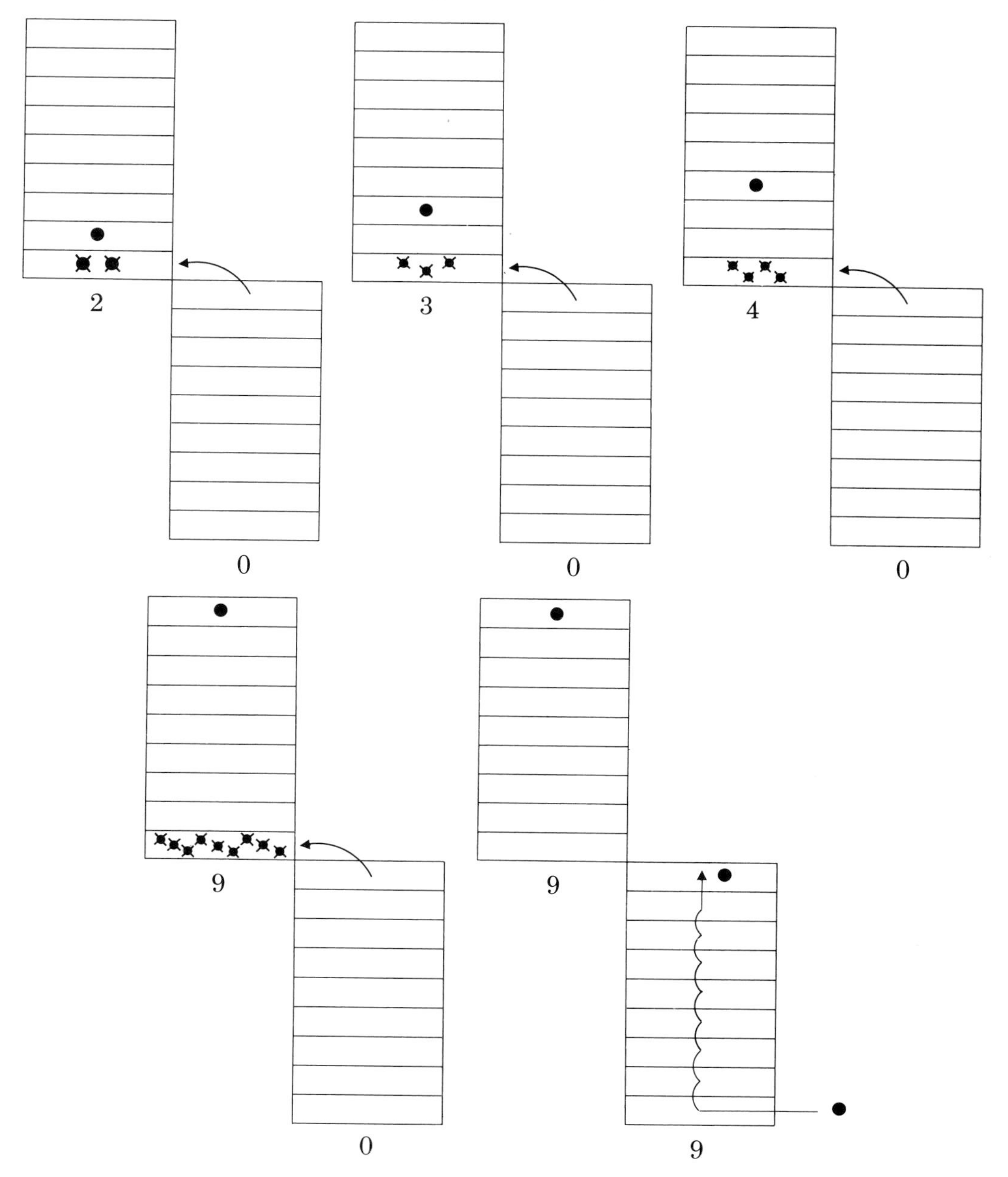

A. Further counting may involve 3-for-1, 4-for-1, etc. exchanges.

B. There are many situations that involve use of numbers of 10 and greater. Recognition tasks will provide practice. For example, give students books with at least 99 pages. Call out page numbers for students to locate.

C. Have students complete tables as shown below.

1	2			5	6			9	10
11	12				16	17	18		20
	22		24	25			28	29	
31		33			36	37			40
		43	44		46		48		50
	52	53		55	56		58		
61			64	65		67	68		70
	72	73		75			78		80
				85	86	87		89	
	92	93							

Activity 6
Recognizing odd and even numbers may serve as a cue in making 2-for-1 exchanges.

A. Review counting by 2s by asking the child to complete a table like the one shown on top of p. 55. Ask her to color all even numbers orange. Then tell her to count by 2s and put an *X* on all even numbers. Ask what number pattern is shown in the units place.

B. Have students use the computation boards to show even and odd numbers (see bottom of p. 55). Even numbers may be shown by 2-for-1 exchanges and can be shown by fewer chips than odd numbers on the computation boards. The 2-for-1 exchanges here must involve 2 chips of the same value, for example, two 2s or two 4s.

0	1	2		4	5				9
10		12	13			16			19
		22		24		26		28	29
		32							
40	41	42	43	44	45	46	47	48	49
				54				58	
60		62		64				68	
70		72		74		76		78	
		82							
90	91	92				96		98	99

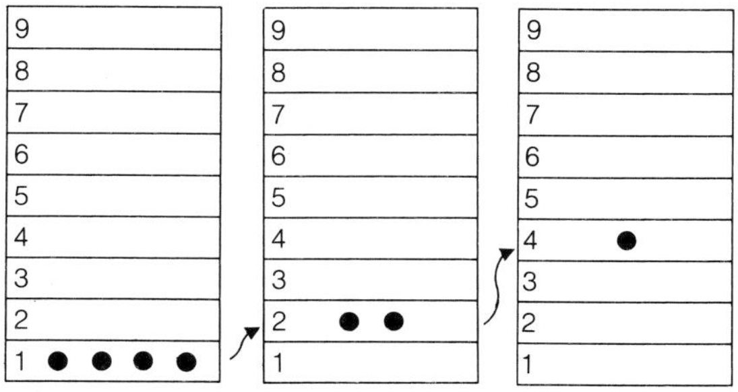

"Four is even."

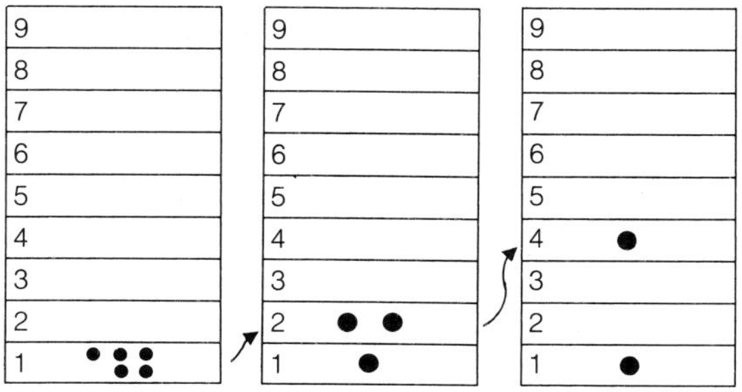

"Five is odd."

C. Paper-pencil activities to show odd and even numbers include asking the child to circle odd and even numbers.

"All the circled numerals are even. All the underlined numerals are odd."

Then show the child pictures like the following, asking him to color all even numbers black and all odd numbers blue.

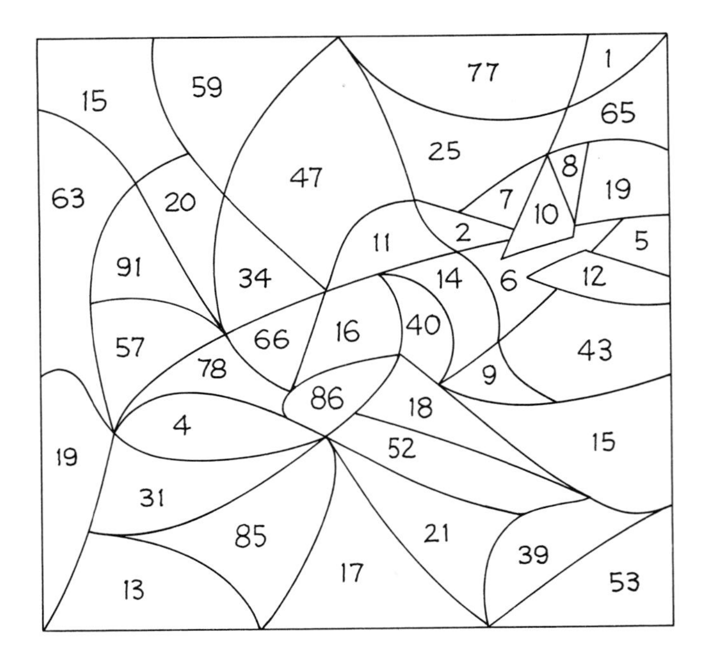

Activity 7
Students may now use their skills with basic addition facts to make 2-for-1 chip exchanges that sum to 10, e.g., 9-space and 1-space, 8-space and 2-space, and 7-space and 3-space (see top of p. 57).

A. Next, have students make exchanges that sum to 10 with chips remaining on the units board. This is a beginning division idea (see bottom of p. 57).

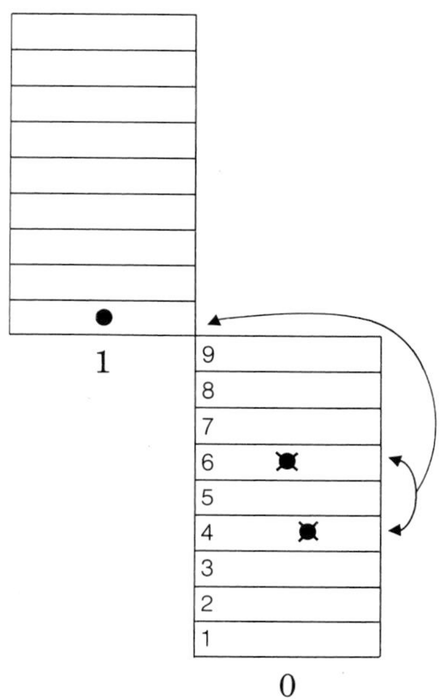

2-for-1 exchange showing
6 units and 4 units as
one 10.

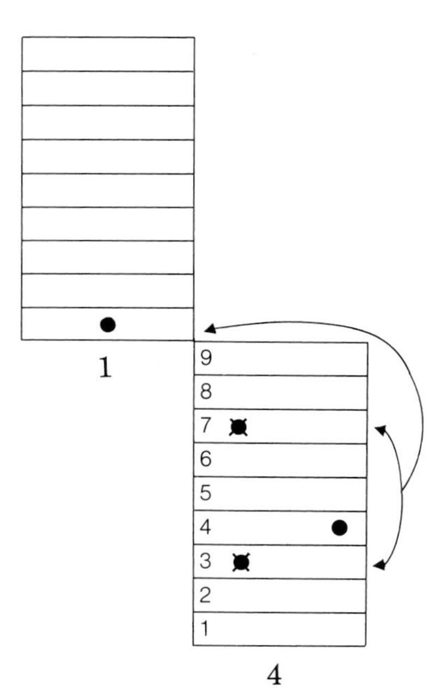

2-for-1 exchange showing
14 units as one 10 and
4 extra units

B. To assess if students are conservers of number, provide activities whereby they must show recognition of different names and physical representations of the same number. For example, have students show the numbers 3, 8, and 9 on the units counting board in several ways. Then, using the tens and units board, have them show numbers greater than 9 in many ways.

Activity 8
Have students show multiplication situations related to many-to-one relationships.

A. After counting 10 units represented on the 1-space, assess whether the student recognizes that the statement "there are ten units" is a multiplication relationship that can be expressed as $10 \times 1 = 10$. This is then extended to "three tens are thirty" expressed as $3 \times 10 = 30$, etc., as shown below.

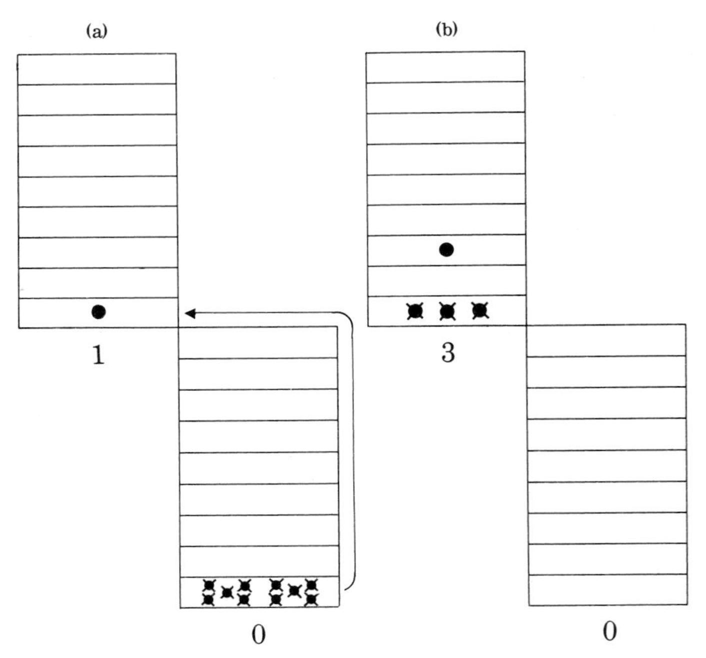

B. Extend multiplication expressed as many-to-one generalization through $9 \times 10 = 90$. Have students show this on the abacus, with pocket chart, and with multibase blocks.

Observe evidence that students notice the two ways multiplication is involved: to generate place value and to indicate the value of a digit (face times place values).

Activity 9
To show awareness of the reflexive relation, have students point out the difference between *absolute* and *similar* interpretation of "the

same as." The absolute meaning refers to identical objects, events, or numbers. This underlies the fact that the physical change of 10 units for 1 ten is a situation where we are really making a 10-for-10 exchange; the number is the same regardless of its representation.

Activity 10

By now, students should be aware that the sequencing by adding 1 to 9 to obtain 10 (used in writing numbers) is not the same as making a 10-for-1 exchange to obtain 1 ten for place value. They must also be aware of the identity idea whereby different physical models are possible to represent the same number. They should notice that in a written number, the place value is unseen, whereas the face value of a digit is seen.

A. Have students complete activities involving exchange of 10 units for a chip on the 1-space of the tens board.

B. Provide experiences involving a combination of addition and multiplication utilizing repeated 2-for-1 moves as in a 9-space and a 1-space exchanged for a chip on the 1-space of the tens board. Then have them make an exchange on the tens board of the three 10-space tokens for a tens board 3-space token to show 30.

C. Provide practice such as showing 13 units as follows: Show a 2-for-1 exchange to represent 10 as 8 + 2 and a forward move to the tens board; the units board shows 3.

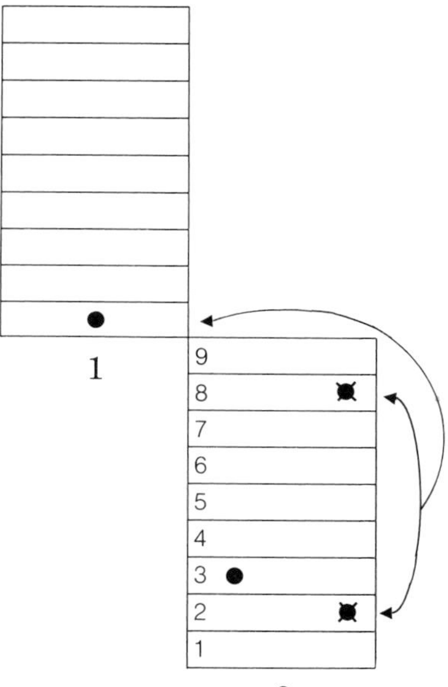

Additional considerations: It should be apparent to the teacher that the exchange model for place value underlies renaming in addition, subtraction, multiplication, and in division when the dividend is renamed. Therefore, the teacher will find it is time well spent to provide students with practice to insure a good understanding of the place value principle.

Note: Mini-lesson 2.2.1 is a prerequisite for Mini-lessons 2.2.4 and 3.1.4.

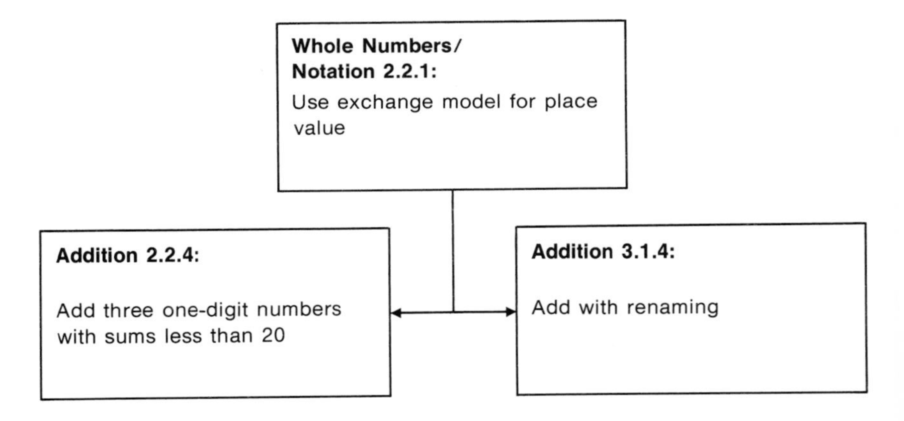

Mini-lesson 4.1.1 Extend Place Value Through Millions

Vocabulary: hundred, thousand, ten thousand, hundred thousand, million, expanded form, expanded notation, periods, comma

Possible Trouble Spots: needs more drill on reading and writing numbers; forgets names of periods; misplaces comma to separate periods, i.e., 431,92 or 4,3192; does not see relationship of expanded form and standard form of numeral, i.e., $125 = 100 + 20 + 5$

Requisite Objectives:

Whole Numbers/Notation 1.3.3: Use count-on place value model

Whole Numbers/Notation 2.2.1: Use exchange model for place value

Whole Numbers/Notation 2.2.2: Extend place value to hundreds place

Addition 2.2.3: Use additive identity

Whole Numbers/Notation 3.1.2: Extend place value through thousands

Multiply with powers of 10 as one of the factors

Whole Numbers/Notation 3.1.3: Write numerals in expanded notation

Activity 1
Extend use of counting boards to hundreds place value by placing a third board in position as shown.

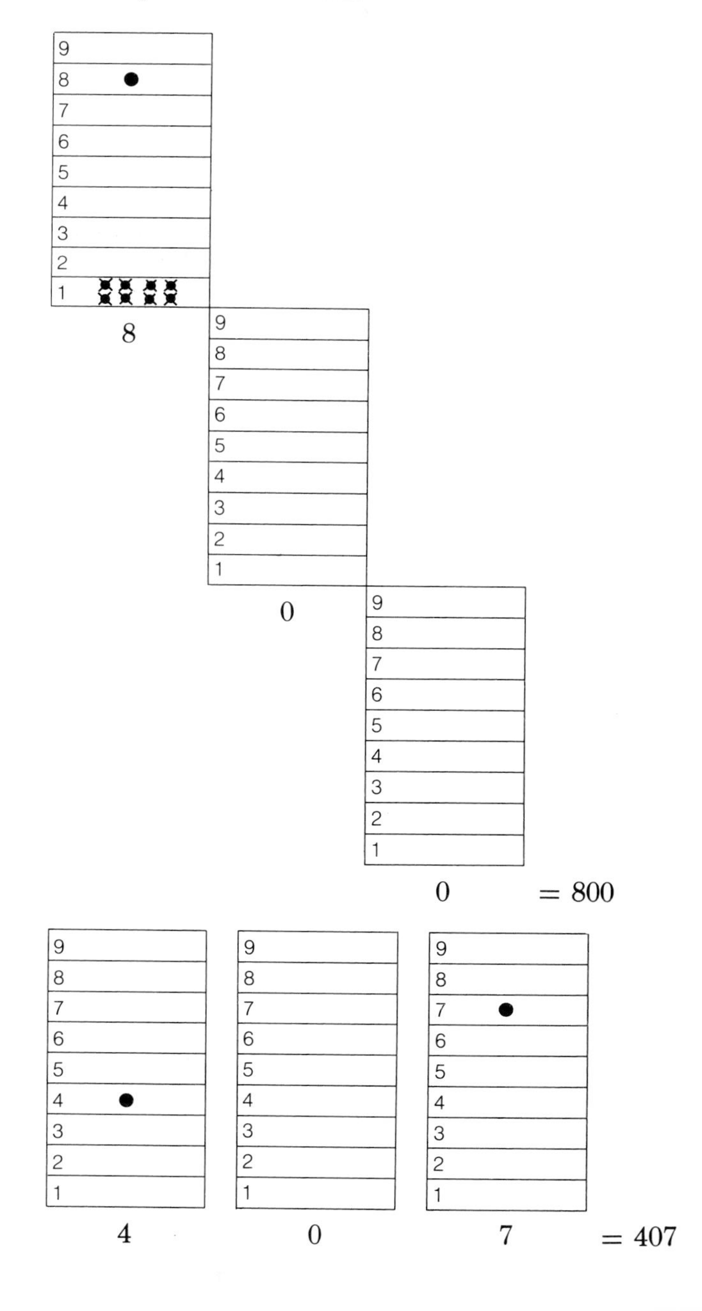

A. Have students make forward moves on the counting boards using 2-for-1 exchanges as shown. They should practice showing numbers to the hundreds place on the boards and record the numbers represented by chips.

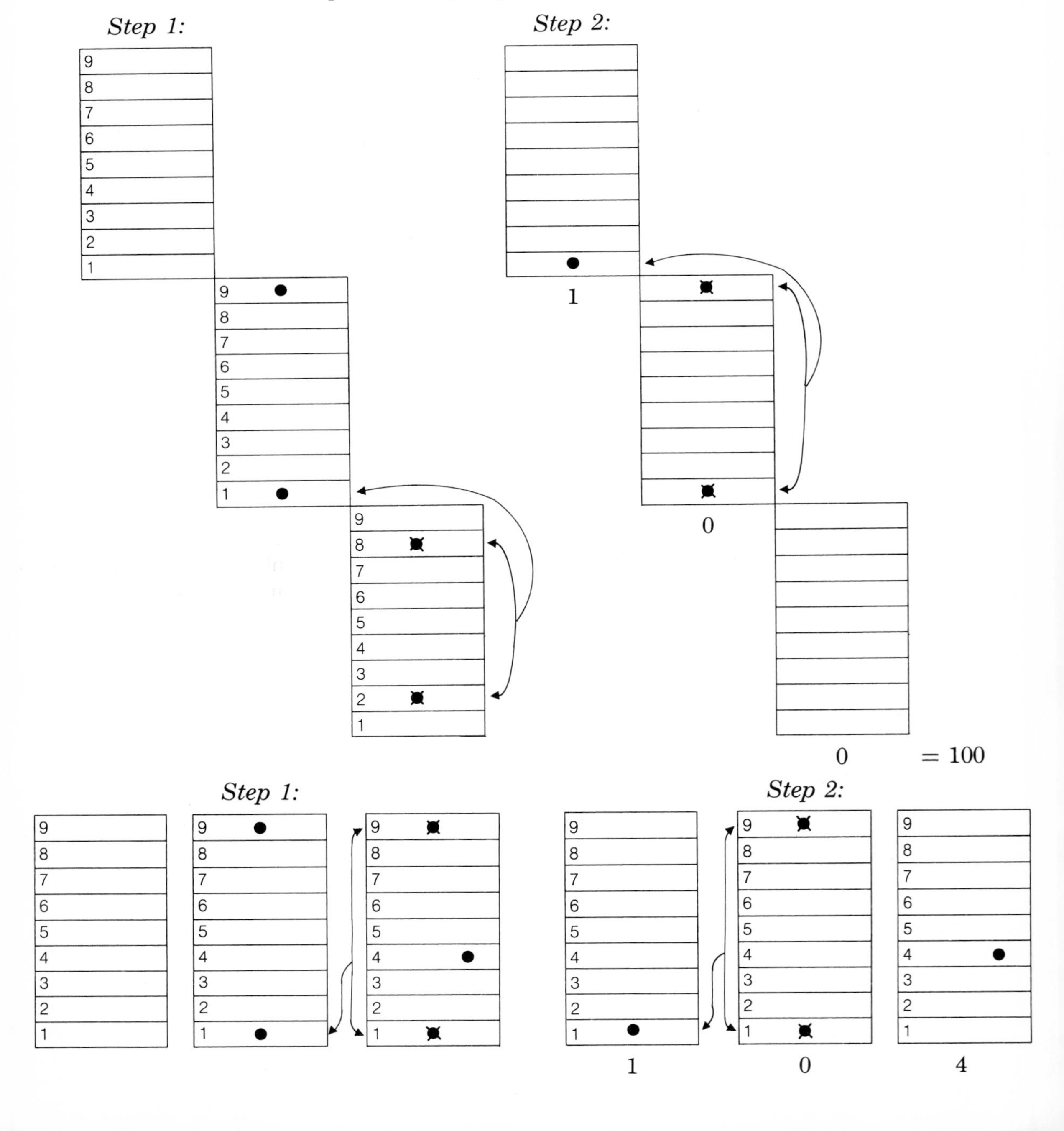

B. Students may also use multibase blocks, abacus, and a place value chart to show hundreds. However, these activities may best be used to represent numbers already shown on the counting boards. In this way, the teacher may check place value by using transfer-type tasks.

Procedure A: Make an abacus by using various colored pipe cleaners bent in a U-shape, with 9 rings on the top loop of each pipe cleaner and each pipe cleaner glued to a base. Have the students compare the loop positions with the counting board positions as they assign place values to the loops. The picture portrays place values through ten thousands.

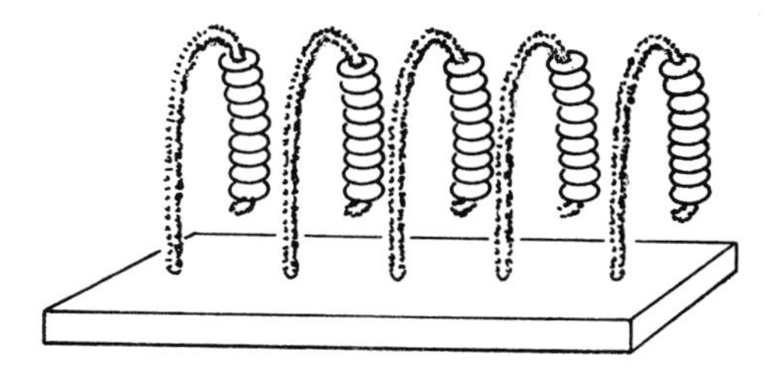

Procedure B: Pull 3 blue rings over the bend in the pipe cleaner and ask how many tens this means and what number 3 tens represents. Pull over 4 red rings and ask what number of singles or units the rings represent. Then ask the child to name the number represented. (Abacus shows 34.)

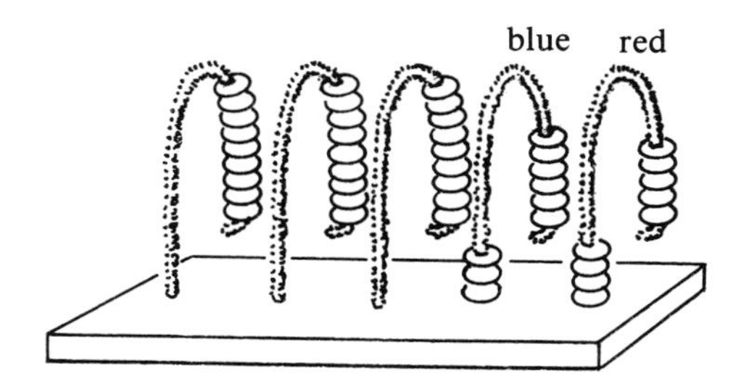

Procedure C: Continue this activity through thousands. If removable loops are used you may control the place values shown.

Activity 2

In order to realize that a numeral is the sum of its component numbers, the student must be aware of the additive identity. For example, to expand 25 as 20 + 5, the student must be able to compute as follows:

$$\begin{array}{r} 20 \\ +\ 5 \\ \hline \end{array} \qquad \begin{array}{r} 400 \\ 30 \\ +\ \ 7 \\ \hline \end{array}$$

Activity 3

Another requisite of writing numbers in expanded form is being able to multiply with powers of ten as one of the factors.

A. Provide examples as follows:

$$2 \times 1 = 2$$
$$2 \times 10 = 20$$
$$2 \times 100 = 200$$
$$2 \times 1000 = 2{,}000$$

B. Have students practice this on a hand calculator and record their answers as follows:

$$1 \times 1 = 1$$
$$1 \times 10 = 10$$
$$1 \times 100 = 100$$
$$1 \times 1{,}000 = 1{,}000$$

by pressing the keys in order as shown: Press 1, then ×, then 1, then =; next, press 1, ×, 1, 0, =; next, 1, ×, 1, 0, 0, =.

Activity 4

Ask the child to complete exercises such as the one on p. 65 for practicing writing numbers in expanded form.

Activity 5

Point out that numbers comprised of more than three place values are read by naming the smallest place value of the three numbers set apart by commas. For example, in a multidigit numeral greater than hundreds, the place values named are *thousand, million, billion, trillion, quadrillion, quintillion, sextillion,* and so on. These are called "periods." Within each period, the numbers are read as those from 1 to 999, followed by the period name. For example, 9,374,681,702,533 is read as "nine trillion, three hundred seventy-four billion, six hundred eighty-one million, seven hundred two thousand, five hundred thirty-three."

A. Have students read numerals given and write numerals that are dictated.

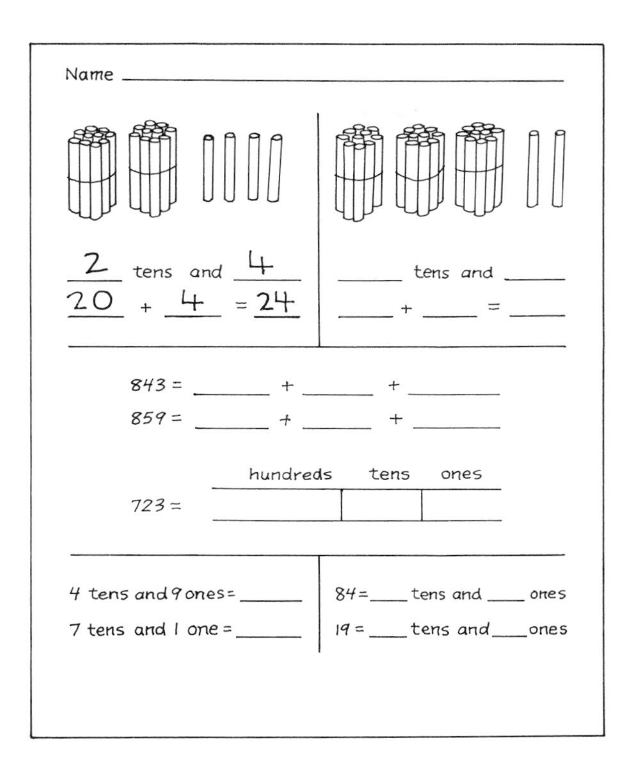

B. Have students complete activities such as the following.

1. Write the numeral: three thousand eight hundred fifty-six: ________.

2. Match each numeral with what we say.

Numeral	*What we say*
3421	four hundred twenty-two
6700	six thousand seven hundred
1080	three thousand four hundred twenty-one
422	one thousand eighty

3. Underline the digit that is in the ones' place: 9843

4. Underline the digit that is in the hundred thousands' place: 4,237,962

Additional considerations: Use of the hand calculator is a helpful drill exercise for teaching notation of large numbers and is especially useful for students with poor eye-hand coordination who have difficulty formulating the digits.

Note: Mini-lesson 4.1.1 is a prerequisite for Mini-lessons 4.1.2 and 4.1.4.

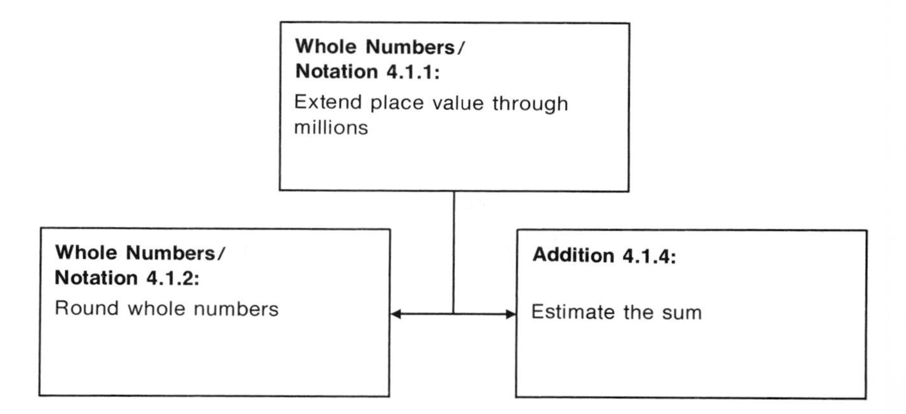

EXERCISES

I. Answer the following questions:
1. Why is our system referred to as the Hindu-Arabic system?
2. Why do we call our system a "decimal numeration system"?
3. In the numeral 3 ③ 1 1̲, how many times as great is the value of the encircled 3 in relation to the value of the underlined 1?
4. If the zero is removed from the numeral 48260, how is the value of each remaining digit affected?

II. Write the following base-ten numerals first in expanded form and then in exponential form:
1. 1,435
2. 796
3. 1,000,431
4. 3,060

III. Explain the difference between the count-on sequence model for teaching place value and the exchange model.

SUGGESTED READINGS

Bidwell, James K. "Mayan Arithmetic." *The Mathematics Teacher* 60 (November 1967): 762–68.

Byrkit, Donald R. "Early Mayan Mathematics." *The Arithmetic Teacher* 17 (May 1970): 387–90.

Cowle, Irving M. "Ancient Systems of Numeration—Stimulating, Illuminating." *The Arithmetic Teacher* 17 (May 1970): 413–16.

Davis, Harold T. "The History of Computation." In *Historical Topics for the Mathematics Classroom,* Thirty-first Yearbook of the National Council of Teachers of Mathematics, edited by Arthur E. Hallerberg, chairman; John K. Baumgart; Duane E. Deal; and Ruth R. Vogeli; pp. 87–117. Washington, D.C.: National Council of Teachers of Mathematics, 1969.

Dienes, Z. P. "Multi-Base Arithmetic." *Grade Teacher* 78 (1962): 97–100.

Freitag, Herta, and Freitag, Arthur. *The Number Story.* Washington, D.C.: National Council of Teachers of Mathematics, 1960.

Ginsburg, Herbert. *Children's Arithmetic: The Learning Process.* New York: D. Van Nostrand, 1977.

Gundlach, Bernard H. "The History of Numbers and Numerals." In *Historical Topics for the Mathematics Classroom,* Thirty-first Yearbook of the National Council of Teachers of Mathematics, edited by Arthur E. Hallerberg, chairman; John K. Baumgart; Duane E. Deal; and Ruth R. Vogeli; pp. 18–36. Washington, D.C.: National Council of Teachers of Mathematics, 1969.

Hess, Adrien L. "A Critical Review of the Hindu-Arabic Numeration System." *The Arithmetic Teacher* 17 (October 1970): 493–97.

Ikeda, Hitoshi, and Ando, Masue. "Introduction to the Numeration of Two-Place Numbers." *The Arithmetic Teacher* 16 (April 1969): 249–51.

Immerzeel, George, and Wiederanders, Don. "Experience in Identification of the Standard Number Symbols." *The Arithmetic Teacher* 18 (December 1971): 577–78.

Moser, James M. "Grouping of Objects as a Major Idea at the Primary Level." *The Arithmetic Teacher* 18 (May 1971): 301–305.

Muente, Grace. "Where Do I Start Teaching Numerals?" *The Arithmetic Teacher* 14 (November 1967): 575–76.

Pincus, Morris. "A Fifth Grade's Revision of Our System of Number Names." *The Arithmetic Teacher* 19 (March 1972): 197–99.

Rudnick, Jesse. "Numeration Systems and Their Classroom Roles." *The Arithmetic Teacher* 15 (February 1968): 138–47.

Smith, David E., and Ginsburg, Jekuthiel. *Numbers and Numerals.* The National Council of Teachers of Mathematics, 1906 Association Drive, Reston, Va. 22091.

Stern, Catherine, and Stern, Margaret B. *Children Discover Arithmetic: An Introduction to Structural Arithmetic.* New York: Harper & Row, 1971. (See pp. 227–59.)

Wolfers, Edward P. "The Original Counting Systems of Papua and New Guinea," *The Arithmetic Teacher* 17 (February 1971): 77–83.

Zaslansky, Claudia. *Africa Counts.* Boston: Prindle, Weber, Schmidt, 1973.

Addition of Whole Numbers

KEY IDEAS FOR ADDITION OF WHOLE NUMBERS

The following are key ideas related to addition of whole numbers:

Numeration

1. Addition is related to enumerating groups of objects to find the cardinality or "how-muchness" of those groups. *Enumeration* refers to meaningful, or rational, counting, which relies upon a one-to-one correspondence of the number names with the objects being counted.

Counting and cardinality

2. The combined number property (cardinality) of two groups of objects may be determined by enumerating the objects in both sets.

What is a sum?

3. Addition is an operation on two numbers that yields the number obtained when the original numbers are combined. Therefore, the expression $3 + 5$ is another name for the number 8. $3 + 5$ is itself a sum. $3 + 5 = 8$ is called an *equation* because both $3 + 5$ and 8 name the same idea, namely, "eightness." The traditional procedure of showing children the labels for numbers added as

$$\text{addend} + \text{addend} = \text{sum}$$

is confusing. A more appropriate visual aid would show that addend plus addend is an indicated sum:

Suggested board work

addend + addend = sum
sum = sum
6 + 3 = 9
9 = 9

The mathematical equation $6 + 3 = 9$ is allowable because of the reflexive property, which states $a = a$ (discussed in Chapter 3).

When two numbers are added, a larger quantity is *not* being created. The values represented by the numbers before they were added were always in existence. For example, consider two sets of apples, belonging to John and Bill. With 5 apples in John's set and 3 apples in Bill's, 8 apples were always in existence. It is only when the two sets are rethought or combined as one set that the quantities 5 and 3 are renamed as 8.

Suggested board work

Another way of looking at the addition idea is to consider the situation in which one quantity matches in a one-to-one relation with two different quantities combined. For example, if John has 5 apples, Bill has 3 apples, and Buddy has the same amount of apples as John's and

Bill's together, then the number of apples John and Bill have together equals the number of apples Buddy has; namely, 5 + 3, or 8. The 8 does not refer to the apples belonging to John and Bill, as it did in the previous example. Here the 8 refers to apples that match up to 8 others in a one-to-one relation.

Suggested board work

Commutative property for addition

4. The order in which two natural numbers are added has no effect on the sum obtained. The sum of 7 and 2 is 9, regardless of whether you think of adding 7 to 2 or 2 to 7; $7 + 2 = 2 + 7$. This is referred to as the Commutative Property for Addition (CPA) and may be represented algebraically as $a + b = b + a$.

Associative property for addition

5. Addition is a binary operation; that is, it is a process that operates on only two numbers. In the addition of any 3 natural numbers a, b, and c, the same sum is obtained regardless of the grouping of the addends. This property is called the Associative Property for Addition (APA) and may be expressed as

$$(a + b) + c = a + (b + c)$$

Young children often refer to this rule as the "grouping law." It is suggested that the teacher use both terms together—"associative, grouping law"—to prevent the children from engaging in meaningless verbalization of the names of the axioms.

Additive identity

6. Additive identity states that

$$a + 0 = 0 + a = a$$

This allows us to perform operations on sets containing a cardinality of zero. Since by definition there is no zero in the set of natural, counting numbers $\{1, 2, 3, \ldots\}$, it is necessary in discussing the concept of additive identity to make use of the set of whole numbers (which includes the number zero).

Closure for addition of whole numbers

7. The closure idea states that for a given set of numbers and a given operation, the number resulting from the operation will be a member of the same set of numbers as the original numbers processed. For example, when two whole numbers are added, the resulting number will also be a whole number. Thus, the set of whole numbers is said to be closed under addition. Is this true for subtraction? In the case of $5 - 3 = 2$, 2 is a whole number, but in $3 - 5 = -2$, -2 is not a whole number. In order to obtain closure under subtraction, the set of integers must be included.

Mini-lesson 1.2.3 Add Numbers with Sums Zero to 9

Vocabulary: sum, addend, plus, join, add, addition, equal, number line, counting-on

Possible Trouble Spots: does not use one-to-one correspondence in counting; does not stop the counting sequence when having to count only part of a group, e.g., when given 8 blocks and asked to count only 6 of them, the student continues until all blocks are counted

Requisite Objectives:
Enumerate objects in a set

Relate combining actions to addition

Whole Numbers/Notation 1.1.5: Indicate the cardinal number property of sets

Whole Numbers/Notation 1.1.7: Write the digits 0 to 9

Sets 1.2.1: Combine disjoint sets

Activity 1
Have students enumerate objects in a set by engaging in the following activity.

Procedure A: Show the child a set of balloons. Ask him how many balloons there are in the set. The last number counted represents the cardinality, or "how-muchness," of the set.

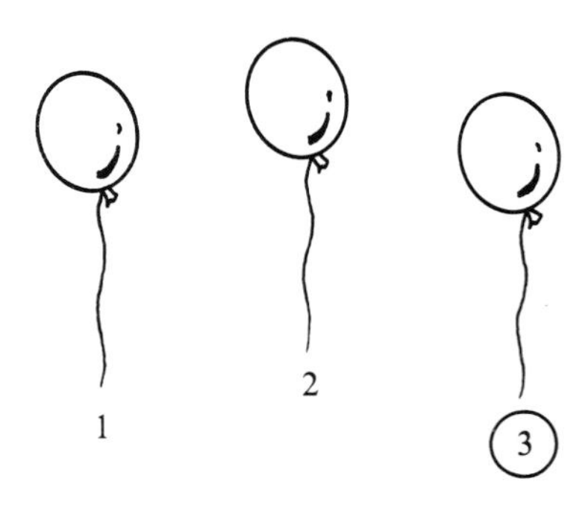

Procedure B: Next show the child two sets of balloons and ask her to count all of the balloons in both sets to find how many balloons there are altogether. (This is called a "counting-on" process.)

Procedure C: Show the child two sets of balloons. Ask him to find the total number of balloons in each set and say a mathematical sentence to show his thinking. The child should respond, "Three plus two is five."

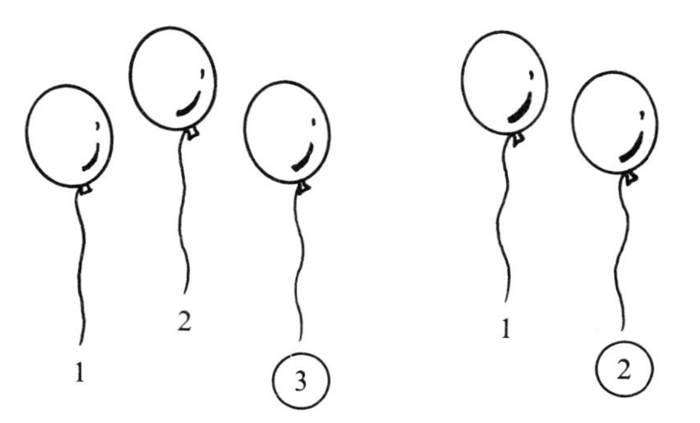

Activity 2
Have the child combine a set of 5 sticks with a set of 4 sticks and record the problem as

$$\begin{array}{r} 5 \\ +4 \\ \hline 9 \end{array}$$

Continue in this manner with problems having sums less than 10.

Activity 3
Have the child show addition with sums less than 10 on the counting boards (shown below is a 2-for-1 exchange) and then record the vertical algorithms as shown previously in Activity 2.

Step 1:		*Step 2:*	
9		9	●
8		8	
7		7	
6		6	
5	●	5	
4	●	4	
3		3	
2		2	
1		1	

Activity 4
Students may use cues to find the standard name of the sum. For example, they may use the counting-on process from one addend to the sum. This may be shown by placing the number of counters for one of the addends vertically beside the board. This is shown below. Point out to the students that the 4 tokens were exchanged for the 4-counter, each token having a value of 1. It would be misleading to show the counting-on process by filling in the 6, 7, 8, and 9-spaces, since a token on a space indicates the value of that space.

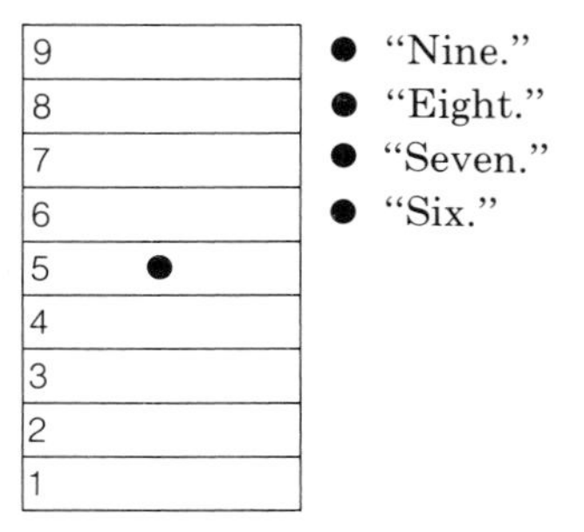

Finally, the students count-on from the addend still shown on the board to the space that names the sum. The 5-space token is then removed, and a token is placed on the 9-space to represent the number 9.

Activity 5
Have the child show addition with sums less than 10 on the abacus and then record vertical algorithms.

Activity 6
Give the child a number line and say to him, "Draw appropriate jumps to show three plus five equals eight."

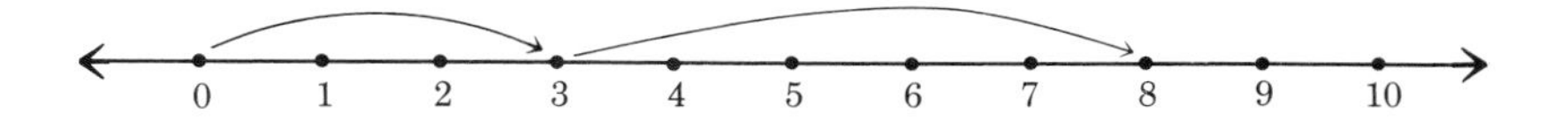

Additional considerations: Students should be instructed to record their activities simultaneously with the manipulations at the concrete level. A mistake to watch for as children use a counting-on procedure to find sums is the pattern of counting some objects more than once.

Note: Mini-lesson 1.2.3 is a prerequisite for Mini-lessons 1.2.5 and 1.2.9.

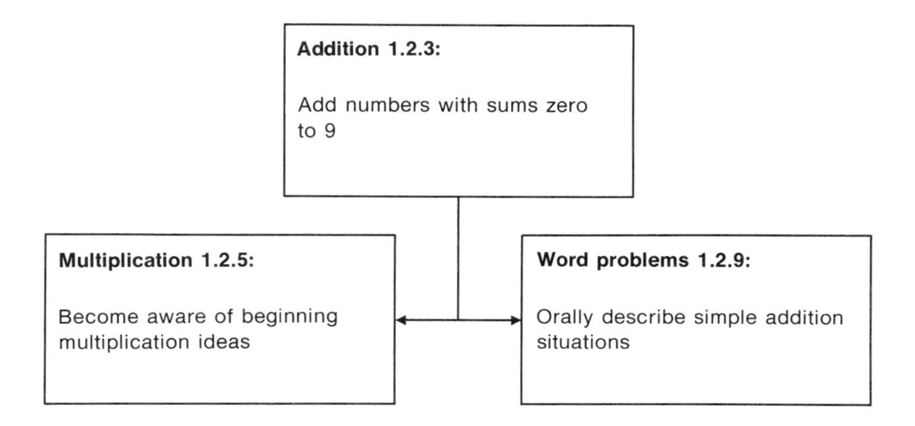

BASIC ADDITION FACTS

A basic addition fact is a sum of two of the digits 0 through 9. The 100 basic addition facts are important tools that children will use throughout their computational experiences in addition. An addition grid as shown on p. 76 is a helpful device for children to use in learning the basic facts. The child who is instructed to circle sums that he or she got wrong may notice a pattern of errors. For example, the student may see that of the 36 sums greater than 10, he or she added 22 incorrectly. By applying the commutative law, this number of errors may be reduced to 11 (if the errors involve the same addends, such as if 7 + 6 and 6 + 7 is recorded as 14). A child is willing to learn 11 basic facts because this becomes a task with success in sight.

Addition Grid

+	0	1	2	3	4	5	6	7	8	9
0	0	1	2	3	4	5	6	7	8	9
1	1	2	3	4	5	6	7	8	9	10
2	2	3	4	5	6	7	8	9	10	11
3	3	4	5	6	7	8	9	10	11	12
4	4	5	6	7	8	9	10	11	12	13
5	5	6	7	8	9	10	11	12	13	14
6	6	7	8	9	10	11	12	13	14	15
7	7	8	9	10	11	12	13	14	15	16
8	8	9	10	11	12	13	14	15	16	17
9	9	10	11	12	13	14	15	16	17	18

If a pupil is having difficulty with particular facts, it is helpful to have him complete grids dealing only with those facts. For example, if a child continually misses sums related to addends of 7, 8, and 9, he or she may be helped by compiling the following grid:

+	7	8	9
0	7	8	9
1	8	9	10
2	9	10	11
3	10	11	12
4	11	12	13

If the pupil correctly completes the cells from 0 to 4, then the teacher may present the grids as shown on p. 77.

+	7	8	9
5	12	13	14
6	13	14	15
7	14	15	16
8	15	16	17
9	16	17	18

+	7	9	8
6	13	15	14
8	15	17	16
9	16	18	17
7	14	16	15
5	12	14	13

These procedures encourage pupils to engage in self-diagnosing activities, which help to build self-responsibility for learning.

The mini-lessons presented in this chapter may be used as drill activities for learning the basic addition facts.

Mini-lesson 1.3.9 Add Two 1-Digit Numbers with Sums to 18

Vocabulary: sum, addend, plus, join, equal, equals, basic fact

Possible Trouble Spots: does not notice patterns; does not relate pattern of "pulling out a group of 10" to finding sums greater than 9, e.g., $7 + 6 = 10 + 3 = 13$; does not conserve number

Requisite Objectives:
Addition 1.2.4: Write the operation sign for addition

Word Problems 1.2.10: Translate simple addition word problems to corresponding basic addition facts and solve

Whole Numbers/Notation 1.3.3: Use count-on place value model

Whole Numbers/Notation 1.3.4: Count and write numbers through 19 in sequence

Recognize equivalent horizontal and vertical addition algorithms.

Activity 1
For basic facts with sums from 10 to 18, show students that forming a group of 10 is helpful. They should be encouraged to engage in self-instruction as a cueing technique, as shown on p. 78 for $5 + 6 = 11$. Ask them to record the problem.

Activity 2
Ask students to show $5 + 6 = 11$ on the counting boards. This is appropriate for teams of students working together and serves as a drill activity for learning basic addition facts.

"I will put five black checkers with six red ones."

"That makes ten checkers and one more. I have eleven checkers. Six plus five equals eleven."

Procedure A: The teacher may model the following cognitive monitoring process:

"Let's see, six and five." (*Step 1*)
"Well, I can change the five to four and one." (*Step 2*)

Step 1:

$$\begin{array}{r} 6 \\ +5 \\ \hline \end{array}$$

Step 2:

$$\begin{array}{rr} 6 & \\ +4 & +1 \\ \hline 10 & +1 \end{array}$$

Step 3:

1

1

"Then, I'll add six and four to get ten. I make a two-for-one exchange. I can show this with a checker on the ten-space and a checker on the one-space. The counting boards show eleven." (*Step 3*)

Procedure B: Then, ask students to simultaneously record the basic addition facts that they show on the boards. For example, in

$$\begin{array}{r} 6 \\ +5 \\ \hline \end{array}$$

the 5 was renamed as 4 + 1 in order to obtain 10 by adding 6 + 4. Of course, some students may prefer to add 5 + 5, renaming the 6 as 5 + 1.

Activity 3
Use of a hand calculator is useful as a related drill activity. As students press the keys 7 + 6 = and see 13 on the display, they are practicing this basic addition fact.

Additional Considerations: This mini-lesson includes only sums to 18 (the basic addition facts) rather than sums to 19 (which is the usual case in most basal textbooks). This emphasizes the need for relating instructional goals to requisite objectives.

Note: Mini-lesson 1.3.9 is a prerequisite for Mini-lessons 1.3.10 and 1.3.12.

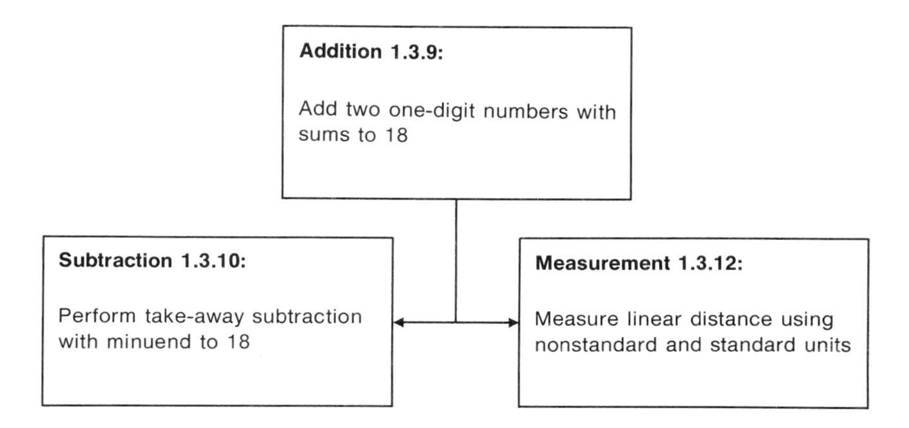

Mini-lesson 2.2.3 Use Additive Identity and Other Properties of Addition

Vocabulary: commutative, order, associative, grouping, identity

Possible Trouble Spots: becomes confused by changes in spatial arrangement of objects; does not distinguish figure from ground in visual tasks

Requisite Objectives:
Sets 1.2.1: Combine disjoint sets

Sets 2.1.1: Represent set relationships and operations at the concrete level

Activity 1

To show that the order of addends does not affect the sum (Commutative Property for Addition), have students engage in the following activities.

A. Mix equal quantities of paint. Point out to the students that if they add blue to yellow or yellow to blue, they obtain green. What about mixing red with blue? Ask if the order of combining colors affects the results.

B. Use a balance and some weights to illustrate that adding two of the same weights to each side but in a different order will balance the scale. This shows that the order of adding equivalent weights does not affect the results.

C. Fit a long rod and a short rod into a boundary that is equivalent in length to the two rods. Change the order of the rods. Both rods fit the boundary, regardless of their positions.

D. Show dominoes or array cards with dot configurations. Students see that no matter how they turn a card or a domino, the number of dots remains the same.

E. Use two sets of measuring cups, including two transparent whole-cups, two half-cups, and two quarter-cups. Fill the half-cups and quarter-cups with colored water. First pour the water from the half-cup and then from the quarter-cup into the whole-cup. Then, by reversing this order, pouring the water from the quarter-cup first and then from the half-cup into the other whole-cup, show that the order of pouring does not affect the measurement—both whole-cups will have three-fourths cup of water. (Marbles may be used instead of water.)

F. Have a child jump for 3 beats of a metronome and clap hands for the next 3 beats. Have another child reverse this sequence at the same time. Guide the children to see that they both ended at the same time even though one child started out jumping and the other started out clapping. Then you might have a child jump 2 times and clap 4 times, while another child claps 2 times and jumps 4 times. (This activity is also helpful in learning how to follow directions.)

G. Place 5 pencils on one side of a table and 3 pencils on the other. Point to 1 set of pencils and ask how many pencils are in the set; repeat this procedure with the other set. Then ask how many pencils there are altogether. Next, reverse the positions of the sets and repeat the above procedure.

Ask if both of the totals are the same. If the child answers incorrectly, check her ability to conserve (discussed in Chapter 22). If the child is able to conserve, you may use smaller numbers to illustrate the commutative property, such as $2 + 1 = 1 + 2 = 3$. Also check for difficulty in visual perception (see Chapter 2).

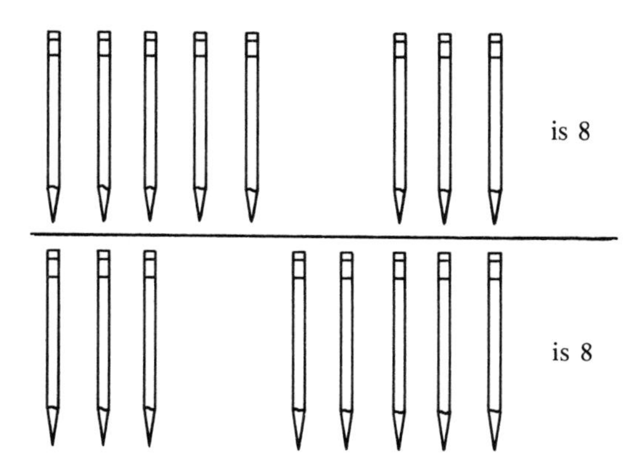

H. Place 7 clothespins on a table and have a child count them. Ask how many different ways the objects can be picked up in two-hand combinations: 6 and 1, 1 and 6, 5 and 2, 2 and 5, 4 and 3, 3 and 4. Emphasize that the entire group is contained in the 2 parts whose total is 7. Write the number of clothespins picked up in each hand for every 2-hand combination.

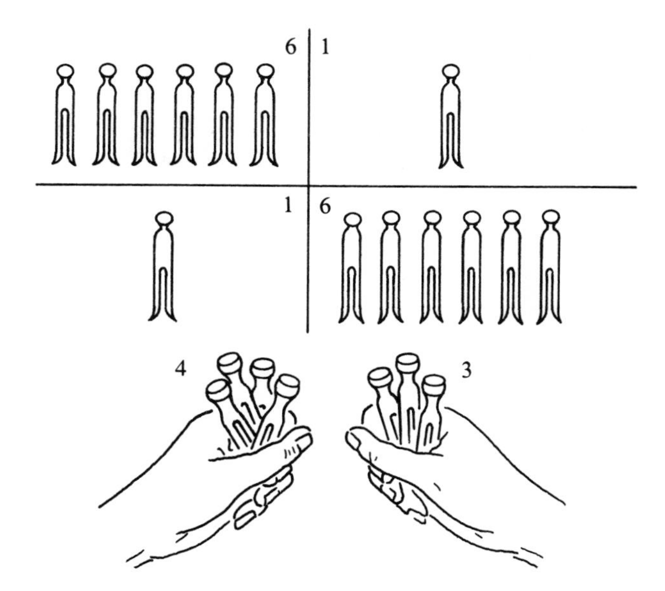

I. Show on a number line the steps involved in solving the equation $2 + 3 = \square$. Then follow the same procedure for $3 + 2 = \square$ and find that answers for both problems are the same number.

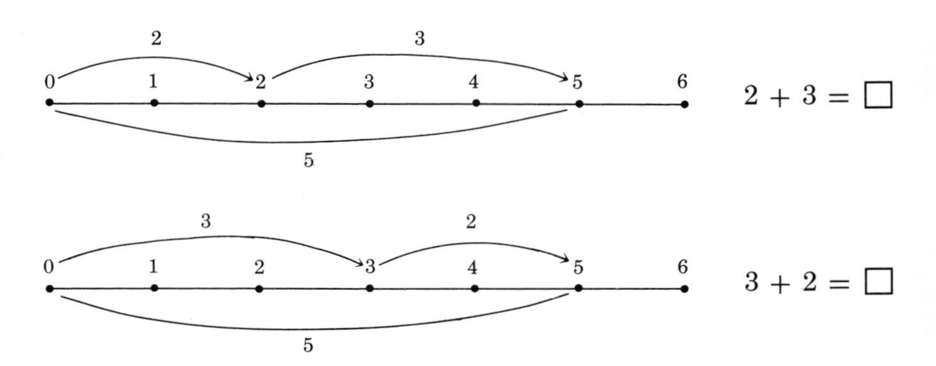

J. When students have a firm grasp of the Commutative Property for Addition, the following might serve as a good practice activity at the symbolic level. Have a child read $3 + 5$ in the first column. Ask which sum in the second column is the same as the sum of $3 + 5$. (Be sure the child understands that "is the same as" and the equals sign have the same meaning.)

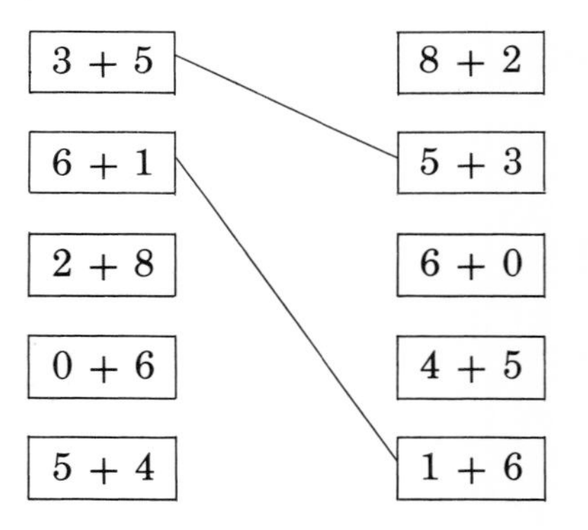

Activity 2
To show that grouping of addends does not affect the sum (Associative Property for Addition), have students engage in the following activities.

A. Give students yellow, red, and blue paint and tell them to mix equal amounts of these colors in different orders. They should notice that there is no change in the resulting color.

B. Using three transparencies, lay the first one on top of the second and then both of them over the third one. Then lay the second transparency over the third and place this combination under the first transparency. Children see that the order (1, 2, 3) and the design are the same, although the transparencies were grouped differently.

C. Obtain a shoe-box cover, 3 red blocks, 2 green blocks, and 4 blue blocks. The box cover represents the addition of numbers in parentheses in the equation $(3 + 2) + 4 = \square$. Have students tell the number of blocks in the shoe-box cover for the indicated sum $3 + 2$. Then regroup blocks as shown. Use this procedure to show that $(3 + 2) + 4 = 3 + (2 + 4)$.

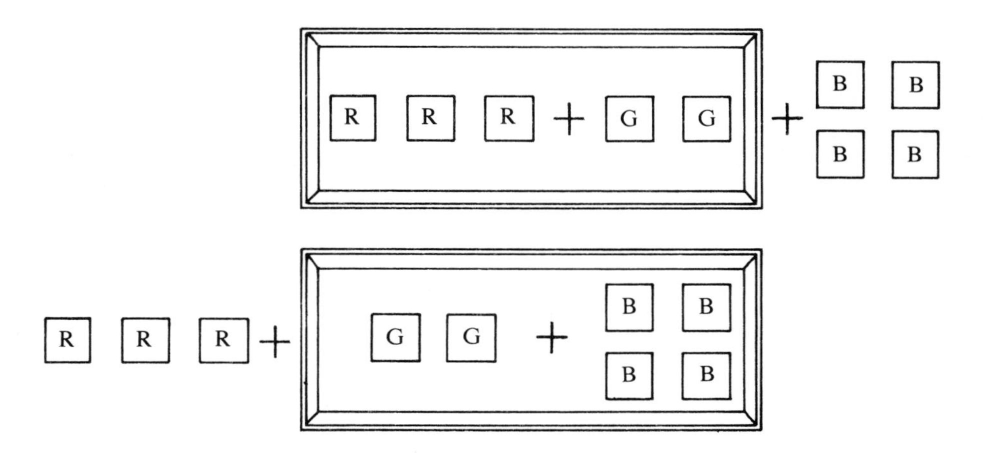

D. Use dots or squares to show associative groupings and write related equations.

$$2 + (1 + 5) = 2 + 6 = 8$$

$$(3 + 1) + 5 = 4 + 5 = 9$$

E. If students seem to have a firm grasp of the grouping concept, present a matching activity on the symbolic level. Have a child read $(2 + 3) + 4$ in the first column. Ask which sum in the second column is the same as that in the first column.

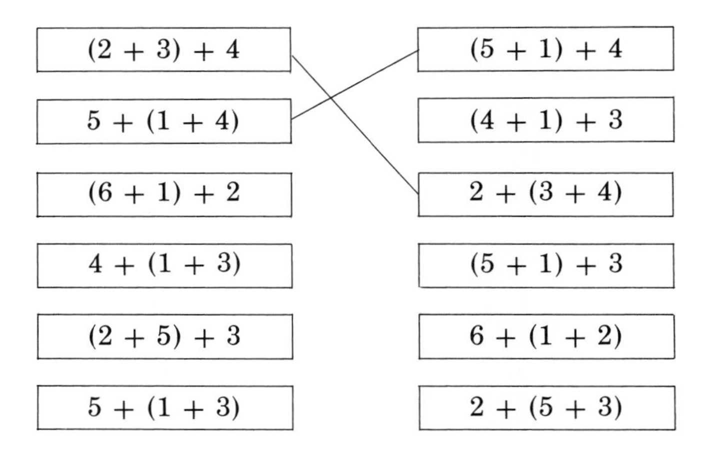

F. Present number lines and ask students to show various ways addends can be grouped for the same numerical problem. These groupings show that the order of grouping does not affect the sum.

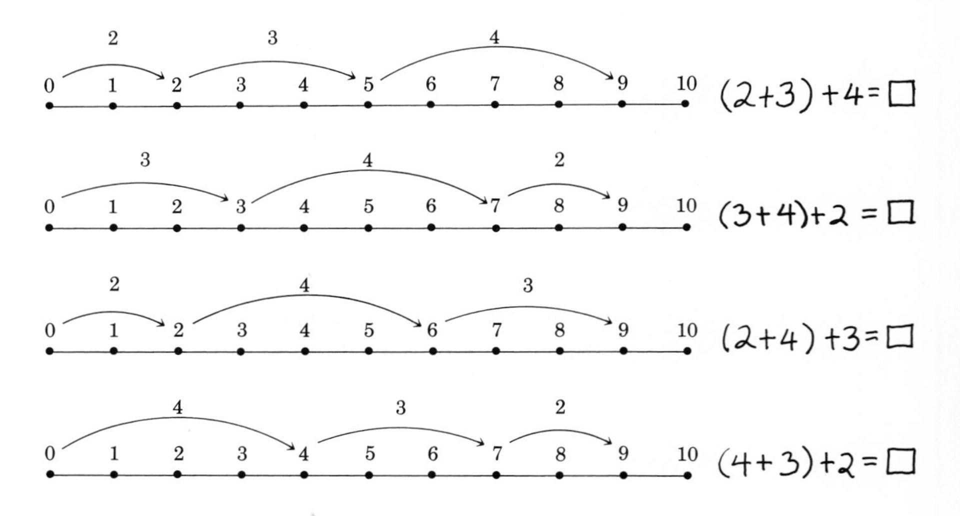

Activity 3
To show that when zero is added to a number, that number is unchanged (Additive Identity), involve students as follows.

A. Show a glass half filled with juice. Tell students you are going to pour some more juice into it. Then, using an empty pitcher, pretend to fill up the glass. Ask if the amount of juice has changed. Finally, pour more juice into the glass. Ask if there is a difference in amounts when you added juice and did not add juice. Follow this same procedure with blocks, cookies, or other appropriate materials. (This activity reinforces the idea that $a + 0 = 0 + a = a$.)

B. Relate the Additive Identity Property ($n + 0 = n$) to the union of a set of objects with an empty set. (See p. 87.)

Activity 4
Have students investigate the Closure Property for Addition by testing addition of odd and even whole numbers. For example, the sum of two even whole numbers is an even number (e.g., $2 + 4 = 6$); this shows that the set of even whole numbers is closed for addition. However, the sum of two odd numbers is not odd; it is an even number ($3 + 5 = 8$). Therefore, the set of odd whole numbers is not closed for addition. If two whole numbers are added (ignoring the aspect of even/odd), the sum is a whole number. Therefore, the set of whole numbers is closed for addition.

Additional considerations: The three properties for addition of whole numbers, namely, commutative, associative, and identity, underlie the addition algorithms used in computation that are discussed next. Encourage students to use these properties as an aid in computation. Analysis of errors students make often reveals patterns that indicate that these properties are not being applied. For example, if a student gets $3 + 4$ correct on one item and misses $4 + 3$ on a later item, he could self-monitor to alleviate this error as follows:

> Let's see. I know that three plus four is seven. Then why did I put that four plus three is nine? That can't be right. I'd better check it. Hmmm. I'll use my fingers. I'll let these fingers stand for the three (holds up three fingers). Four, five, six, seven. OK. That's right. So I'll change this answer to seven.

The teacher may need to model self-monitoring as described above for many of the students. Encourage students to think aloud as they engage in selected tasks.

Note: Mini-lesson 2.2.3 is a prerequisite for Mini-lessons 2.2.4 and 2.2.7.

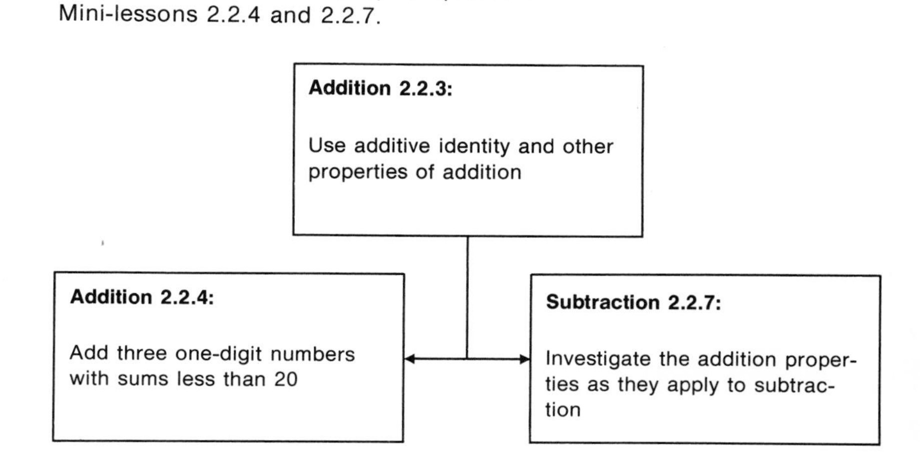

EXERCISES

I. Identify the property for each of the following:

1. $n + 0 = n$
2. $3 + 8 + 4 = 11 + 4 = 3 + 12$
3. $13 + 4 = 17$
4. $(3 + 4) + 2 = 2 + (3 + 4)$

II. Answer true or false for the following statements:

1. The number 1 is the identity element for addition.
2. The set of odd counting numbers is closed for addition.
3. The set of even whole numbers is closed for addition.
4. The set of whole numbers has an additive inverse for addition.

SUGGESTED READINGS

Granito, D. "Number Patterns and the Addition Operation." *The Arithmetic Teacher* (October 1976) 23: 432–34.

Sanders, Walter J. "Cardinal Numbers and Sets." *The Arithmetic Teacher* 13 (January 1966): 26–29.

FOR FURTHER STUDY

1. A set may be equivalent to a given set, contain more elements, or contain fewer elements. Another way to compare sets is to look at the nature of their elements. Two sets may have no elements in common, some elements in common, or all elements in common. When two sets have no elements in common, they are said to be *disjoint* sets.

2. *Union* is a binary operation on sets. A *binary operation on sets* relates two sets to produce a third, just as a binary operation on numbers relates two numbers to produce a third number. The *union* of two sets A and B, denoted by $A \cup B$, is the set of all elements that belong *either* to set A or to set B or to both. $A \cup B$ is read "A union B."

Elements that belong to more than one set are listed only once in the union.

If $A = \{1, 2, 3\}$ and $B = \{2, 3, 4, 5\}$ then
$A \cup B = \{1, 2, 3, 4, 5\}$

If $A = \{1, 2\}$ and $B = \{a, b, c\}$ then
$A \cup B = \{1, 2, a, b, c\}$

If $A = \{a, b, c\}$ and $B = \{a, b\}$ then
$A \cup B = \{a, b, c\}$ or, in this case,
$A \cup B = A$.

The union of any set A and the empty set is A itself:

$$A \cup \varnothing = A.$$

3. The concept of addition can be shown as the cardinal number property of the union of two disjoint sets. For two disjoint sets A and B,

$$n(A \cup B) = n(A) + n(B)$$

(read "the number of the union of sets A and B equals the number of set A plus the number of set B"). For example, if the cardinal number property of set A is 7 and that of set B is 2, and if A and B are disjoint sets (no member of set B is a member of set A), then the cardinal number property of the union of these 2 sets is 9 ($7 + 2 = 9$). (Notice that union ($\cup$) is an operation on sets, while addition is an operation on numbers.)

4. Addition of numbers is related to the union of disjoint sets. Physical joining of disjoint sets forms a model for finding the sum of the cardinal number of the combined sets.

Since for any two disjoint sets A and B, if $A \cup B = B \cup A$, then $n(A \cup B) = n(B \cup A)$.

5. Then how can the cardinal number property of the union of three or more disjoint sets be found, $n(A \cup B \cup C)$? It may be shown that $(A \cup B) \cup C = A \cup (B \cup C)$. This supports the assumption that $n(A \cup B) + n(C) = n(A) + n(B \cup C)$.

Addition Algorithms

TASK ANALYSIS FOR ADDITION ALGORITHM OF WHOLE NUMBERS WITH RENAMING

The following 11-step analysis of the addition algorithm may be used as a model for the reader to identify prerequisite concepts necessary to do the computations involved.

What is involved in order to reach this sum?

$$\begin{array}{r} 387 \\ 98 \\ 3 \\ 1009 \\ +\ \ 430 \\ \hline 1927 \end{array}$$

Place value

1. The "adder" has to know place value so that the addends can be properly positioned—units under units, tens under tens, hundreds under hundreds, and so on.

2. Place value is also involved in writing the numbers 430 and 1,009.

Basic addition combinations

3. Basic addition combinations must be known so after learning these at the principle level, they become stimulus-response learning.

Application of commutative and associative properties for addition

4. An application of the commutative and associative properties for addition is necessary. These properties allow grouping numbers to form tens, as in adding 7 + 3 and then adding 8 + 9 to that sum, instead of proceeding in order: 7 + 8 = 15, 15 + 3 = 18, 18 + 9 = 27.

The mature adder uses both of these techniques within a single problem and thus needs to have learned both strategies for adding. (The second method, adding in sequence as the numerals are listed, is called "adding by endings.")

Absence of a digit

5. The adder must know that the absence of a digit in a column acts as though zero were the addend in that position.

Addends added vertically

6. The addends are added vertically in each column. Some children do not automatically bridge from processing numerals in the horizontal form (13 + 7 + 22) to the vertical algorithm:

$$\begin{array}{r} 13 \\ 7 \\ +22 \\ \hline \end{array}$$

Some ignore place value and compute "3 + 7 = 10," "10 + 2 = 12," continuing to the tens column, "12 + 1 = 13," "13 + 2 = 15."

Renaming based on place value

7. The renaming of the appropriate numeral is based on the place-value idea. The child who adds the units column correctly and then writes the tens number in the units position is not applying the place-value idea. For example, in the sample problem above, the child may get 12 for the sum of the units but may write "1" and "carry" the 2, adding the 2 to the tens column numbers. The child is, therefore, renaming the wrong part of the numeral for 12. The pupil is saying that 2 ones is the same as 2 tens instead of 10 ones equals 1 ten.

Restructuring the problem

8. Some children ignore the addition sign and subtract instead of adding. This sometimes occurs when there are two addends, as in

$$\begin{array}{r} 39 \\ +\ 6 \\ \hline \end{array}$$

and renaming is involved. The child who has difficulty with renaming at the symbolic level, for example 9 + 6 = 15, may restructure the problem so that he or she *can* handle it, thus ignoring the addition sign and instead subtracting as follows: 9 − 6 = 3 in the units column.

One digit per column

9. It is necessary to know the convention that only one digit is written in a column. In the example

$$\begin{array}{r} 39 \\ +\ 6 \\ \hline 315 \end{array}$$

the child who writes the 15 in the units column does not realize that this makes the entire sum read as "three hundred fifteen." For this reason, an intermediate algorithm is helpful. This less-efficient algorithm may be written in expanded notation as in the examples below:

$$\text{(a)}\quad \begin{array}{r} 39 = 30 + 9 \\ +\ 6 = \quad\ +\ 6 \\ \hline 30 + 15 \end{array} \quad \begin{array}{r} 30 \\ +15 \\ \hline 45 \end{array} \qquad \text{(b)}\quad \begin{array}{r} 39 \\ +\ 6 \\ \hline 15 \\ +30 \\ \hline 45 \end{array}$$

Similar place values added

10. Only similar place-value numbers are added. The child who adds the example

$$\begin{array}{r} 13 \\ +\ 2 \\ \hline 35 \end{array}$$

as "3 + 2 = 5" and "1 + 2 = 3" is not applying the place-value principle.

For some pupils, moving from the expanded form of a numeral to the standard name may foster this confusion. When a child is told 3 tens plus 9 ones equals 39, or $(3 \times 10) + (9 \times 1) = 30 + 9 = 39$, he or she needs a good grasp of place value and the associative property in order to see that the problem $13 + 2 = \square$ involves adding $10 + 3 + 2$, not adding 2 to both the 1 and the 3 in 13 to obtain 35, or adding 2 only to the 1 of the 13 to obtain 33. Renaming a number from expanded notation to its standard name is seen by some children as the same as adding units numbers to tens numbers. The child may try to solve the problem $30 + 9 = \square$ by thinking "three plus zero is three and three plus nine is twelve."

Algorithm written legibly

11. Writing the algorithm legibly is a necessary skill that depends upon the maturity of fine-motor skills. The use of large-cell graph paper helps some children keep the place values one under the other. Illegible writing often clouds mathematical thinking, so children should be encouraged to write neatly.

The above analysis was presented to aid the teacher in identifying possible errors in computation and in hypothesizing causes for such errors.

The mini-lessons presented next lead to the end goal of adding whole numbers with renaming.

Mini-lesson 2.2.6 Add Numbers with Sums to 999, with No Renaming

Vocabulary: compute, algorithm

Possible Trouble Spots: does not know basic addition facts; reverses 2-digit numbers when copying from textbook or chalkboard; does not organize algorithm so that numbers are lined up within place values; writes so messy that digits are misinterpreted; organizes work on paper so that algorithms become intermingled and errors result

Requisite Objectives:
Addition 2.1.6: Add two numbers, with no renaming, sums to 99, and check

Word Problems 2.1.15: Translate simple addition word problems to corresponding algorithm and solve

Addition 2.2.4: Add three 1-digit numbers with sums less than 20

Addition 2.2.5: Add three numbers, sums to 99, with no renaming

Activity 1
To record addition algorithms with sums greater than 100, first review algorithms for smaller sums.

A. Have the child combine a set of 6 sticks with a set of 12 sticks and record the problem as

$$\begin{array}{r} 12 \\ +\ 6 \\ \hline 18 \end{array}$$

Then tell the child to exchange 10 of the sticks for a single ten-stick. Ask the child to describe the 12 sticks as they are regrouped. She should respond that she has 1 ten and 2 ones. Then ask her how many ones she has altogether. The correct response is 8 ones. Point out that ones are added to ones.

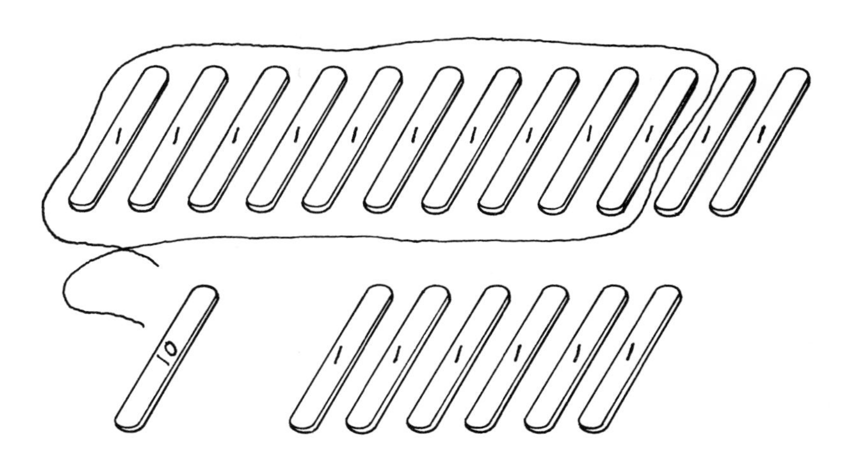

Ask children in grades 3 and up to use a simple proof to record the grouping action.

$12 + 6 = (10 + 2) + 6$	Expanded form
$= 10 + (2 + 6)$	Associative Property for Addition
$= 10 + 8$	Basic addition fact
$= 18$	Addition table for whole numbers

Activity 2
To record addition algorithms with sums greater than 100, show the child a homemade abacus reading 3 hundreds, 5 tens, 6 ones. Add markers representing 6 hundreds, 3 tens, 2 ones. Ask the child to write the addition shown on the abacus, with and without the aid of a place value chart as shown below.

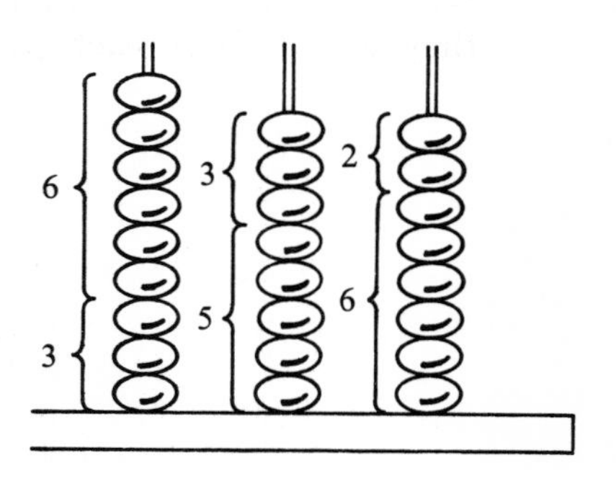

Adding within a place-value chart

h	tens	ones
3	5	6
6	3	2
9	8	8

Adding without a place-value chart

$$\begin{array}{r} 356 \\ 632 \\ \hline 988 \end{array}$$

Activity 3
Use the counting boards. For example, have students show their computation on the boards and record their work for 356 + 632. Let ●'s represent 356 and let ■'s make up 632. Place a hundreds board to the left of the tens board.

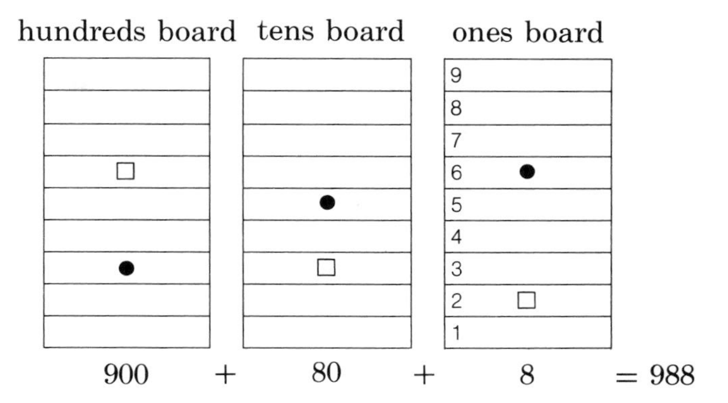

Additional considerations: For students who are not sure of their basic addition facts, encourage them to check each partial sum obtained with a hand calculator. For example, in the previous problems, they would check each partial sum: 6 + 2, 5 + 3, and 3 + 6.

Note: Mini-lesson 2.2.6 is a prerequisite for Mini-lessons 2.2.8 and 2.2.14.

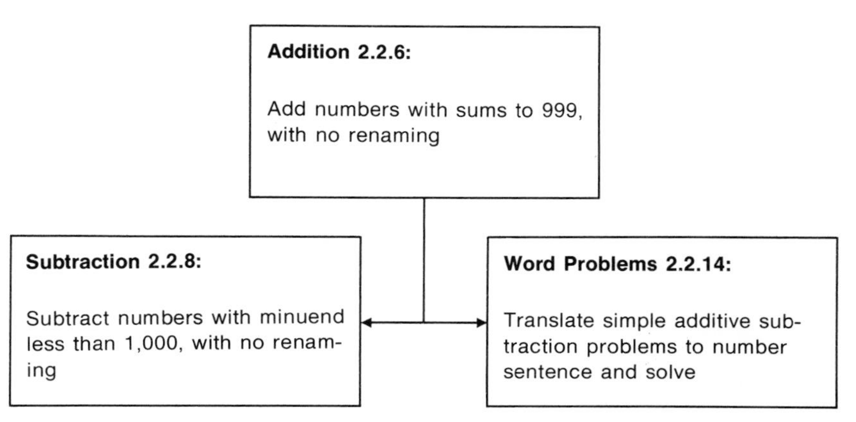

Mini-lesson 3.1.4 Add with Renaming

Vocabulary: regroup, rename, forward move

Possible Trouble Spots: does not understand place value relationships; lacks ample concrete experience with exchange model for place value; is not a conserver of number and therefore, does not understand the transformation involved in renaming 10 units as 1 ten

Requisite Objectives:
Whole Numbers/Notation 2.2.1: Use exchange model for place value
See task analysis, pages 27–29
Addition 2.3.2: Add two numbers with sums to 999, one renaming
Addition 2.3.3: Add with sums to 999, two renamings

Activity 1
Demonstrate how to add two 2-place numbers with renaming on the counting boards. For the example 26 + 47, have the child enter 26 and then 47 on the boards (Steps 1 and 2 on p. 95). Next, direct him to make forward moves so that the sum is shown with the fewest number of tokens (Steps 3 and 4).

Activity 2
Have the child show the addition for the example 35 + 48 on a homemade abacus.

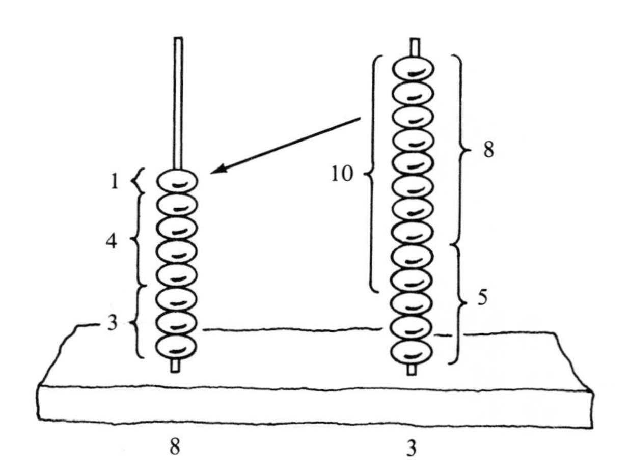

Activity 3
Present the child with addition problems requiring renaming and ask her to solve them using the following algorithms.

(a) $\begin{array}{r} 65 \\ +29 \\ \hline \end{array}$

tens	ones
6	5
2 1↖	9
9	①4

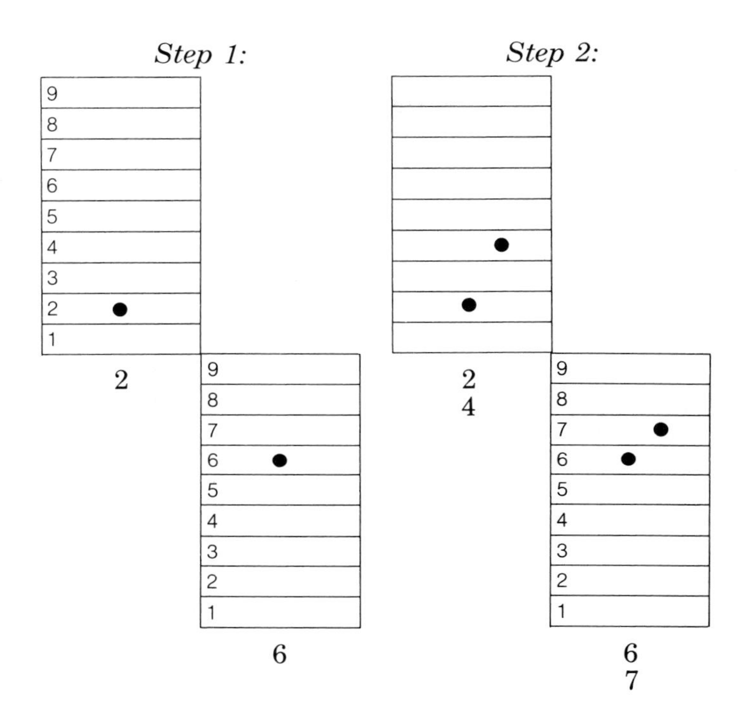
Step 1:
9
8
7
6
5
4
3
2
1
2
9
8
7
6
5
4
3
2
1
6
Step 2:
2
4
9
8
7
6
5
4
3
2
1
6
7

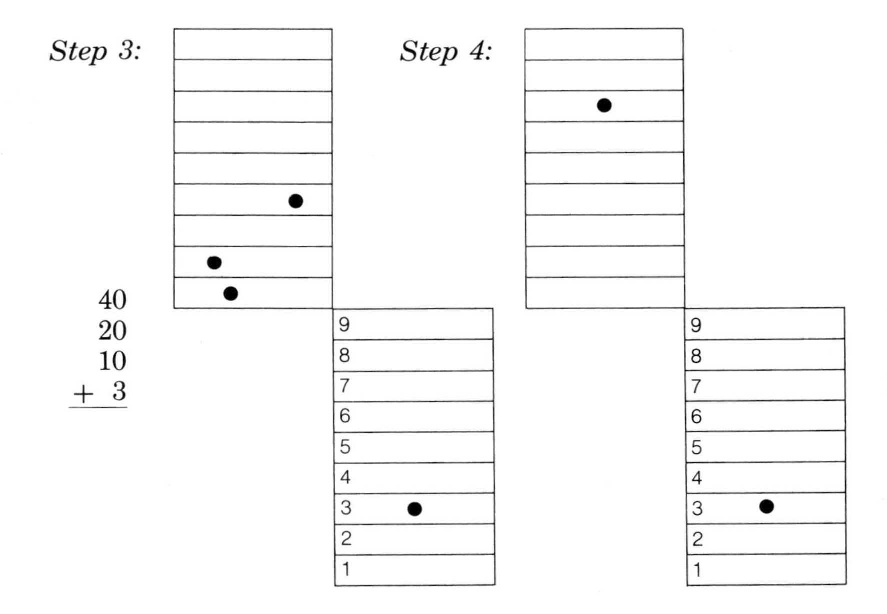
Step 3:
40
20
10
+ 3
9
8
7
6
5
4
3
2
1
Step 4:
9
8
7
6
5
4
3
2
1

(b)
$$\begin{array}{r} 726 \\ 1092 \\ 434 \\ +\ 873 \\ \hline 23{,}125 \end{array}$$

th	h	tens	ones
	7	2	6
1	0	9	2
	4	3	4
2 ↖	8 2 ↖	7 1 ↖	3
3	② 1	② 2	① 5
↓	↓	↓	↓
3	1	2	5

(c)
$$\begin{array}{r} 65 \\ +29 \\ \hline 14 \\ 80 \\ \hline 94 \end{array}$$

Add downward (5 + 9).
Units are blocked
out with zero;
tens are added.

(d)
$$\begin{array}{r} 1\ \textcircled{\not{1}4} \\ 65 \\ +29 \\ \hline 94 \end{array}$$

Add upward (9 + 5).

14 is renamed as 1 ten and 4 ones;
4 ones is written below and the 1 ten above;
the associative property is used to add the tens,
1 + 6 + 2 = 9.

(e)
$$\begin{array}{r} 65 \\ +29 \\ \hline 94 \end{array}$$

Add (5 + 9).
4 is written. Think "one ten."
Think "one plus six plus two"; the sum is
written in the tens place.

Activity 4

Present to the child an addition problem requiring renaming and ask him to solve it using mental computation.

Additional considerations: For those students who need a great deal of structure in organizing their written work, the following algorithm has been helpful. Each cell in the algorithm is divided diagonally so that a 2-place number may be entered. The addition is performed from right to left as shown by the arrows in the example on p. 97.

It should be apparent that the algorithm just described is also helpful in diagnosing how well a student knows the basic addition facts,

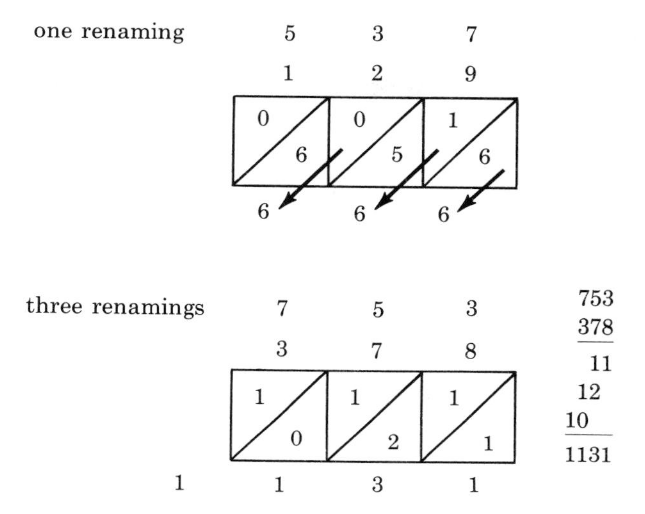

since each cell represents the sum for a basic fact, e.g., 7 + 9, 3 + 2, and 5 + 1.

For students who do not seem able to grasp the place value principle that underlies renaming, this algorithm circumvents having to understand place value. Then, perhaps after practice with this algorithm, these students may be helped to use the conventional algorithm by comparing the two, side by side, as shown in the example 378 + 753.

Note: Mini-lesson 3.1.4 is a prerequisite for Mini-lessons 3.1.5 and 3.1.15.

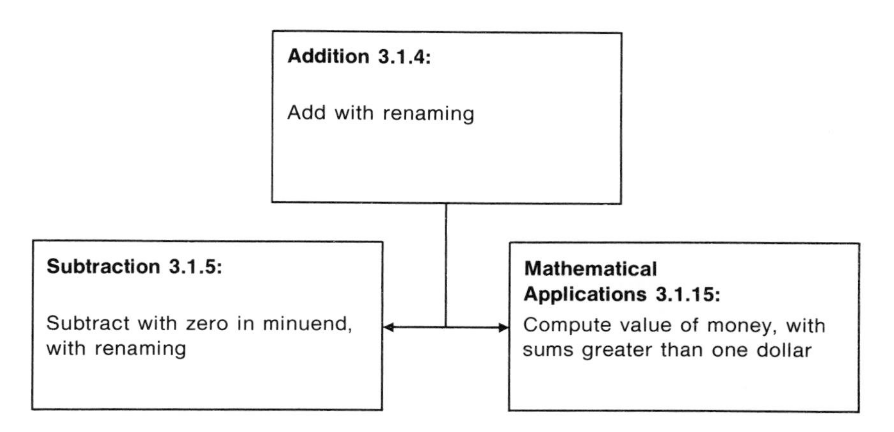

EXERCISES

I. Each cell in the illustration below is divided diagonally to allow for 2-digit numerals to be recorded. Tell how it is similar to the ancient retrograde method of adding.

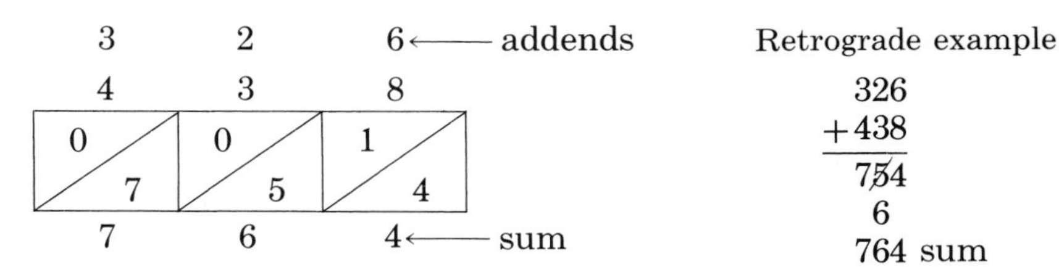

II. Identify the steps in the following proof:

13 + 25	= 10 + 3 + 20 + 5	1. ______
	= 10 + (3 + 20) + 5	2. ______
	= 10 + (20 + 3) + 5	3. ______
	= (10 + 20) + (3 + 5)	4. ______
	= 30 + 8	5. ______
	= 38	6. ______

III. Identify the possible reasons for errors in the following addition algorithms:

(1) $\begin{array}{r} 306 \\ +\ 50 \\ \hline 300 \end{array}$ (2) $\begin{array}{r} 3 \\ 4 \\ +6 \\ \hline 7 \end{array}$ (3) $\begin{array}{r} 137 \\ +\ 9 \\ \hline 132 \end{array}$ (4) $\begin{array}{r} 306 \\ +\ 19 \\ \hline 415 \end{array}$

IV. What property is used in the following algorithm?[1]

$\begin{array}{r} 3987 \\ 1564 \\ +3981 \\ \hline 9532 \end{array}$

th	h	tens	units
2 + 3 = 5	2 + 9 = 11	1 + 8 = 9	
5 + 1 = 6	1 + 5 = 6	9 + 6 = 15	7 + 4 = 11
6 + 3 = 9	6 + 9 = 15	5 + 8 = 13	1 + 1 = 2

SUGGESTED READING

King, I. "Giving Meaning to the Addition Algorithm." *The Arithmetic Teacher* 19 (May 1972): 345–48.

1. Suggested by F. Hutchins, formerly at the University of Maryland.

Multiplication of Whole Numbers

It may bother some that multiplication rather than subtraction is presented at this point. However, children are expected to use multiplication ideas in kindergarten and first grade, although traditional curriculum placement of multiplication comes at the end of second grade or in third grade.

Early multiplication experiences include learning to multiply face value by place value in determining the value of a numeral, identifying time on a clock face (the digits on a clock face indicate multiples of 5); and participating in activities dealing with many-to-one correspondence (which also involves multiplication). In fact, children learn the threads of all four operations—addition, multiplication, subtraction, and division—contiguously, but they are often taught as unique steps on a hierarchy.

A child may develop only a rote parroting of the names of the places—ones, tens, hundreds—if he or she is not aware of the multiplying involved in generating these names, which represent products. Perhaps, then, the reason why first and second graders have difficulty in understanding place-value concepts stems from the fact that underlying multiplication ideas are not taught until third grade. Also, many children become confused as to the job of the 12 numerals on a clock face. A typical incorrect response when a clock face shows five minutes after an hour is "one after" or "one minute after." The child does not realize that the 1 on the clockface, in addition to representing one o'clock, indicates one group of 5 minutes, or 1×5. The child also needs to realize that if he or she is given one cookie on three consecutive occasions, he or she will have received 3×1, or 3, cookies.

Thus, the discussion of multiplication precedes that of subtraction to emphasize the necessity of focusing on basic multiplication relationships in the early grades.

MULTIPLICATION AS A MANY-TO-ONE IDEA

Ratio idea underlying multiplication

Our numeration system employs a many-to-one multiplicative idea. The names of the places are in a 10-to-1 relation in descending order from left to right: 1000 to 100; 100 to 10; 10 to 1; 1 to 1/10; 1/10 to 1/100; and so on. These may be expressed as ratios: 1000:100, 100:10, 10:1, and so forth. Thus, there is a ratio idea underlying multiplication.

Division as prerequisite to multiplication

The relationship between the ratio idea of multiplication and the ratio idea of division is thereby established. (This will be discussed in Chapter 11.) The ratio idea that underlies multiplication raises the point that division is a prerequisite notion to multiplication rather than the other way around.

Experiences in one-to-one correspondence

Consider the very basic idea of one-to-one correspondence. Some activities designed to provide experiences in one-to-one correspondence resemble the following:

1. *Establishing a one-to-one correspondence between two sets.* The child is asked to give a napkin to each of 6 children.

2. *Continuing one-to-one correspondence.* The child is asked to give a juice cup to each of the same 6 children.

3. *One-to-one correspondence becomes a many-to-one correspondence.* The child is given 18 cookies and is asked to distribute them to 6 children so that each child may have the same number of cookies.

A young child, who is probably unaware that each of the 6 children will receive 3 cookies, will distribute them as a one-to-one action until all of the cookies have been given away. The idea behind this activity is the same as that in partitive division, in which the number of the whole group (18 in this case) and the number of groups (6 children) are known, and the number sought is how many will be in each group (in this case, how many cookies each of the 6 children will receive).

MULTIPLICATION AS THE UNION OF EQUIVALENT DISJOINT SETS

Multiplication is the operation that determines the number of objects in the union of several disjoint sets, with each set having the same cardinal number. If diagnostic pretesting indicates that the child knows the meaning of disjoint sets and of union and is able to use the union of equivalent disjoint sets as a physical model for the addition operation, this model may be used for the multiplication operation.

Multiplication as shortcut addition

One concept of multiplication of whole numbers is that it is shortcut addition. Showing children that multiplication can be a time saver serves as a motivation for them to learn this way of operating on numbers. For example, the teacher may compare addition and multiplication operations to solve the problem of how many children there are in 6 families, with each family having 2 children. The addition method is as follows:

Suggested board work or transparency

A generalized associative property allows the addition of all of the 2s as follows:

$$\begin{aligned} & (2 + 2) + (2 + 2) + (2 + 2) \\ = & (4 + 4) + 4 \\ = & 8 + 4 \\ = & 12 \end{aligned}$$

This procedure involves writing 32 separate symbols, including parentheses as well as addition and equal signs. Instead of listing each addend separately, in multiplication the addend needs to be listed only once. Then, it is necessary to show how many times the addend is used.

Suggested board work

The 2 is used as an addend 6 times. The notation to show this is 6×2, read "6 times 2" or "six twos." The notation $6 \times 2 = 12$ involves writing only six separate symbols.

The following verbal models for both addition and multiplication may be used as a summary activity in the later stages of instruction.

Suggested board work

addend + addend + addend = sum
sum = sum
number of addends × addend = product
product = product

Terminology

The name *product* refers to the number property of the union of the equivalent disjoint sets in multiplication (in addition, the name *sum* refers to the number property of the union of two disjoint sets).

Multiplication is a binary operation; that is, it is used to process two numbers to produce a third number. The term *multiplier* (from the Latin *numerus multiplicans,* meaning "multiplying number") refers to the number that indicates the number of addends. *Multiplicand* (from the Latin *numerus multiplicandus,* meaning "number to be multiplied") is the number that tells the cardinal number property of the equal addends. Thus, another verbal model is as follows:

$$\underbrace{\text{multiplier} \times \text{multiplicand}}_{\text{product}} = \text{product} \updownarrow \text{product}$$

During the "modern math" era, the term *factor* became more popular than *multiplier* and *multiplicand* and has remained the preferred terminology.

$$\underbrace{\text{factor} \times \text{factor}}_{\text{product}} = \text{product} \updownarrow \text{product}$$

Multiplication to be considered in context

From a mathematical view, as stated in the Commutative Property for Multiplication, the product will be the same, regardless of the order of the factors. But in the physical world, there is a difference, of course. The food and clothing bills would be quite different for 6 families with 2 children each than for 2 families each having 6 children, although a total of 12 children is still involved.

Mini-lesson 3.1.6 Show Multiplication as the Union of Several Equivalent Disjoint Sets

Vocabulary: disjoint sets, union, equivalent

Possible Trouble Spots: does not generalize well; does not notice salient aspects of situations; does not recognize equivalent sets

Requisite Objectives:
Multiplication 1.2.5: Become aware of beginning multiplication ideas

Whole Numbers/Notation 1.3.5: Show many-to-one and one-to-many representations of the same number

Whole Numbers/Notation 1.3.6: Count by 2s, 5s, and 10s

Mathematical Applications 1.3.16: Compare the value of penny, nickel, and dime

Whole Numbers/Notation 2.1.2: Apply many-to-one generalization to exchange model of place value

Multiplication 2.1.8: Show multiplication situations related to many-to-one relationships

Sets 3.1.1: Combine disjoint equivalent sets

Activity 1
Assess that students recognize that two groups are disjoint, i.e., do not overlap.

A. Assign classroom helpers: e.g., John, Martin, and Bill are the set of students who erase the board; Mary and Alice are the set of students who straighten the library corner; John, Mary, and Alice feed the fish. Have the students discuss which of the sets have the same members (not disjoint) and which of the sets of helpers do not have members in common (disjoint).

B. Discuss with the class other examples where two sets do not have elements in common, e.g., boys/girls, animals/boats, and glass marbles/wooden beads.

Activity 2
Provide 7 students with 5 pieces of candy each. Tell the class that the candy costs a penny for each piece. Ask how much money is needed to buy the candy using nickels. Class should state they would use 7 nickels, or a total of 35 cents.

Activity 3
Show the child two sets of 5 triangles each on the flannel board, overhead projector, or chalkboard. Ask the child how many sets there are, the number of triangles in each set, and the number of triangles in all. Continue using other groups of equivalent disjoint sets. Each time you present them, stress the number of sets, the equivalence of each set, and then the number of the union of all the sets.

△△△△△ △△△△△

5 + 5 = 10

"Two fives equal ten."

Activity 4
Ask the child to take 6 steps of 2 spaces each on a number line. Record the child's steps on the chalkboard and say, "Six twos equal twelve; six times two equals twelve."

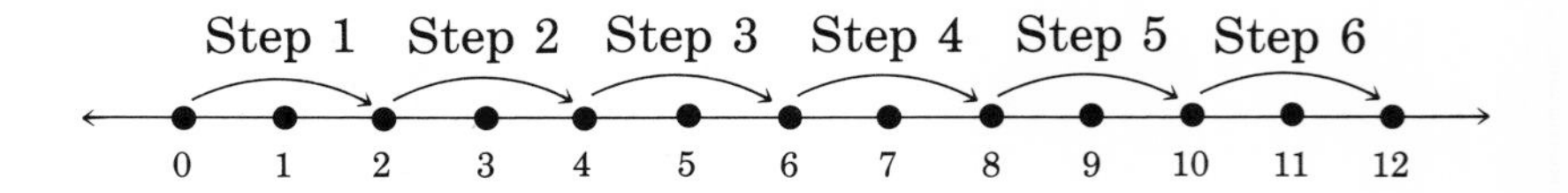

Some children will count "one" before they take a step, as shown on the enlarged portion of the number line below. Emphasize that they are counting steps taken, and, therefore, while they are still standing on zero, they have not as yet taken a step. (This is an important prerequisite to telling time. An error often made by children learning to tell time is to start counting "one" at zero, saying it is 6 minutes after an hour rather than 5).[1]

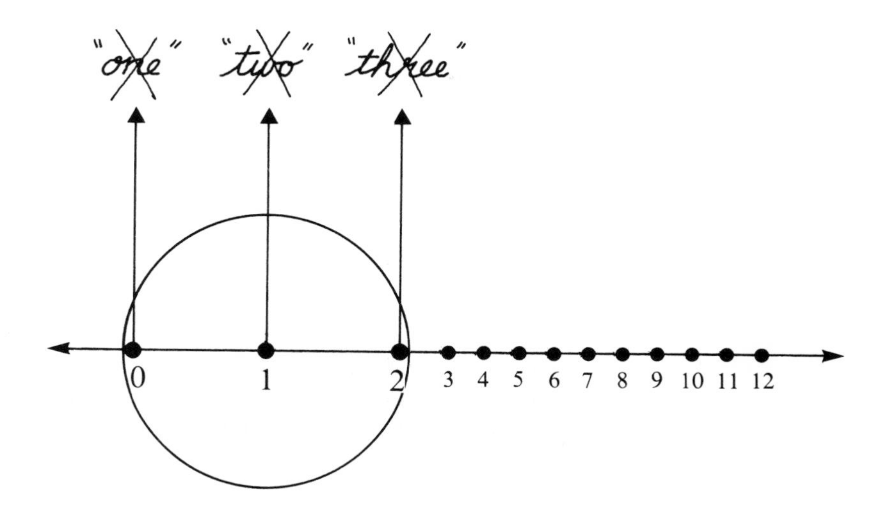

Draw a number line labeled 0 to 18. Ask the child to draw jumps on the number line to correspond with various multiplication problems.

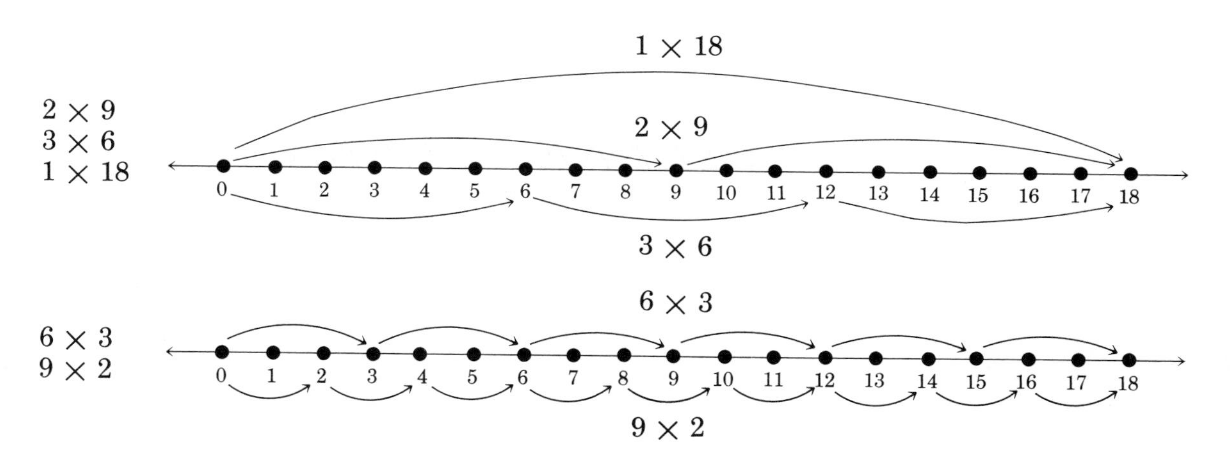

Additional considerations: Use of many examples at the concrete level provides a foundation for understanding the goal of this mini-lesson. However, it is important that the teacher record what is shown at the concrete level. For example, in Activity 2, the teacher should

1. Fredricka K. Reisman, "Children's Errors in Telling Time and a Recommended Teaching Sequence," *The Arithmetic Teacher* 18 (1971): 152–55.

record 5 + 5 + 5 + 5 + 5 + 5 + 5 simultaneously with the students' conclusion that they would use 7 nickels.

Note: Mini-lesson 3.1.6 is a prerequisite for Mini-lessons 3.1.15 and 3.1.18.

Multiplication 3.1.6:

Show multiplication as the union of several equivalent disjoint sets

Mathematical Applications 3.1.15:
Compute value of money, with sums greater than one dollar

Word Problems 3.1.18:

Solve simple word problems involving multiplication

THE CROSS-PRODUCT APPROACH

Imagine a situation in which two nonequivalent sets are to be paired and the number of pairs is asked for. For example, if Lisa has 2 skirts (blue and green) and 3 tops (red, yellow, and white), how many different outfits can she wear? This may be set up as follows:

Suggested board work

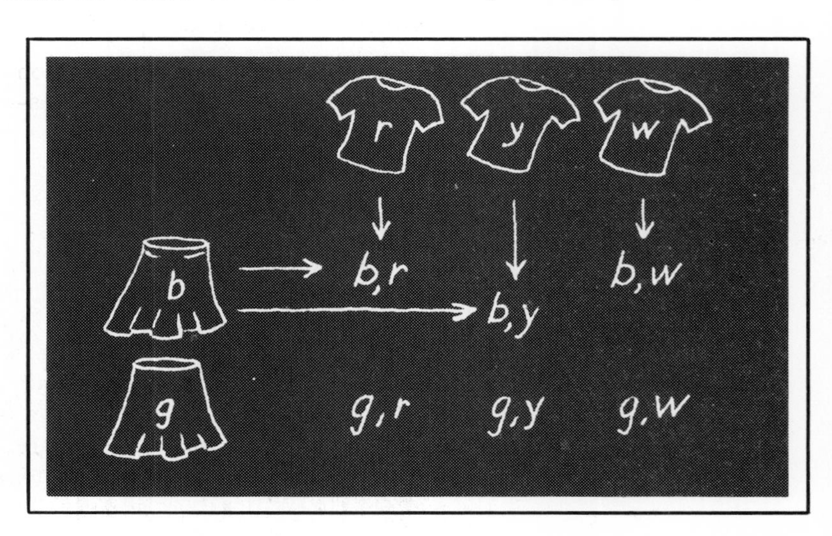

Counting the combinations shows that Lisa will have 6 outfits. The set of pairs is called the *cross product* of the given sets. The symbol for

cross product (×) is read "cross" not "times." Although the symbol for cross product looks like the symbol for multiplication, it does not denote multiplication. Multiplication is an operation on numbers and cross product is an operation on sets; just as addition is a number operation and union is a set operation. Sometimes a large, darker × is used to symbolize cross product.

Elements of cross product are ordered pairs

Notice that in the illustration the placement of the elements from each set is shown where the arrows meet as in a matrix or grid. Also notice that the colors of the skirts are listed first. There is a reason for this. Cross product (or Cartesian product, named after Descartes) is a set whose elements are *ordered pairs*. This means that the individual elements of each set making up the cross product are arranged in order. The colors of the skirts are listed first and then the colors of the tops, resulting in ordered pairs.

The cross product of Lisa's outfits would be noted as follows:

Given: set of skirts, $S = \{b, g\}$
set of tops, $T = \{r, y, w\}$
set of the cross product, $S \times T$
$= \{(b, r), (b, y), (b, w), (g, r), (g, y), (g, w)\}$

The notation places the ordered pairs in parentheses with a comma between each element within an ordered pair and between the ordered pairs. The elements of the cross product, $S \times T$, are the ordered pairs. So there are six elements in the set $S \times T$, read "S cross T."

Number of $S = n(S) = 2$
Number of $T = n(T) = 3$
Number of the cross product $= n(S \times T) = 6$

Another example might be a combination of sandwiches and beverages. If a choice is given of 3 sandwiches—hamburger, salami, and tuna—and 2 drinks—milk and coffee—how many pairs of sandwiches and drinks can there be?

Suggested board work

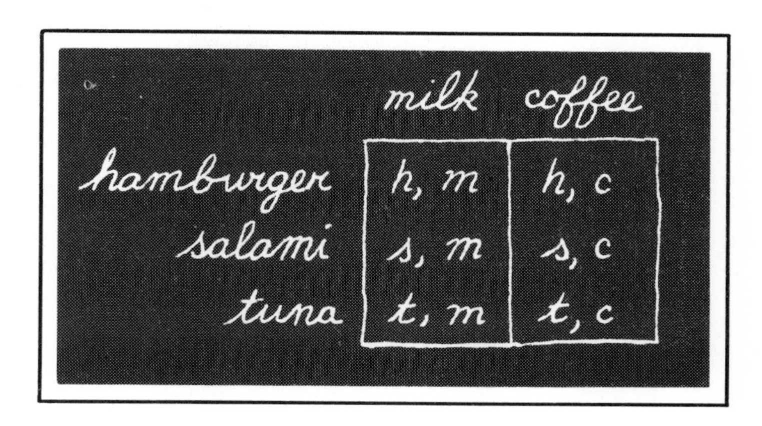

$$S \times D = \{(h, m), (h, c), (s, m), (s, c), (t, m), (t, c)\}$$
$$n(S) = 3$$
$$n(D) = 2$$
$$n(S \times D) = 6$$

This may also be set up as follows;

Suggested board work

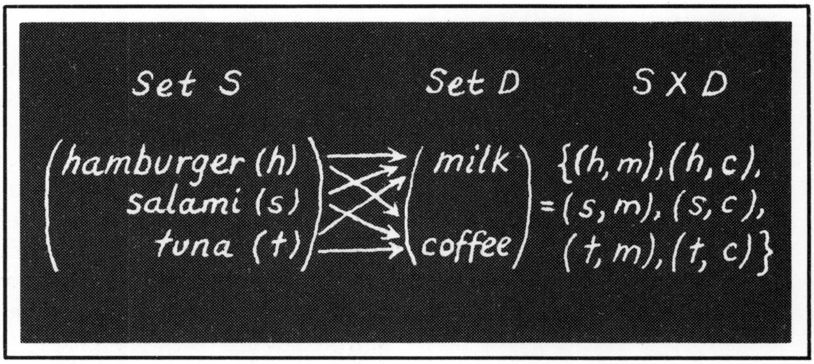

Set S | Set D | S X D

(hamburger (h), salami (s), tuna (t)) → (milk, coffee) = {(h, m), (h, c), (s, m), (s, c), (t, m), (t, c)}

A summary definition for cross product may state: If $S = \{s_1, s_2, s_3, \ldots\}$ and $D = \{d_1, d_2, d_3, \ldots\}$, the cross product of the two sets, $S \times D$, is the set of all possible pairs whose first member comes from the first set and whose second member comes from the second set. Also, if $S \times D$ represents the set of all ordered pairs of elements in which an element from S is paired with one and only one element from D, then

$$n(S \times D) = n(S) \times n(D) = n(D) \times n(S) = n(D \times S)$$

One advantage of this definition of multiplication is that it no longer requires that the sets used as a physical model be disjoint, as in the model based on the union of equivalent disjoint sets. In fact, it specifies a product even when $S = D$. For example,

if $S = \{1, 2\}$ and $D = \{1, 2\}$,
then $S \times D = \{(1, 1), (1, 2), (2, 1), (2, 2)\}$,

and the number operation of multiplication would still hold: $n(S) = 2$, $n(D) = 2$, $n(S \times D) = 4$. The number product, $n(S \times D)$, does not depend upon the nature of the elements comprising either set S or set D. The assignment of a product (a third number) to a pair of numbers is what the multiplication operation means.

Mini-lesson 4.2.4 Extend Multiplication Involving Any Whole Numbers Using Cross-Product Approach

Vocabulary: cross product, multiplication, array, ordered pair, grid

Possible Trouble Spots: difficulty in generalizing; difficulty in dealing with complexity; does not understand set operations.

Requisite Objectives:
Interpret arrays

Understand ordered pairs

Graph and Functions 4.1.17: Interpret and use a grid

Activity 1
To show that cross product may be used to define multiplication of whole numbers without reference to addition, review with the children sets and the set operation of union before beginning this activity.

Using cutout felt objects of stars and circles, display one star and one circle on a flannel board. Call them an ordered pair. Stress that you do not change the order of an ordered pair. The ordered pair of a star and a circle is different from that of a circle and a star. Then display another pair of objects, for example, a square and a triangle.

Show the cross product of the two sets by choosing the first element of the ordered pair from the first set and the second element from the second set. For example, the cross product of sets A and B, where $A = \{☆, ○\}$ and $B = \{□, △\}$ is $\{(☆, □), (☆, △), (○, □), (○, △)\}$.

Then discuss with the children that $n(A) = 2$ and $n(B) = 2$, $n(A \times B) = 4$ (which is a model for $2 \times 2 = 4$).

Activity 2
Use a 2-by-2 array to show the cross product of ordered pairs. The elements on the left side of each ordered pair indicates the first element of the ordered pair. Once again, this demonstrates the multiplication idea of $2 \times 2 = 4$.

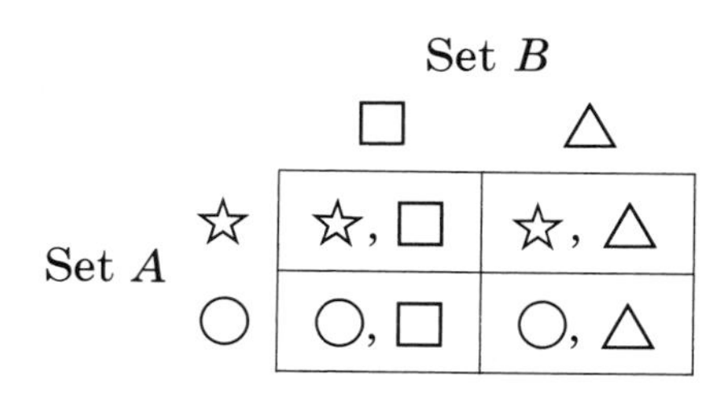

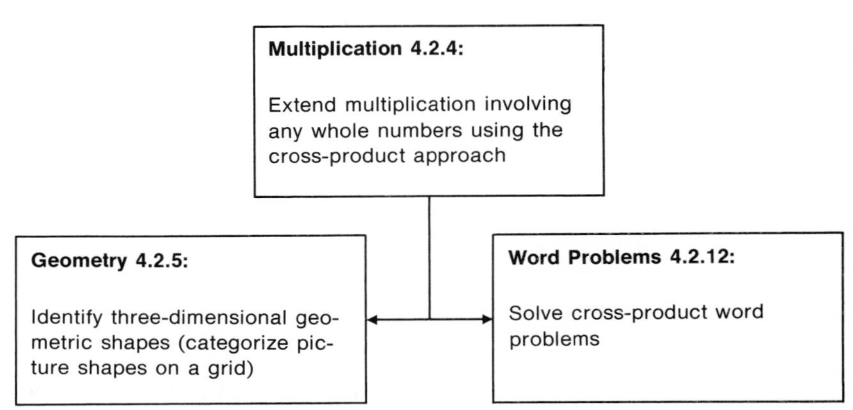

CROSS PRODUCT AND UNION OF EQUIVALENT DISJOINT SETS COMPARED

Models as involving different thought processes

The two models for multiplication presented thus far—the union of equivalent disjoint sets and cross product—are useful for classroom instruction. The two methods yield the same product for any pair of factors, but they involve different thought processes. For example, the union of equivalent disjoint sets model may be used in solving the problem: "Nick planted 3 rows of corn, using 8 seeds in each row. How many seeds did he plant?" On the other hand, the cross-product model for multiplication would be more appropriate in solving the problem: "If 3 children play trumpet and 8 other children play piano, how many trumpet-piano duets can be formed?" The learner should be aware that these problems are different but that both situations involve multiplication.

Models as related to the child's development

The model for multiplication as the union of equivalent disjoint sets appears more appropriate for children in the earlier stages of cognitive development. Take the following example:

> Seven-year-old children were individually given 3 plastic bags with 5 candy kisses in each bag. The children were told that each bag had 5 pieces of candy in it (this was obvious as the bags were transparent). Placed before each child were a counting abacus, paper and a pencil, and 6 each of the flannel numerals 3, 5, and 1. The children were asked how many candy kisses there are altogether.

This procedure was repeated with several 7-year-old second graders. Their strategies for finding the total number of candy kisses are described below in an ascending order of maturity of mathematical thinking:

Johnathan dumped the candy out of all of the bags and counted the pieces of candy.

Larry emptied only one of the bags to verify that it had 5 pieces of candy. He then asked, "Are you sure the others have five, too?" He received an affirmative answer. He said, "Well, five, ten, fifteen. There are fifteen pieces."

Tommy placed a numeral 5 by each bag and wrote on the paper

$$\begin{array}{r} 5 \\ 5 \\ +\ 5 \\ \hline 15 \end{array}$$

Joseph simply stated, "Three fives are fifteen." When asked how this is written, he wrote, "3 fives are 15," and asked if he could write this still another way, he wrote, "3 5s are 15."

In further interviews, it was found that the children seemed to think of the factors as a concrete mode. For example, Joseph explained that he "really" was thinking "three five-candy-kisses" and that the "three was really the bags."

Tommy made the transfer from his addition algorithm to the oral sentence "three fives are fifteen." He was then shown a further translation of this word sentence to a word-numeral sentence, "three 5s are 15," making use of the visual appearance of his own algorithm

$$\begin{array}{r} 5 \\ 5 \\ +\ 5 \\ \hline 15 \end{array}$$

Joseph and Tommy were then told that their solutions could be expressed as a multiplication sentence that was very similar to Joseph's sentence "3 5s are 15."

Tommy was asked to read Joseph's sentence and said, "Thirty fives are fifteen." Joseph said, "No, that says three fives are fifteen." They were then asked if they knew a sign that could be used to separate the 3 and the 5 and show what Joseph's sentence meant. Larry, who had also been listening, wrote $\times$ on the paper. The children were told that $\times$ represents multiplication and that "$3 \times 5 = 15$" is read as "three times five equals fifteen."

Notice that both the concrete level and symbols were used in these interviews. The children could also have been tested at the picture level: Pictures of the bags could have been given to the children.

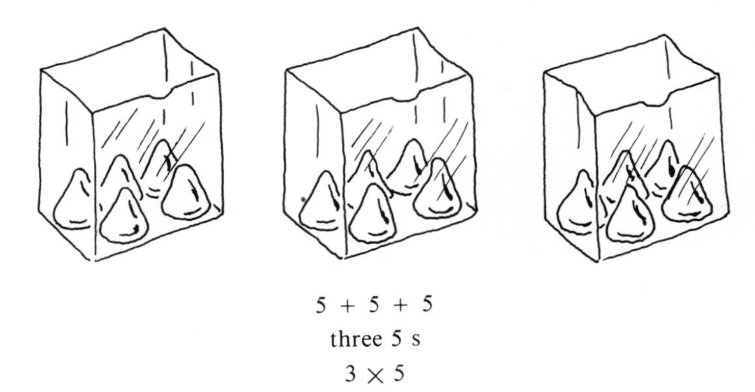

USE OF ARRAYS

Both models for multiplication discussed above may be represented by an array, which is a specific arrangement of elements of two sets by rows (represented horizontally) and by columns (represented vertically, like the standing marble columns of buildings). The convention is to state the number of rows first and then the number of columns.

A rectangular array having 3 rows and 4 columns would be called a "3-by-4 array" and is noted "3 × 4." This is pictured below. The upper picture focuses on the three horizontal rows. The lower picture emphasizes the four vertical columns. If the two rectangular regions were superimposed, the result would represent a 3-by-4 array.

Suggested transparencies

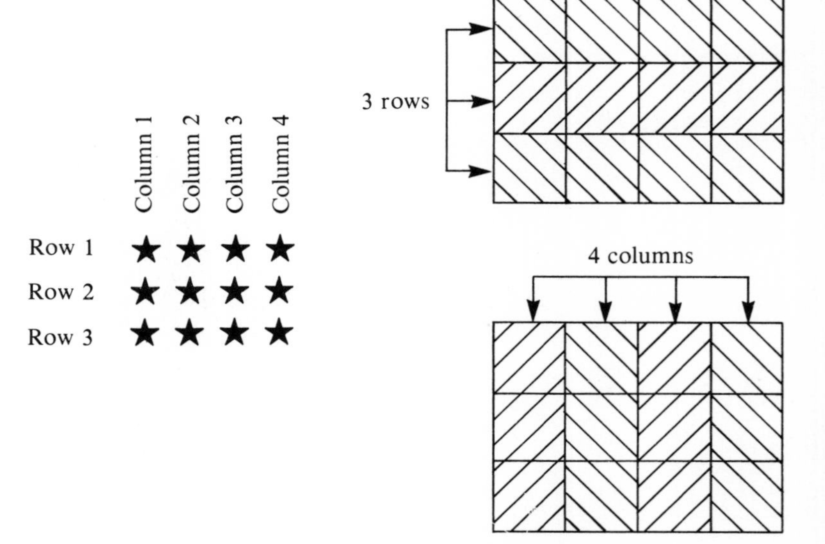

By counting the number of elements in an array, we arrive at the product of the number of elements in the rows and the number of elements in the columns. Thus, a 3-by-4 array has 12 elements.

PROPERTIES OF MULTIPLICATION

The set of whole numbers is used to illustrate the multiplication properties discussed below.

1. *Commutative Property for Multiplication* (CPM) $a \times b = b \times a$. It was discussed previously that when the product of two whole numbers is obtained, the order in which they are multiplied is immaterial. This was shown as $n(S \times D) = n(S) \times n(D) = n(D) \times n(S) = n(D \times S)$ and is called the Commutative Property for Multiplication.

2. *Associative Property for Multiplication* (APM): $(a \times b) \times c = a \times (b \times c)$. Multiplication, like addition, is a binary operation and can be used to work on exactly two numbers at a time. If three numbers are to be multiplied, the product of two of the numbers is multiplied by the third number. For example, in the problem $1 \times 2 \times 3$, we can begin by either multiplying 1×2 or 2×3, renaming the problem to either 2×3 or 1×6, respectively. The final product 6 is obtained in both cases.

The following board work helps young children in distinguishing between the associative property and the commutative property. Note that there is no change in the order of the factors from one side of the equals sign to the other; rather, there is a change of grouping. The location of the numbers are numbered "one," "two," and "three" to illustrate this point.

Suggested board work

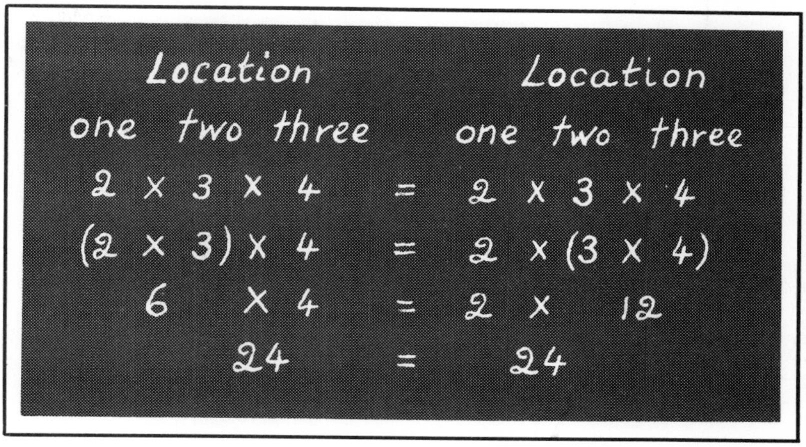

Location		Location
one two three		one two three
$2 \times 3 \times 4$	=	$2 \times 3 \times 4$
$(2 \times 3) \times 4$	=	$2 \times (3 \times 4)$
6×4	=	2×12
24	=	24

3. *Closure for Multiplication:* For two numbers a and b in a set of numbers, N, it is true that $a \times b$ is a unique element of N. For the whole numbers 3 and 4, $3 \times 4 = 12$. Since 12 is also a whole number, the set of whole numbers is said to be closed under multiplication (unless someone comes up with an exception, which has not occurred as yet and is not likely to happen).

What about the even and odd whole numbers? The following examples are instances verifying that the set of whole numbers is closed under multiplication for both odd and even numbers.

$3 \times 5 = 15$	Two odd whole numbers multiplied yield an odd whole number.
$2 \times 4 = 8$	Two even whole numbers multiplied yield an even whole number.

4. *Distributive Property for Multiplication over Addition* (DPMA): $a \times (b + c) = (a \times b) + (a \times c)$. This property states that multiplication distributes over addition. This means that to find a product of a sum of two whole numbers by multiplying a third whole number, the individual may add first and then multiply or multiply first and then add. In the following example, let $a = 3$, $b = 8$, and $c = 2$.

$$\begin{aligned} a \times (b + c) &= (a \times b) + (a \times c) \\ 3 \times (8 + 2) &= (3 \times 8) + (3 \times 2) \\ 3 \times 10 &= 24 + 6 \\ 30 &= 30 \end{aligned}$$

The distributive property relates multiplication and addition of whole numbers. It underlies the multiplication algorithm. The distributive property also underlies some important ideas in algebra. For example, since it allows the multiplication of ones, tens, hundreds, and so on separately, the following arithmetic and algebraic algorithms are possible:

(a)

$$\begin{array}{rl} 32 & \\ \times\ 3 & \\ \hline 6 & (3 \times 2) \\ 90 & (3 \times 30) \\ \hline 96 & (90 + 6) \end{array}$$

(b)

$$\begin{aligned} & 3(32) \\ &= 3(30 + 2) \\ &= (3 \cdot 30) + (3 \cdot 2) \end{aligned}$$

(c)

$$a(b + c)$$
$$(ab) + (ac)$$

(d)

$$\begin{array}{r} b + c \\ \times \quad a \\ \hline ab + ac \end{array}$$

Unless the pupil has a grasp of the Distributive Property of Multiplication over Addition, multiplication algorithms will hold no meaning for the child, and he or she will have difficulty bridging from arithmetic algorithms to algebraic problems such as $(a + b)(c + d)$.

5. *Multiplicative Identity* (also called Identity Element): 1 is a unique natural number such that $1 \times a = a \times 1 = a$ for any natural number a.

This property may be shown by a "1-by-a" array. Let a in the equation $1 \times a = a$ equal 2, 3, and 8. Then,

$$1 \times 2 = 2 \cdot \cdot$$
$$1 \times 3 = 3 \cdot \cdot \cdot$$
$$1 \times 8 = 8 \cdot \cdot \cdot \cdot \cdot \cdot \cdot \cdot$$

Now let us commute the factors to read $a \times 1 = a$

$$2 \times 1 = 2 \; :$$
$$3 \times 1 = 3 \; \vdots$$
$$8 \times 1 = 8 \; \vdots$$

Using cross product is an alternate method for showing the multiplicative identity:

	Set B □	△
Set A ∗	∗, □	∗, △

$A = \{*\}$ $A \times B = \{(*, □), (*, △)\}$ $n(A) = 1$

$B = \{□, △\}$ $n(B) = 2$

$n(A \times B) = 2$

Mini-lesson 3.1.7 Use Multiplication Properties

Vocabulary: multiplicative identity, commutative, associative, distributive property of multiplication over addition, zero as a factor

Possible Trouble Spots: does not attend to salient aspects of situations; does not apply generalizations to new situations; does not seem aware of relationships; has difficulty labeling situations; has poor memory skills; has visual discrimination difficulties

Requisite Objective:
Addition 2.2.3: Use additive identity

Activity 1
To teach the multiplicative identity property that may be shown as $1 \times a = a$ or $a \times 1 = a$, use the following procedures.

A. Grids may be used as a drill activity for students to show the identity property for multiplication as well as the basic facts. Instruct the class to complete only the row and column that shows the products for $n \times 1$.

X	0	1	2	3	4	5	6	7	8	9
0										
1										
2										
3										
4										
5										
6										
7										
8										
9										

B. Collect as many objects as possible and allow children to make one group (stress the one group concept). The group may vary in number of objects.

C. The number wheel can be used. The factor is in the center. The outer ring is for the product.

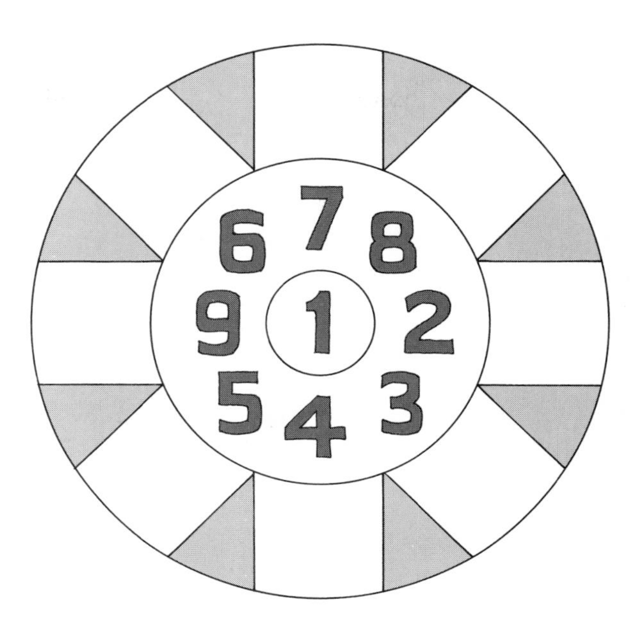

Activity 2
To show multiplication when zero is a factor, have the child participate in the following activities.

A. Use a number line and include 2×0 as one of the multiplication problems. Demonstrate that no jumps can be made; therefore, the answer is zero.

B. Emphasize that when the number property of equivalent disjoint sets is zero, sets are being combined and, therefore, the number property must be zero. Ask the child, "If you combine three empty bags of candy, how much candy do you have?" The child should answer, "Three zeros equal zero. Three times zero equals zero."

Activity 3
To recognize the Commutative Property for Multiplication, draw sets of figures on the chalkboard while students simultaneously use counters or draw diagrams at their seats such as follows:

$2 \times 3 = 6$ ○○○ ○○○
2 sets of 3

$3 \times 2 = 6$ △△ △△ △△
3 sets of 2

Activity 4
To recognize the Associative Property for Multiplication, have students use a hand calculator to show examples, such as the following:

$$(3 \times 6) \times 2 = 3 \times (6 \times 2)$$
$$18 \quad \times 2 = 3 \times \quad 12$$
$$36 \quad = \quad 36$$

Activity 5
To show the Distributive Property of Multiplication over Addition, use the following procedures.

A. Give the child three pieces of graph paper, the first ruled to show a 4-by-12 array; the second, a 4-by-10 array; and the third, a 4-by-2 array. Guide the child to see that there are 48 squares, regardless of whether the array is a single 4-by-12 arrangement or if it is divided into a 4-by-10 and a 4-by-2 arrangement.

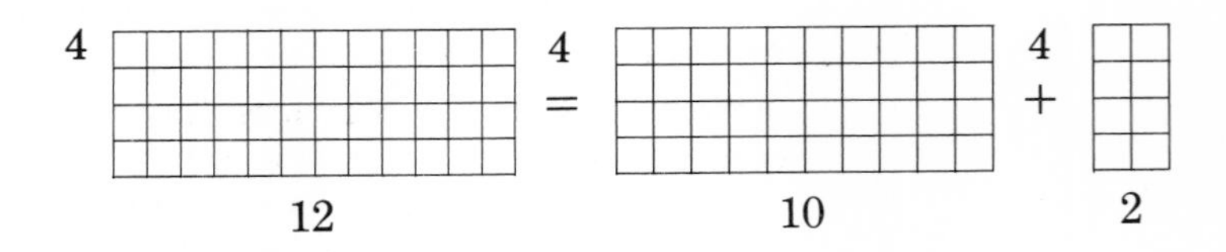

$$\begin{aligned} 4 \times 12 &= 4 \times (10 + 2) \\ &= (4 \times 10) + (4 \times 2) \\ &= 40 + 8 \\ 48 &= 48 \end{aligned}$$

B. Use rectanglar arrays as follows.

$3 \times 8 = \square$

$(3 \times 5) + (3 \times 3) = \square + \square = \square$

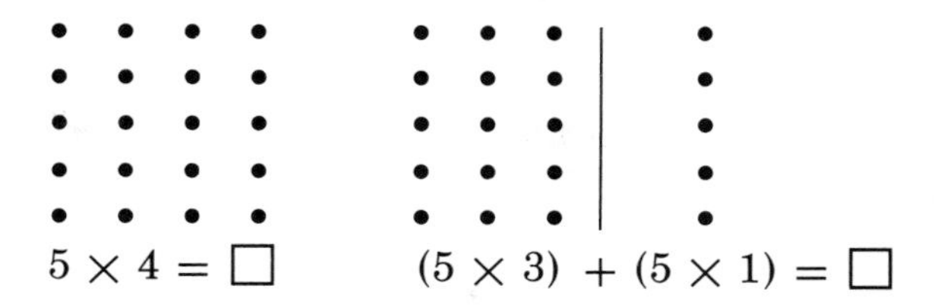

Note: Mini-lesson 3.1.7 is a prerequisite for Mini-lessons 3.1.8 and 3.1.18.

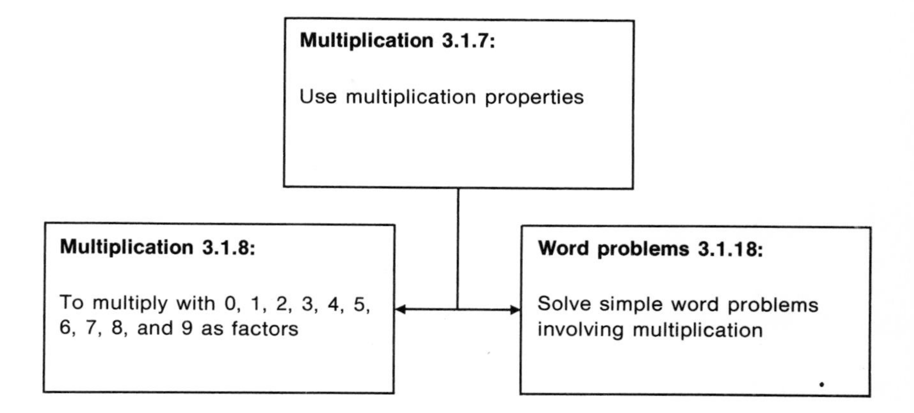

BASIC MULTIPLICATION FACTS

Basic multiplication facts are products formed by one-digit factors, for example, $3 \times 9 = 27$ or $7 \times 8 = 56$. Products that have as their factors the number 5 or less may be shown on an abacus or at the concrete level by using the union of equivalent disjoint sets model. Then grids may be used, as in addition, as a drill activity for the pupil to learn some of the properties of multiplication as well as the basic facts.

Mini-lesson 3.1.9 Compute Basic Multiplication Facts

Vocabulary: basic fact, factor, product

Possible Trouble Spots: difficulty seeing patterns and forming series; need for repetition due to difficulty in remembering

Requisite Objectives:
Complete multiplication grids

Multiplication 3.8: Multiply with 0, 1, 2, 3, 4, 5 as factors; with 6, 7, 8, 9 as factors

Activity 1
Make a multiplication grid designed to check the child's knowledge of the 2s and 5s basic facts. The child should notice the counting by 2s and counting by 5s patterns as she completes the grid.

×	0	1	2	3	4	5	6	7	8	9
2	0	2	4	6	8	10	12	14	16	18
5	0	5	10	15	20	25	30	35	40	45

Activity 2
Make a multiplication grid, selecting factors whose products are 25 or less and ask the child to complete the grid. If the child fills in an incorrect product, circle that product.

×	2	3	4
2	4	(5)	(6)
3	6	9	(7)
4	8	12	16

Ask the child how he got the encircled answers. Guide him to see patterns of counting. In the first column (in which all the answers are correct), there is a counting by 2s pattern. But in the second and third columns, there are no such patterns. He has ignored counting by 3s and then by 4s in the columns that are completed incorrectly.

Guide the child to look for patterns within rows. Remind him of the commutative property. Ask him if $3 \times 2 = 6$ in the first column, how can $2 \times 3 = 5$ in the second column? Raise the same question for 3×4 and 2×4.

Activity 3
Make a multiplication grid including 0, 1, and 5 as factors and ask the child to complete the grid.

Since 0 and 1 as factors are trouble spots in multiplication, recording them with colored chalk or pencil helps point up the following patterns:

$0 \times n = n \times 0 = 0$ Zero times a number is zero.
$1 \times n = n \times 1 = n$ One times a number is that number.

×	0	1	2	3	4	5
0	0	0	0	0	0	0
1	0	1	2	3	4	5
2	0	2	4	6	8	10
3	0	3	6	9	12	15
4	0	4	8	12	16	20
5	0	5	10	15	20	25

Activity 4
Give the child a partially completed 0-to-9 multiplication grid and ask him to fill in the blank spaces.

×	0	1	2	3	4	5	6	7	8	9
0	0	0	0	0	0	0	0	0	0	0
1	0	1	2	3	4	5	6	7	8	9
2	0	2	4	6	8	10	12	14	16	18
3	0	3	6	9	12	15	18	21	24	27
4	0	4	8	12	16					
5	0	5	10	15	20	25				
6	0	6	12	18	24	30	36			
7	0	7	14	21	28	35	42	49		
8	0	8	16	24	32	40	48	56	64	
9	0	9	18	27	36	45	54	63	72	81

Ask the child the following questions:

1. How many basic facts are there for multiplication? (100)
2. When considering the commutative property, how many basic facts are zero facts? (10)
3. How many of them are commutative (eliminate square numbers)? (45)
4. When considering the commutative property, how many basic facts involve the multiplicative identity? (9)
5. After eliminating facts obtained from using the commutative property, the 10 zero facts and the 9 multiplicative identity facts, how many basic facts remain to be learned? (36); $100 - (45 + 10 + 9) = 100 - 64$.

Guide the children to look for the following patterns:

1. $0 \times n = 0 \quad n \times 0 = 0$
2. $1 \times n = n \quad n \times 1 = n$
3. Square numbers are found on the top left to bottom right diagonal (4, 9, 16, 25, 36, 49, 64, 81).
4. The Commutative Property for Multiplication is shown by reversing factors.
5. Going down the "2" column shows counting by 2s (also across the "2" row).
6. Going down the "3" column shows counting by 3s (also across the "3" row).
7. Each of the following columns show counting patterns: "4," "5," "6," "7," "8," and "9."
8. As you go down columns headed by 2 or multiplies of 2, all of the cells contain even numbers. How about the rows headed by even numbers? What do you notice about the columns and rows headed by odd numbers?

Activity 5
Ask the child to complete simple open mathematical sentences by filling in the two *single-digit* factors.

$\square \times \Diamond = 15$	3×5 or 5×3
$\triangle \times \square = 27$	3×9 or 9×3
$\_\_ \times \_\_ = 36$	6×6, 4×9, or 9×4
$\_\_ \times \_\_ = 42$	7×6 or 6×7

Activity 6
Use the Distributive Property for Multiplication over Addition in learning basic multiplication facts.

Thus, the 8 is renamed as $8 = 3 + 5 = 2 + 6 = 4 + 4 = 7 + 1$. The child may use the distributive property as a strategy for renaming one of the factors so that she may deal with lower-level basic multiplica-

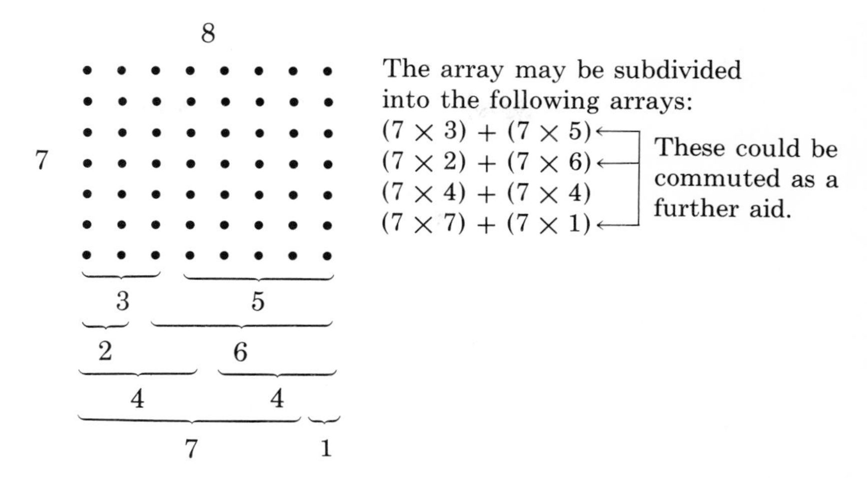

tion facts whose products she can add to find the product of the original basic fact.

Additional considerations: Properties other than the distributive property can be used to teach basic multiplication facts. For example, make use of the associative property by asking the child to name products as follows:

$$4 \times 4 = \square$$
$$(2 \times 2) \times 4 = \square$$
$$2 \times (2 \times 4) = \square$$
$$2 \times 8 = \square$$

Encourage children to use the properties in this manner for drill activities.

Note: Mini-lesson 3.1.9 is a prerequisite to Mini-lessons 3.1.12 and 3.2.6.

Multiplication 3.1.9:

Compute basic multiplication facts

Measurement 3.1.12:

Write time shown on a clock face

Multiplication 3.2.6:

Multiply two-digit by one-digit numbers, with no renaming

EXERCISES

I. Give examples of children in kindergarten and first grade using multiplication ideas.

II. Give two models for teaching multiplication.

III. Discuss why the phrase "factor $\times$ factor $=$ product" is an accurate mathematical statement but cannot necessarily be applied to physical world situations.

IV. Tell when the cross-product model for multiplication is more appropriate than the union of equivalent disjoint sets.

V. Write word problems that show thought processes for the two models of multiplication.

VI. Draw an array to show 2×4.

VII. State the properties for multiplication of whole numbers and give an example of each.

VIII. Identify the following properties for multiplication:

1. $(74 + 8) \times 97 = 97 \times (74 + 8)$
2. $35 \times (7 \times 9) = (35 \times 7) \times 9$
3. $3 \times 12 = 36$
4. $6 \times 24 = 144$
5. $35 \times 17 = (30 \times 10) + (30 \times 7) + (5 \times 10) + (5 \times 7)$
6. $132 \times 1 = 132$

IX. What are products formed by one-digit factors called?

X. Write the product shown in the array below as pairs of factors using the Distributive Property of Multiplication Over Addition.

● ● ● ●
● ● ● ●
● ● ● ●
● ● ● ●

SUGGESTED READINGS

Cacha, F. B. "Understanding Multiplication and Division of Multi-digit Numbers." *The Arithmetic Teacher* 19 (May 1972): 349–54.

Gardner, M. "Tricks of Lightning Calculators," In *Mathematical Carnival.* New York: Knopf, 1975. Pp. 77–88.

Multiplication Algorithms

EARLY METHODS OF MULTIPLICATION

Quarter-Squares Method

The Arabs used a process of multiplication called the "quarter-squares" method. The product of two numbers is obtained by subtracting the square of half their difference from the square of half their sum. For example, in considering the two factors 3 and 9, half the difference of 3 and 9 is 3; the square of 3 is 9. This is shown in the illustration below:

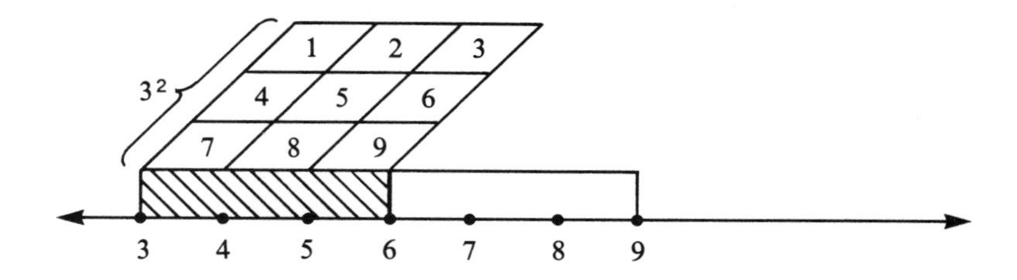

Half the difference of 3 and 9 is 3. $\qquad \dfrac{9-3}{2} = 3$

The square of 3 is 9. $\qquad 3^2 = 9$

Half the sum of 3 and 9 is 6. The square of 6 is 36. This is shown in the construction below. The product of our two original factors, then, is 27, or 36 − 9.

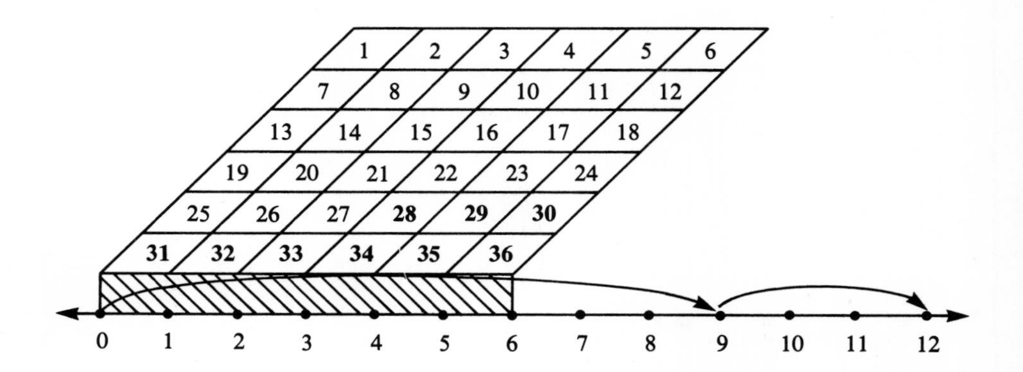

Half the sum of 3 and 9 is 6. $\frac{9+3}{2} = 6$

The square of 6 is 36. $6^2 = 36$

$$\therefore 36 - 9 = 27, \text{ or } 3 \times 9 = 27$$

The Sluggard's Rule for Multiplying

The Sluggard's Rule for Multiplying involves a thinking process similar to that used in the quarter-squares method and states: "Subtract each digit from 10, and write down the difference; multiply these differences together, and add as many tens to their product as the first digit exceeds the second difference, or the second digit the first difference."[1]

If, for example, the learner wants to multiply 6 × 7 but cannot remember the product, he or she may subtract as follows:

$$10 - 6 = 4$$
$$10 - 7 = 3$$

Then the product 3 × 4 = 12 is obtained. Next, it is necessary to find the difference between one of the original factors (7) and the complement to 10 of the other factor (10 − 6, or 4). The difference is 3, or 7 − 4. Then 3 tens, or 30, is added to 12, yielding 42, the product of 6 × 7.

1. Edward Brooks, *Philosophy of Arithmetic* (Lancaster, Pa.: Normal Publishing Co., 1880), p. 53.

Example: 7×8

$$\begin{array}{rr} 10 - 7 = & 3 \\ 10 - 8 = & \times\ 2 \\ \hline & 6 \\ & +50 \\ \hline & 56 \end{array}$$

Example: 8×9

$$\begin{array}{rr} 10 - 8 = & 2 \\ 10 - 9 = & \times\ 1 \\ \hline & 2 \\ & +70 \\ \hline & 72 \end{array}$$

Network Method

The Network method of multiplication was very popular in the East and was adopted by the Arabs, who termed it *shabacah,* meaning "network," because of its reticulated appearance. (This process is the underlying principle of multiplying using Napiers rods). The illustration below shows the network method of multiplication.

One factor (135) is written across the top of the grid and the other factor (12) is written vertically along the right side of the grid. Partial products are placed in the corresponding cells and are then summed along each diagonal.

Example: 12×135

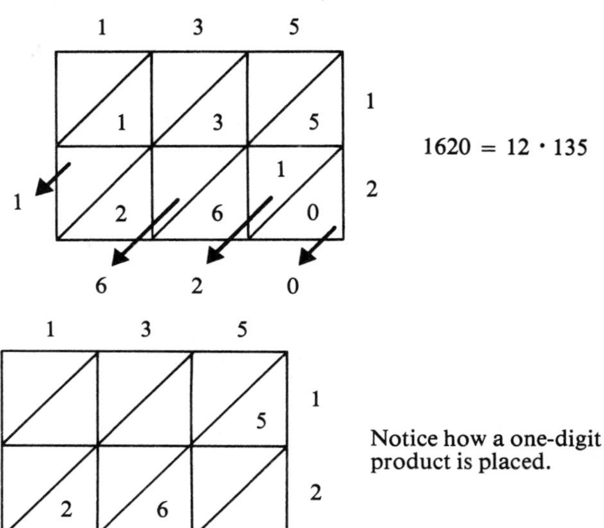

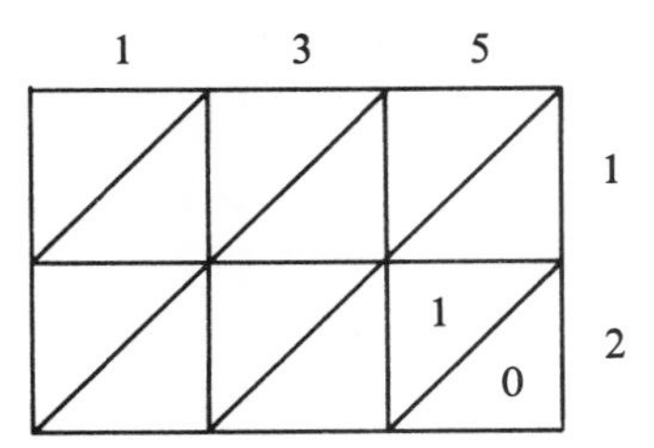

Notice how a two-digit product is placed.

Napier's Bones

In 1617, a Scotsman named Napier invented a calculating machine that was based on the Network method of multiplication. "Napier's bones," as they have come to be called, were described by Brooks as 10 separate thin strips of bone, metal, ivory, or wood "about two inches in length and a quarter of an inch in breadth."[2] Modern-day versions of Napier's bones are constructions made of a heavy poster paper (see Figure 7.1). The advantage of this tool over the abacus is that the "bones" or "rods" show a number that is "carried," whereas an abacus does not.

Figure 7.1. Napier's Bones

1	2	3	4	5	6	7	8	9	0	R
1	2	3	4	5	6	7	8	9	0	I
2	4	6	8	1/0	1/2	1/4	1/6	1/8	0	II
3	6	9	1/2	1/5	1/8	2/1	2/4	2/7	0	III
4	8	1/2	1/6	2/0	2/4	2/8	3/2	3/6	0	IV
5	1/0	1/5	2/0	2/5	3/0	3/5	4/0	4/5	0	V
6	1/2	1/8	2/4	3/0	3/6	4/2	4/8	5/4	0	VI
7	1/4	2/1	2/8	3/5	4/2	4/9	5/6	6/3	0	VII
8	1/6	2/4	3/2	4/0	4/8	5/6	6/4	7/2	0	VIII
9	1/8	2/7	3/6	4/5	5/4	6/3	7/2	8/1	0	IX

2. Brooks, *Philosophy of Arithmetic,* p. 160.

In using Napier's bones to solve the problems 5×738 and 15×738, the 7, 3, and 8 rods would be placed side by side next to the guiding rod (R) as shown below. The product is found by adding in a downward movement within a diagonal in the section of squares that is in a row with the multiplier in the R column.

7	3	8	R
/7	/3	/8	I
1/4	/6	1/6	II
2/1	/9	2/4	III
2/8	1/2	3/2	IV
3/5	1/5	4/0	V
4/2	1/8	4/8	VI
4/9	2/1	5/6	VII
5/6	2/4	6/4	VIII
6/3	2/7	7/2	IX

$$5 \times 738 = 3690$$
$$15 \times 738 = 11070$$

Example: 5×738

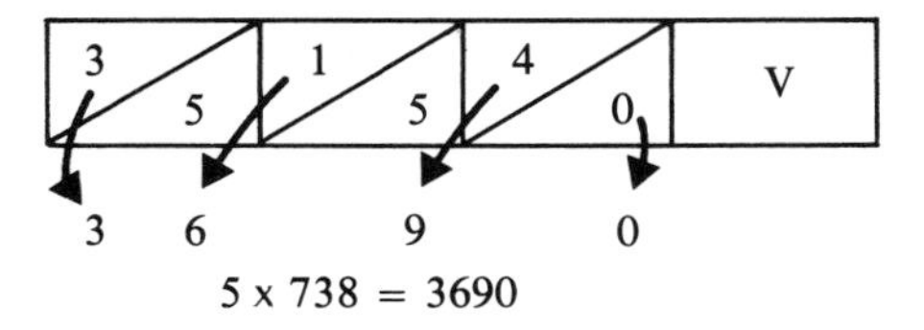

In the problem 15×738, the distributive property is employed in renaming 15 as $10 + 5$. The product of row I is multiplied by 10 and then added to the product in row V.

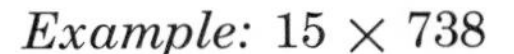

Example: 15 × 738

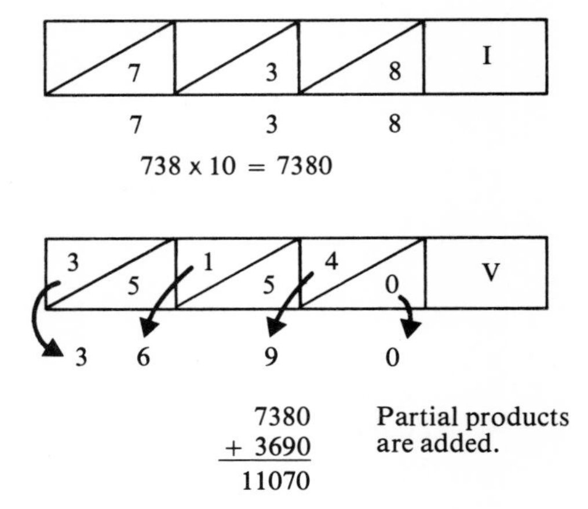

MODERN ALGORITHMS FOR MULTIPLICATION OF WHOLE NUMBERS

Distributive property as basis for modern algorithm

The algorithms used today depend upon the Distributive Property of Multiplication over Addition (as did the *shabacah* [network] and *schacherii* [chessboard] methods in particular). The distributive property states that for any three natural numbers (a, b, c), it is true that $a \times (b + c) = (a \times b) + (a \times c)$.

This property was also used in the method of multiplication by duplation. Referring to the duplation example 15 × 37, the factor 15 was expressed as 8 + 4 + 2 + 1. The product was therefore expressed as (8 + 4 + 2 + 1) × 37, which is the same as the following:

$$\begin{aligned} & 37 \times (8 + 4 + 2 + 1) \\ = & (37 \times 8) + (37 \times 4) + (37 \times 2) + (37 \times 1) \\ = & \quad 296 \quad + \quad 148 \quad + \quad 74 \quad + \quad 37 \\ = & 555 \end{aligned}$$

Modern multiplication algorithms below show the partial products in their respective place value positions.

For the example 156 × 349, the horizontal algorithm and properties justifying it are as follows:

$156 \times 349 = 156 \times (300 + 40 + 9)$ Renaming in expanded form
$= (156 \times 300) + (156 \times 40) + (156 \times 9)$ Distributive Property of Multiplication over Addition
$= 46800 + 6240 + 1404$ Table for multiplication
$= 54444$ Table for addition

The vertical algorithms for 156 × 349 include:

(a)

```
   349
 ×156
    54   (6 × 9)
   240   (6 × 40)
  1800   (6 × 300)
   450   (50 × 9)
  2000   (50 × 40)
 15000   (50 × 300)
   900   (100 × 9)
  4000   (100 × 40)
 30000   (100 × 300)
 54444
```

(b)

```
   349
 ×156
 17450
 34900
 54444
```

An intermediate step is to block out place values with zeros so that the next partial product is "pushed over" to its proper position. Child thinks for 50 × 9, "tens times ones gives me tens, so I block out the units place"; for 100 × 9, the child thinks, "hundreds times ones are hundreds, so block out units and tens places."

(c)

```
    24
    25  (crossed out)
   349
 ×156
  2094
 1745
 349
 54444
```

Shortcut using aids to show renaming

Aids for positioning partial products

For children having difficulty with writing partial products in the correct position, Table 7.1 (which they should build) has been found to be helpful. Teachers should allow children to refer to this table until they no longer need a visual aid. Some children may need this aid throughout several grades.

Checking answers

In checking the answers, the factors are simply interchanged, as in changing 156 × 349 to 349 × 156. The product should be the same, regardless of the order of the factors.

Table 7.1. Aids in Determining Positions of Partial Products in the Vertical Multiplication Algorithm

Factor place values	*Product place values*
$1 \times 1 =$	1
$1 \times 10 =$	10
$1 \times 100 =$	100
$10 \times 1 =$	10
$10 \times 10 =$	100
$10 \times 100 =$	1000
$100 \times 1 =$	100
$100 \times 10 =$	1000

```
    11
    2̸2̸
    5̸5̸
   156
  ×349
  ----
  1404
  6240
 46800
 -----
 54444
```

Casting out 9s

Casting out 9s is a procedure that detects errors in computation if the two excesses of 9s do not agree. The excess of 9s in the product is equal to the excess of 9s in the product of excesses of the multiplicand and multiplier. However, this is a method of verification that an answer is correct rather than a checking procedure. If an incorrect product of 45444 were obtained, the casting-out 9s method would not detect the error.

	Sum of the digits	*Excess of 9s*	
349	$3 + 4 + 9 = 16 = 7$	$3 \times 7 = 21; \quad 2 + 1 = 3$	*Excess of 9s in product of excesses*
$\times$156	$1 + 5 + 6 = 12 = 3$		
54444	$5 + 4 + 4 + 4 + 4 = 21 = 3$	*Excess of 9s in product*	

Arrays and the multiplication algorithm

Children must understand the relationship between the Distributive Property of Multiplication over Addition and the multiplication algorithm in order to perform such multiplications as 6×35 either mentally or on paper. Using arrays helps to show the distributive property as related to the multiplication algorithm.

Mini-lesson 3.2.5 Use the Distributive Property of Multiplication over Addition as Related to the Multiplication Algorithm

Vocabulary: distributive property of multiplication over addition, factor, multiple, rename

Possible Trouble Spots: cannot rename a two-digit number as a sum; mixes up the + and × signs, e.g.
$3 \times 25 = 3 + (20 \times 5)$ *or* $3 \times (20 + 5) = (3 + 20) \times (3 + 5)$

Requisite Objectives:
Rename a number such as 24 as a sum, e.g., $20 + 4$

Whole Numbers/Notation 2.2.1: Use exchange model for place value

Multiplication 3.1.9: Compute basic multiplication facts

Multiply when one factor is a multiple of 10

Activity 1
Ask the child to build the following individual arrays.

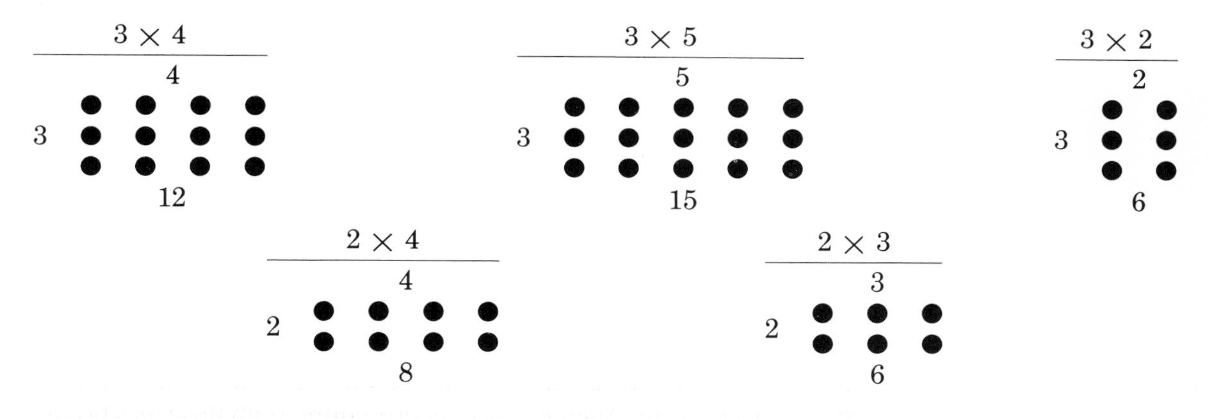

Activity 2
Ask the child to build a 3-by-6 array.

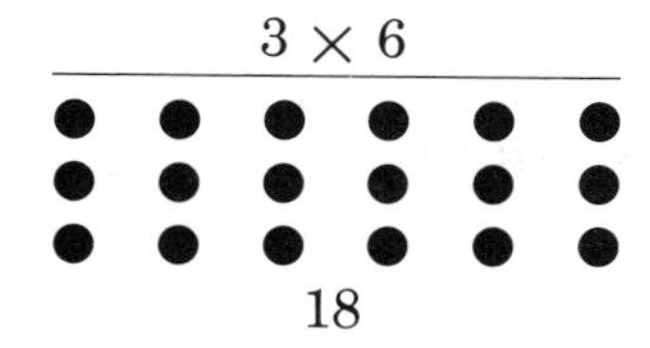

Suggest that he select from the arrays in Activity 1 two sets of arrays whose unions have 18 dots. When he selects the 3-by-2 and the 3-by-4 arrays, ask him to draw a vertical line through the 18-dot array to show a 3-by-2 array. Ask him to label each part.

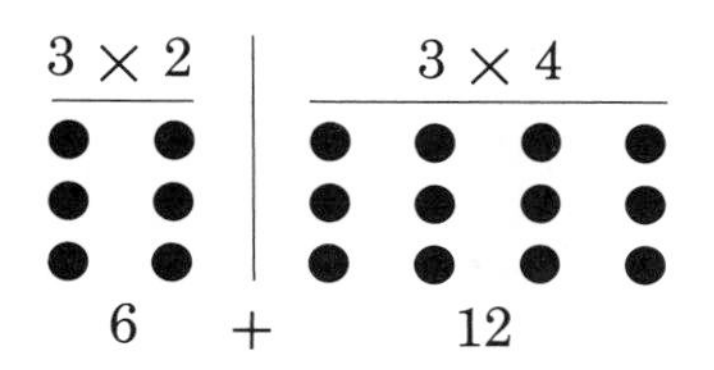

Activity 3
Ask the child to combine two other sets of arrays from Activity 1 to form a single multiplication problem, such as $(2 \times 3) + (2 \times 4) = 2 \times 7 = 14$.

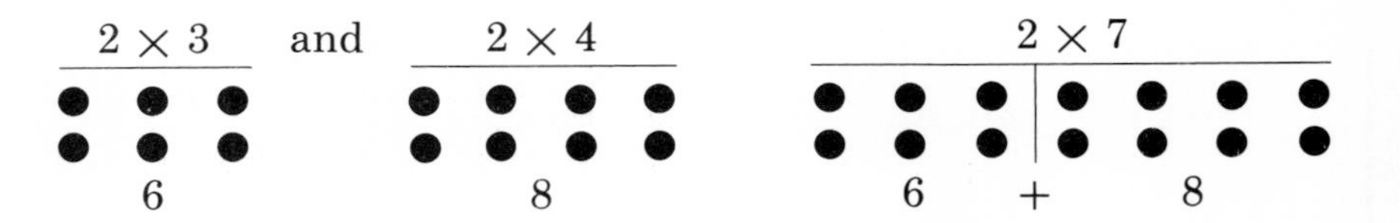

Next ask him how the 7 is related to putting the 2-by-3 and 2-by-4 arrays together. If the child knows the basic addition facts, he will associate the pair 3, 4, with the number 7. Point to the 3 and to the 4 if necessary.

Then introduce the algorithm showing the Distributive Property of Multiplication over Addition.

$(2 \times 3) + (2 \times 4)$	Distributive Property of Multiplication
$= 2 \times (3 + 4)$	over Addition
$= 2 \times 7$	Basic addition fact
$= 14$	Basic multiplication fact

Additional considerations: Use of hand calculators should be allowed, especially for the longer computations. Computing with the calculator is a good drill activity for learning the basic facts. The children should compute and then check their results with the calculator. Hand calculators allow slow learners in mathematics to participate successfully in activities that previously were trouble spots (especially for those who find it difficult to memorize).

For the student with visual perception problems that interfere with arrangement of the multiplication algorithm, use should be made of the historical algorithms. The Network method is particularly helpful in structuring the placement of the partial products. These algorithms, as well as Napier's bones, serve as cues for students who do not understand place value because there is no need to rename from one place to another and then to add in the "carried" number. Furthermore, these algorithms do not require that the student be able to place the partial products in the correct column according to place value. The format of the algorithm facilitates correct placement of partial products.

Instruction to write the multiplication problem within a place value chart provides structure for using face value and place value as shown for the problem 13×21.

Example: 13×21

hundreds	tens	units	*Face values*	*Place values*
	2	1		
	1	3		
		3	$3 \times 1 = 3$	*units* × *units* = *units* $(1 \times 1 = 1)$
	6	0	$3 \times 2 = 6$	*units* × *tens* = *tens* $(1 \times 10 = 10)$
	1	0	$1 \times 1 = 1$	*tens* × *units* = *tens* $(10 \times 1 = 10)$
2	0	0	$1 \times 2 = 2$	*tens* × *tens* = *hundreds* $(10 \times 10 = 100)$
2	7	3	*Sum of partial products*	

Note: Mini-lesson 3.2.5 is a prerequisite for Mini-lessons 3.2.6 and 3.2.22.

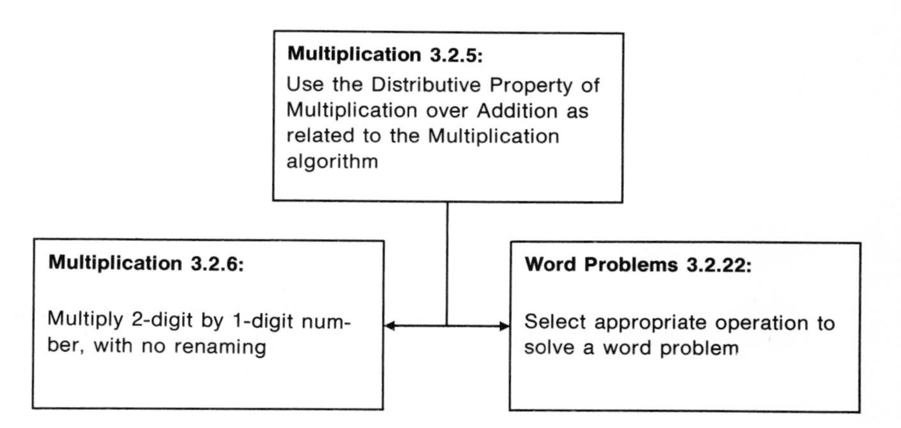

MULTIPLICATION EXAMPLES AS SAMPLES FOR DIAGNOSTIC TEACHING

The following examples serve as a structure for assessing multiplication computation. The teacher may use these items to identify students' competence in multiplying.

1. Basic facts:

$$8 \times 9 \qquad \begin{array}{r} 9 \\ \underline{\times 8} \end{array}$$

2. Eleven as a factor; number 0 to 9 as the other factor:

$$\begin{array}{r} 11 \\ \underline{\times\ 6} \end{array} \qquad \begin{array}{r} 11 \\ \underline{\times\ 9} \end{array}$$

3. Twelve as a factor; number 0 to 9 as the other factor:

$$\begin{array}{r} 12 \\ \times\ 6 \\ \hline \end{array} \qquad \begin{array}{r} 12 \\ \times\ 9 \\ \hline \end{array}$$

4. One-place factor times two-place factor, without renaming:

$$\begin{array}{r} 21 \\ \times\ 4 \\ \hline \end{array} \qquad \begin{array}{r} 34 \\ \times\ 2 \\ \hline \end{array}$$

5. One-place factor times three-place factor, without renaming:

$$\begin{array}{r} 123 \\ \times\ \ 3 \\ \hline \end{array} \qquad \begin{array}{r} 324 \\ \times\ \ 2 \\ \hline \end{array}$$

6. One-place factor times two-place factor, with renaming in the units place:

$$\begin{array}{r} 13 \\ \times\ 7 \\ \hline \end{array} \qquad \begin{array}{r} 29 \\ \times\ 2 \\ \hline \end{array}$$

7. One-place factor times two-place factor, with renaming in the units and tens places:

$$\begin{array}{r} 45 \\ \times\ 3 \\ \hline \end{array} \qquad \begin{array}{r} 79 \\ \times\ 6 \\ \hline \end{array}$$

8. Ten as a factor times one-place factor and the commuted form:

$$10 \times 3 \qquad \begin{array}{r} 10 \\ \times\ 3 \\ \hline \end{array}$$

9. One hundred as a factor times one-place factor and the commuted form:

$$100 \times 3 \qquad \begin{array}{r} 100 \\ \times\ \ 3 \\ \hline \end{array}$$

10. One thousand as a factor times one-place factor and the commuted form:

$$1000 \times 3 \qquad \begin{array}{r} 1000 \\ \times\ \ \ 3 \\ \hline \end{array}$$

11. Ten as a factor times two-place factor:

$$10 \times 23 \qquad \begin{array}{r} 23 \\ \times 10 \\ \hline \end{array}$$

12. One hundred as a factor times two-place factor:

$$100 \times 35 \qquad \begin{array}{r} 35 \\ \times 100 \\ \hline \end{array}$$

13. One hundred as a factor times three-place factor:

$$100 \times 985 \qquad \begin{array}{r} 985 \\ \times 100 \\ \hline \end{array}$$

14. Two-place factor times two-place factor, without renaming:

$$\begin{array}{r} 32 \\ \times 13 \\ \hline \end{array} \qquad \begin{array}{r} 21 \\ \times 14 \\ \hline \end{array}$$

15. Two-place factor times n-place factor, with renaming:

$$\begin{array}{r} 16 \\ \times 14 \\ \hline \end{array} \qquad \begin{array}{r} 16 \\ \times 41 \\ \hline \end{array} \qquad \begin{array}{r} 76 \\ \times 14 \\ \hline \end{array} \qquad \begin{array}{r} 76 \\ \times 34 \\ \hline \end{array} \qquad \begin{array}{r} 786 \\ \times\ 35 \\ \hline \end{array} \qquad \begin{array}{r} 9187 \\ \times\ \ 14 \\ \hline \end{array}$$

16. Factors with zeros in the tens place:

$$\begin{array}{r} 504 \\ \times\ 27 \\ \hline \end{array} \qquad \begin{array}{r} 9803 \\ \times\ 139 \\ \hline \end{array} \qquad \begin{array}{r} 702 \\ \times 103 \\ \hline \end{array}$$

TRANSFERRING FROM THE HORIZONTAL ALGORITHM TO THE VERTICAL ALGORITHM

Children are likely to encounter the problem of transferring from the horizontal algorithm to the vertical algorithm. When multidigit factors are involved, the horizontal algorithm is difficult for use in computation. The vertical algorithm is easier for computing three-place numbers by two-place numbers or larger. However, horizontal mathematical sentences are the most appropriate algorithm for translating word problems and thus are not likely to be discarded.

The word problem "Gilbert had 13 interviews and each interview was to last 7 minutes. How long would it take him to finish if he adhered to this schedule?" This implies 13 sevens and is shown horizontally in the form 13×7. The 13 is the multiplier telling how many groups, and the 7 is the multiplicand specifying the size of each group.

Vertical form displays reversal of thinking

At some stage of development, the child is asked to write this in vertical form. The traditional vertical form shows the multiplicand atop the multiplier as shown:

$$\begin{array}{r} 7 \\ \times 13 \\ \hline \end{array}$$

This is read "thirteen times seven." But when writing "thirteen times seven," the child usually goes from top to bottom and so has to reverse his thinking to "seven times thirteen":

$$\begin{array}{ccc} \textit{Step 1} & \textit{Step 2} & \textit{Step 3} \\ 7 & 7 & 7 \\ & \times & \times 13 \\ \end{array}$$

or to "seven, thirteen times":

Step 1	*Step 2*	*Step 3*
7	7	7
	13	$\underline{\times 13}$

Suggestions for transferring from horizontal to vertical algorithm

Following are some suggestions for helping the child to bridge the gap between arranging factors in a horizontal algorithm and placing them in a vertical algorithm:

1. Suggest to the child that the vertical algorithm be written in an upward direction (instead of from top to bottom) so that the motor sequence involved in writing the algorithm parallels the verbal sequence of "thirteen times seven."

Step 1	*Step 2*	*Step 3*	*Step 4*
		7	7
13	$\times 13$	$\times 13$	$\underline{\times 13}$

In the case of multiplication word problems, the learner is often involved with the multiplicand being a concrete, or denominate, number. A "denominate" number is one that expresses a specific kind of unit—minutes, for example. Thus, 7 minutes is a denominate number, while 13, without reference to concrete units, is referred to as "abstract." The vertical multiplication algorithm is often arranged so that the denominate number is placed above the abstract number.

2. Reverse the convention for writing the algorithm so that it meshes more closely with the verbal-motor chaining (just as algorithms historically have undergone changes, modern changes can also occur). The horizontal algorithm coincides with the verbal-motor coordination of saying (either aloud or to oneself) "thirteen sevens equal ninety-one" and simultaneously writing $13 \times 7 = 91$. The same could be accomplished by placing the abstract number atop the denominate number (7 minutes from the example above).

Step 1	*Step 2*	*Step 3*	*Step 4*
13	13	13	13
	$\times$	$\times\ 7$	$\underline{\times\ 7}$

3. Suggest to the child that he or she say, "seven multiplied by thirteen" instead of "thirteen times seven." This language fits the vertical algorithm:

$$\begin{array}{r} 7 \\ \underline{\times 13} \end{array}$$

EXERCISES

I. State the justification for each step:

1. $4 \times 16 = 4 \times (10 + 6)$ Renaming
   $= (4 \times 10) + (4 \times 6)$
   $= 40 + 24$
   $= 64$
2. $25 \times 76 = (20 + 5) \times (70 + 6)$
   $= (20 \times 70) + (20 \times 6) + (5 \times 70) + (5 \times 6)$
   $= 1400 + 120 + 350 + 30$
   $= 1900$

II. Identify the properties of multiplication of whole numbers that underlie the following algorithm for 12×135:

```
  135      540
×   4    ×   3
-----    -----
   20      120
  120     1500
  400     ----
  ---     1620
  540
```

III. Use the Distributive Property of Multiplication over Addition to perform the following computations. Use both horizontal and vertical algorithms.

1. $3(a + b)$	4. $(7 + 1)(8x + 3)$
2. $3x(a + b)$	5. $(7x + 1)(8x + 3)$
3. $4x(x + 8)$	6. $(3x + y)(2y + 4x)$

IV. Choose the relevant hypotheses below that appear helpful in determining possible causes for error in the following algorithm.

```
 35
×62
---
831
 10
---
841
```

1. does not know basic addition facts
2. does not know basic multiplication facts
3. computes from left to right
4. does not understand place value
5. cannot rename in multiplication
6. cannot rename in addition
7. is left-handed and this may cause left-to-right procedure
8. in first grade, this child reversed letters and numerals but no longer does this
9. child is in fifth grade
10. child has recent I.Q. of 101 on WISC
11. child is from Spanish-speaking home

V. Following is a typical error pattern made by children when computing in multiplication. Choose items from the diagnostic information below that appear helpful in determining possible causes for the errors.

$$\begin{array}{r} 26 \\ \times 42 \\ \hline 8 \\ 24 \\ 12 \\ \hline 314 \end{array}$$

1. does not know basic multiplication facts
2. does not know basic addition facts
3. computes from left to right
4. does not understand place value
5. cannot rename in multiplication
6. cannot rename in addition
7. is left-handed
8. reverses words, such as *top* to *pot; was* to *saw*
9. in first grade, she wrote ∂ for 6, ƨ for 5, and Ɛ for 3.

VI. Complete the arrow graphs:

1.

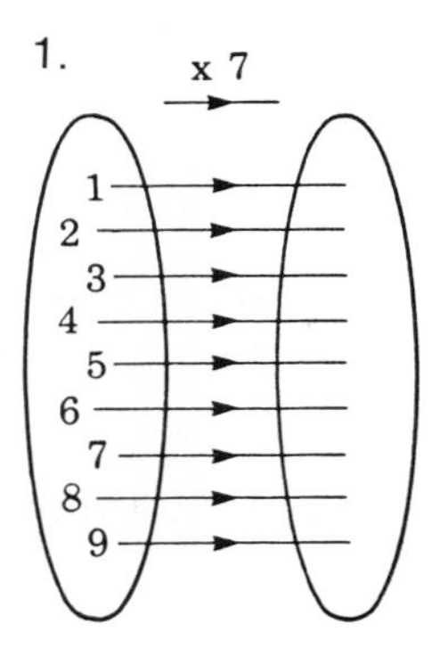

2.

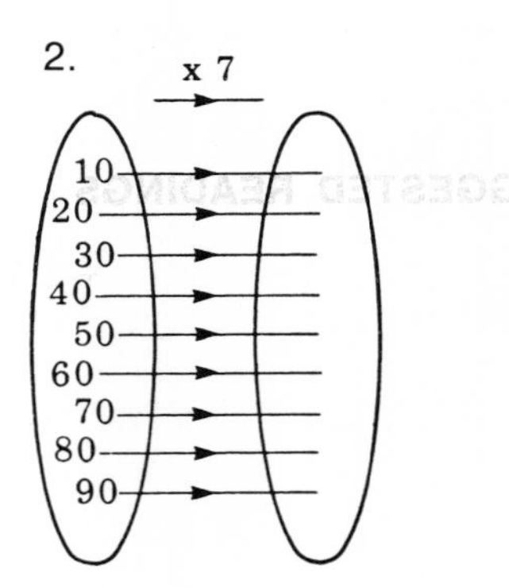

VII. After you complete the arrow graphs in number VI above, use them to find the answer for 93×7.
(*Hint:* The two completed arrow graphs are used to find the product of 38×7 as shown in the following algorithms):

$$\begin{array}{r} 30 \times 7 = 210 \\ 8 \times 7 = 56 \\ \hline 38 \times 7 = 266 \end{array} \qquad \begin{array}{r} 38 \\ \times\ 7 \\ \hline 210 \\ 56 \\ \hline 266 \end{array} \qquad \begin{array}{r} 38 \\ \times\ 7 \\ \hline 56 \\ 210 \\ \hline 266 \end{array} \qquad \begin{array}{r} 8 \times 7 = 56 \\ 30 \times 7 = 210 \\ \hline 38 \times 7 = 266 \end{array}$$

VIII. Using letters, a two-digit number may be written $10m + n$.
1. What values can m have?
2. What values can n have?
3. Using the same idea, write a model for a three-digit number.

IX. What property is used if the following grid is completed without doing any multiplication?

×	1	4	9	16	25
1	1		9		25
4	4	16		64	
9		36	81		225
16			144	256	
25		100		400	625

X. Use the Network method of multiplication for multiplying algebraic quantities. Multiply $x + 5$ by $x + 3$ as shown:

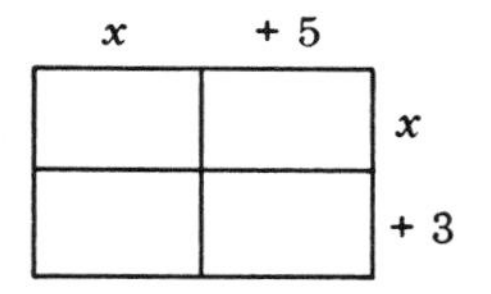

SUGGESTED READINGS

Boykin, W. E. "The Russian-Peasant Algorithm: Rediscovery and Extension." *The Arithmetic Teacher* 20 (January 1973): 29–32.

Traub, R. G. "Napier's Rods: Practice with Multiplication." *The Arithmetic Teacher* 16 (May 1969): 363–64.

Extending Whole Numbers to Negative Numbers

The numbers we call *integers* consist of positive numbers, the number zero, and negative numbers. Two common subgroupings of integers are *whole numbers* and *counting numbers*. This is illustrated below.

$$\text{Integers} = \{\ldots, {}^{-}3, {}^{-}2, {}^{-}1, 0, 1, 2, 3, \ldots\}$$
$$\text{Whole Numbers} = \{0, 1, 2, 3, \ldots\}$$
$$\text{Counting Numbers} = \{1, 2, 3, \ldots\}$$

Notice that the whole numbers are contained within the integers and that the counting numbers are contained within the whole numbers.

NEGATIVE NUMBERS

Negative numbers were used by the Chinese around 200 B.C. The Chinese used red computing rods to represent monetary credits and gains. Black computing rods represented debits. Thus, red rods symbolized positive numbers, and black rods symbolized negative numbers. The convention today is just the reverse. We talk about "being in the red" when we are in debt, and "being in the black" when we are solvent.

SIGNED NUMBERS

The term *signed numbers* refers to the positive and negative numbers. Signed numbers allow measurements in both directions on a scale from a fixed position. In measuring temperature, for example, warmer weather is expressed by positive numbers, whereas cold weather, below zero on the Celsius scale, is expressed by negative numbers. In launching rockets, signed numbers are used to state time relative to the moment of blast-off, which is designated as zero-time. For example, "minus ten and counting" means that it is 10 minutes to launch time and that the clock is counting in reverse ($^-9$, $^-8$, $^-7$, . . . minutes). The time after the launch is counted positively ($^+1$, $^+2$, $^+3$, . . . , minutes). Signed numbers are also used to describe elevation. The usual reference point for measuring altitude is sea level. Points below sea level are designated by negative numbers, points above sea level by positive numbers. Signed numbers, also called *directed numbers,* can be represented on a number line as shown below:

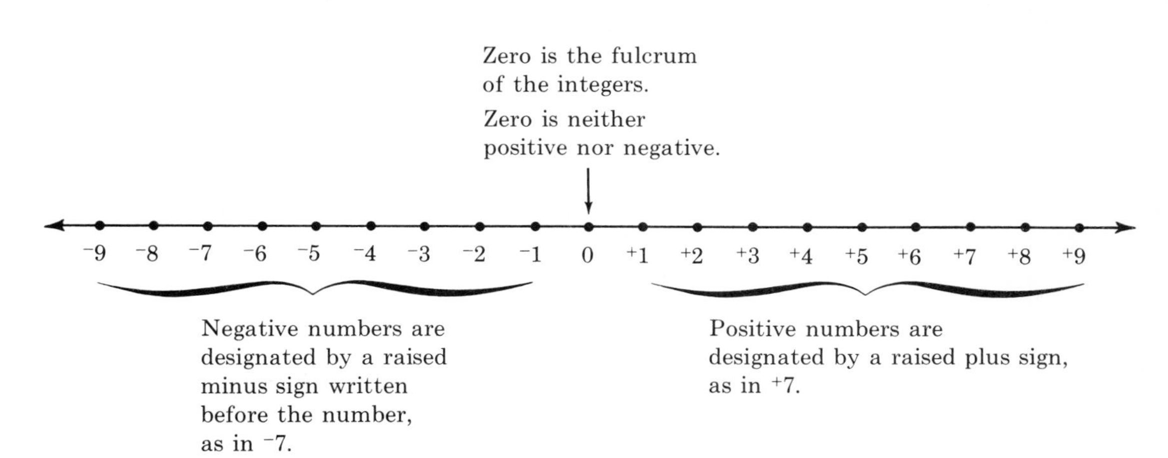

THE USE OF VERTICAL NUMBER LINES

Scanning is a developmental skill; that is, children scan vertically before they scan horizontally. Some children with learning disabilities involving perceptual problems thus find it easier to read sentences in which the words are arranged vertically. The same is true in working with number lines.

Vertical number lines facilitate learning place value notation

An advantage of using vertical number lines is that they facilitate the learning of place value notation. I have heard students through the fifth grade explain how confused they were when simultaneously they learned that, in place value, numbers greater in value or amount are to the left, while on a number line, larger numbers are to the right. This is further confused by use of the usual place value chart and the number line, as shown below.

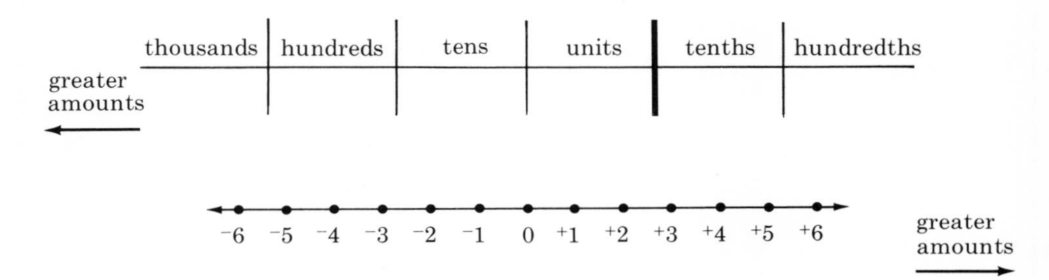

Mini-lesson 3.3.1 Use Inverse Relationship Between Addition and Subtraction

Vocabulary: inverse, positive, negative, earn, debit, net

Possible Trouble Spots: does not know basic addition facts; does not know basic subtraction facts; number line errors: counts one too many; counts one too few; begins counting "1" at zero; counts points rather than jumps or spaces

Requisite Objectives:
Addition 1.2.3: Add numbers with sums zero to 9

Sets 1.3.2: Remove a subset from a given set

Subtraction 1.3.10: Perform take-away subtraction with minuend to 18

Mathematical Applications 2.3.10: Compute value of coins

Graphs and Functions 3.2.20: Use a number line

Activity 1

Have students show addition example on a vertical number line as shown:

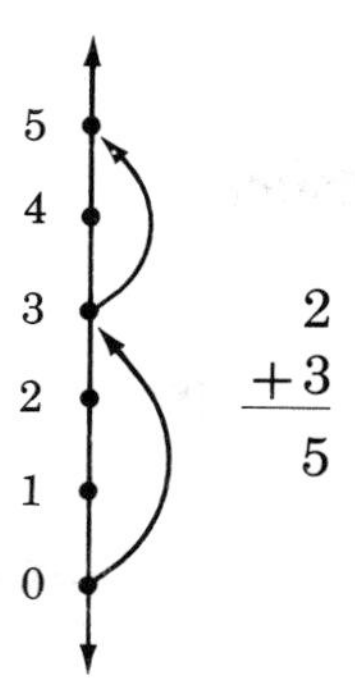

Activity 2
Show simple take-away subtraction on a vertical number line.

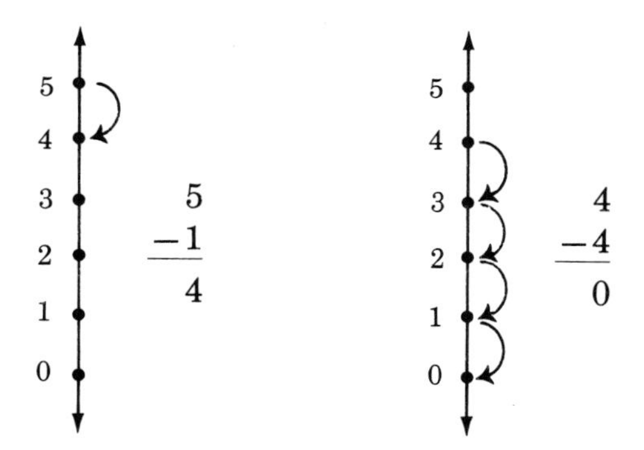

Activity 3
Set up situations using values of money greater than one dollar in order to avoid computations with decimals. The inverse relationship between addition and subtraction may be represented by money earned (addition) and money owed (subtraction). The net difference will be negative if more is owed than was earned. For students who need more structure, use play dollars to represent money earned and give them a bill for services rendered or for goods bought to represent what is owed. The student will then be left with either money or a bill as his or her net worth after the transactions are completed. You may make the bill for services rendered or for goods bought out of red paper to represent being "in the red," or in debt.

earned	owed	net
1	2	−1
5	3	+2
5	7	−2
3	2	+1
10	8	+2

Additional considerations: Students' facility using the number line should be observed. If necessary, walk with a child as you both count aloud. Point out that steps are counted and that at position zero, you have not yet moved, and so there is nothing to count. For some students, it may help to have them commence counting with "zero." Thus, for $3 + 2 = 5$ they would count "zero, one, two, three; zero, one, two" until they arrive at 5 on the vertical number line in Activity 1 above. To show the inverse relation, have the students back up a designated number of steps on the number line.

For practice at the paper-pencil level, the following exercise is suggested. Using a vertical number line from $^{-}10$ to $^{+}10$, have students name the number at the end of each move that you present as follows:

a. Start at 0. Move $^{+}4$.

b. Start at 0. Move $^{-}3$.

c. Start at $^{+}7$. Move $^{+}3$.

d. Start at $^{+}7$. Move $^{-}4$.

e. Start at $^{-}9$. Move $^{+}16$.

f. Start at $^{-}2$. Move $^{-}7$.

g. Start at $^{+}10$. Move $^{-}10$.

After students have completed these exercises—either independently or with a partner—lead the class as a whole in correcting their papers. This wind-up activity is important in making students aware of the inverse relationship between adding and subtracting as shown by integers. Because they may need models for cognitive monitoring (See Chapter 1), probes such as the following should be encouraged: "Start at $^{-}2$. Move $^{-}7$. Let's see, I'm supposed to start at $^{-}2$ and move 7 steps in the negative direction. That means my answer must be a negative number. Okay, I start at $^{-}2$. Then, counting seven more into the negatives, let's see where I end up. Negative nine. That's reasonable."

Note: Mini-lesson 3.3.1 is a prerequisite for Mini-lessons 3.3.7 and 4.1.5.

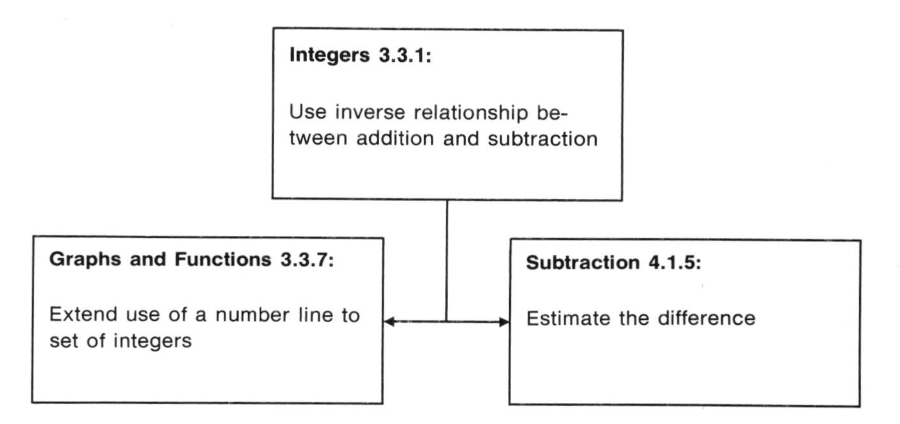

Mini-lesson 3.3.2 Identify the Additive Inverse of a Number

Vocabulary: additive inverse, opposite, signed number

Possible Trouble Spots: does not notice symmetry in situations

Requisite Objective:
Mathematical Applications 1.2.8: Recognize symmetry in nature and in own body

Activity 1
Have students analyze a set of integers in terms of balance. They should notice that each whole number has its opposite as does each negative number. They are mirror images.

A diagram like the following will help point up the idea of a number and its opposite.

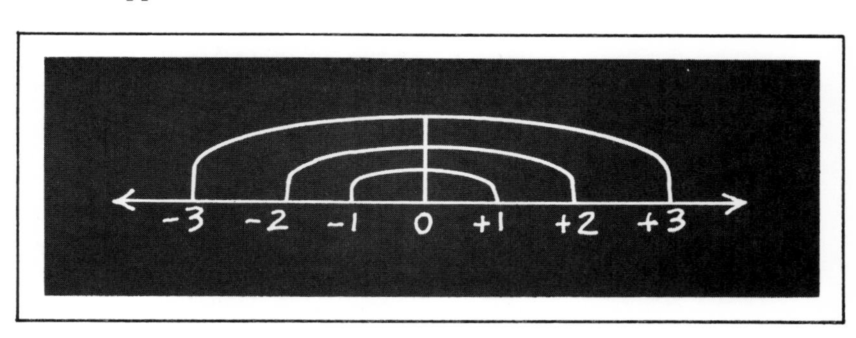

Activity 2

A large number line placed on the floor will allow students to develop a sense of balance of the set of integers. Paint a number line on the floor or on paper from $^{-}5$ to $^{+}5$. Have one student stand on $^{-}4$ and have another student stand on $^{+}4$, both facing in the same direction. Have them simultaneously count steps until they meet at zero. Then ask them to explain where the idea of "inverse" was shown. The student on the negative number had to back up.

Pair students in teams and provide time for students to test each other by asking what number on a number line is the same distance from zero in the opposite direction. Students should practice making these comparisons so that they can immediately give the opposite of a given number.

Activity 3

This activity is appropriate as a total class exercise. Ask "What number will get you to zero on the number line, given the following situations? You know the present temperatures and in some cases the number of temperature change in degrees. (See p. 149.) Use the number line as an aid to help complete the table." If necessary, point out that on the vertical number line, a downward movement indicates a negative direction and an upward movement indicates a positive direction.

Additional considerations: Students with learning disabilities that involve disorientation in space may have difficulty with number-line activities requiring their own movement, such as relating a number's inverse to a sequence of backward steps. Activity 1 above would be more appropriate for such students, even though it is considered more abstract than walking along a number line.

Mini-lesson 3.3.3 Add Inverses to Obtain the Additive Identity (Zero)

Vocabulary: zero, additive inverse

Possible Trouble Spots: may not realize that the additive inverse of a negative number is its positive mirror image; does not become aware that the identity element for an operation is obtained by performing the particular operation on inverse elements within the operation; that is, adding opposites yields the number zero, which is the identity element for addition, just as we obtain the multiplicative identity (1) by multiplying a number and its reciprocal.

Requisite Objectives:

Geometry 1.1.8: Show topological relationships: proximity, enclosure, separateness, order

Geometry 2.1.9: Show rigid transformations, e.g., slides, turns, flips

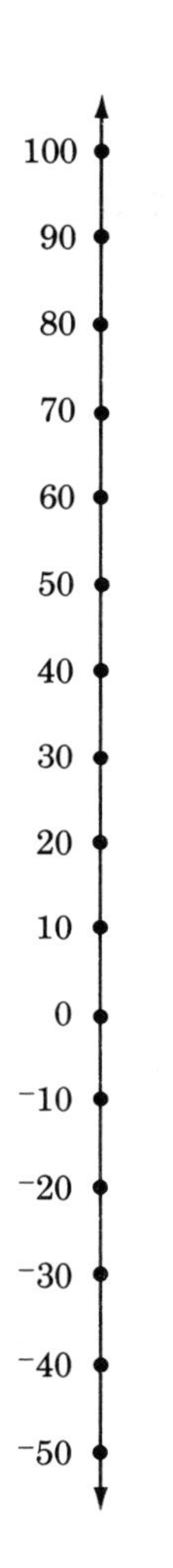

Present Temperature	Degree Change
32°	⁻32°
48°	______
⁻11°	⁺11°
⁻50°	______
17°	______
27°	______
⁻27°	______
______	⁻13°
______	⁺55°
______	⁻11°

Activity 1

Represent negative numbers on the counting boards with red tokens and positive numbers with black. Have the students form empty board values by making *null matches*.[1] They show a null match by considering placement of a red and a black token on the 4-space, for example, as representing a battle, where each match knocks out the opponent, leaving an empty space.

1. The term *null match* was used by Frederique Papy, *Les Enfants et la Mathematique 1* (Bruxelles-Montreal-Paris: Didier, 1970).

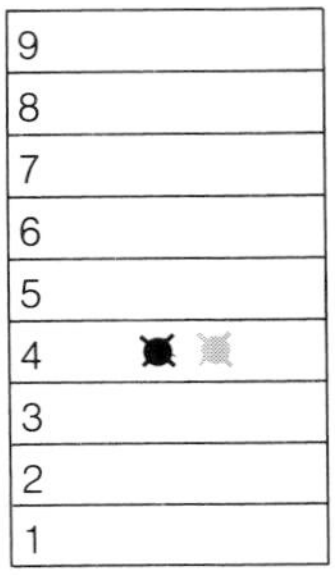

9
8
7
6
5
4
3
2
1

Record this as: $^{-}4 + 4 = 0$. Continue this with other numbers on the board to show that a number nullifies its opposite.

Activity 2
Using the counting boards, children may count up and then down the board by the same number. For example, count onto the board to 6; then count off the board by taking 6 steps down and off, thus showing that a value of zero on the board was obtained by combining inverse movements of the same number of steps. Students should agree that a red and a black chip on the same space cancel each other. Also, that n number of steps up the counting board is cancelled by ^{-}n, or the same number of steps down the counting board.

Activity 3
If you owe 25 cents to your friend, you are in debt. This may be expressed as $^{-}25$. Then you do an errand and receive 25 cents, thus creating a situation that may be expressed as follows on the counting boards:

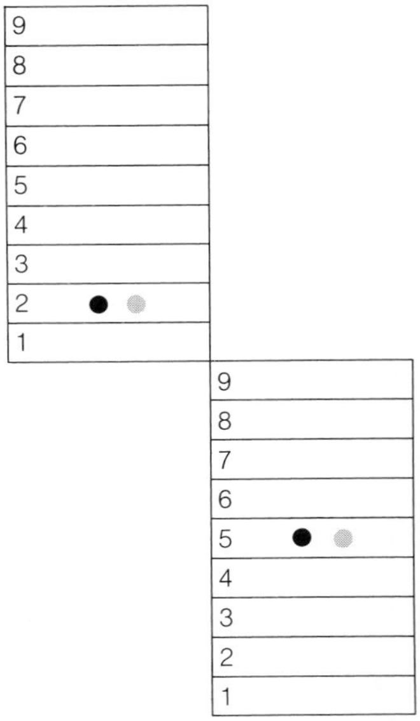

9
8
7
6
5
4
3
2
1

9
8
7
6
5
4
3
2
1

Point out that the debit may be represented by the red checker, by writing the number in red, or by recording the debit as a negative number: $^{-}25 + 25 = 0$.

A second situation might involve having 35 cents and receiving a bill for 35 cents. As soon as you receive the bill for 35 cents, your net worth is zero. The bill may be represented by a red checker or as a negative number: $35 + {}^{-}35 = 0$.

Now, reverse the situation. You have 35 cents and you spend 35 cents. You are left with zero cents, or $35 - 35 = 0$. This is an example of take-away subtraction.

These last two examples form the basis for pointing out that subtracting a number yields the same results as adding a negative number.

Additional considerations: The relationship between the inverse operation and the number zero, which is the identity element for addition, involves 2 of the 11 basic properties that elementary school children should know. Students should be reminded that the sum of inverse numbers is the additive inverse zero. Supplement your explanation with bulletin board displays, verbal daily reminders, investigations on hand calculators, and practice in computing such examples as $^{+}3 + {}^{-}3 = \square$, $^{-}2 + {}^{+}2 = \square$, $^{+}98 + {}^{-}98 = \square$, $^{-}49 + {}^{+}49 = \square$, and so on.

Note: Mini-lesson 3.3.3 is a prerequisite for Mini-lessons 3.3.6 and 4.2.11.

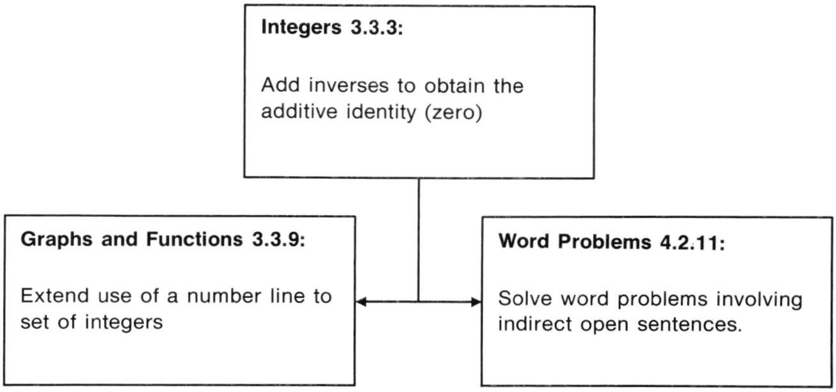

Integers 3.3.3:

Add inverses to obtain the additive identity (zero)

Graphs and Functions 3.3.9:

Extend use of a number line to set of integers

Word Problems 4.2.11:

Solve word problems involving indirect open sentences.

Mini-lesson 8.2.1 Apply Whole Number Properties to Integers

Vocabulary: inverse, commutative, associative, distributive, closure

Possible Trouble Spots: counts one too many on the counting boards; is not aware of properties as applied to whole numbers; does not compute integers correctly in one or more of the operations

Requisite Objectives:
Addition 2.2.3: Use additive identity

Subtraction 2.2.7: Investigate the addition properties as they apply to subtraction

Multiplication 3.1.7: Use multiplication properties

Addition 8.1.2: Add integers

Subtraction 8.1.3: Subtract integers

Multiplication 8.1.4: Multiply integers

Division 8.1.5: Divide integers

Activity 1
Using the counting boards, have students work in teams of two to practice computing with integers. For example, to show that $^{-}7 + {}^{-}2 = {}^{-}9$, use red checkers to indicate an original debit of 7 dollars and an additional bill for food purchased in the amount of 2 dollars. Making a 2-for-1 exchange of the chips, show a debit for 9 dollars on the 9-space with a red checker. If necessary, the teacher should demonstrate first, and then after modeling computation of integers on the counting board, the students should discuss their moves as they compute.

To show that addition of integers is commutative, direct one student to show $^{-}7 + {}^{-}2$ on the units board and the teammate to show $^{-}2 + {}^{-}7$ on his or her board. When the two boards are compared, students should state that both show the same number indicated by a red checker on the 9-space to indicate $^{-}9$. One of the students should record the findings to show that $^{-}7 + {}^{-}2 = {}^{-}9$ and that $^{-}2 + {}^{-}7 = {}^{-}9$, or that $^{-}7 + {}^{-}2 = {}^{-}2 + {}^{-}7 = {}^{-}9$.

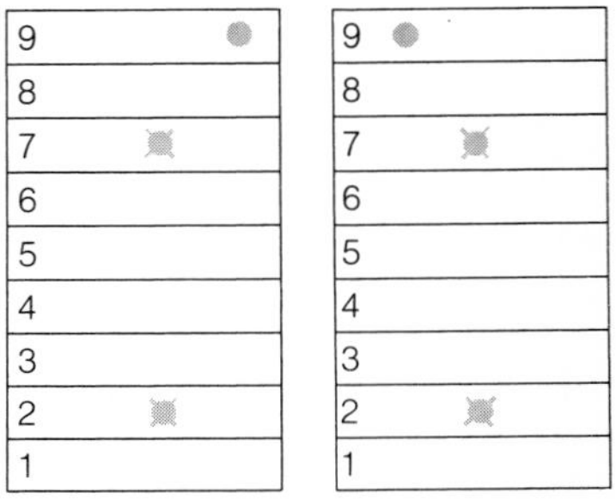

9	9
8	8
7	7
6	6
5	5
4	4
3	3
2	2
1	1

$$^{-}2 + {}^{-}7 \quad = \quad {}^{-}7 + {}^{-}2$$

Activity 2
Show that addition of integers is associative on the counting boards as shown below for the example $({}^{-}2 + {}^{-}3) + {}^{-}4 = {}^{-}2 + ({}^{-}3 + {}^{-}4) = {}^{-}9$.

Starting with the bottom boards, follow the vertical sequence of computations that represent moves with the checkers.

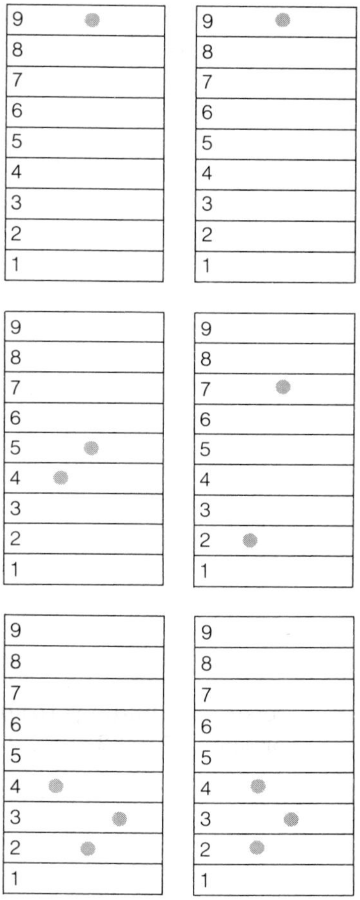

$$
\begin{array}{cc}
{}^{-}9 & {}^{-}9 \\
\uparrow & \uparrow \\
{}^{-}5 + {}^{-}4 & {}^{-}2 + {}^{-}7 \\
\uparrow & \uparrow \\
({}^{-}2 + {}^{-}3) + {}^{-}4 & {}^{-}2 + ({}^{-}3 + {}^{-}4)
\end{array}
$$

Activity 3

Show the Distributive Property of Multiplication over Addition by having students compute both sides of examples, such as the following:

$$
\begin{array}{ccc}
{}^{-}2 \times ({}^{+}4 + {}^{-}6) & = & ({}^{-}2 \times {}^{+}4) + ({}^{-}2 \times {}^{-}6) \\
{}^{-}2 \times {}^{-}2 & = & {}^{-}8 + {}^{+}12 \\
{}^{+}4 & = & {}^{+}4
\end{array}
$$

Additional considerations: It is important that number properties be seen as necessary tools for developing basic mathematics skills. They are of little use if knowledge of the properties is acquired in isolation and students do not use them to simplify computations. Because students do not intuitively apply the properties to whole numbers even when they know them,[2] the teacher must demonstrate how the various properties can be helpful in performing basic computational skills. For example, begin a group discussion as follows: "Class, we have been applying the commutative and associative properties to adding and multiplying integers. Let us talk about how we can use these laws to make computing easier." Then write the following on the board.

a. $+2 \times \square = +2$
b. $-12 \times 0 = \square$
c. $(-59 \times -2) \times +5 = \square$
d. $-333 \times 18 = \square \times -333$
e. $(-17 + -16) + -4 = \square + -20$
f. $-\frac{5}{6} \times \square = -1$

Continue the discussion as follows:

"What property helps you with example *a*?" (Students should respond that the multiplicative identity is involved; $\square = 1$.)

"What is the answer to example *b*? What information helps you to know that?" (Students state that any number times zero is zero.)

"What property is apparent in example *c*? Someone tell me your thinking for *c*." (Student responds that by using the Associative Property for Multiplication, you get 10×59, which is 590 and then just have to decide the sign. Negative two times $^{+}5 = {}^{-}10$. Then $^{-}59$ times $^{-}10$ equals 590.)

"What property is shown in example *d*?" (Student states that the

2. *Mathematical Knowledge and Skills,* Report No. 90-MA-02, 77-78 Assessment. (Denver, Colo.: National Assessment of Educational Progress, Education Commission of the States, 1979).

Commutative Property for Multiplication is pointed out in this example. The order of factors will not affect the product.

"How about *e*?" (Student volunteers that the Associative (or grouping) Law for Addition allows adding $^-16 + ^-4$, which equals the $^-20$ and that leaves the $^-17$, which goes in the box.)

"Who would like to tell their thinking for example *f*?" (Student responds that $^+6/5$ goes in the box and that illustrates the inverse law for multiplication. The sign is positive to obtain $^-1$.)

Those who are suggesting that teachers model self-instructions are recommending this type of analysis.[3] This cognitive monitoring or thinking about one's thinking is a cueing strategy. The related ideas of cognitive-behavioral therapy applied to teaching were discussed in Chapter 1.

Mini-lesson 8.2.2 Simplify Expressions of Integers (Order of Operations)

Vocabulary: simplify, grouping symbols

Possible Trouble Spots: may not remember the order of simplifying; leaves out some of the computations

Requisite Objectives:
Whole Numbers/Notation 7.2.1: Simplify expressions involving order of operations

Addition 8.1.2: Add integers

Subtraction 8.1.3: Subtract integers

Multiplication 8.1.4: Multiply integers

Division 8.1.5: Divide integers

Activity 1
Show students how to simplify within grouping symbols beginning with the innermost computation. Since the order of simplifying involves an arbitrary convention, a didactic approach whereby the teacher demonstrates is most economical and probably most effective. Examples to use include the following:

a. $^-3 \times (6 - 10) = \square$

b. $42 - (^-24 \div 6) = \square$

c. $\dfrac{(^-8 \div 2) + ^-6}{^-8 + ^+3} = \square$

3. J. Flavell and H. Wellman, "Metamemory." In R. Kail and J. Hagen, eds., *Perspectives on the Development of Memory and Cognition* (Hillsdale, N.J.: Lawrence Erlbaum, 1977); D. Meichenbaum, "Enhancing Creativity By Modifying What Subjects Say to Themselves," *American Educational Research Journal* 12 (1975): 129–45.

Solutions: a. ${}^{-}3 \times {}^{-}4 = {}^{+}12$ or 12 (No sign indicates a positive number.)

b. $42 - {}^{-}4 = {}^{+}46$

c. $\dfrac{({}^{-}4) + {}^{-}6}{{}^{-}5} = \dfrac{{}^{-}10}{{}^{-}5} = 2$

Activity 2

In the absence of grouping symbols, use the following order: multiplication and division in order from left to right; then additions and subtractions in order from left to right. This is illustrated as follows:

a. $10 - 8 \times {}^{-}2 = 10 - {}^{-}16 = 26$
b. ${}^{-}4 \times 3 - 7 = {}^{-}12 - 7 = {}^{-}19$
c. $24 - {}^{-}18 \div {}^{-}9 = 24 - {}^{+}2 = 22$

Additional considerations: This lesson is inappropriate for students who cannot handle highly abstract and complex mathematics tasks. Other aspects of student behavior to watch out for are weakness in remembering sequences of occurrences and difficulty attending to salient aspects of a situation. Cues may be helpful, such as writing the parts of the problem in different sizes with the simplification proceeding from largest number to smallest (or vice versa, so long as agreement is obtained and the use of this cue is consistent). An example of this is shown in the following:

Problem: $[9 \times (3 + {}^{-}8)] + {}^{-}3$

Step 1: $[9 \times {}^{-}5] + {}^{-}3$

Step 2: ${}^{-}45 + {}^{-}3$

Solution: ${}^{-}48$

Note: Mini-lesson 8.2.2 is a prerequisite for Mini-lessons 8.2.4 and 8.3.4.

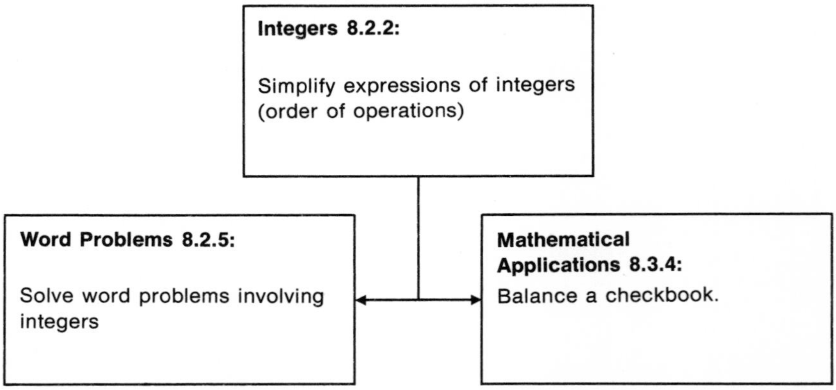

Integers 8.2.2:

Simplify expressions of integers (order of operations)

Word Problems 8.2.5:

Solve word problems involving integers

Mathematical Applications 8.3.4:

Balance a checkbook.

EXERCISES

Simplify the following expressions:

1. $(^-2)^2 + {}^-4 \times {}^-1$
2. $24 - (^-9 \div {}^-3)$
3. $\dfrac{^-3 + (10 \div {}^-5)}{^-4 + 5}$
4. $12 - (4 \times {}^-2)$
5. $(4 \times (6 - {}^-2)) \div 8$
6. Choose those properties that hold for integers.
   a. Addition of integers is commutative.
   b. Addition is associative.
   c. Subtraction is commutative.
   d. Subtraction is associative.
   e. Multiplication is commutative.
   f. Division is commutative.
   g. Multiplication distributes over addition.
7. Tell two trouble spots that number lines present as an instructional tool.

SUGGESTED READINGS

Flavell, J., and Wellman, H. "Metamemory." In *Perspectives on the Development of Memory and Cognition,* edited by R. Kail and J. Hagen, Hillsdale, N.J.: Lawrence Erlbaum, 1977.

Gardner, M. "The Concept of Negative Numbers and the Difficulty of Grasping It." *Scientific American* 236 (June 1977): 131–35.

Mathematical Knowledge and Skills, Report No. 90-MA-02, 77–78 Assessment. Denver, Colo.: National Assessment of Educational Progress, Education Commission of the States, 1979.

Meichenbaum, D. "Enhancing Creativity by Modifying What Subjects Say to Themselves." *American Educational Research Journal* 12 (1975): 129–45.

Subtraction

THREE INTERPRETATIONS OF SUBTRACTION

Subtraction may be thought of as the inverse of addition. In addition, numbers are united to find a sum; in subtraction, numbers are separated to find a difference.

Subtraction can be interpreted as being of three different types, depending upon its application in a given circumstance. The first type, *take-away subtraction,* has to do with taking away a part of a number and retaining the rest, thus decreasing the value of the number. *Comparison subtraction* involves finding how much larger or how much smaller one number is than another. The third type, *additive subtraction,* involves analyzing a number in terms of its component parts. The teacher needs to understand these three interpretations of subtraction in order to select relevant instructional activities for the pupils.

Take-Away Subtraction

Take-away subtraction involves taking away a part of a set, which results in a decrease in the number of elements.

Mini-lesson 1.3.10 Perform Take-away Subtraction with Minuend to 18

Vocabulary: take-away, minus, less, more, difference, left over, part of a set, equal

Possible Trouble Spots: difficulty forming concepts, e.g., cardinality; needs more concrete experiences; needs to record the take-away subtraction algorithm simultaneously with manipulative activity

Requisite Objectives:
Whole Numbers/Notation 1.3.3: Use count-on place value model

Subtract numbers with minuends to 9

Activity 1
To find the cardinal number property of the remaining subset when the number of the original set and of the subset taken away is known, ask the child to solve the following problem: "Mary had 6 gumdrops. She gave 2 to Betsy. How many gumdrops does Mary have left?" ($6 - 2 = \square$)

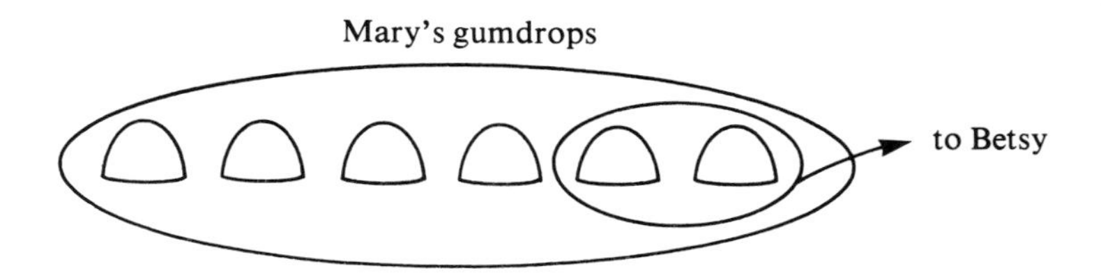

Activity 2
To find the number of the subset taken away when the number of the original set and of the remaining subset is known, ask the child to solve the following problem: "Mary had 6 gumdrops. She gave some to Betsy and has 4 left. How many gumdrops did she give to Betsy?" ($6 - \square = 4$, which is solved by using the related equation $6 - 4 = \square$.)

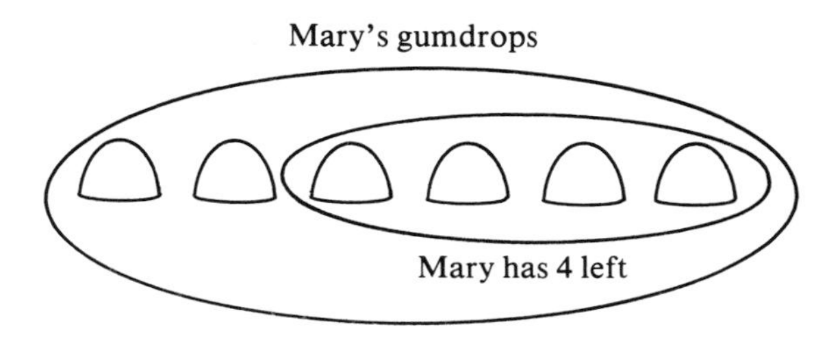

Activity 3
Show the child pictures that illustrate the idea of take-away subtraction and have the child complete the corresponding equations.

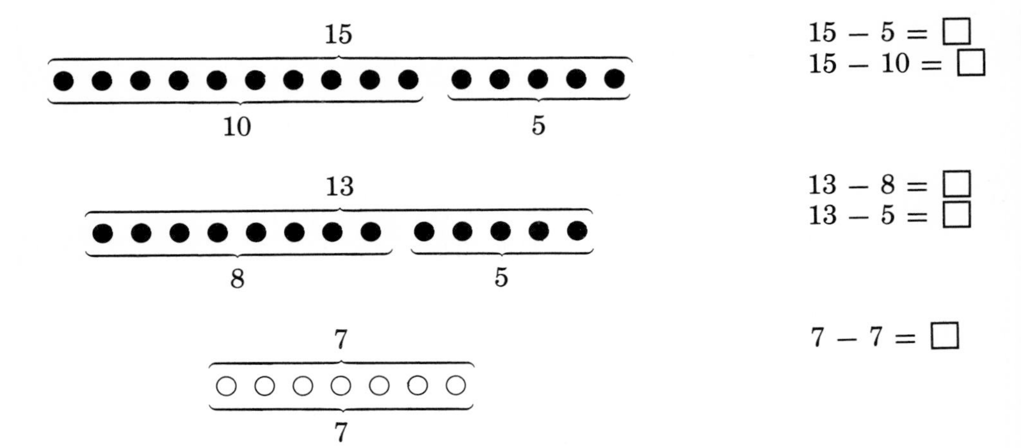

Additional considerations: Certain children may have difficulty in going from the picture level to the symbolic level in take-away subtraction. A pupil may not see the difference, for example, between physically removing a part of the universal set and erasing the part of the equation that stands for removing the set. Consider Activity 1, in which it was stated that Mary had 6 gumdrops and gave 2 to Betsy. We may show this on the chalkboard by drawing 6 gumdrops (△△△△△△) and then erasing 2 gumdrops, leaving 4 gumdrops (△△△△). Unless the child understands that 6 − 2 stands for the number 4, the child may erase the − 2 (to coincide with erasing Betsy's 2 gumdrops), being left with 6 as an answer. This erroneous concept then leads a child to compute as follows: "6 − 2 = 6."

Diagnostic teaching involves questioning the child about both correct and incorrect responses. Say to the child, "I can't see your think-

---

Note: Mini-lesson 1.3.10 is a prerequisite for Mini-lessons 1.3.11 and 1.3.18

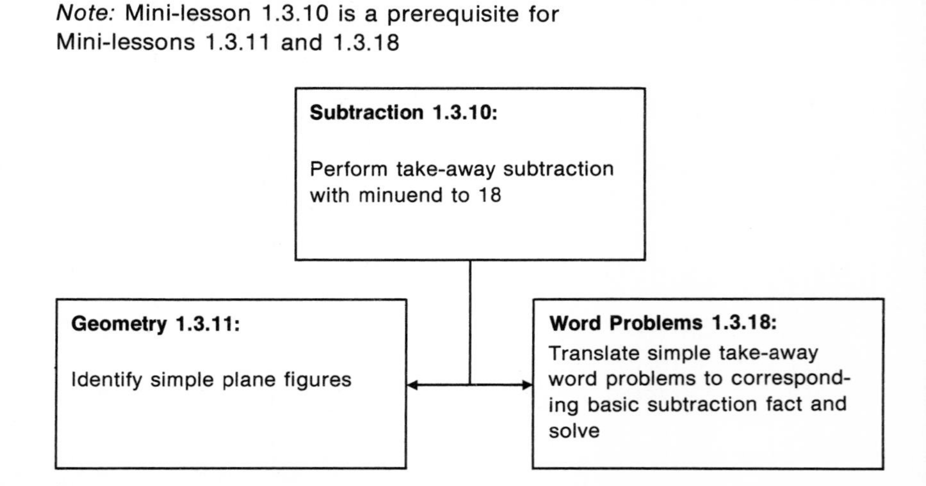

ing; you'll have to tell me." "Tell me your thinking here." "What were you thinking when you wrote six minus two equals six?"

Comparison-Type Subtraction

Thinking strategies in comparison situations involve finding how much larger or how much smaller one number is than another. This usually entails finding (a) the number of the difference between two disjoint sets or (b) the number of the smaller set (or the larger set) when the number of one set and the difference is given.

Mini-lesson 3.3.7 Perform Comparison Subtraction

Vocabulary: smaller, larger, difference, fewer, more

Possible Trouble Spots: does not understand how to use a number line; does not understand greater than/less than comparison

Requisite Objectives:
Subtraction 1.3.10: Perform take-away subtraction with minuend to 18

Graphs and Functions 3.2.20: Use a number line

Subtraction 2.1.7: Subtract numbers with minuend less than 100, with no renaming

Activity 1
To find the number of the difference between two disjoint sets ask the child to solve the following problem: "Laura has 2 sisters; Lisa has none. How many more sisters does Laura have than Lisa?" or "How many fewer sisters does Lisa have than Laura?" (2 – 0 = ☐)

Activity 2
To find the number of the smaller set (or the larger set) when the number of one set and the difference is given, ask the child to solve the following problem: "Laura has 5 sisters, Lisa has 3 fewer. How many sisters does Lisa have?" (5 – 3 = ☐)

Activity 3
Present the following example to the child: "Laura has 5 marbles in her sack; Lisa has 3 fewer. How many marbles are in Lisa's sack?" Allow the child to verify that all of the marbles have the same weight, as do both empty sacks. Have her put 5 marbles into Laura's sack and place the sack on the balance. Place Lisa's closed sack on the other side of the balance. Tell the child that Lisa's sack has 3 marbles fewer than Laura's sack. Ask the child to balance the scale to find how many marbles are in Lisa's sack.

When the number of the smaller set is asked for, simply have the child remove 3 marbles from Laura's sack. Guide the child to notice that the 2 remaining marbles are now balancing Lisa's sack. The child should conclude that Lisa's sack contains 2 marbles.

The problem might also read: "Laura has 5 marbles in her sack, Lisa has 2 in hers. How many more marbles are in Laura's sack?"

Additional considerations: This topic lends well to consumer-type problems involving price comparisons, including unit pricing, fuel costs, and comparisons of gas mileage for various automobiles.

Note: Mini-lesson 3.3.7 is a prerequisite for Mini-lessons 4.1.5 and 4.1.19.

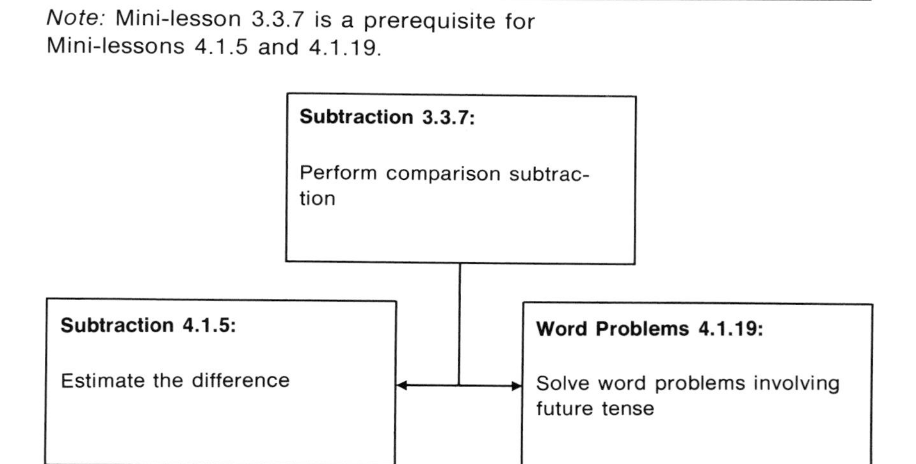

Additive Subtraction

The additive idea of subtraction involves thinking of a number as a sum of component parts.

Mini-lesson 3.3.8 Perform Additive Subtraction

Vocabulary: complement

Possible Trouble Spots: does not see cause-effect relationships; does not understand inverse relationship between addition and subtraction; is not aware of the three types of subtraction

Requisite Objectives:

Subtraction 1.3.10: Perform take-away subtraction with minuends to 18

Subtraction 2.2.7: Investigate the addition properties as they apply to subtraction

Integers 3.3.1: Use inverse relationship between addition and subtraction

Activity 1

To find the number of the complement set when the number of a subset and of the universal set is known ask the child to solve the following problem: "Lisa has a dog named Chuteka. Chuteka had some puppies. Lisa has 4 dogs altogether with Chuteka's puppies. How many puppies did Chuteka have?" ($1 + \square = 4$)

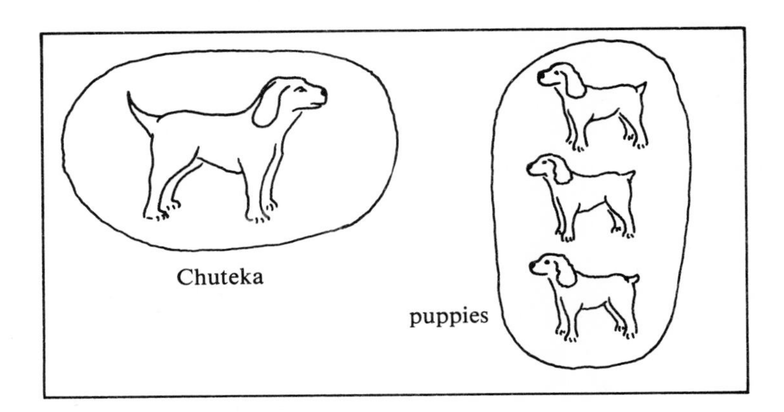

Activity 2

Procedure A: Ask the child to solve the following problem: "Libby has walked 3 miles. She would like to walk 5 miles. How many more miles must she walk?" ($3 + \square = 5$)

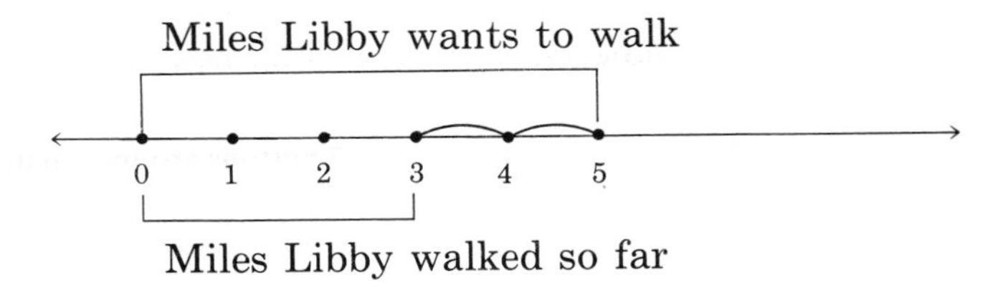

Procedure B: Ask the child to use two rulers to solve the word problem in Procedure *A* ($3 + \square = 5$). Have him place ruler *a* over the number of miles walked (3) on ruler *b*. Ruler *a* becomes the counter for the number of miles needed to go to get to 5. Ruler *a* shows that Libby still has 2 miles to walk.

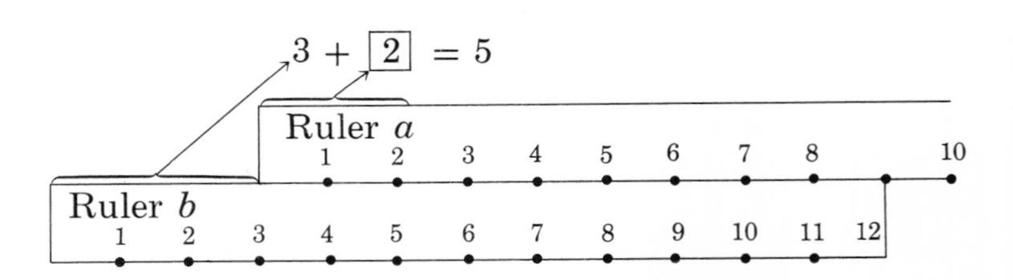

Activity 3
Ask the child to solve the following problem: "Libby walked some miles. Then Laura joined her and together they walked 2 miles, making a total of 5 miles for Libby: How many miles had Libby walked alone?" ($\square + 2 = 5$)

Suggested board work

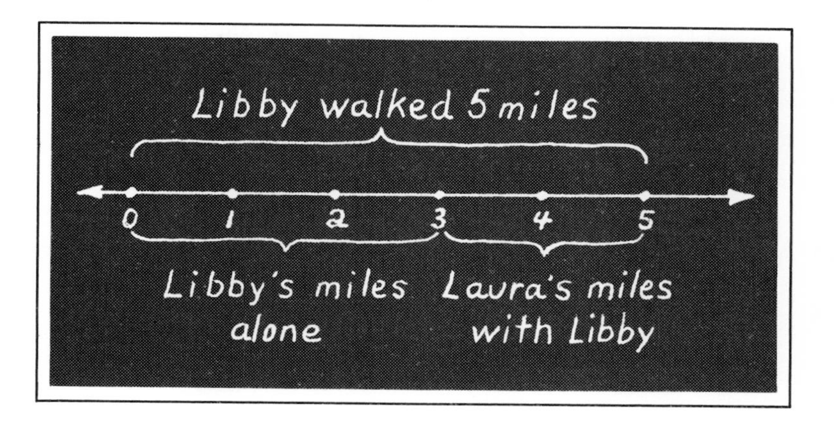

Activity 4
To separate a set into two subsets, place a set of objects on a table. Tell the child to separate the objects into two sets. Then ask her to make a different arrangement of two sets. For instance, in a set of 7 objects, possible arrangements are 3 and 4, 2 and 5, 1 and 6, 0 and 7.

7 buttons grouped in different ways

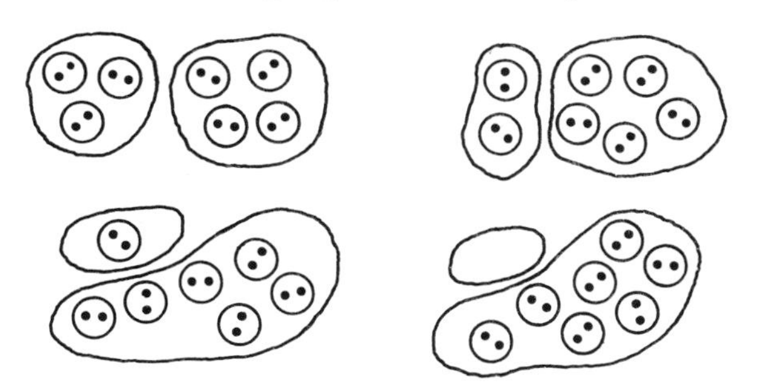

Activity 5
Give the child pieces of paper with 8 objects drawn on each sheet. Tell him to cut the paper into two pieces (if necessary, show the child how to use a pair of scissors), making sets with different numbers of elements. Help him to record the related addition/subtraction sentences.

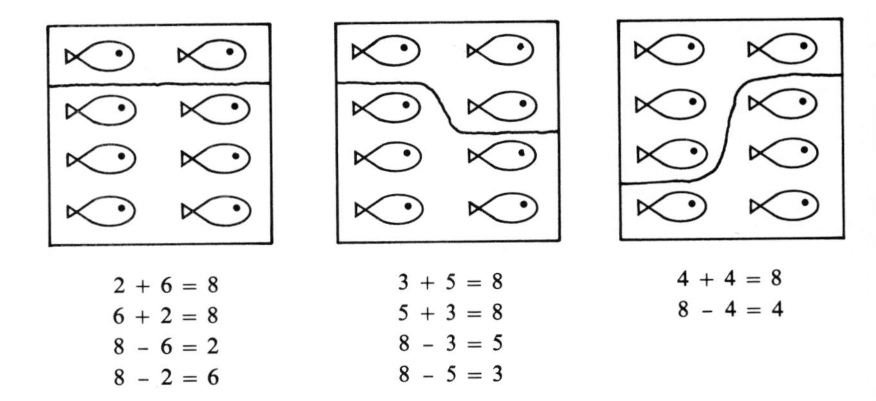

Additional considerations: Since additive subtraction is the most abstract of the three subtraction models, it should be avoided in the first grade (and removed from first grade textbooks).

Note: Mini-lesson 3.3.8 is a prerequisite for Mini-lessons 3.3.9 and 4.1.2:

Subtraction 3.3.8:

Perform additive subtraction

Graphs and Functions 3.3.9:

Extend use of a number line to set of integers

Whole Numbers Notation 4.1.2:

Round whole numbers

SUBTRACTION AND ADDITION AS INVERSE OPERATIONS

As mentioned in the previous chapter, subtraction is the inverse of addition; one undoes the other.

Although activities dealing with the inverse idea can be introduced to children in kindergarten or in first grade, the teacher must keep in mind that children at this level of conceptualization will be capable of only the barest understanding of the inverse idea. It is not until third grade that most students recognize the invariance of events in the physical world, regardless of visual-spatial rearrangements. The child should be allowed to informally investigate the inverse relationship through various activities without being forced into premature verbalization or computation. When the child is allowed to describe the inverse idea without having to use labels of adult language, such as "addition and subtraction are inverse operations" or "subtraction undoes addition," he or she is given an opportunity to develop understandings of concepts related to the inverse relationship.

Mini-lesson 2.2.7 Investigate the Addition Properties as They Apply to Subtraction

Vocabulary: addend, sum, difference

Possible Trouble Spots: difficulty constructing generalizations; does not conserve number

Requisite Objectives:
Addition 1.2.3: Add numbers with sums zero to 9

Addition 1.3.9: Add two 1-digit numbers with sums to 18

Graphs and Functions 2.1.12: Select situations showing the reflexive relation

Activity 1
Give the child a candy bar that is grooved for division into 10 pieces. Ask her to divide the bar into those 10 pieces and place the pieces in the original positions. Then ask her to count all the pieces and to tell how many pieces are in the whole candy bar. Take away part of the candy bar, for example, 4 pieces, and ask the pupil what part of the candy bar is left. She should reply that there are 6 pieces left. Tell her to put the parts of the candy bar back together to form a whole candy bar.

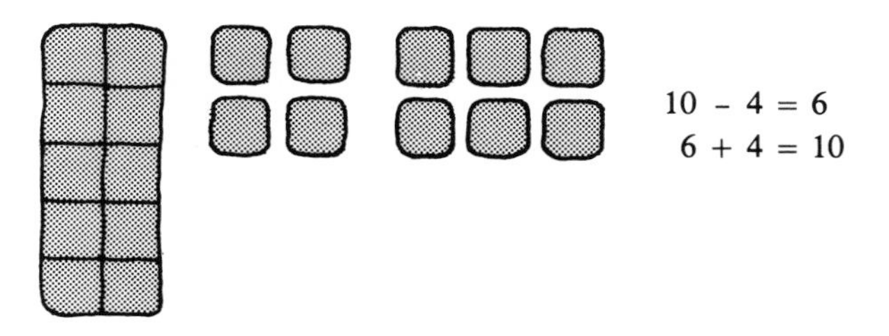

Continue the investigation by taking away different numbers of pieces and asking the child to write corresponding equations.

Activity 2
Buy or make a plastic number line one foot wide divided into units one foot long. Ask the child to walk over the number line to show

$3 + 4 = \square$. The child starts at 0 and then takes 3 steps. She stops and stands still and then takes 4 more steps. Guide her to notice that the number she ends up on represents the total number of steps she has taken.

Ask her to work a subtraction problem such as $7 - 4 = \square$ by beginning at 7 and taking 4 steps backward. Point out that the number she ends up on represents the difference.

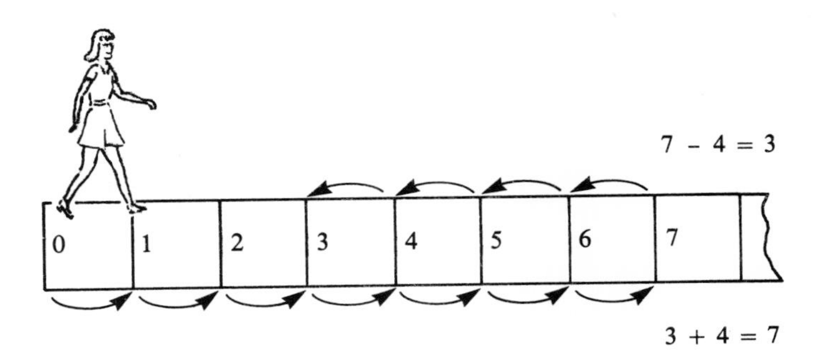

Activity 3
Make task cards containing open sentences for the operations of addition and subtraction. Ask the child to use a number line numbered from 0 to 20 to solve the problems and to write her answers in the appropriate boxes.

Activity 4
Show the child how to make a slide rule from cardboard. Mark off the slide rule with consecutive integers. Ask the child to use the slide rule to solve the problem $3 + 5 = \square$. He should line up the zero mark on the inside scale with the 3 mark on the outside scale to find the sum 8, shown below the 5 on the inside scale. Then ask the child to take 6 from 9. Tell him to place the 6 above the 9 to find the difference below the left end of the inside scale.

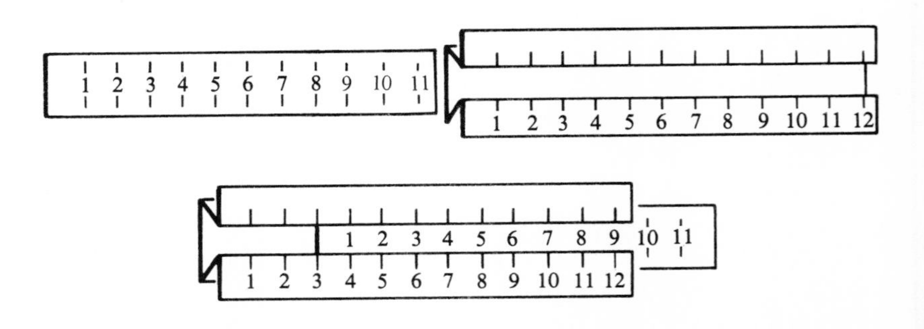

Activity 5

Ask the child to use an addition/subtraction grid to solve $2 + 3 = \square$. Tell her to locate 2 on the left side of the grid and then move over to the cell under the 3 at the top of the grid. Ask her to write the sum (5) in the cell where the arrows originating from 2 and from 3 meet.

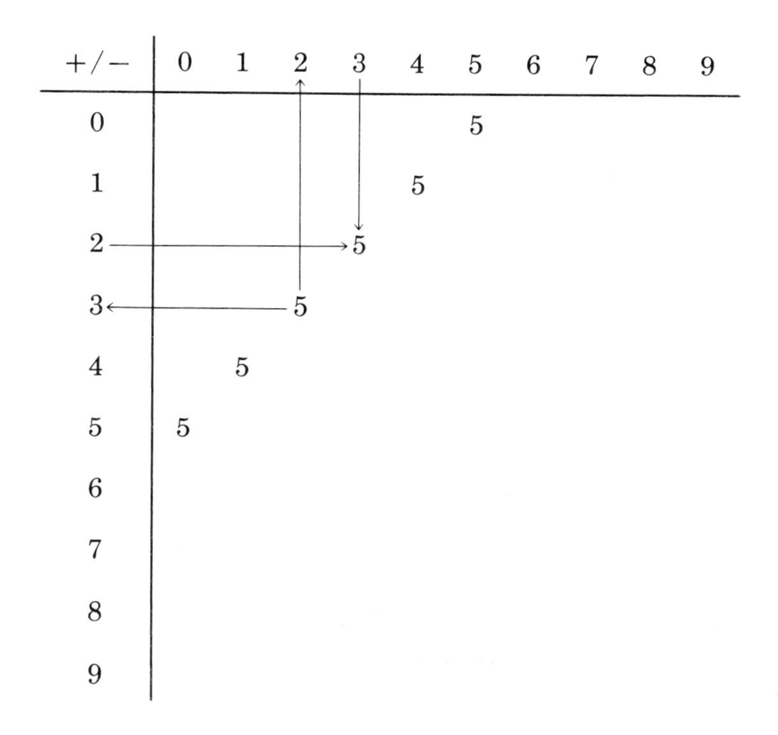

Ask the child to use the grid to solve $5 - 3 = \square$. Tell her to find a sum 5 and move along the 5 diagonal until finding a 3 as one of the addends. Show her that the other addend is seen to be 2.

After the child tries to add all of the possible combinations, circle any incorrect responses. Ask her to notice those sums she couldn't fill in and those that are circled. Such a diagnosis may show the child that out of the 100 possible sums on the grid, she missed perhaps 24 answers. If these twenty-four combinations involved the commutative property, she needs to learn only twelve combinations. A child will not be overwhelmed by the task of learning twelve combinations. Also, the child's self-esteem is raised when she realizes that she knows quite a few of the combinations.

After the child has learned how to use the grid, give her portions of the grid for practice or drill.

Activity 6
Select sets of numbers to investigate using the addition/subtraction grid. For example, use as sets of addends combinations like 0, 1, 2, 3; 4, 5, 6, 7; 6, 7, 8, 9; 1, 3, 5, 7, 9; 0, 2, 4, 6, 8; 4, 7, 8, 9; 3, 7, 6, 2, 9.

+/−	4	7	8	9
4	8	⑭	12	13
7	11	14	⑭	16
8	12	15	16	17
9	13	16	17	18

Encourage children to look for patterns in the grid to help them correct errors in their sums. Portions of the above grid are presented below to illustrate this point.

(a)

7	8	9
⑭	12	13
14	⑭	16
15	16	17
16	17	18

(b)

7	11	14	⑭	16
8	12	15	16	17
9	13	16	17	18

The child should be guided to notice that his incorrect answers do not fit the patterns in the sequences.

Additional considerations: There are 10 axioms (rules or principles) that underlie addition and multiplication:

1. *The order of addends does not affect the sum.* (Commutative Property for Addition)

$$a + b = b + a$$

2. *The grouping of addends does not affect the sum.* (Associative Property for Addition)

$$(a + b) + c = a + (b + c)$$

3. *When zero is added to any number, the sum is that number.* (Additive Identity, also called Identity Element for Addition)

$$n + 0 = n$$

4. *The order of factors does not affect the product.* (Commutative Property for Multiplication)

$$ab = ba$$

5. *The grouping of factors does not affect the product.* (Associate Property for Multiplication)

$$(ab)c = a(bc)$$

6. *When any number is multiplied by 1, the product is that number.* (Multiplicative Identity, also called Identity Element for Multiplication)

$$1 \times n = n$$

7. *Factors may be renamed as sums, which, in turn, become factors, and the resulting products are then summed.* (Distributive Property for Multiplication over Addition)

$$a(b + c) = ab + ac$$

or

$$(a + b)(c + d) = ac + ad + bc + bd$$

8. *For each number $n (n \neq 0)$, there is a number $1/_n$* (Multiplicative Inverse) *such that their product is 1.*

$$n \times \frac{1}{n} = \frac{n}{n} = 1(n \neq 0)$$

9. *For each number n, there is a number $-n$* (Additive Inverse) *such that their sum is zero.* If n is a negative number then its additive inverse is positive.

$$n + -n = 0$$

10. *Given a set of numbers and an operation on these numbers, if the answer (resulting number) is a member of the same set as the original numbers, the set is closed for that operation.* (Closure Principle)

   a. For all a and b, if a and b are in set X, then $a + b$ is in set X. (Closure of Addition)

   b. If a and b are in set X, then ab is in set X. (Closure of Multiplication)

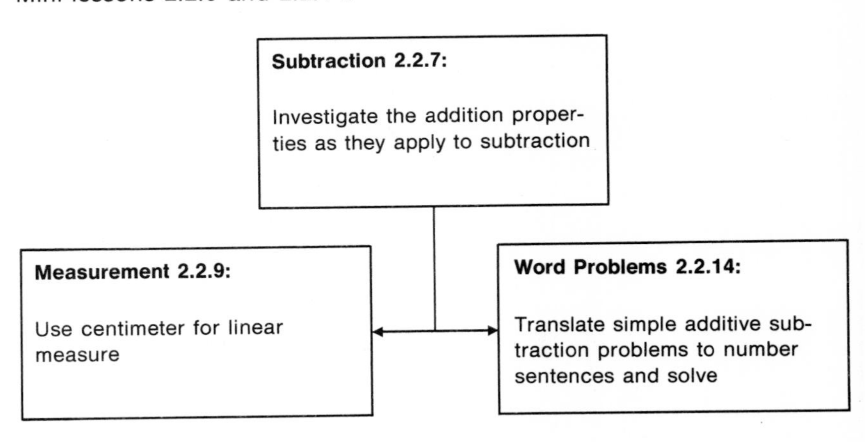

PROPERTIES OF SUBTRACTION

Four properties were applied to addition of whole numbers:

1. Commutative property ($a + b = b + a$)
2. Associative property $(a + b) + c = a + (b + c)$
3. Identity element ($a + 0 = 0 + a = a$)
4. Closure (for any two whole numbers, a and b, $a + b = c$ such that c is also a whole number).

Testing for commutativity

Is subtraction commutative? Will the order of the numbers affect the difference? Let us test for commutativity using the whole numbers 7 and 4. Does $7 - 4 = 4 - 7$? No, because $7 - 4 = 3$, but $4 - 7 = -3$. Thus, $a - b \neq b - a$ and subtraction is not commutative.

Testing for associativity

Is subtraction associative? Do differences in groupings of three or more numbers affect the difference? Let us use the whole numbers 7, 4, and 2 to find out.

$$\begin{aligned} (7 - 4) - 2 &\stackrel{?}{=} 7 - (4 - 2) \\ 3 \quad - 2 &\stackrel{?}{=} 7 - \ 2 \\ 1 &\neq 5 \end{aligned}$$

Subtraction is not associative.

Testing for identity element

Is there an identity element for subtraction? It is true that $a - 0 = a$ for all a, and $a - a = 0$; but

$$0 - a = {}^{-}a$$
$$\text{so} \quad a - 0 \neq 0 - a$$

and the identity element 0 does not hold for subtraction. Zero in subtraction satisfies the identity requirement only when placed to the right of the subtraction sign. (We shall encounter this concept again in Chapter 12 when testing for the Distributive Property of Division over Addition.)

Testing for closure

Does the closure property apply to the set of whole numbers in the subtraction operation? Is it true that for any two whole numbers a and b, there exists a whole number c such that $a - b = c$? In order to answer yes to this question, there may be no exceptions. What if $a = 5$ and $b = 8$? There exists no whole number c such that $5 - 8 = c$. The closure property does not apply to the set of whole numbers when a is less than $b (a < b)$, for there is no whole number equal to c that will satisfy the equation $5 - 8 = c$. Thus, in the set of whole numbers, it is possible to subtract only when a is greater than or equal to $b (a \geq b)$ in such examples as $a - b = c$.

EXERCISES

I. Solve the following problems in the set of whole numbers:
1. $8 = 3 + \square$.
2. $17 = 8 + \square$.
3. $53 = 16 + (4 + \square)$.
4. $83 = 35 + (7 + \square)$.

II. The binary operation of subtraction, defined in terms of addition, is $a - b = c$ (a, b, c represent whole numbers) only if ____.

III. The set of whole numbers (is/is not) closed for subtraction.

IV. The commutative property (can/cannot) be applied to subtraction of whole numbers.

V. $(7 - 3) - 2 (=/\neq) 7 - (3 - 2)$.

VI. The associative property (can/cannot) be applied to subtraction of whole numbers.

VII. $6 - 0 = 6$ since $6 =$ ____.

VIII. For any whole number n, $n - 0 = \square$.

IX. Classify the following subtraction word problems as (a) comparison, (b) take away, or (c) additive:

1. Glenn has 4 pieces of candy and Pat has 7. Tell the difference in the number of pieces of candy the children have.
2. John has 4 dogs and Jennifer has 7 cats. How many more pets does Jennifer have than John?
3. Marie has 3 dogs; Barbara has 2 fewer. How many dogs does Lisa have?
4. Rosalind had 3 paintings for sale. She sold 2. How many does she have left?

5. Susan had 7 books to sell. She sold some and has 2 left. How many did she sell?

6. Chuteka had 9 puppies. There were 4 black ones. The others were brown. How many were brown?

7. Lisa's dog, Chuteka, had some puppies. Lisa now has 6 dogs, since she has Chuteka and Chuteka's puppies. How many puppies did Chuteka have?

8. Joan has 4 yards of material. She needs 6 yards to sew a long dress. How many more yards does she need?

9. Nan read some comic books. Sara came to her house, and they read 3 comic books together. This made a total of 7 comic books read in one day for Nan. How many did she read alone?

10. Find the number of the smaller set when the number of one set and the difference are given.

11. Jane has 6 cookies, and Mabel has 2 fewer. How many cookies does Mabel have?

12. Lottie has 5 dresses. She gave her sister 2 dresses. How many dresses does Lottie have left?

13. Find the number of the complement set when the number of the universal set and that of the subset are known.

14. Find the number of the subset needed to complete a given number of a universal set when the number of the other subset is known.

X. What is meant by saying that addition and subtraction are inverse processes?

SUGGESTED READINGS

Bennett, A. B., Jr., and Musser, G. L. "A Concrete Approach to Integer Addition and Subtraction," *The Arithmetic Teacher* 25 (May 1976): 332–36.

Bradford, J. W. "Methods and Materials for Learning Subtraction." *The Arithmetic Teacher* 25 (February 1978): 18–20.

Burns, M. "Ideas." *The Arithmetic Teacher* 22 (January 1975): 34–46.

Locke, F. M. *Math Shortcuts*. New York: John Wiley & Sons, 1972.

Werner, M., "The Case for a More Universal Number-Line Model of Subtraction." *The Arithmetic Teacher* 20 (January 1973): 61–64.

Zakariya, Norma; McClung, Margo; and Winner, Alice-Ann. "The Calculator in the Classroom." *The Arithmetic Teacher* 27 (March 1980): 12–16.

Subtraction Algorithms

Just as subtraction is the inverse of addition, so the subtraction algorithm is the inverse of the addition algorithm, as shown below:

Addition algorithm		*Subtraction algorithm*		
Addend	6	Sum	11	11
+Addend	+5	−Addend	− 5	− 6
Sum	11	Addend	6	5

The subtraction algorithm is so closely related to the addition algorithm (especially when two addends are being processed) that they are usually taught side by side. In fact, since all four arithmetic operations—addition, multiplication, subtraction, and division—occur in the everyday lives of children, the teacher should take advantage of the natural settings and situations that give rise to opportunities for mathematics instruction.

> *Note:* The word *regrouping* is used in the example involving concrete objects; but at the symbolic level, the word *renaming* is used. When concrete objects are used, they are regrouped; numerals are renamed.

THE ABACUS AS AN AID IN RECORDING THE SUBTRACTION ALGORITHM

The decomposition take-away method seems to provide optimal transfer from the concrete level to the symbolic level. For example, an abacus representing the concrete level may be used to show a subtraction computation and, at the same time, the process may be recorded with the subtraction algorithm.

By this approach of combining an activity at the concrete level (computing on the abacus) with that at the symbolic level (writing the algorithm), the child is recording, step by step, the physical representation of the algorithm.

Mini-lesson 2.3.6 Subtract with Minuends Less than 1,000, with One Renaming

Vocabulary: rename, regroup

Possible Trouble Spots: does not understand place value; does not apply place-value principle to renaming; does not conserve number

Requisite Objectives:
Subtraction 2.1.7: Subtract numbers with minuend less than 100, with no renaming

Word Problems 2.1.15: Translate simple addition word problems to corresponding algorithm and solve

Whole Numbers/Notation 2.2.1: Use exchange model for place value

Subtraction 2.2.8: Subtract numbers with minuend less than 100, with no renaming

Subtraction 2.3.5: Subtract with minuends less than 100, with renaming

Activity 1
Write the example

$$\begin{array}{r} 328 \\ -\ 49 \\ \hline \end{array}$$

Give the child an abacus and a pile of digit symbols (wooden rings, spools, beads, Life Savers, or similar objects). Ask the child, "What number is the whole that you start out with?" He should respond that the number is 328. Then ask him to show 328 on the abacus. This is shown in Step 1.

Activity 2
There are many variations on the next step. You may simply ask the child to show 328 − 49 on the abacus, leaving the procedure unstruc-

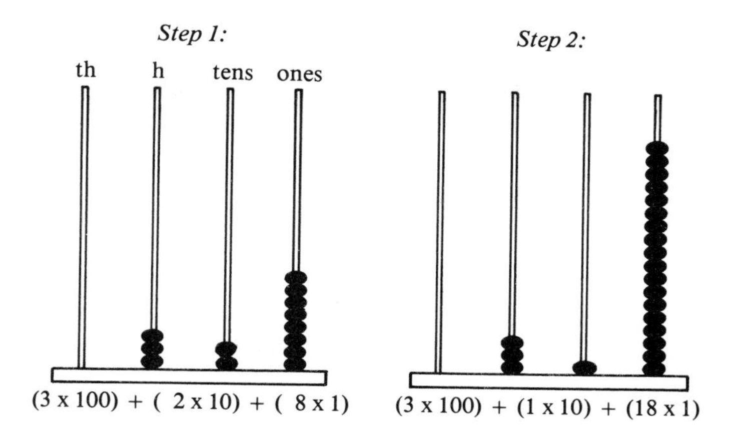

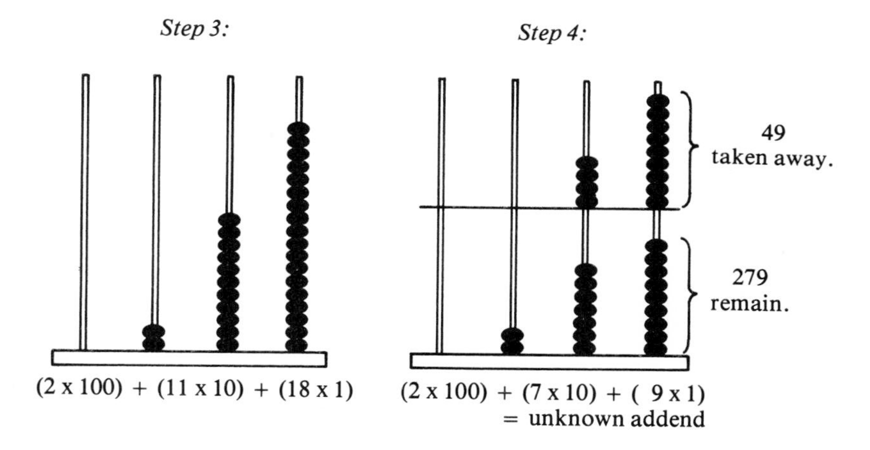

tured for the child. Or you may guide him to perform *all* necessary groupings first before subtracting. If the latter step is preferred, ask the child if he has enough ones to do the subtracting. The child, seeing that he must remove 9 digit symbols and that he has only 8, should realize that he does not have enough ones. Point to the units digit symbols and ask the child how much each is worth. He should realize that each symbol is worth 1.

Then point to the tens-place value and ask how much each symbol is worth. The child should respond "ten." Then ask him what he must do to get more beads (digit symbols) in order to subtract 9 ones. If the child cannot explain that he can exchange 1 ten for 10 ones and put them with the other ones, it may prove helpful to elaborate on the process by saying to the child, "Do you remember in addition, when you had more than ten in the units column, how you pulled out groups of ten units and exchanged each group for one ten? Well, now you can undo that. You may go the other way—from the tens place to the ones place."

Once the child has exchanged 1 ten for 10 units and summed the units, he may either do the unit subtraction (18 − 9 = 9) or proceed to other necessary exchanges. In the latter case, a renaming of 1 of the hundreds as 10 tens, yielding 11 tens, is necessary.

The recording of the algorithm may accompany the abacus activities so that the recordings by step would be as follows:

Step 1	*Step 2*
328	118 3~~2~~~~8~~
Step 3	***Step 4***
211 ~~1~~ 18 ~~3~~~~2~~~~8~~	211 ~~1~~ 18 ~~3~~~~2~~~~8~~ − 49 279

Additional considerations: The ability to relate the place-value principle, the renaming principle, and the fact that a number may be represented in different ways is crucial to the goal of this mini-lesson. Class discussion should clarify the interrelatedness of these three complementary principles.

Note: Mini-lesson 2.3.6 is a prerequisite for Mini-lessons 2.3.8 and 2.3.9.

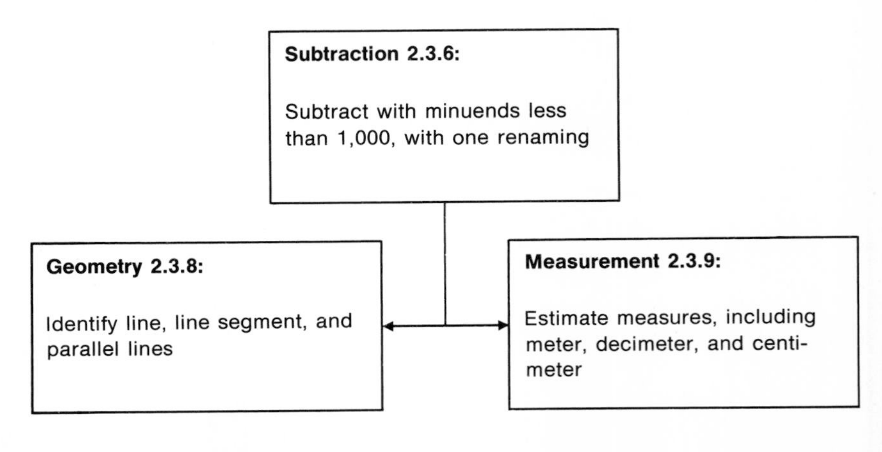

RENAMING PROCESS AS A POTENTIAL TROUBLE SPOT

Numeration system ideas as prerequisites to algorithms

Unless the child understands that in base ten the position values are based on groupings of ten, he or she may move one bead from the tens place to the units place. Some children, on the other hand, become confused as to why they must add 10 units when they may need only 1 or 2 units to allow for the subtraction. They do not understand that *the exchange must involve the entire set, whose cardinal number property is determined by the base.* It is to avoid this confusion that activities involving numeration systems are an important prerequisite to the algorithms. A historical review of both nonplace-value and place-value numeration systems helps children see that, in our numeration system, exchanges may be made as based upon groupings of the base. In renaming, numeration-system ideas are even more important than concepts about the properties for the operation, since renaming involves groupings by the base.

Readiness activities

The teacher also needs to provide the children with readiness activities for the subtraction-with-renaming algorithm. These activities may include bundling units into tens and then tens into hundreds, or unbundling a group of 10 tens (100) and then unbundling one of these tens to get 10 units.

Diagrams to reflect renaming process

Combining the exchanged 10 units with the units already present is not always understood by the children. Drawings sometimes hinder the renaming process. To avoid this, use the same symbol to represent tens and units, as shown below.

Step 1: $\begin{array}{r} 60 \\ -49 \\ \hline \end{array}$

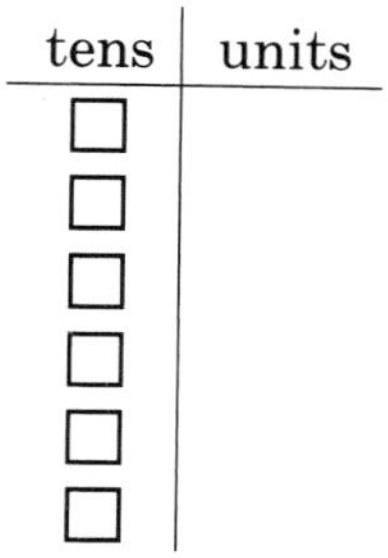

tens	units
□	
□	
□	
□	
□	
□	

Note that there are no symbols for 49.

Step 2: $\begin{array}{r} 60 \\ -40 \\ \hline 20 \end{array}$

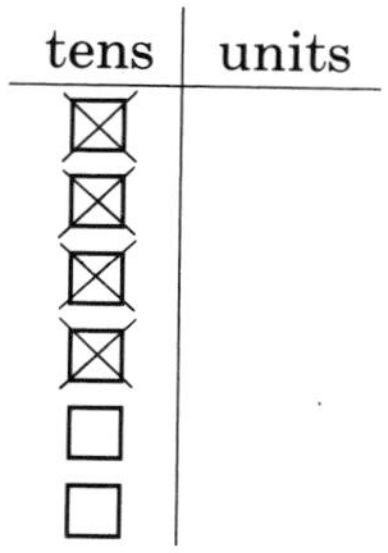

tens	units
⊠	
⊠	
⊠	
⊠	
□	
□	

The 4 tens to be subtracted are crossed out. The number of remaining tens are recorded.

Step 3:

$$\begin{array}{r} 60 \\ -40 \\ \hline \overset{1}{\not 2}\overset{10}{\not 0} \end{array}$$

tens	units

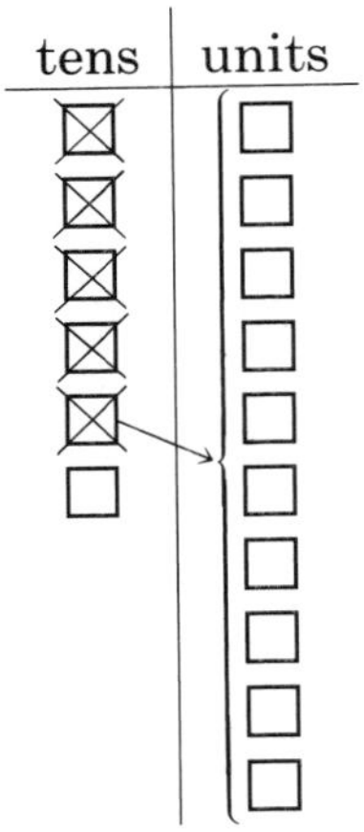

1 ten is exchanged for 10 ones.

Step 4:

$$\begin{array}{r} \overset{1}{\not 2}\overset{10}{\not 0} \\ -\ 9 \\ \hline 11 \end{array}$$

tens	units

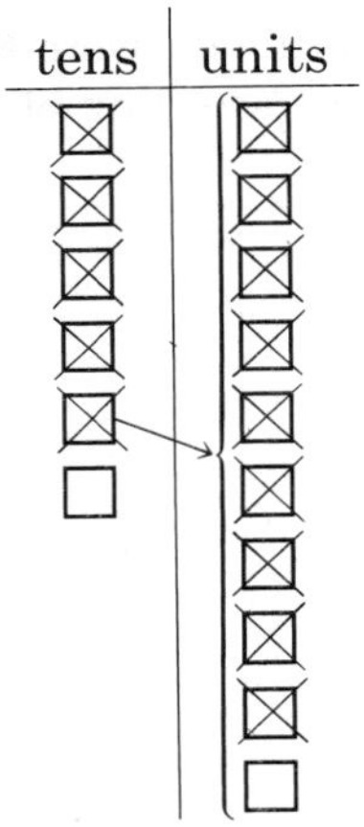

1 ten and 1 unit remain. Thus 60 − 49 = 11.

The sequence could have been illustrated as follows:

Step 1:

$$\begin{array}{r} \overset{5}{\not 6}\overset{10}{\not 0} \\ -49 \\ \hline \end{array}$$

tens	units

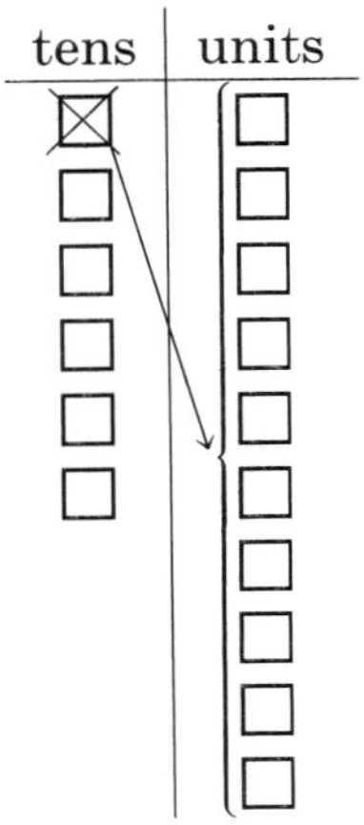

1 ten is renamed as 10 units.

Step 2:

$$\begin{array}{r} \overset{5}{\not6}\overset{10}{\not0} \\ -49 \\ \hline 11 \end{array}$$

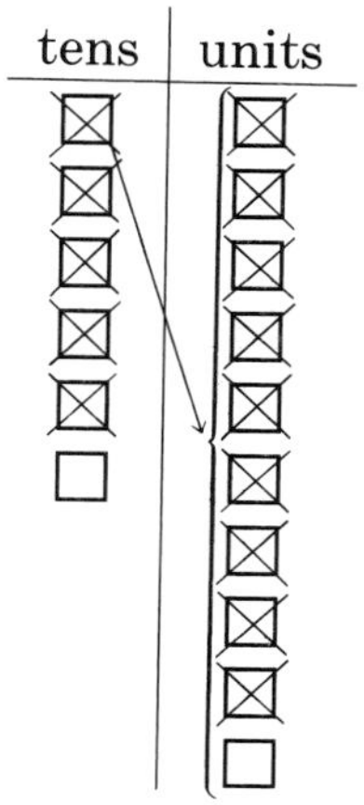

tens	units

Subtraction is done.

If the problem were 63 − 49 = □, the teacher should encourage immediate recording of the exchanged ten. For example,

Step 1:

$$\begin{array}{r} 63 \\ -49 \\ \hline \end{array}$$

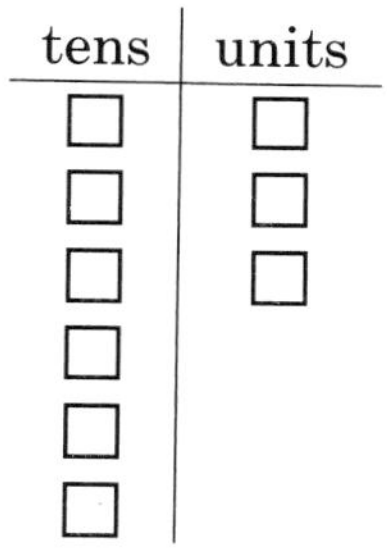

tens	units

Step 2:

$$\begin{array}{r} \overset{5}{\not6}\overset{13}{\not3} \\ -49 \\ \hline \end{array}$$

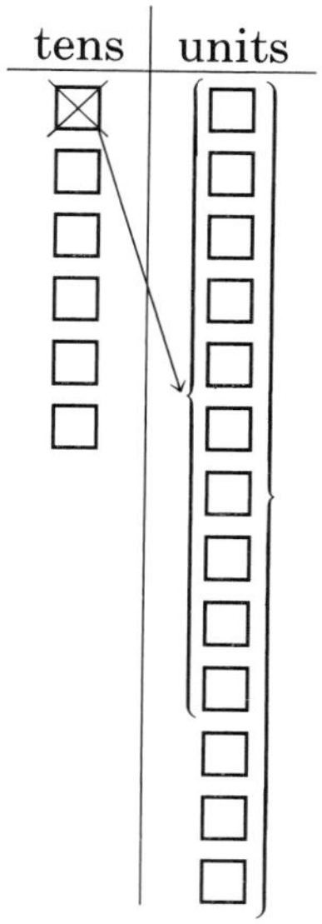

tens	units

While drawing the renaming symbols, the child thinks, "One, two, three, four, five, six, seven, eight, nine, ten, one, two, three" and then, recounting all of the units, One, two, three, . . . , thirteen."

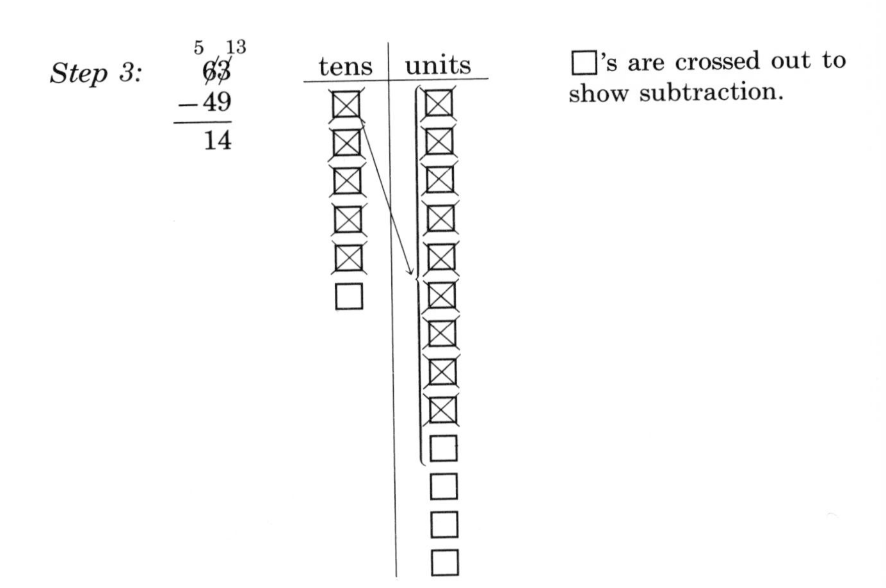

If it appears that the child needs more experience at the concrete level using bundles of Popsicle sticks or tongue depressors, a place-value chart may be introduced. However, there are several activities that should precede working with a place-value chart. A task analysis of place-value activities shows that these activities should be presented in sequence as shown next in mini-lesson 3.1.5.

Mini-lesson 3.1.5 Subtract with Zero in Minuend, with Renaming

Vocabulary: minuend, renaming

Possible Trouble Spots: does not conserve number; has visual-perceptual difficulty; does not understand the renaming process; applies count-on principle, rather than the exchange model of place value

Requisite Objectives:
Group objects by tens with no extra units

Group objects by tens when there are units left over

Exchange a group of 10 objects for an object that symbolizes 1 ten

Show the exchange of 10 units for a single symbol for 10 by moving over one place

Enumerate using a place-value chart

Activity 1
To group objects by tens with no extra units, present arrays or dot cards to the child and ask her to draw a boundary around each group of 10 dots.

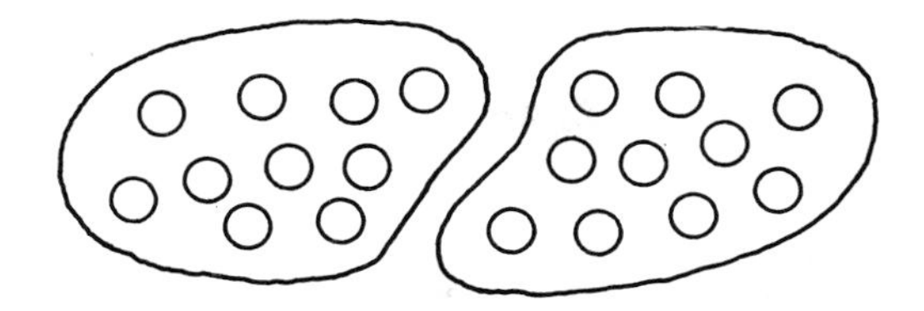

Activity 2
To group objects by tens when there are units left over, have the child place objects in groups of 10 and some extra. (Note that this is a division idea.)

"Two tens and three ones."

Activity 3
To exchange a group of 10 objects for an object that symbolizes 1 ten, ask the child to exchange 10 pennies for a dime, or 10 natural-colored tongue depressors for a colored one.

Activity 4
To show the exchange of 10 units for a single symbol for 10 by moving over one place, place two cards marked "tens" and "ones," respectively, on a table. Give the child a bundle of tongue depressors and ask him to show representations of various numbers by placing the tongue depressors under the appropriate columns. He should recognize that a single tongue depressor in the units column means 1 unit but that the same stick in the tens position means 1 ten.

	tens	units
"Show one."		0
"Show two."		0 0
"Show three."		0 0 0
"Show nine."		0 0 0 0 0 0 0 0 0
"Show thirteen."	0	0 0 0
"Show ten."	0	
"Show twenty-three."	0 0	0 0 0

Activity 5

To enumerate using a place-value chart, ask the child to make a place-value chart by folding a piece of paper (as shown), stapling the ends, and naming the places.

A. Ask the child to show counting from 1 to 13 using the place-value chart. The child inserts 1 stick in the units place for 1, a second stick in the units place for 2, and so on, until she has placed 9 sticks in the units place. She then *takes out the 9 sticks in the units place and, as 10 is counted, places a single stick in the tens position.* Then she inserts 1 stick in the units place for "11," a second stick in the units place for "12," and a third stick in the units place for "13."

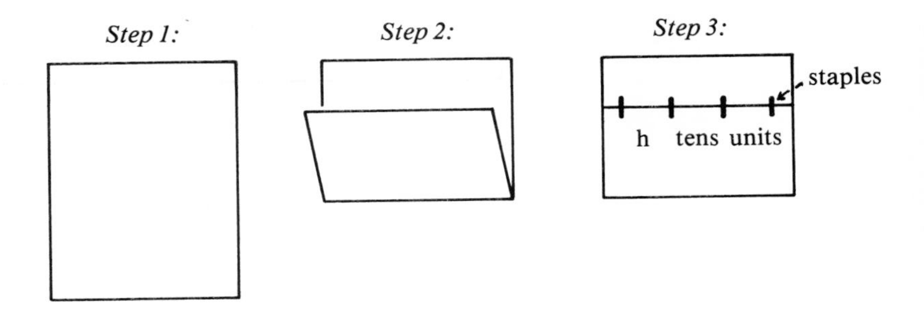

B. To subtract without renaming using the place-value chart ask the child to solve the problem 43 − 12 = ☐ using a place-value chart. The child inserts 4 sticks in the tens place and 3 sticks in the units place. Then he takes out 1 ten and 2 units, leaving 3 tens and 1 unit.

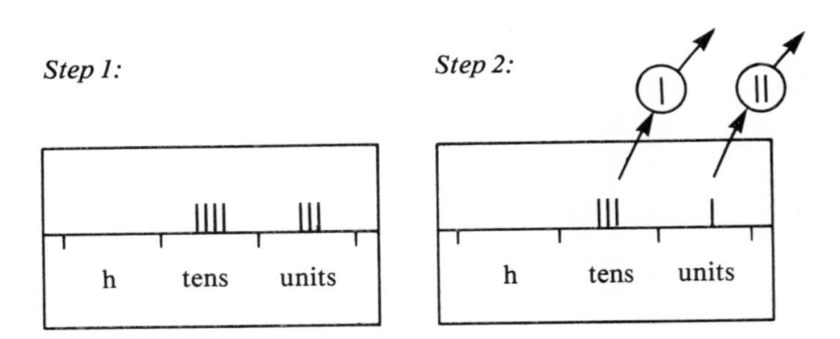

Have the child record his actions:

$$\begin{array}{r} 43 \\ -12 \\ \hline 31 \end{array}$$

C. To subtract with renaming using the place-value chart for the example $43 - 19 = \square$, guide the child to show subtraction in the units place first. The child inserts 4 sticks in the tens place and 3 sticks in the units place. The child then starts subtracting 9 sticks but can only go as far as 3 sticks.

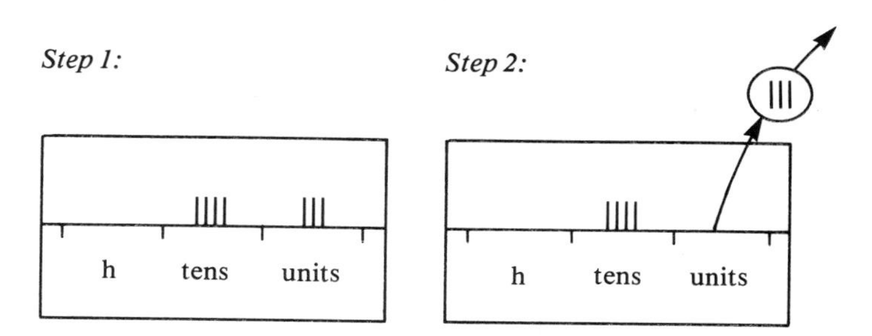

He may take out the 3 sticks and remember that he owes 6 more $(9 - 3 = 6)$. Then he exchanges 1 stick in the tens position for 10 units, placing them in the units position. He then takes out the 6 sticks he owes, leaving 4 units, and subtracts the tens $(3 - 1)$, leaving 2 tens.

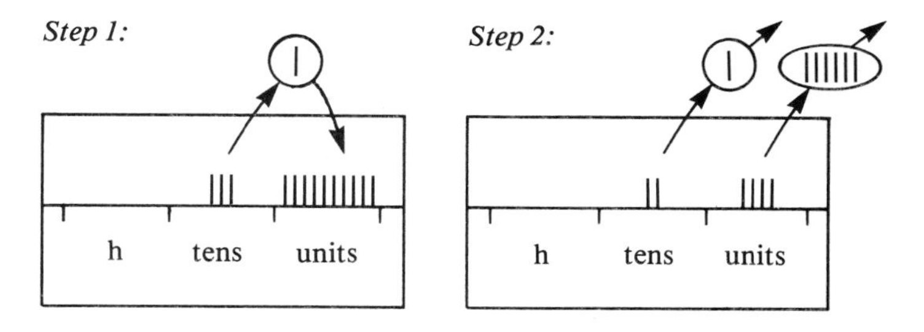

Or he may regroup first, obtaining 3 tens and 13 units, and then subtract

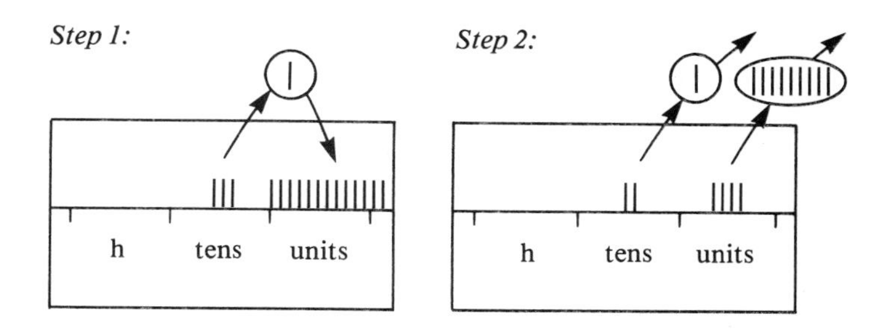

Ask the child to record his actions:

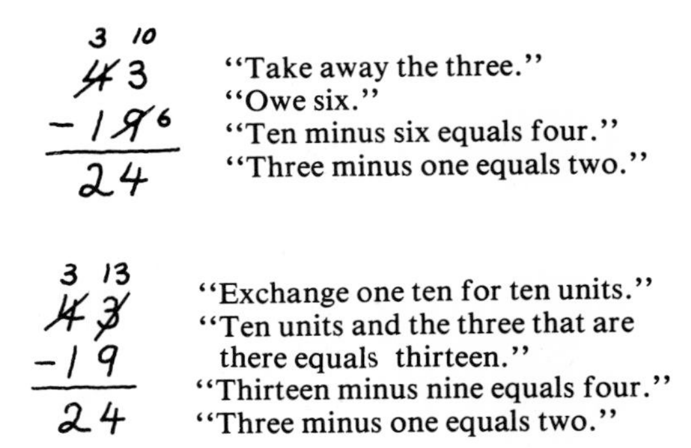

Activity 6

Use the counting boards for examples such as the following.

A. To solve 90 − 18, use checkers that are black on one side and red on the other side. Show the minuend (90) in black. Then, rename 90 so that the subtrahend (18) is apparent on the boards. Then show 18 with the red side of the checkers. The remaining black checkers represent the answer as shown next.

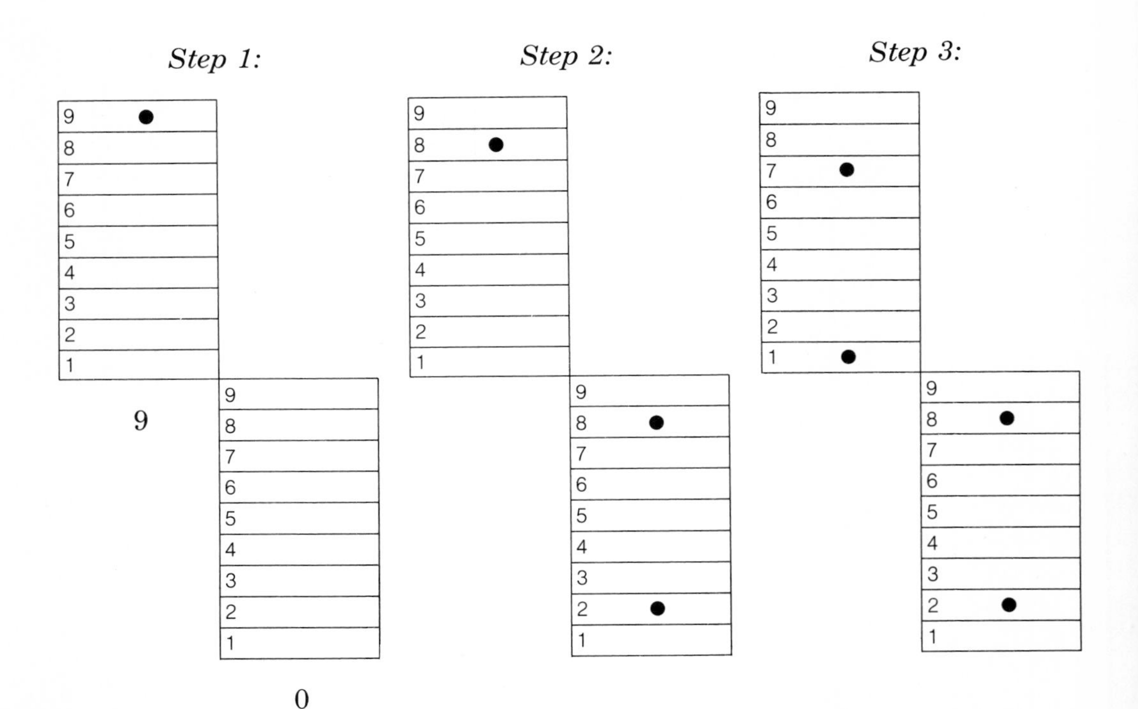

Step 4:

$$\begin{array}{r} 90 \\ -18 \\ \hline 72 \end{array}$$

B. If only black checkers are available, then, after renaming to show the subtrahend, take away those checkers that represent the subtrahend. The value of the remaining checkers represents the difference. This procedure is shown below for the example 302 − 191:

Step 1:

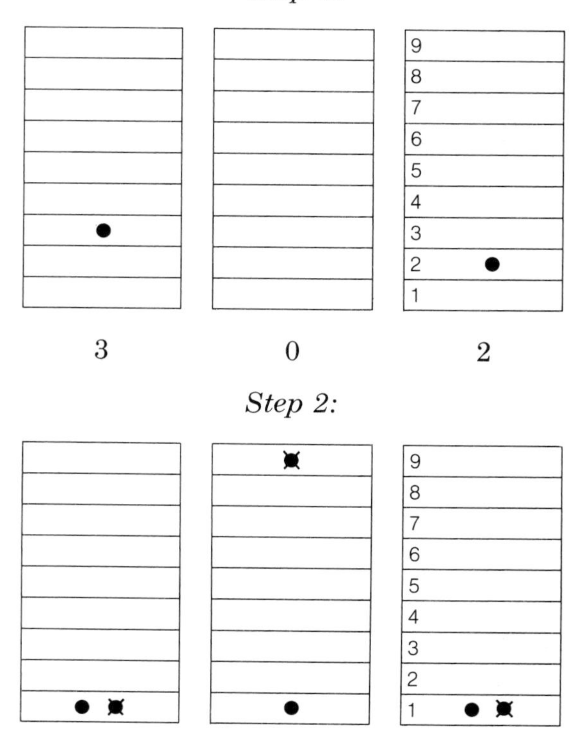

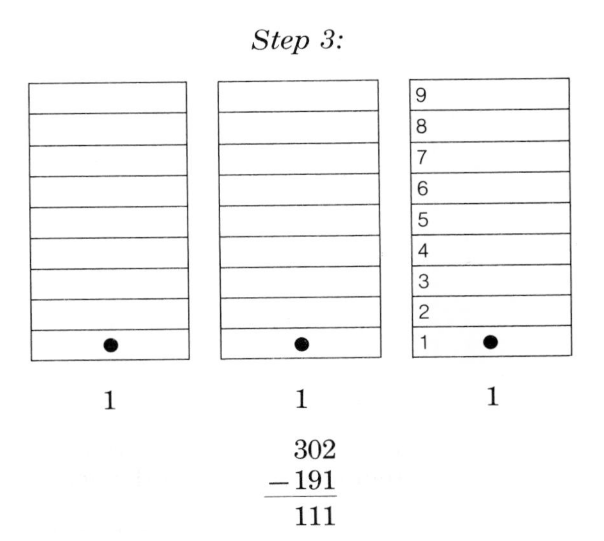

Additional considerations: I have found that the counting boards serve as a bridge from more concrete activities to the subtraction algorithm. The boards are a physical representation of processing the subtraction with renaming algorithm.

Note: Mini-lesson 3.1.5 is a prerequisite for Mini-lessons 3.1.10 and 3.1.16.

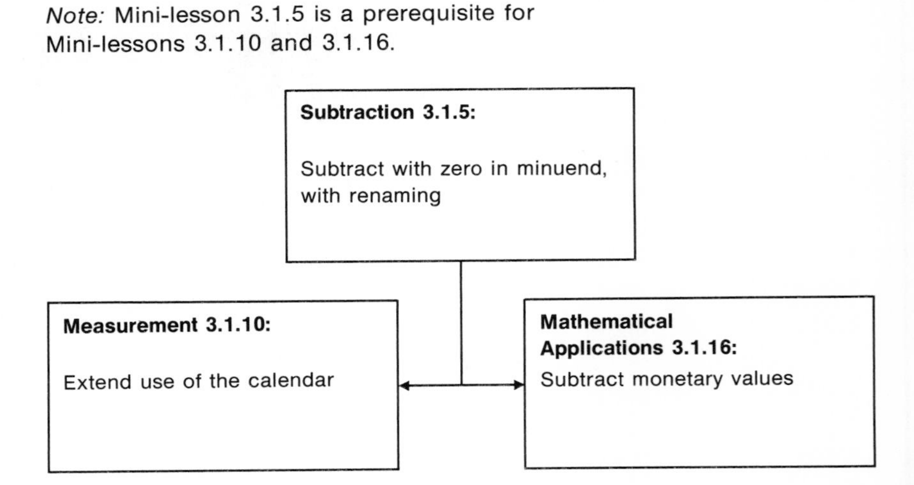

AIDS IN SUBTRACTION

"When do you get rid of an aid?" "When your leg is better, you get rid of the crutch." There are many crutches in mathematics. Counting on fingers (or toes), counting cracks in the ceiling, counting positions on numerals such as

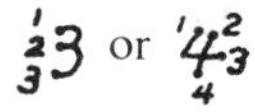

$\overset{1}{\underset{3}{2}}3$ or $'4^{2}_{3}$ (4)

Writing the renamed number in algorithms, using the addition/subtraction or multiplication/division grids, using concrete materials to represent symbols—all are crutches, or aids. When should the aid be discarded? When the user is ready to discard it. The teacher should gently guide the learner to the most mature level of performance so the aids will no longer be needed (some persons may not reach this level until adulthood).

There are some variations on the subtraction algorithm that have been effective in helping children understand renaming in the algorithm. They involve showing the renaming as the learner crosses over from one position to the next.

Suggested board work

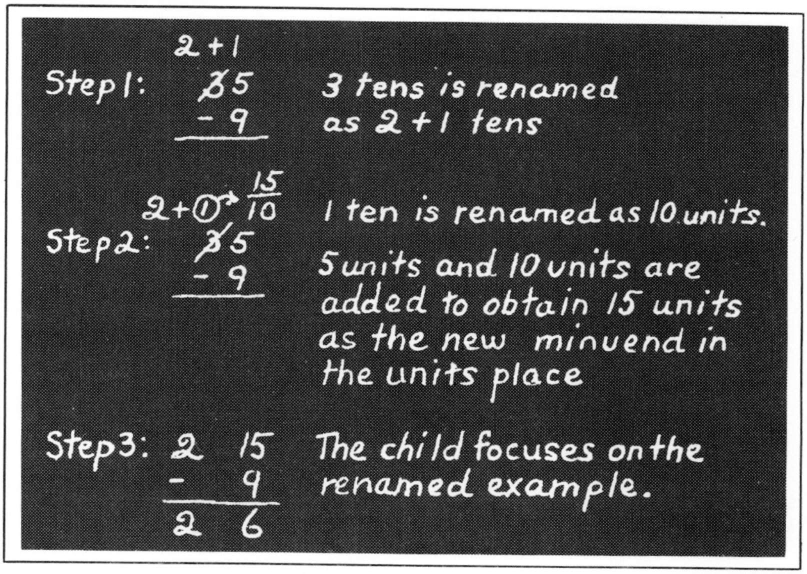

Step 1:
```
 2+1
  3̸5
 - 9
```
3 tens is renamed as 2 + 1 tens

Step 2:
```
        15
 2+(1)→ 10
      3̸5
     - 9
```
1 ten is renamed as 10 units.

5 units and 10 units are added to obtain 15 units as the new minuend in the units place

Step 3:
```
  2 15
 -   9
  2  6
```
The child focuses on the renamed example.

Notice that in Step 3 above, the minuend was not written in expanded notation as 2 tens + 15 units. This was purposely avoided because children become confused when expanded forms are used.

The following variation is appropriate for children who can compute mentally the new names for the minuend numbers.

```
   2 15
   3̸ 5̸
 -   9
   2 6
```
Renaming is shown immediately.

CAUTION AGAINST USING EXPANDED NOTATION FOR THE SUBTRACTION ALGORITHM

As mentioned in Chapter 9, the expanded form of an addition algorithm may lead children to add tens and ones. For example, they may process 1 ten + 3 ones + 2 ones (13 + 2) as 1 + 3 = 4 and 4 + 2 = 6.

Expanded form in textbooks may confuse children

An expanded form of the subtraction algorithm has also caused confusion. Textbooks show the expanded form of a subtraction problem like

$$\begin{array}{r} 77 \\ -42 \\ \hline \end{array}$$

as follows:

$$\begin{array}{rcr} 77 & = & 70 + 7 \\ -42 & = & 40 + 2 \\ \hline & & 110 + 9 \end{array}$$

The child follows the direction of the sign.

Another form used in textbooks is

$$\begin{array}{rcr} 77 & = & 7 \text{ tens} + 7 \text{ ones} \\ -42 & = & 4 \text{ tens} + 2 \text{ ones} \\ \hline & & 3 \text{ tens} + 9 \text{ ones} \end{array}$$

This child followed the operation signs in a different way.

The following forms are also used "as aids" in renaming:

(a)

$$\begin{array}{rcrcr} 72 & = & 70 + 2 & = & 60 + 12 \\ -49 & = & 40 + 9 & = & 40 + 9 \\ \hline & & & & 100 + 21 \end{array}$$

Logical response, isn't it?

(b)

$$\begin{array}{rcr} 72 & = & 7 \text{ tens} + 2 \text{ ones} \\ -49 & = & 4 \text{ tens} + 9 \text{ ones} \\ \hline & & 11 \text{ tens} + 11 \text{ ones} \end{array}$$

This, too, is a logical error.

The correct expanded form should be written as

$$\begin{array}{rcr} 77 & = & 70 + 7 \\ -42 & = & -40 - 2 \\ \hline \end{array} \quad \text{or} \quad \begin{array}{r} 70 + 7 \\ -(40 + 2) \\ \hline \end{array}$$

for $^-42 \neq 42$, which is what the textbook algorithm shows. The mathematically equivalent form

$$\begin{array}{r} 70 + 7 \\ -40 - 2 \\ \hline \end{array}$$

to represent 77 − 42 in expanded notation confuses young children, who, without knowledge of multiplying positive and negative integers, could not intuitively grasp that $^{-}42 = ^{-}(40 + 2) = ^{-}40 + (^{-}2) = ^{-}40 - 2$. To a young child, $^{-}40 - 2$ might be $^{-}38$, or just 38.

Avoid expanded form in early grades

It is therefore suggested that the expanded notation form be avoided in teaching the subtraction algorithm in the early grades. Instead, the teacher may use the abacus, a place-value chart, or the computation boards to emphasize place value in the subtraction algorithm.

ADDITIVE SUBTRACTION AND NEGATIVE NUMBERS

Including integers in the mathematics program

Since addition and multiplication are usually restricted to the set of whole numbers in elementary school mathematics, chapters dealing with these operations were concerned with whole numbers only. It is not until the middle-grade years that operations with negative numbers become prevalent. At that time, the sets of numbers studied include the set of integers, which may be thought of as the set of whole numbers plus the negative counting numbers (set of integers = $\{\ldots ^{-}3, ^{-}2, ^{-}1, 0, 1, 2, 3, \ldots\}$).

Introducing children to negative numbers

Young children may develop beginning thoughts on the idea of negative numbers. For example, children encounter such situations as the temperature falling below zero, owing more money than they have and, perhaps, scoring less than zero on tests that are graded according to a strict point system. Since children do experience such situations, they need not be told that "you can't take six away from two." Rather, it is more sensible to tell young children that "there is a set of numbers that allows you to subtract a larger number from a smaller number, but for now we'll work with situations that do not involve those numbers."

The computation boards serve as a model for moving from subtraction of whole numbers to subtraction of integers.

Mini-lesson 8.1.3 Subtract Integers

Vocabulary: positive, negative, integer

Possible Trouble Spots: has difficulty understanding computation on a vertical number line

Requisite Objectives:

Addition 1.3.9: Add two 1-digit numbers with sums to 18

Subtraction 1.3.10: Perform take-away subtraction with minuend to 18

Subtraction 2.2.7: Investigate the addition properties as they apply to subtraction

Integers/Notation 3.3.1: Use inverse relationship between addition and subtraction

Activity 1
To show that the sum of additive inverses equals the additive identity, ask the child to solve the following problem: "Mary and Beth each won 5 games. What is the difference in their winnings?" Tell the child that the red checkers represent Mary's wins and the black checkers stand for Beth's wins. He places red checkers to indicate a value of 5. Explain that every one-to-one match is considered a null match; that is, one of Mary's wins and one of Beth's wins cancel each other out and neither of them receives a point.

The child then cancels Mary's 5 points, leaving the board empty. Therefore, he should realize that the answer is zero. This is a concrete representation for $5 + {}^{-}5 = 0$.

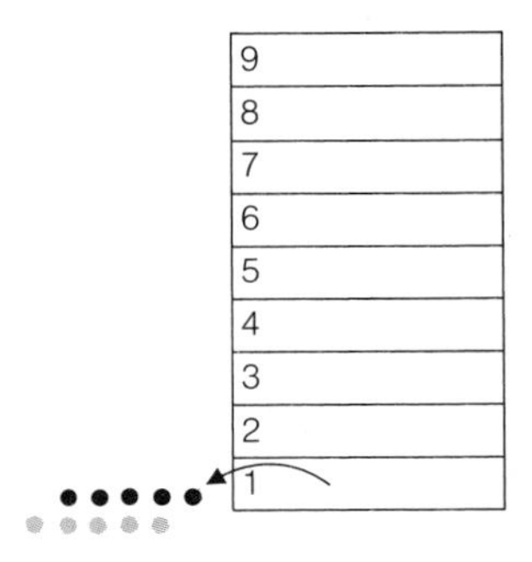

Activity 2
To show comparison subtraction, give the child red and black checkers and help her solve the following problem: "Mary and Beth played a game 10 times. Mary won 6 times and Beth won 4. How many more times did Mary win than Beth?" Explain to the child that the red

checkers represent Mary's wins and the black checkers stand for Beth's wins. The child places red checkers to indicate a value of 6. Beth's 4 wins cancel out 4 of Mary's 6 wins, and the board shows a red checker whose value is 2.

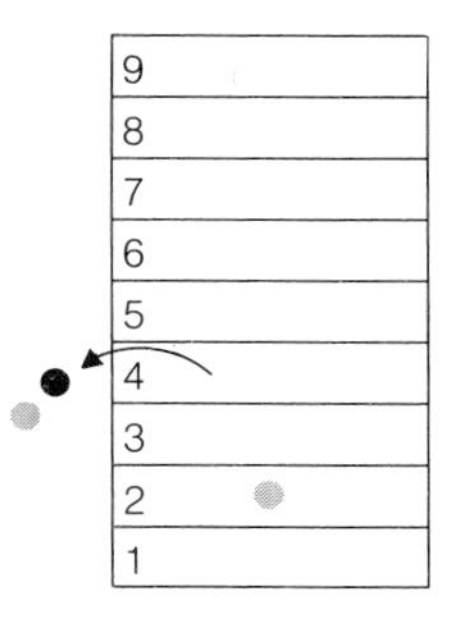

Activity 3
Solve the problem $7 + {}^{-}4 = \square$.

The black checkers represent positive numbers and the red checkers represent negative numbers. Black checkers show the value of 7.

Then cancel a black checker whose value is 4 with a red checker whose value is ${}^{-}4$.

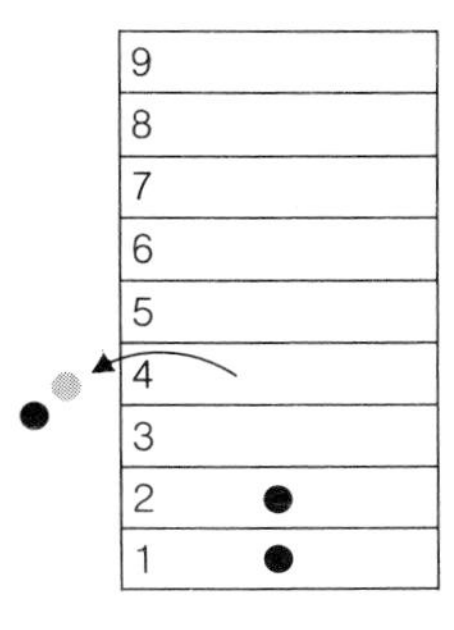

The board shows $+3$, so $7 + {}^-4 = +3$. Color coding the equation is a useful aid. This may be done by writing the 7 and the $+3$ in black and the $^-4$ in red in the equation $7 + {}^-4 = +3$.

Activity 4
Solve the problem, $^-9 + 6 = \square$.

The black checkers represent positive numbers and the red checkers stand for negative numbers.

Red checkers show the value $^-9$.

9	●
8	
7	
6	
5	
4	
3	
2	
1	

Regroup $^-9$ as $^-4 + {}^-2 + {}^-2 + {}^-1$ so that the number 6 is apparent.

9	
8	
7	
6	
5	
4	●
3	
2	● ●
1	●

The child then cancels red checkers whose value is $^-6$ with black checkers whose value is 6. The board shows $^-3$, so $^-9 + 6 = {}^-3$.

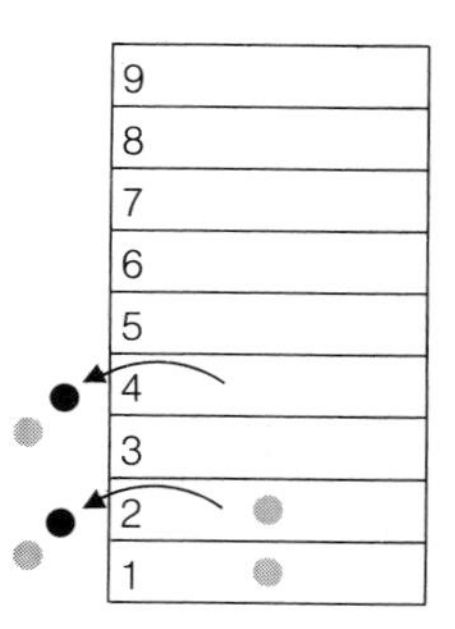

Additional considerations: Computation with integers is particularly useful in consumer mathematics. For example, signed numbers are used in credits and debits, balance of trade, and balance of payments. Subtraction of integers, especially the take-away model, should be kept in mind by the teacher before telling a primary grade student that examples like $3 - 5$ *cannot* be done. Erroneous generalizations are far-reaching in confusing students; maybe the confusion does not show up until years later.

Note: Mini-lesson 8.1.3 is a prerequisite for Mini-lessons 8.2.2 and 8.2.5.

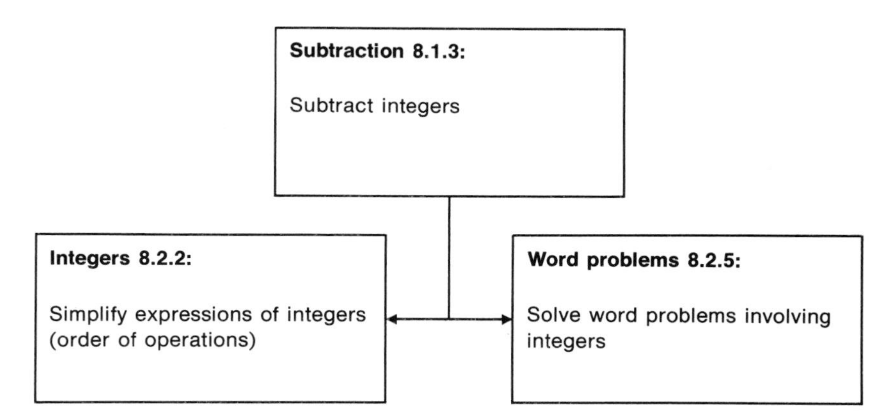

CHECKING SUBTRACTION

Adding the difference to subtrahend

Checking subtraction problems involves adding the difference to the subtrahend as shown below:

$$\begin{array}{r} 873 \\ -549 \\ \hline 324 \\ \hline 873 \end{array}$$

Casting out 9s

Another procedure for verifying answers is casting out 9s (which was described for checking multiplication in Chapter 7).

$$\text{(a)}\quad \begin{array}{r} 873 \\ -549 \\ \hline 324 \end{array} \quad \begin{array}{r|r} 8+7+3=18 & 18 \div 9 = 2 \text{ rem. } 0 \\ 5+4+9=18 & 18 \div 9 = 2 \text{ rem. } 0 \\ \hline 3+2+4=\ \ 9 & 9 \div 9 = 1 \text{ rem. } 0 \end{array}$$

$$\text{(b)}\quad \begin{array}{r} 8639 \\ -5473 \\ \hline 3166 \end{array} \quad \begin{array}{r|r} 8+6+3+9=26 & 26 \div 9 = 2 \text{ rem. } 8 \\ 5+4+7+3=19 & 19 \div 9 = 2 \text{ rem. } 1 \\ \hline 3+1+6+6=16 & 16 \div 9 = 1 \text{ rem. } 7 \end{array}$$

In subtraction, the face values of the digits in each number are summed. This sum is then divided and the remainders of these divisions observed. If the minuend remainder minus the subtrahend remainder equals the difference remainder, this is evidence that the subtraction in the original problem is correct.

EXERCISES

I. Draw and use the Reisman computation boards to solve the following problems:

1. $\begin{array}{r} 35 \\ -14 \\ \hline \end{array}$

2. $\begin{array}{r} 14 \\ -\ 8 \\ \hline \end{array}$

3. $\begin{array}{r} 43 \\ -29 \\ \hline \end{array}$

4. $\begin{array}{r} 346 \\ -243 \\ \hline \end{array}$

II. Describe the errors represented in the following subtraction algorithms:

1. $\begin{array}{r} 38 \\ -13 \\ \hline 24 \end{array}$

2. $\begin{array}{r} 248 \\ -\ 27 \\ \hline 21 \end{array}$

3. $\begin{array}{r} 108 \\ -\ 37 \\ \hline 101 \end{array}$

4. $\begin{array}{r} 45 \\ -11 \\ \hline 45 \end{array}$

5. $\begin{array}{r} 483 \\ -\ 2 \\ \hline 261 \end{array}$

6. $\begin{array}{r} 92 \\ -\ 7 \\ \hline 95 \end{array}$

7. $\begin{array}{r} 3471 \\ -\ 598 \\ \hline 3127 \end{array}$

III. Compute and check the following problems by casting out 9s:

1. $\begin{array}{r} 385 \\ -168 \\ \hline \end{array}$

2. $\begin{array}{r} 8736 \\ -2578 \\ \hline \end{array}$

SUGGESTED READINGS

Asimov, Isaac. *How Did We Find Out About Numbers?* New York: Walker Publishing Co., 1973.

Barnard, Douglas St. Paul. *It's All Done by Numbers.* New York: Hawthorn Books, 1968.

Bendick, Jeanne, and Levin, Marcia. *Mathematics Illustrated Dictionary.* New York, McGraw-Hill Co., 1965.

Schlossberg, Edwin, and Brockman, John. *The Pocket Calculator Game Book.* New York: William Morrow and Co., 1975.

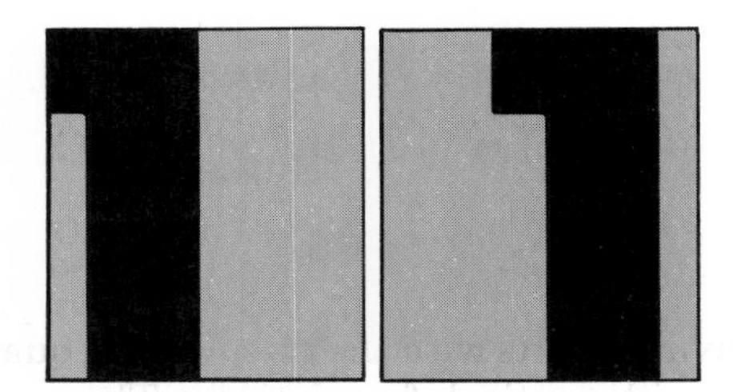

Division

CHILDREN'S EARLY EXPERIENCES WITH THE DIVISION OPERATION

Children are performing division very early in life. For example, division is involved in one-to-one correspondence when a child is presented with the problem of distributing 9 napkins to 6 children so that each child has at least 1 napkin. The child is told that he or she has a specific number of napkins to use and must see that 6 children each get 1 napkin. This tells the child how large the whole is (9 napkins) and how many groups of napkins he or she needs to make (6 groups).

Partitive and measurement types of division

If you ask the child how many napkins each child may receive, he or she will probably answer, "Each child will get one napkin and I'll have three extra" or "Each child will get one napkin and then three of them can get an extra one." *Partitive division* is involved since the question implied is "how many in each group?" *Measurement division,* on the other hand, involves the question, "how many groups?"

Young children confronted with "uneven" division

Notice in the above example that the child is confronted with division with a remainder. It is artificial to introduce only "even" division when children are handling "uneven" division in their daily lives from the time they have to share something with another child or with a group of children. Young children, then, should be encouraged to describe their experiences with partitive division.

Now consider a many-to-one example. A 4-year-old child, having gone to the family doctor for a checkup, is given a handful of 7 lollipops by the kindly physician who says, "Now, honey, share these with your sister and brother." The child has been presented with an uneven partitive division problem. He or she may have a mental picture of the division operation (picture level) and may think, "one for me, one for sister, one for brother." But the child realizes there are some left, so he or she decides to distribute further, thinking, "one for me again, one for sister, one for brother." Now there is only one left, so the child eats it on the way home and feels good about how he or she solved that problem.

Parents can help provide "readiness" experiences

Preschool, kindergarten, and primary teachers must analyze their pupils' environments with the goal of using quantitative situations for building mathematical foundations. These situations can become "readiness" experiences and must be capitalized on by both teachers and parents. When parents ask teachers what they can do to help their child in mathematics, specific suggestions for creating experiences in adding, multiplying, subtracting, and dividing (like the napkin and lollipop situations) may be described by the teacher. It should be emphasized that no notation was used in either case. The teacher should be careful not to equate providing early quantitative situations with early *formal* instruction in computing.

TEACHING THE DIVISION OPERATION

Just as a physical model for multiplication is the union of equivalent disjoint sets, a model for division may be the "disunion" or "taking apart the union" and partitioning it into equivalent disjoint sets with and without a remainder.

Mini-lesson 3.2.11 Identify Measurement and Partitive Division Situations

Vocabulary: measurement, partitive, partition

Possible Trouble Spots: does not relate the question "how many groups?" with measurement division; does not relate the question "how many in a group?" with partitive division.

Requisite Objectives:
Sets 3.2.1: Remove all disjoint equivalent sets from a given set
Division 3.2.10: Become aware of the division operation

Activity 1
To show measurement division, give the child 12 pennies (real money or play money, depending upon your financial condition!). Tell the child to exchange them for nickels. Ask, "How many nickels will you receive?" (Note that the child needs to know that 5 pennies are equal in number to a nickel.) The child should make two piles of 5 pennies each and exchange these piles for 2 nickels. He will have 2 pennies extra. (This is an example of measurement division: The child knew the size of each group [5 pennies], and he was asked to find "how many groups" [how many nickels he would receive].)

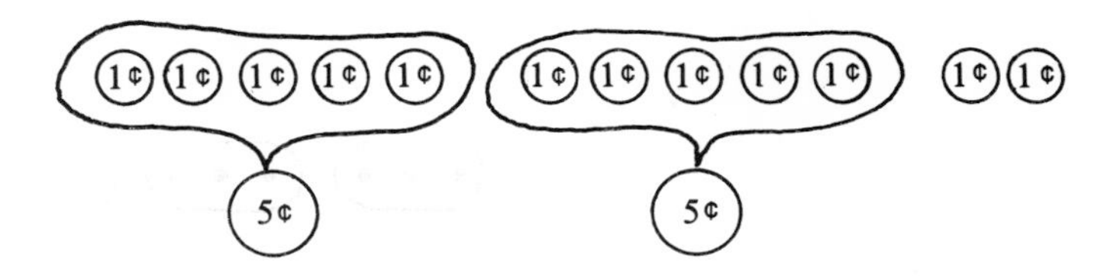

Activity 2

Counting steps backward on a number line

To show measurement division as the inverse of multiplication, show the child a number line and ask her how many 2-space steps must be taken to get from 6 to 0. The child should start at 6 and take steps of 2 spaces each to show subtraction and division. Point out to the child that the size of the whole (6) and the size of each group (2) are given and that the question asked is "how many groups?" Therefore, this problem involves measurement division and demonstrates that division can be shown by counting equal-sized steps backward on a number line (and as such is related to the idea of repeated subtraction). This is inverse to multiplication, which can be shown by counting equal-sized steps forward on a number line.

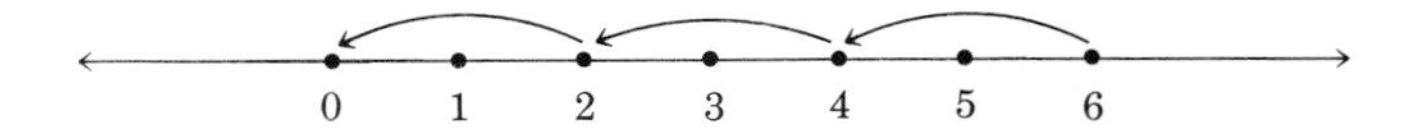

Activity 3
To show measurement division as repeated subtraction of equivalent disjoint sets as a model for division, have the child participate in the following activities.

A. Give the child a group of 21 objects and ask him to take out sets of 7 objects. Ask the child how many sets of 7 he could get out of 21 and if he has any objects remaining. The child should respond that he can make 3 sets with no objects remaining.

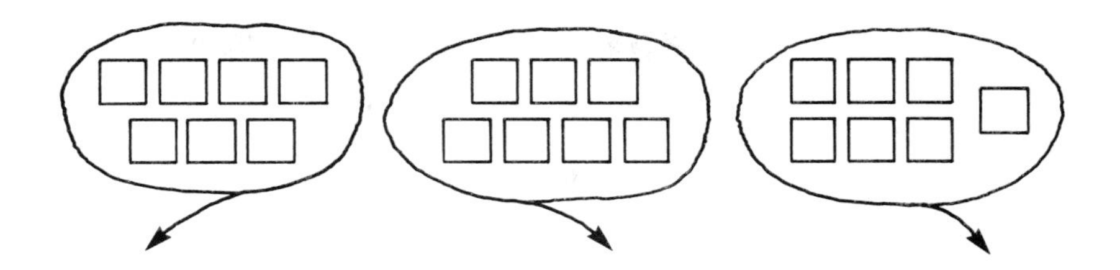

B. Show the child a card with 13 dots on it. Ask the child to circle sets of 3 dots. Then ask her to tell how many sets of 3 dots can be circled and if she has any extra. The child should respond that she can circle 4 sets of 3 dots and has 1 extra. She has a remainder of 1.

Activity 4
To show partitive division, have the child engage in the following activities.

A. Give the child 12 pennies and ask him to make two equal-sized groups. The child should distribute the pennies into two piles to find he has 6 in each group. (This is an example of partitive division: The problem specifies the whole and the number of groups, and the learner is asked to find "how many in each group.")

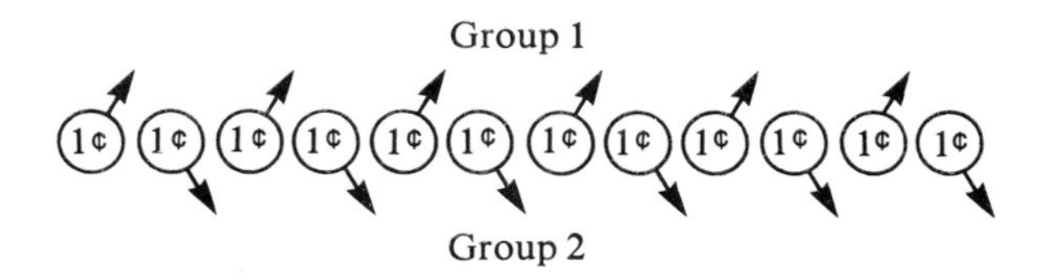

B. To show different groupings of a number, have the child show that 36 objects may be regrouped as two 18s, three 12s, four 9s, six 6s, nine 4s, twelve 3s, and eighteen 2s. Then have the child complete the following equations.

$2 \times 18 = \square$ $\quad 36 \div 2 = \square$ $\quad 4 \times 9 = \square$ $\quad 36 \div 9 = \square$
$18 \times 2 = \square$ $\quad 36 \div 18 = \square$ $\quad 9 \times 4 = \square$ $\quad 36 \div 4 = \square$

$3 \times 12 = \square$ $\quad 36 \div 3 = \square$ $\quad 6 \times 6 = \square$ $\quad 36 \div 6 = \square$
$12 \times 3 = \square$ $\quad 36 \div 12 = \square$

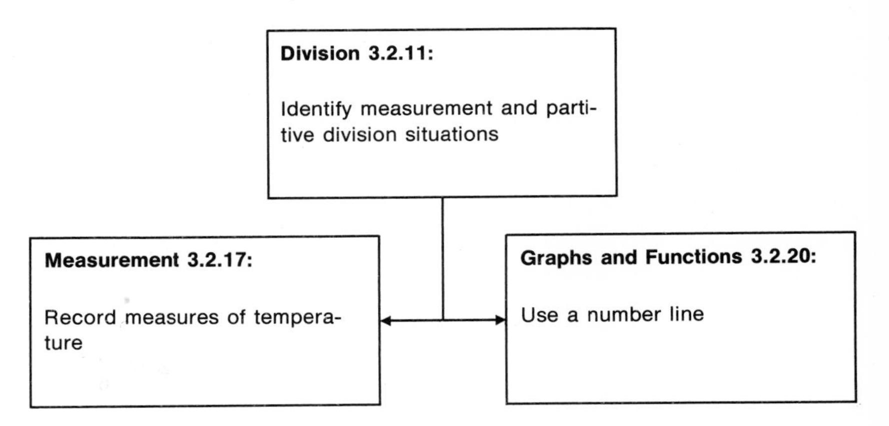

Remainders as related to multiples

The division operation is essentially a "taking-out-of" process or a simultaneous separation of a set into equivalent disjoint sets with or without a remainder. The existence of a remainder depends upon whether or not the dividend is a multiple of one of the equivalent disjoint sets. The term *multiple* is characteristic of the set of natural numbers.

The multiples of a natural number are defined as those natural numbers that result from multiplying the number by all possible natural numbers. The multiples of the natural number 2 include 2 (1×2), 4 (2×2), 6 (3×2), 8 (4×2), 10 (5×2), and so forth, and the multiples of the natural number 3 are 3, 6, 9, 12, 15, 18, 21, 24, 27, 30, 33, and so on.

As shown above, if a number (m) is a multiple of a natural number (n), then that number (m) is divisible by n. When m is divided by n, there is no remainder (or there is a remainder of zero). For example, let m equal a multiple of n: $m = 21$, $n = 7$. The statement "m is divisible by n" or "21 is divisible by 7" implies that there is no remainder. Some children find it easier to think of this case as having a remainder of zero.

TESTING PROPERTIES AS APPLIED TO DIVISION OF WHOLE NUMBERS

Testing for closure

Is division of whole numbers closed? No, there is no whole number that satisfies the following division examples:

$$3 \div 4 = \square$$
$$4 \div 3 = \square$$

Testing for commutativity

Does the commutative property apply to division of whole numbers? To answer this question, the teacher should allow the children to investigate examples like the following:

$$6 \div 3 \stackrel{?}{=} 3 \div 6$$
$$2 \neq \frac{3}{6}$$

Testing for associativity

Is division of whole numbers associative? The teacher should help the children explore examples like the following:

$$(12 \div 3) \div 2 \stackrel{?}{=} 12 \div (3 \div 2)$$
$$4 \div 2 \stackrel{?}{=} 12 \div \frac{3}{2}$$
$$2 \neq \frac{12}{\frac{3}{2}}\text{, or } 8$$

> *Note:* See Chapter 16 for the process of solving $12 \div \frac{3}{2}$.

MULTIPLICATION/DIVISION GRID

Testing for associativity

Is division of whole numbers associative? The teacher should help the children explore examples using the multiplication/division grid. In the grid below, find a product in a cell and the divisor along one axis ($8 \div 2$). The quotient is found by traversing the other axis as shown above.

Multiplication/Division Grid

×/÷	0	1	2	3	4	5	6	7	8	9
0	0	0	0	0	0	0	0	0	0	0
1	0	1	2	3	4	5	6	7	8	9
2	0	2	4	6	8	10	12	14	16	18
3	0	3	6	9	12	15	18	21	24	27
4	0	4	8	12	16	20	24	28	32	36
5	0	5	10	15	20	25	30	35	40	45
6	0	6	12	18	24	30	36	42	48	54
7	0	7	14	21	28	35	42	49	56	63
8	0	8	16	24	32	40	48	56	64	72
9	0	9	18	27	36	45	54	63	72	81

Testing for distributive property

Children may complete equations such as the following *to test for the Distributive Property of Division over Addition:*

(a)

$$\begin{aligned} 36 \div 4 &\stackrel{?}{=} (20 + 16) \div 4 \\ &\stackrel{?}{=} (20 \div 4) + (16 \div 4) \\ &\stackrel{?}{=} \quad 5 \quad + \quad 4 \\ 9 &= \quad\quad 9 \end{aligned}$$

(b)

$$\begin{aligned} 36 \div 4 &\stackrel{?}{=} 36 \div (2 + 2) \\ &\stackrel{?}{=} (36 \div 2) + (36 \div 2) \\ &\stackrel{?}{=} \quad 18 \quad + \quad 18 \\ 9 &\neq \quad\quad 36 \end{aligned}$$

Notice the distributive property worked only in example *a,* where the number distributed over the parts of the renamed number was placed on the right. For this reason, it is said that *only right distribution over addition holds.*

Testing for identity element

Is there an identity element for division? The teacher may provide examples like the following for investigation:

$$4 \div 1 = 4$$
$$93 \div 1 = 93$$
$$5 \div 5 = 1$$

The teacher should point out to the children that *there is a right identity only.* The multiplicative identity element (1) serves as an identity element for division when 1 is the divisor. Any number divided by itself is 1. However, for the example $1 \div 6 = \square$, 1 is not considered to be an identity element.

> *Note:* Notice the similarity between 1 as a right identity element for division of whole numbers and zero as a right identity element for subtraction of whole numbers. See Chapter 9 on subtraction of whole numbers.

THE ZERO IN DIVISION

An inverse model for division in relation to multiplication is presented below:

Multiplication

factor $a \times$ factor b = product
factor $b \times$ factor a = product

Division

product $\div$ factor a = factor b
product $\div$ factor b = factor a

Substituting whole numbers in the above models, the following conditions may occur.

a. $8 \times 2 = 16$ — $16 \div 8 = 2$; $16 \div 2 = 8$

b. $5 \times 1 = 5$ — $5 \div 5 = 1$; $5 \div 1 = 5$

c. $9 \times 0 = 0$ — $0 \div 9 = 0$; but $0 \div 0 \neq 9$

d. $0 \times 18 = 0$ — $0 \div 18 = 0$; but $0 \div 0 \neq 18$

$a/0$ and $0/0$ are not defined

There is no counting number that results when zero is divided by zero. In fact, there is no counting number that results when any number is divided by zero. In mathematical language, one states that *division by zero is not defined.* Therefore, zero divided by zero is also not defined. If $0/0 = n$, then $n \times 0 = 0$. But n could be any number; a unique number could not be obtained for n.

For the examples $8 \div 0$, $15 \div 0$, $25 \div 0$, no partitioning or measuring of the whole is apparent. If, for example, no cookies are pulled out of a group of 8 cookies, then the original set of 8 cookies is intact. For $8 \div 0 = \square$, no answer exists because there is no number $\square$ such that $0 \times \square = 8$.

For the examples $0 \div 8$, $0 \div 15$, $0 \div 25, \ldots$, if you start out with an empty set of cookies, how can you partition no cookies? You end up with zero cookies in each case. For the case $0 \div n = \square$, $\square = 0$.

EXERCISES

I. Indicate by marking yes or no which properties apply to the sets of numbers listed.

	Closure				Associativity				Commutativity				Identity element				Inverse element				Distributive property
	+	×	−	÷	+	×	−	÷	+	×	−	÷	+	×	−	÷	+	×	−	÷	
Natural numbers																					
Whole numbers																					
Integers																					

II. Write a division equation for the following illustration:

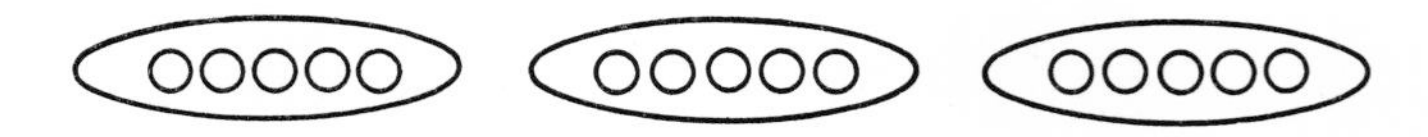

III. Classify the following situations as measurement or partitive division:

1. John has 36 objects to distribute equally into 3 boxes. How many objects must he put into each box?

2. Lisa had 100 pennies and wanted to give 25 cents to each of some neighborhood children. How many children could receive 25 cents?

IV. Show $4 \div 2$ on a number line.

V. In whole numbers, $2 \div 8 = n$ implies that n is ______.

VI. Write a division statement to show that 5 divided by 3 is 1, remainder 2.

VII. How are multiplication and division related?

VIII. How can division be checked by using subtraction?

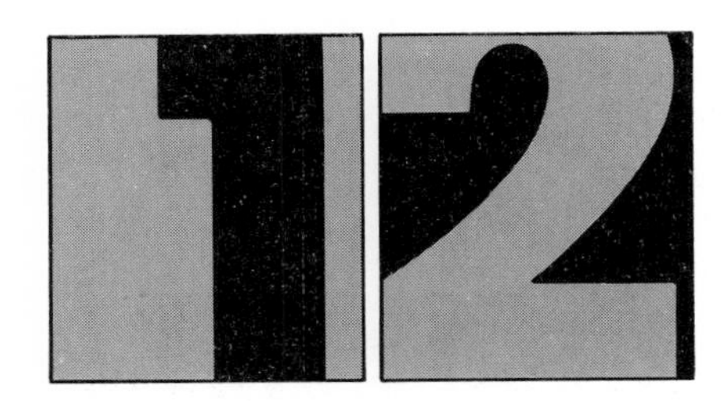

Division Algorithms

Situations involving division are expressed by a variety of words like *share, distribute, give, pull out, times, combine,* and *get from.* Children have expressed division ideas in the following ways:

> Tom (9 years old): Eight of us have seventy-two pennies to *share.* How much money did each of us get?
>
> Maribeth (13 years old): I *distributed* thirty-five lollipops equally, giving each seven. How many received lollipops?
>
> Loretta (7-1/2 years old): I *gave away* twelve pieces of candy to four friends. They each got the same amount. How many did I give to each friend?
>
> Eddie (7-1/2 years old): If I have a quarter, how many nickels can I *pull out?*
>
> Sammy (10 years old): Nine *times* what is seventy-two?
>
> Jennifer (6 years old): How many nickels do I *add together* to make a quarter?
>
> Billy (12 years old): How many fifteen-inch-long lengths can I *get from* a piece of wire sixty inches long?

Each of these comments may be expressed by a division algorithm.

Tom used repeated subtraction by ones. He gave a penny to each boy and one to himself in succession until he had distributed all of the pennies. Each child received 9 cents. He used the partitive division idea. This may be expressed as $72 \div 8 = \boxed{9}$.

Maribeth used repeated subtraction by groups of 7. Five children received lollipops.

Loretta and Eddie also used repeated subtraction by groups. Loretta used partitive division. Eddie used measurement.

Sammy made use of the inverse relation between division and multiplication.

$$9 \times \square = 72$$
$$72 \div 9 = \square$$

Jennifer used a repeated addition idea.

$$⑤ + ⑤ = 10$$
$$10 + ⑤ = 15$$
$$15 + ⑤ = 20$$
$$20 + ⑤ = 25$$

DIVISION ALGORITHMS—LONG OR SHORT FORM?

Both the long and short form algorithms for division are included in the elementary school curriculum. Thinking processes that underlie the long form parallel those in repeated subtraction. In using the long division algorithm, the child may check each step along the way, thus making it easier to identify where an error in computation may have occurred.

What prerequisites do children need in order to use the division algorithms? Children must be able to subtract with renaming; to multiply; to add, while using the basic addition, subtraction, and multiplication facts; and to estimate quotients.

Long form algorithms

Algorithm *a* relies on mental computation and correct placement of digits according to their place values.[1]

(a)

```
       228 rem. 29
   ________
59 | 13481
     118
     ---
      168
      118
      ---
       501
       472
       ---
        29
```

1. See Fredricka K. Reisman, *Diagnostic Teaching of Elementary School Mathematics: Methods and Content* (Chicago: Rand McNally College Publishing Co., 1977), p. 288, for other algorithm forms.

Algorithm *b* deals the least with estimation and bridges from the subtraction algorithm directly to the division algorithm, as shown below for 20 ÷ 4. The subtraction algorithm shows that five 4s are taken out of 20.

(b)

20		4	20	
− 4	✓		− 4	1
16			16	
− 4	✓		− 4	1
12			12	
− 4	✓		− 4	1
8			8	
− 4	✓		− 4	1
4			4	
− 4	✓		− 4	1
0			0	5

Procedure *c* points up the relation of division to repeated subtraction.

(c)

59	13481			13481
	−5900	100	(× 59)	−5900
	7581			7581
	−5900	100	(× 59)	−5900
	1681			1681
	− 590	10	(× 59)	− 590
	1091			1091
	− 59	1	(× 59)	− 59
	1032			1032
	− 118	2	(× 59)	− 118
	914			914
	− 590	10	(× 59)	− 590
	324			324
	− 118	2	(× 59)	− 118
	206			206
	− 118	2	(× 59)	− 118
	88			88
	− 59	1	(× 59)	− 59
	29	228		29

Repeated subtraction

Short form algorithms

The short form of the division algorithm may be used by the very able student. Most children should use the short division algorithm only with divisors no larger than 11. The short form has little meaning, as illustrated by the following problem:

$$\begin{array}{r} 1\ 2 \quad \text{rem. } 21 \\ 29\overline{)36^{7}9} \end{array}$$

When teaching the short form of the division algorithm, a single-digit divisor should be used first. Practice with divisors no larger than 9 is a helpful drill activity that aids the child in becoming comfortable with the short form algorithm and in learning basic division facts. (The short form algorithm is useful in finding prime numbers, as shown in Chapter 13.)

BASIC DIVISION FACTS

The basic division facts in the short form algorithm are shown in Chart 12.1.

Learning basic facts through practice

After children have been allowed to investigate the basic division facts at the concrete level, they should participate in a planned practice or a drill program. Memorization of the basic facts will facilitate

Chart 12.1. Basic Division Facts

$\overset{1}{1\overline{)1}}$	$\overset{2}{1\overline{)2}}$	$\overset{3}{1\overline{)3}}$	$\overset{4}{1\overline{)4}}$	$\overset{5}{1\overline{)5}}$	$\overset{6}{1\overline{)6}}$	$\overset{7}{1\overline{)7}}$	$\overset{8}{1\overline{)8}}$	$\overset{9}{1\overline{)9}}$
$\overset{1}{2\overline{)2}}$	$\overset{2}{2\overline{)4}}$	$\overset{3}{2\overline{)6}}$	$\overset{4}{2\overline{)8}}$	$\overset{5}{2\overline{)10}}$	$\overset{6}{2\overline{)12}}$	$\overset{7}{2\overline{)14}}$	$\overset{8}{2\overline{)16}}$	$\overset{9}{2\overline{)18}}$
$\overset{1}{3\overline{)3}}$	$\overset{2}{3\overline{)6}}$	$\overset{3}{3\overline{)9}}$	$\overset{4}{3\overline{)12}}$	$\overset{5}{3\overline{)15}}$	$\overset{6}{3\overline{)18}}$	$\overset{7}{3\overline{)21}}$	$\overset{8}{3\overline{)24}}$	$\overset{9}{3\overline{)27}}$
$\overset{1}{4\overline{)4}}$	$\overset{2}{4\overline{)8}}$	$\overset{3}{4\overline{)12}}$	$\overset{4}{4\overline{)16}}$	$\overset{5}{4\overline{)20}}$	$\overset{6}{4\overline{)24}}$	$\overset{7}{4\overline{)28}}$	$\overset{8}{4\overline{)32}}$	$\overset{9}{4\overline{)36}}$
$\overset{1}{5\overline{)5}}$	$\overset{2}{5\overline{)10}}$	$\overset{3}{5\overline{)15}}$	$\overset{4}{5\overline{)20}}$	$\overset{5}{5\overline{)25}}$	$\overset{6}{5\overline{)30}}$	$\overset{7}{5\overline{)35}}$	$\overset{8}{5\overline{)40}}$	$\overset{9}{5\overline{)45}}$
$\overset{1}{6\overline{)6}}$	$\overset{2}{6\overline{)12}}$	$\overset{3}{6\overline{)18}}$	$\overset{4}{6\overline{)24}}$	$\overset{5}{6\overline{)30}}$	$\overset{6}{6\overline{)36}}$	$\overset{7}{6\overline{)42}}$	$\overset{8}{6\overline{)48}}$	$\overset{9}{6\overline{)54}}$
$\overset{1}{7\overline{)7}}$	$\overset{2}{7\overline{)14}}$	$\overset{3}{7\overline{)21}}$	$\overset{4}{7\overline{)28}}$	$\overset{5}{7\overline{)35}}$	$\overset{6}{7\overline{)42}}$	$\overset{7}{7\overline{)49}}$	$\overset{8}{7\overline{)56}}$	$\overset{9}{7\overline{)63}}$
$\overset{1}{8\overline{)8}}$	$\overset{2}{8\overline{)16}}$	$\overset{3}{8\overline{)24}}$	$\overset{4}{8\overline{)32}}$	$\overset{5}{8\overline{)40}}$	$\overset{6}{8\overline{)48}}$	$\overset{7}{8\overline{)56}}$	$\overset{8}{8\overline{)64}}$	$\overset{9}{8\overline{)72}}$
$\overset{1}{9\overline{)9}}$	$\overset{2}{9\overline{)18}}$	$\overset{3}{9\overline{)27}}$	$\overset{4}{9\overline{)36}}$	$\overset{5}{9\overline{)45}}$	$\overset{6}{9\overline{)54}}$	$\overset{7}{9\overline{)63}}$	$\overset{8}{9\overline{)72}}$	$\overset{9}{9\overline{)81}}$

work with the division process. The teacher may test the children on selected sets of basic division facts in order to diagnose which facts they know and which they have yet to learn.

It should be remembered that children vary in their ability and motivation to memorize. Some learn the basic facts quickly and need no extensive drill. The majority can develop facility with the basic facts through practice. A few may never be able to remember all of the basic facts. For this last group in particular, hand calculators may prove to be a helpful aid.

Avoid senseless repetition of basic facts

It is again emphasized that a child first understands what he or she is doing. Activities presented in the chapters dealing with the nature of the operations and of practical applications in the physical world should precede work with division algorithms and drill activities on the basic facts.

Patience and time are needed by the teacher to ensure that these facts are not merely symbols or sounds to the child. Senseless repetition should be avoided; repetition should be natural and not done in a singsong voice. When a child repeats "thirty-five divided by seven is five," he or she should be told to picture the underlying meaning. Each fact should be known by itself: The child should not have to chant through an entire sequence of the basic fact table in order to state a particular fact. Drill must therefore be arranged so that each basic fact can be stated with speed and accuracy.

The following mini-lesson is appropriate for children who need to talk their way through learning experiences. These children include those who may have neurologically based learning disabilities or those who are slow learners. The activities can also be used by the entire class for learning the basic facts.

Mini-lesson 3.2.12 Compute Basic Division Facts

Vocabulary: share, distribute, pull out, divide

Possible Trouble Spots: has poor memory skills; needs concrete activities relating repeated take-away subtraction to division

Requisite Objectives:
Subtraction 1.3.10: Perform take-away subtraction with minuend to 18

Division 3.2.10: Become aware of the division operation

Auditory and speech-related activities

Activity 1
Ask the child to say the table of basic division facts in a left-to-right sequence, starting with "twenty-seven divided by three, nine; twenty-four divided by three, eight; twenty-one divided by three, seven; eighteen divided by three, six; fifteen divided by three, five."

Activity 2
Ask the child to state basic division facts with odd divisors only, such as "threes out of three, one; threes out of six, two; threes out of nine, three; threes out of twelve, four." Then ask the child to repeat this process with even divisors.

Activity 3
Have the boys say the basic division facts with odd divisors and the girls say the facts with even divisors. Then ask the boys to say the evens and the girls to say the odds.

Activity 4
Have the child say a basic division fact, such as "twenty-four divided by three equals eight." Ask another child to say the related fact, "Twenty-four divided by eight equals three."

Visual activities for practicing basic division facts

Activity 5
Have the children build their own charts or tables as basic division facts are learned. The entire class may compile one large chart and then place it so that it can be seen from all parts of the classroom. Topics for such charts may include "Division Facts with Dividends No Greater than 18" and "Which Division Facts in a Row Are Related?" (see Chart 12.2). For the latter topic, each cluster of problems may contain a fact and its twin ($2\overline{)12}$ and $6\overline{)12}$). Each cluster may include easy and difficult facts.

Chart 12.2. Which Division Facts in a Row Are Related?

Row 1	$2\overline{)12}$	$4\overline{)0}$	$3\overline{)21}$	$6\overline{)6}$	$6\overline{)12}$	$7\overline{)21}$	$1\overline{)6}$
Row 2	$4\overline{)8}$	$3\overline{)24}$	$5\overline{)0}$	$1\overline{)7}$	$2\overline{)8}$	$2\overline{)4}$	$7\overline{)7}$
Row 3	$3\overline{)12}$	$1\overline{)8}$	$2\overline{)18}$	$6\overline{)0}$	$3\overline{)9}$	$4\overline{)12}$	$8\overline{)8}$
Row 4	$2\overline{)0}$	$4\overline{)16}$	$7\overline{)0}$	$6\overline{)18}$	$1\overline{)9}$	$3\overline{)18}$	$3\overline{)15}$

Activity 6
Fasten on cards buttons, shells, beads, macaroni, or similar objects to show number patterns. Write corresponding number facts on the backs of the cards. These may be used for drill in multiplication as well as in division.

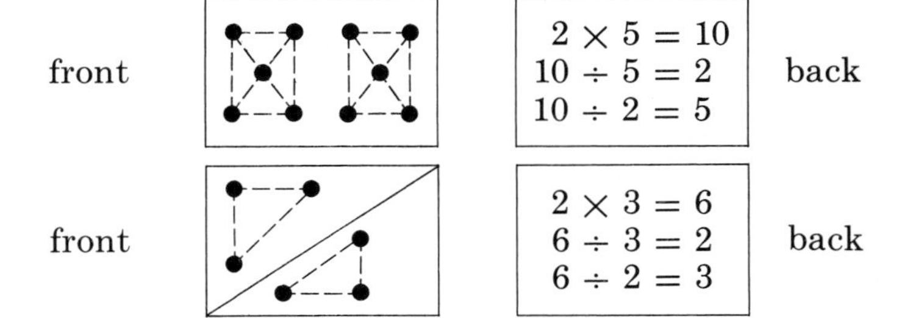

Activity 7
Give the child a multiplication division grid and ask him to circle those numbers that show patterns of basic division facts.

×/÷	2	4	6
3	6	12	18
4	8	16	24
5	(10)	(20)	(30)

Child selects:
$10 \div 5 = 2$
$20 \div 5 = 4$
$30 \div 5 = 6$

Activity 8
Have each of two children select a set of flash cards. The sets should be equal in number and have related facts. Child A has the question side up; child B has the answer side up. The cards are shuffled. A question is shown first. Then an answer card is placed on top. If a child recognizes a valid relationship between the problems on the cards, he should say "Snap!" to win all of the cards in the pile. If the child who says "Snap!" is wrong, the pile goes to the other child. The winner is the child who has the most flash cards. An example follows:

Child A's question cards	*Child B's answer cards*
$9 \div 3$	$6 \div 6 = 1$
$8 \div 2$	$12 \div 3 = 4$ "Snap!"

Since both $8 \div 2$ and $12 \div 3$ equal 4, child B, who recognizes the relationship, takes all.

The game continues:

$24 \div 8$	$16 \div 4 = 4$
$32 \div 4$ "Snap!"	$32 \div 8 = 4$

Child A recognizes the relationship among 32, 4, and 8 and takes all of the cards.

The play continues:

$12 \div 12$ $9 \div 9 = 1$ "Snap!"

Child B recognizes that both answers are 1, so she takes the cards.

Remember that the child who says "Snap!" must point out a valid relationship in order to win. Some relationships are more subtle than others.

Activity 9
Draw a clock face on the chalkboard or on cardboard. Choose a number for a dividend and write that number in the center of the clock face.

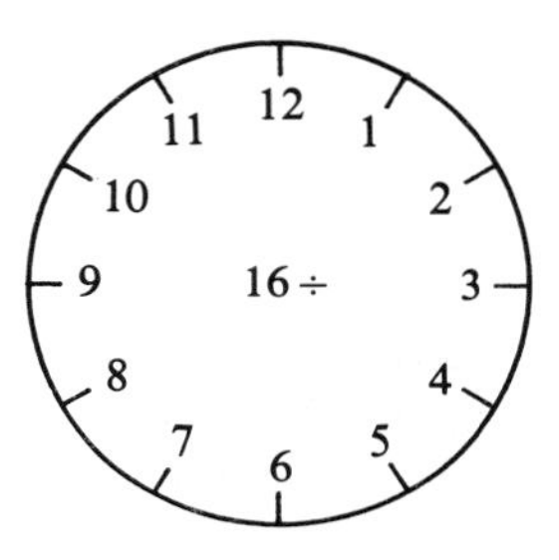

Point to a numeral on the clock face and ask the child to divide the dividend by that number. Sometimes place a zero as the dividend. (This activity involves not only basic division facts but also division with remainders.)

Activity 10
Give half of the class question cards in the form $\boxed{12 \div 4 = \square}$ and the other half of the class answer cards in the form $\boxed{3}$. Tell the children that those who correctly match a pair will each receive a new card. The three children who obtain the greatest number of new cards at the end of a specified time are the winners.

Activity 11

Relate basic facts to children's experiences

The teacher should take advantage of everyday situations that lend themselves to teaching the basic division facts. For example, questions such as the following give children the opportunity to learn the basic facts as well as to recognize when to divide and to understand why they should divide:

1. There are 21 days until the end of school. How many weeks is that?
2. We are inviting another class in to see a play. The principal is lending us some benches. There are 24 pupils in the other class and 1 bench seats 8 pupils. How many benches do we need to borrow?

Additional considerations: The repeated subtraction model for division can be shown on a hand calculator. This serves as an alternative model for comparing repeated subtraction with division. Estimation of the expected quotient should be encouraged.

Note: Mini-lesson 3.2.12 is a prerequisite for Mini-lessons 3.2.21 and 3.3.4.

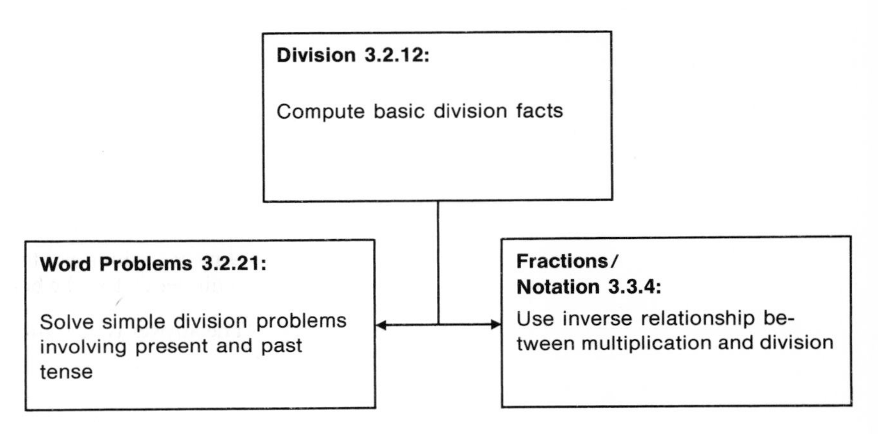

SUGGESTED INSTRUCTIONAL SEQUENCE FOR DIVISION ALGORITHMS

Following are examples of division problems arranged in a suggested sequence for instruction. This sequence commences with division examples containing one basic division fact with a remainder. Transcriptions of taped interviews accompany many of the examples to show the various language patterns used by the children as they were computing. The sequence of computations within each example is shown by the varying widths of the numerals.

Contains basic division fact

	tens	units
		8
6	5	2
	4	8
		4

Nancy (8 years old): "I know six eights are forty-eight. Forty-eight from fifty-two leaves four extra. It's eight, remainder four."

Remainders in the tens and the units places

	tens	units
	3	9
2	7	9
	6	0
	1	9
	1	8
		1

(Graph paper was used and "tens" and "units" were marked by the teacher.)

Jon (8½ years old): "First I'll divide the tens place. I can get three twos out of seven." (Jon writes 3 above the 7 in the dividend.)

"Three twos are six." (Jon writes 6 under the 7 in the dividend.)

"But that's really six tens." (Jon writes 0 under the 9 to make 60.)

"Sixty from seventy-nine, hmmm, nine minus zero, nine; six from seven, one. I can get nine twos out of nineteen." (Jon writes 18 under the 19 and 9 in the units place of the quotient.)

"Nineteen minus eighteen, one." (Jon writes remainder of 1 at the bottom of the algorithm.)

"The answer is thirty-nine, remainder one."

Remainder in the tens place

	tens	units
	1	4
3	4	2
	3	0
	1	2
	1	2

Amy (8½ years old): "I can take one three out of four. It's really four tens, so I write the one in the tens place." (Amy writes 1 in the tens place of the quotient.)

"Thirty from forty, ten. Two minus zero, two." (Amy writes 12 as the new dividend.)

"Four threes are twelve." (She writes 4 in the units place of the quotient and 12 under the 12 below.)

"There are fourteen threes in forty-two."

Remainder in the units place

	tens	units
	2	2
3	6	7
	6	
		7
		6
		1

Dan (9 years old): "There are two threes in six." (Dan writes the 2 in the tens place of the quotient and 6 below.)

"It was really sixty, so sixty-seven minus sixty is seven." (He writes 7 below.)

"There are two threes in seven. Two threes are six." (Dan writes 6 below the 7 and subtracts.)

"The answer is twenty-two, remainder one."

No remainders

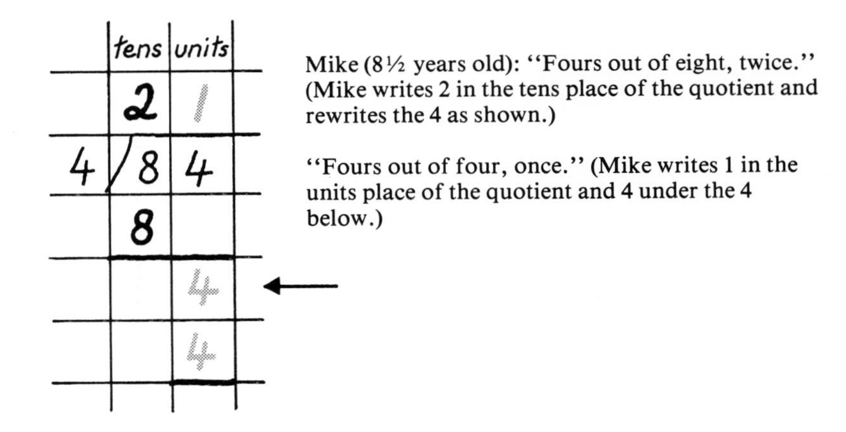

Mike (8½ years old): "Fours out of eight, twice." (Mike writes 2 in the tens place of the quotient and rewrites the 4 as shown.)

"Fours out of four, once." (Mike writes 1 in the units place of the quotient and 4 under the 4 below.)

The teacher asked Mike why he rewrote the 4 next to the arrow above. He responded, "I'm really subtracting eighty from eighty-four."

The algorithm Mike used may seem easier than the previous examples since there are no remainders. However, if children begin with this type of problem, they often have difficulty transferring to such problems as 3913 ÷ 13, which would be computed by using the long form. Also, many children would not think about place-value concepts if the learning sequence commenced with the algorithm that Mike used. The sequence presented here gives the children experience with the long division form, which involves computing the tens first and then the units. Also, this sequence postpones examples with zeros in the quotient, such as 81 ÷ 4 or 417 ÷ 4.

The following example is a division problem with zero in the units place of the quotient and a remainder.

Zero in the units place of quotient; remainder

	tens	units
	2	0
4	8	3
	8	
		3
		0
		3

"The answer is twenty and there are three extra."

The following problem involves remainders in the hundreds, tens, and units places.

Remainders in the hundreds, the tens, and the units places

	h	tens	units
	2	3	8
3)	7	1	6
	6		
	1	1	
		9	
		2	6
		2	4
			2

The following problems involve remainders in two places only.

Remainders in two places only

	h	tens	units
	2	5	4
3)	7	6	2
	6		
	1	6	
	1	5	
		1	2
		1	2

	h	tens	units
	2	4	2
3)	7	2	8
	6		
	1	2	
	1	2	
			8
			6
			2

	h	tens	units
	1	2	5
3)	3	7	6
	3		
		7	
		6	
		1	6
		1	5
			1

Next is presented a problem with zero in the quotient and zero in the tens place.

Zero in the quotient and in the tens place

	h	tens	units
	1	0	4
4)	4	1	7
	4		
		1	7
		1	6
			1

The following example has zero in the hundreds place of the quotient.

Zero in the hundreds place of the quotient

	h	tens	units
	0	7	3
5)	3	6	8
	3	5	
		1	8
		1	5
			3

Note that placing 0 in the hundreds column ensures that 7 goes in the right place. When children become comfortable with this type of problem, they may omit the zero in the hundreds place. The answer is then written "73 rem. 3" instead of "073 rem. 3." However, 073 = 73.

The next example involves zero in the tens place of the dividend.

Zero in the tens place of the dividend

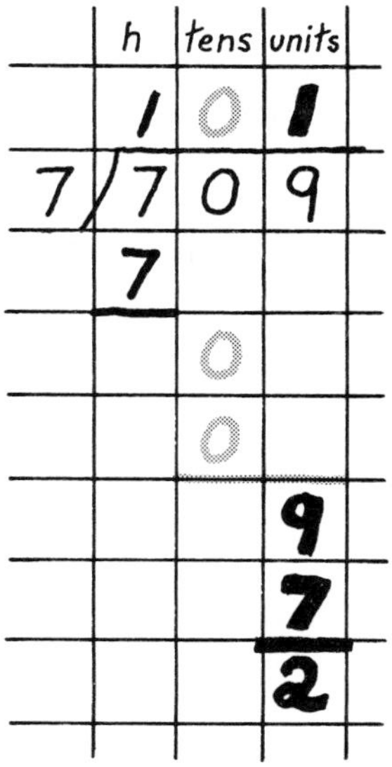

h	tens	units
1	0	1
7) 7	0	9
7		
	0	
	0	
		9
		7
		2

USING THE PLACE-VALUE SLIDE

A device called a "place-value slide" is helpful for activities involving dividing by 10 and multiplying by 10. This device is also useful in studying decimals. It is made from four pieces of thick cardboard. The following dimensions are recommended: 24-by-6-inch base, 24-by-2-inch place-value name strip, 24-by-3-inch slide, and 24-by-1-inch bottom strip.

The place-value strip and the bottom strip are glued to the base, leaving a groove through which the slide may be moved. The top of the strip should be marked "thousands," "hundreds," "tens," and "units."

To find the answer to the problem 329 ÷ 10, the child writes the digits under the appropriate place values and then moves the slide one section to the right. The 3 appears in the tens column and the 2 in the units column, and the 9 is extra. The answer, then, is shown to be 32 and 9 extra. This may easily be extended to division by 100 by moving the slide two places to the right. Extension to decimals may occur as the quotient is written "32.9."

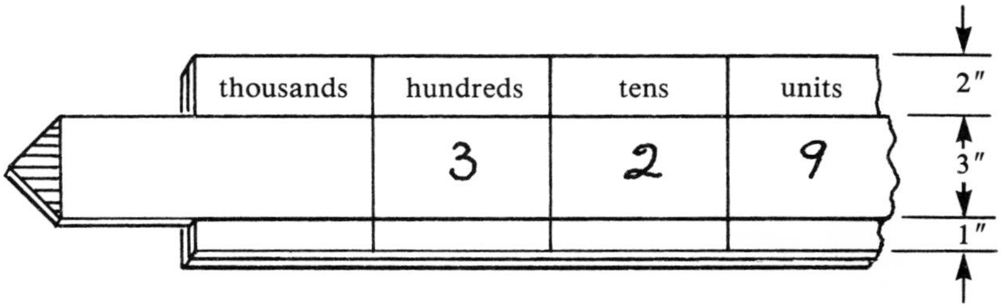

thousands	hundreds	tens	units
	3	2	9

2″ 3″ 1″

NOTICING PATTERNS IN MULTIPLYING BY 10 AND POWERS OF 10

It is also helpful for the learner to have facility in multiplying by 10 and powers of 10. Analyzing patterns such as the following grow out of activities that involve generating place values.

Multiplicative identity

$$1 \times 10 = 10 \qquad 10 \times 1 = 10$$
$$1 \times 100 = 100 \qquad 100 \times 1 = 100$$
$$1 \times 1000 = 1000 \qquad 1000 \times 1 = 1000$$

A teacher may ask, "What do you notice about the number of zeros in two factors for a product and the number of zeros in that product?" The teacher may then help the child analyze the following pattern:

$$10 \times 10 = 100$$
$$10 \times 100 = 1000$$
$$10 \times 1000 = 10000$$
$$100 \times 100 = 10000$$
$$1000 \times 100 = 100000$$

The next step is to look for patterns involving digits greater than 1.

$30 \times 10 = 300$	2 zeros in factors, 2 zeros in product $3 \times 1 = 3$ (product of nonzero factors)
$300 \times 10 = 3000$	3 zeros in factors, 3 zeros in product
$300 \times 1000 = 300000$	5 zeros in factors, 5 zeros in product
$30 \times 200 = 6000$	3 zeros in factors, 3 zeros in product $3 \times 2 = 6$ (product of nonzero factors)

The generalization regarding the number of zeros in products and factors becomes a necessary prerequisite in division. For division, the teacher should question the children to help them notice the reverse pattern: "What do you notice about the number of zeros in the dividend and divisor? How does this relate to the number of zeros in the quotient?"

$60\emptyset\emptyset \div 2\emptyset\emptyset = 30$	The number of zeros in the divisor (2) is subtracted from the number of zeros in the dividend (3), leaving 1 zero in the quotient of nonzero numbers. $6 \div 2 = 3$ (quotient of nonzero numbers)

$6000\!\!\!/0\!\!\!/ \div 30\!\!\!/0\!\!\!/ = 200$

The number of zeros in the divisor (2) is subtracted from the number of zeros in the dividend (4), leaving 2 zeros in the quotient of nonzero numbers.

$6 \div 3 = 2$ (quotient of nonzero numbers)

ESTIMATION AND THE DIVISION ALGORITHM

At this point, the children should feel comfortable using the long division algorithm and should be able to describe selected computational steps used in the long division algorithm. However, since forcing a child to describe step-by-step procedures while he or she is computing hinders learning for some children, it is *not* suggested here that children must "talk their way through" the algorithm. Rather, it is suggested that the teacher observe the child's computations and, after the child completes a particular problem, ask such diagnostic questions as "Why did you rewrite this?" or "How did you get this?"

Now the introduction of divisors greater than 12 is suggested. This is an important step, because for these problems, a knowledge of the basic division facts is not enough. It is at this step that estimation comes into play.

Prerequisites for division using estimation

A helpful prerequisite activity for estimation is working with two-digit dividends that are grouped by decades, such as 20, 21, 22, . . . ; 30, 31, 32, . . . ; 90, 91, 92, ... Also helpful is a review of counting by 10s and its use in dividing by 10. Then, the child may be asked to count by 20s—"twenty, forty, sixty, eighty, ..." The children may then be guided to use this counting to divide by 20. The subtractive division algorithm lends itself to this approach, as shown below:

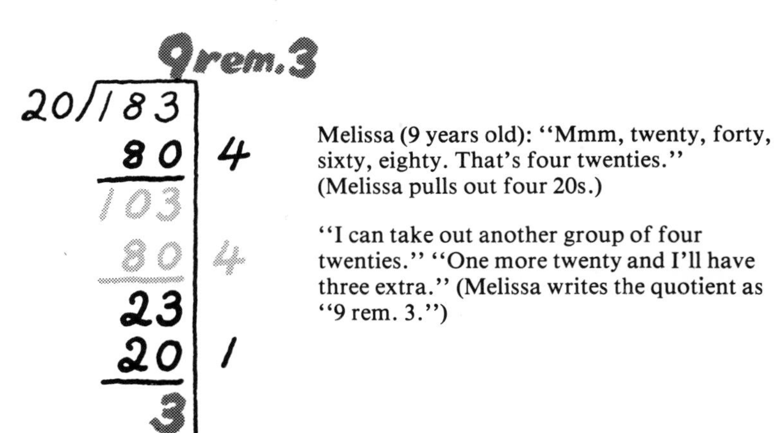

Melissa (9 years old): "Mmm, twenty, forty, sixty, eighty. That's four twenties." (Melissa pulls out four 20s.)

"I can take out another group of four twenties." "One more twenty and I'll have three extra." (Melissa writes the quotient as "9 rem. 3.")

Charlie approached this same problem in a different manner, as shown below in the long division algorithm. He counted by 20s, keeping a tally on his fingers.

20 40 60 80 100 120 140 160 180

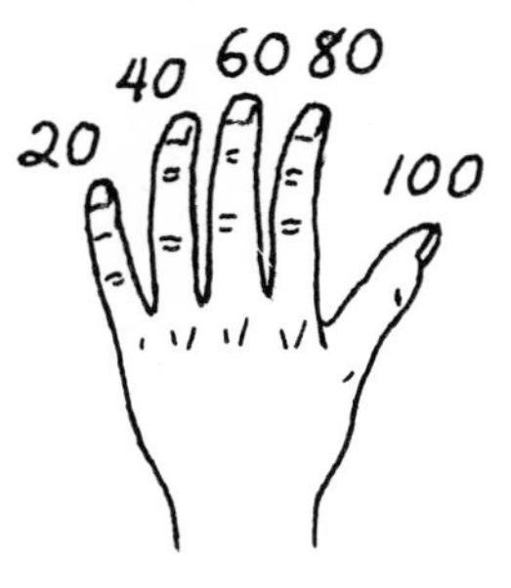

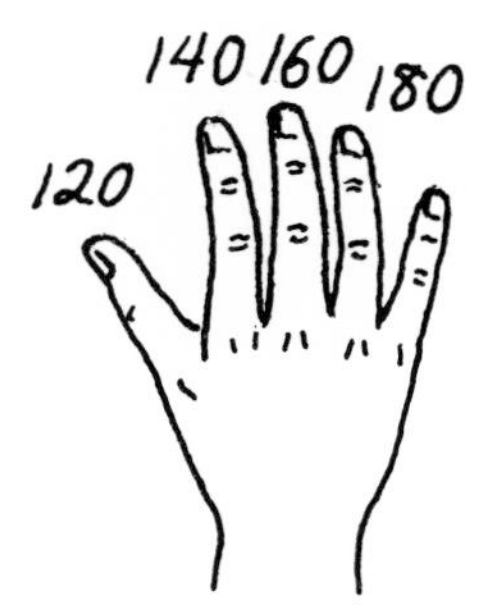

	h	tens	units
			9
20)	1	8	3
	1	8	0
			3

Charlie (10 years old): "Twenty, forty, sixty, eighty, one hundred, one hundred twenty, one hundred forty, one hundred sixty, one hundred eighty. That's all the twenties I can get out. That's nine twenties and three extra."

This approach may then be extended to counting by 30s and 40s.

Division using estimation

Division using estimation is the next step. The same example will be used, but this time the children are guided to use the 2 in 20 to see how many times 20 can be pulled out of 183.

> *Note:* Division is a separating operation, so the teacher should ask, "How many fours can you pull out of twenty?" instead of "How many fours go into twenty?" The latter is referred to as the "gazinta syndrome."

The teacher should point out to the child that the 2 is 2 tens and the 18 is 18 tens. Then, the teacher may ask, "How many two tens can you get out of eighteen tens?" Finally, the child may be asked, "How many twos can you get out of eighteen?" The teacher may find it helpful to use colored chalk for the numbers that are wider in the algorithm below.

	h	tens	units
			9
20)	1	8	3
	1	8	0
			3

For the example 95 ÷ 30, the teacher may ask the child to count by 30s until reaching a multiple of 30 closest to but less than the dividend (95).

	h	tens
		3
30)	9	5
	9	0
		5

"Thirty, sixty, ninety."

Therefore, 30 will come out of 95 three times, and there will be 5 extra. Estimation occurs in thinking, "Thirty will come out of ninety-five about three times."

The teacher should present to the child an example in which the first estimate, although logical, does not work. For the example 81 ÷ 21, to estimate that the quotient is 4, since 2s come out of 8 four times, yields a quotient that is too large.

$$\begin{array}{r} 4 \\ 21\overline{)81} \\ \underline{84} \end{array} \quad \leftarrow \text{This is too large, so try this} \rightarrow \quad \begin{array}{r} 3 \\ 21\overline{)81} \\ \underline{63} \\ 18 \end{array}$$

Children may now work with problems with two-digit divisors by "rounding up" the divisor to the next multiple of 10 or "rounding down" the divisor to the previous multiple of 10. A vertical number line is a helpful teaching aid for making the estimation. A portion of a number line that applies to a particular problem may be used, as shown below:

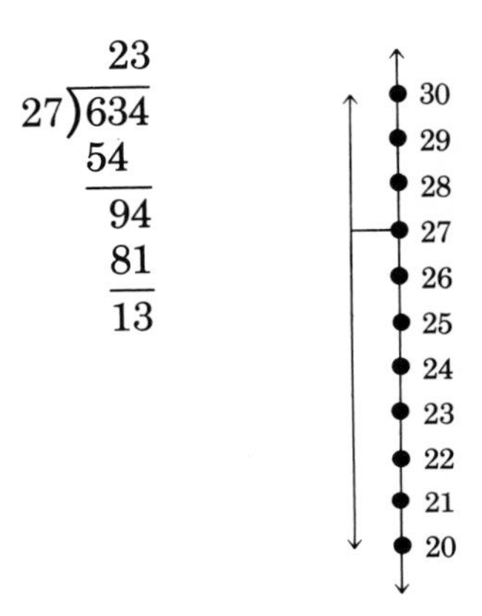

Using the number line, the child sees that 27 is closer to 30 than to 20, so he or she would estimate how many 30s can come out of 60. Then, the child would estimate how many 30s can come out of 90.

The same procedure may be used for the example below, but this time the estimate yields a quotient that is too small.

$$\begin{array}{r} 6 \\ 19\overline{)137} \\ \underline{114} \\ 23 \end{array}$$

Nineteen is closer to 20 than to 10, so the child would estimate how many 20s can come out of 130. The child tries 6 but finds that the remainder is too big. Since another divisor can be pulled out, a correction is necessary. The answer is 7, remainder 4.

$$\begin{array}{r} 7 \\ 19\overline{)137} \\ \underline{133} \\ 4 \end{array}$$

After many experiences with the subtractive division algorithm, the teacher may wish to guide the learner to estimating the largest multiples of the divisor that may be subtracted from the dividend. There are essentially three procedures, as shown for the example $989 \div 43 = 23$. In the round-down method, the child looks to the largest place-value digit in the divisor as a guide (4 in the numeral 43). Forty-three is said to be rounded down to 40, and the problem becomes $40 \times \square = 989$. The child estimates that there are about twenty 43s in 989. (This procedure often produces an estimate that is too large.)

$$\begin{array}{r l}
\left.\begin{array}{r} 3 \\ 20 \end{array}\right\} 23 & \\
43\overline{)989} & \text{"There are about three forties} \\
\underline{860} & \text{in one hundred twenty-nine."} \\
129 & \\
\underline{129} &
\end{array}$$

The round-up method increases the largest place-value digit in the divisor by 1. The problem, then, becomes $50 \times \square = 989$. The child estimates that there are about eighteen 50s in 989.

$$\begin{array}{r l}
18 & \\
50\overline{)989} & \\
\underline{900} & \text{"Too small."} \\
89 &
\end{array}$$

This method often produces an estimate that is too small, so it works well with the subtractive division algorithm.

$$\begin{array}{r r}
43\overline{)989} & \\
\underline{-430} & 10 \\
559 & \\
\underline{-344} & 8 \\
215 & \\
\underline{-172} & 4 \\
43 & \\
\underline{-43} & \underline{1} \\
 & 23
\end{array}$$

The "two-rule process" combines the round-down and round-up methods. The divisor is rounded to the nearest multiple of 10; for example, 43 is rounded to 40, and 48 is rounded to 50. If the divisor ends in 5, either multiple of 10 may be chosen.

COMMON ERRORS MADE WHEN USING THE LONG DIVISION ALGORITHM

The following analysis of errors children make when using the long division algorithm is presented as a diagnostic inventory. Since the division algorithm employs all four operations, this analysis may be used as a model for creating your own diagnostic inventories for addition, subtraction, multiplication, and division algorithms.

1. Not using all digits in the dividend
2. Error in basic division fact
3. Error in basic multiplication fact
4. Rewriting ("bringing down") the wrong number of digits (usually too many digits)

> *Note:* The term *bring down* to express rewriting is confusing. How can you move down chalk configurations on a chalkboard? This saying probably has its roots in the operations performed on counting boards, on which the apices were indeed "brought down." The teacher might instead say, "Rewrite."

5. Quotient figure too small, so the remainder is greater than the divisor
6. Quotient figure reflecting error in place value
7. Quotient figure ignores placement of zero to show a place value
8. Failure to subtract to obtain remainder
9. Expressing remainder as "rem. 2/3."[2]
10. Error in subtraction[3]

---

2. The division algorithm is defined as follows: For any given integers a and b, with $b > 0$, there exist unique integers q and r such that $a = q \cdot b + r$ and $0 \leq r < b$. The phrase $0 \leq r < b$ means that the remainder (r) may be zero (which is referred to as even division), or it may be greater than zero but it must be less than the divisor (b). The division algorithm yields an integral remainder that may be either greater than or equal to zero. Therefore, when using the notation "quotient, remainder number," where the quotient (q) equals 3 and the remainder is 4, as in the example $79 \div 25$, the child should understand that this means 75 may be separated into 3 groups with 25 in each group, and there will be 4 extra. However, to express the quotient as "3 rem. 4/25" is incorrect. The remainder is an integer and may be expressed as a fraction only in the form 4/1. The notation "3 4/25" is correct; it is a complete quotient obtained by dividing 79 by 25 since $3\ 4/24 \times 25 = 79$. This latter notation, however, has to do with the set of rationals (discussed in Chapter 14).

3. See Reisman, *Diagnostic Teaching,* pp. 308–309, for an analysis of subtraction errors made by children (aged 8 to 10) during clinical interviews with the author.

CHECKING DIVISION

Using the multiplication/division relation

A popular method of checking the accuracy of the quotient is to use the multiplication/division relation. Since product ÷ factor = factor, then factor × factor = product.

For example:

```
    3 rem. 4    Check
25)79             25
   75           ×  3   Quotient
   --           ----
    4             75
                +  4   Remainder
                ----
                  79
```

This may be expressed as $(3 \times 25) + 4$, showing the inverse relation of division to multiplication.

Casting out 9s

For the more able mathematics pupil, the teacher may introduce the casting out 9s method. Casting out 9s in division is similar to casting out 9s in multiplication so long as the quotient in division has a remainder of zero.

```
     2    Sum of the digits    Excess of 9s
19)38       19→1 + 9 = 10           1
   38     × 2 ──────────→2         ×2
   --   ────────────────────────────────
            38→3 + 8 = 11           2
```

If the dividend is not a multiple of the divisor, the remainder is subtracted from the dividend before the multiplication is done.

```
    20 rem. 17
19)397
   38        397
   --      -  17
   17      -----
             380
```

The remainder (17) is subtracted from the dividend (397) to obtain a multiple of the divisor.

```
Excess of 9s
  19  ──→   1
 ×20  ──→  ×2
 ---       --
 380        2
```

The procedure is the same as in checking multiplication.

EXERCISES

I. Describe the errors in the following problems:

1. 27 ÷ 3 = 7

2.
```
   60
7)430|
  420|
  ---
```

```
3.      70 rem. 13
     7)503|
       490|
       ---
        13|

4.      40 rem. 1
     7)321|
       280|   40
       ---
         1|

5.      56 rem. 1
     7)413|
       350|   50
       ---
        43|
        42|    6
       ---
         1|

6.       32
     4)1208|
       1200|   30
       ----
          8|    2
```

II. Find the quotient for $168 \div 8$ by using the right distributive property for division.

III. The fact that $5 \div 3$ is undefined in the set of whole numbers shows that division of whole numbers is not ________.

IV. For all whole numbers n such that $n \neq 0$, $n = n \times 1$ and $n = 1 \times n$. Thus, $n \div 1 =$ ________ and $n \div n =$ ________ by definition.

V. If $m = 0$ and $n \neq 0$, then m $(= / \neq)$ $m \cdot n$.

VI. Since $0 \div 0$ is not unique, division of zero by zero (is/is not) defined for the set of whole numbers.

SUGGESTED READINGS

Denmark, Tom, and Kepner, Henry S., Jr. "Basic Skills in Mathematics: A Survey." *Journal for Research in Mathematics Education.* 11 (March 1980): 104–123.

Swart, W. L. "Teaching the Division-By-Subtraction Process." *The Arithmetic Teacher* 19 (January 1972): 71–75.

Tucker, B. F. "The Division Algorithm." *The Arithmetic Teacher* 20 (December 1973): 639–46.

Zweng, M. J. "The Fourth Operation Is Not Fundamental." *The Arithmetic Teacher* 19 (December 1972): 623–27.

Prime and Composite Whole Numbers

Composite number

A *composite number* is a whole number a that can be expressed as a product $b \cdot c$, where both b and c are whole numbers other than 1 or zero. The integers b and c are factors of a. Numbers like 18, 50, and 121 are composite numbers.

Prime number

A *prime number* is a whole number greater than 1 that has no factors other than itself and 1—for example, 2, 3, 5, 7, 11.

Procedures for prime factorization

A composite number can be factored into its primes. The prime numbers of the composite numbers 18, 50, and 121 are shown below as the bottom row of the "factor trees."

18	50	121
2×9	2×25	11×11
$2 \times 3 \times 3$	$2 \times 5 \times 5$	(11^2)
$(2 \cdot 3^2)$	$(2 \cdot 5^2)$	

Another process for finding primes involves the short form of the division algorithm, as shown below:

$$\begin{array}{r} 3 \\ 3\overline{)9} \\ 2\overline{)18} \end{array} \qquad \begin{array}{r} 5 \\ 5\overline{)25} \\ 2\overline{)50} \end{array} \qquad \begin{array}{r} 11 \\ 11\overline{)121} \end{array}$$

This procedure involves using the smallest prime as a divisor. The individual continues dividing by that prime until the dividend is no longer its multiple. Then, the new dividend is divided by the next largest prime that is a factor of that dividend. This procedure is continued until the dividend itself is a prime number.

USING THE SIEVE OF ERATOSTHENES

Eratosthenes (c. 270–190 B.C.) was a mathematician of Alexandria whose interests included the study of number theory. He devised a number pattern that is a helpful aid in identifying those numbers that are prime. He wrote the sequence of counting numbers on parchment and then punched out the composite numbers; the remaining numbers, then, are prime numbers. The parchment with holes resembled a sieve; thus, the term *Sieve of Eratosthenes.*

Mini-lesson 5.1.4 Identify Prime Numbers

Vocabulary: prime, composite, factor tree, greatest common factor (GCF), least common multiple (LCM)

Possible Trouble Spots: poor memory skills; difficulty making judgments; inattentive to salient aspects of tasks; poor problem-solving strategies.

Requisite Objectives:
Multiplication 3.1.9: Compute basic multiplication facts

Division 4.1.10: Divide any whole number by a 1-digit number

Whole Numbers/Notation 5.1.3: Compare prime and composite numbers

Activity 1
Give the child a ditto sheet showing the Sieve of Eratosthenes with the numbers 1 to 100. Discuss the fact that the number 1 is neither prime nor composite but has the special job of being the multiplicative identity. Have the child start with the first prime, 2, and mark out all multiples of 2 with an X; all multiples of the next prime, 3, with a right-to-left diagonal line; then multiples of 5 with a right-to-left diagonal line; and finally multiples of 7 with a dash line. It is not necessary to continue, for it is obvious that all of the composites have already been eliminated, since they are multiples of smaller primes.

Activity 2
Ask the pupils to look for patterns in the sequence of prime numbers identified in the previous activity. The class may place on the bulletin

1	2	3	~~4~~	5	~~6~~	7	~~8~~	~~9~~	~~10~~
11	~~12~~	13	~~14~~	~~15~~	~~16~~	17	~~18~~	19	~~20~~
~~21~~	~~22~~	23	~~24~~	~~25~~	~~26~~	~~27~~	~~28~~	29	~~30~~
31	~~32~~	~~33~~	~~34~~	~~35~~	~~36~~	37	~~38~~	~~39~~	~~40~~
41	~~42~~	43	~~44~~	~~45~~	~~46~~	47	~~48~~	~~49~~	~~50~~
~~51~~	~~52~~	53	~~54~~	~~55~~	~~56~~	~~57~~	~~58~~	59	~~60~~
61	~~62~~	~~63~~	~~64~~	~~65~~	~~66~~	67	~~68~~	~~69~~	~~70~~
71	~~72~~	73	~~74~~	~~75~~	~~76~~	~~77~~	~~78~~	79	~~80~~
~~81~~	~~82~~	83	~~84~~	~~85~~	~~86~~	~~87~~	~~88~~	89	~~90~~
~~91~~	~~92~~	~~93~~	~~94~~	~~95~~	~~96~~	97	~~98~~	~~99~~	~~100~~

board "Tommy's pattern," "Jane's pattern," and so on. Then scan the following list and ask the children if some of these more complicated generalizations regarding prime numbers are true.

1. All primes except 2 are odd numbers (see Chapter 3 for a discussion of odd and even numbers). Are all odd numbers prime?
2. All prime numbers except 2 and 5 terminate with 1, 3, 7, or 9; all other numbers are composite.
3. Every prime number except 2, if either increased or lessened by 1, is divisible by 4. (Every prime except 2 can be expressed as $4n \pm 1$.)
4. Every prime except 2 and 3, if either increased or lessened by 1, is divisible by 6. (Every prime except 2 and 3 can be expressed as $6n \pm 1$.)

Additional considerations: The analysis of a composite number to identify its prime factors is a good activity for developing planned problem solving. It also provides practice on basic multiplication and division facts. For example, to find the prime factors of 36, the students test their knowledge of dividing, of multiplying, of trying various divisibility rules, of analyzing, and of synthesizing as they multiply to check their factorization. This is shown below:

a. $36 = \{1, 2, 3, 4, 6, 9, 12, 18, 36\}$
b. $36 = 1 \times 36,\ 2 \times 18,\ 3 \times 12,\ 4 \times 9,\ 6 \times 6$
c. $36 = 2 \times 2 \times 3 \times 3$

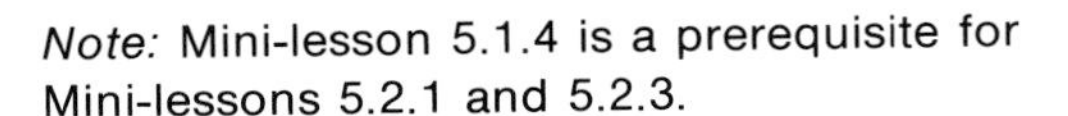

Note: Mini-lesson 5.1.4 is a prerequisite for Mini-lessons 5.2.1 and 5.2.3.

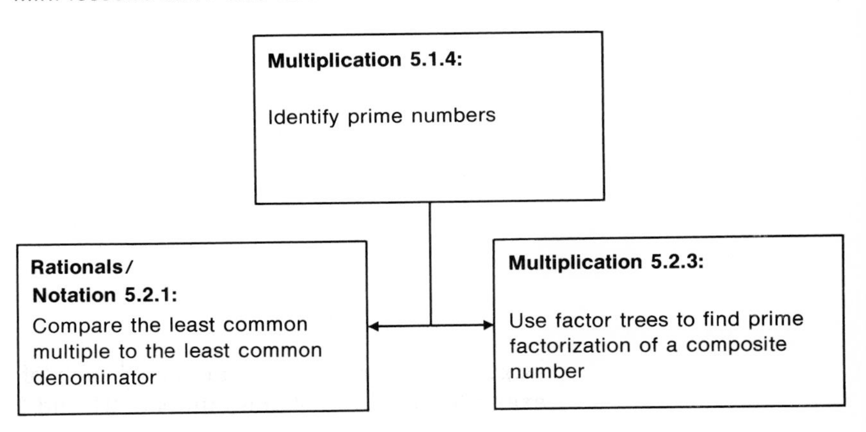

DIVISIBILITY RULES

A helpful prerequisite for determining if a prime number is a factor of another number is a knowledge of divisibility rules. The rules that follow may be presented to children by using a guided discovery method or by presenting the generalization and asking them to verify the statement.

1. *Divisibility by 2.* A number is divisible by 2 if the units-place number is divisible by 2.

> *Note:* Two statements often used are "is a factor of" and "is divisible by." Zero can be a factor of a number (including itself), but no number is divisible by zero. This appears to be the only difference between the meaning of these two statements.

2. *Divisibility by 3.* A number is divisible by 3 if the sum of its digits is a number divisible by 3.
3. *Divisibility by 5.* An integer is divisible by 5 if its units digits is zero or 5.

APPLICATIONS OF THE FUNDAMENTAL THEOREM OF ARITHMETIC

Any whole number greater than 1 may be written as the product of primes in one and only one way if the order of the factors is disregarded. This is called the Fundamental Theorem of Arithmetic. The prime factorization for a number is the same regardless of the factoring sequence. Following are applications of the Fundamental Theorem of Arithmetic to finding the greatest common factor and the least common multiple of two or more numbers.

Greatest Common Factor

The *greatest common factor* (GCF) of two or more numbers is the largest number that is a factor of each of the given numbers. For example, the GCF of the numbers 24 and 36 is 12, as shown below:

$$\begin{array}{r} 3 \\ 2\overline{)6} \\ 2\overline{)12} \\ 2\overline{)24} \end{array} \qquad \begin{array}{r} 3 \\ 3\overline{)9} \\ 2\overline{)18} \\ 2\overline{)36} \end{array}$$

$$36 = 2 \cdot 2 \cdot 3 \cdot 3$$
$$24 = 2 \cdot 2 \cdot 3 \cdot 2$$

Both numbers contain the prime factors, $2 \times 2 \times 3$, or 12, which is the GCF.

Using GCF to find simplest name of a fraction

The concept of GCF, sometimes called the "greatest common divisor," is useful in finding the simplest name for a fraction (this is discussed more in depth in Chapter 16).

$$\frac{24}{36} = \frac{2 \times 2 \times 3 \times 2}{2 \times 2 \times 3 \times 3} = \boxed{\frac{2}{2}} \times \boxed{\frac{2}{2}} \times \boxed{\frac{3}{3}} \times \frac{2}{3}$$
$$= 1 \times 1 \times 1 \times \frac{2}{3} = \frac{2}{3}$$

> *Note:* The teacher should point out to the children that the notations $(2 \times 3)/(3 \times 3)$ and $(2/3) \times (3/3)$ as well as $(6 \div 3)/(9 \div 3)$ and $(6/9) \div (3/3)$ are equivalent and merely different ways of writing the same mathematics idea. Children are often unaware of this point.

In renaming 24/36 as 2/3, the GCF was "pulled out" of both the numerator and the denominator, employing the related idea that any number divided by 1 is that number:

$$\frac{24}{36} \div \boxed{\frac{12}{12}} = \frac{2}{3}$$

When the numerator and denominator of a fraction are multiplied or divided by the same nonzero number, the resulting fraction value is the same as the original fraction value, and they are called "equivalent fractions."

Finding the GCF using sets

The GCF may also be taught by using the set of all factors, rather than only prime factors. To find the GCF of 24 and 36 using sets, let A = the set of all factors of 24 and B = the set of all factors of 36. Then,

$$A = \{1, 24, 2, 12, 3, 8, 4, 6\} \quad \text{or}$$

$$\{1, 2, 3, 4, 6, 8, 12, 24\}$$

$$B = \{1, 36, 2, 18, 3, 12, 4, 9, 6\} \quad \text{or}$$

$$\{1, 2, 3, 4, 6, 9, 12, 18, 36\}$$

Notice that $n(A)$ is an even number and $n(B)$, which is a square number, is odd.

The intersection set comprises the set of common factors:

$$A \cap B = \{1, 2, 3, 4, 6, 12\}$$

The largest common factor is 12, which is the greatest common factor.

Now let us consider another set (C). If set C = the set of all factors of 12, then $C = \{1, 12, 2, 6, 3, 4\}$ or $\{1, 2, 3, 4, 6, 12\}$. The intersection of sets A, B, and C can be expressed as

$$A \cap B \cap C = \{1, 2, 3, 4, 6, 12\}$$

The GCF, then, for sets A, B, and C (or 24, 36, and 12) is 12.

Least Common Multiple

The *least common multiple* (LCM), which is the least common denominator when applied to fractions, is the smallest multiple of two or more numbers. A simple procedure for determining the LCM of 6 and 8, for example, is to find successive multiples of each number until reaching the number that is a multiple of both 6 and 8.

6, 12, 18, (24)
8, 16, (24)

To find the LCM of three numbers, the same procedure may be used. For example, for 6, 8, and 9, the LCM is found to be 72, as shown below.

6, 12, 18, 24, 30, 36, 42, 48, 54, 60, 66, (72)
8, 16, 24, 32, 40, 48, 56, 64, (72)
9, 18, 27, 36, 45, 54, 63, (72)

The mini-lesson that follows illustrates that prime factorization is an alternative procedure for finding the lowest common multiple.

Mini-lesson 5.1.2 Find the Least Common Multiple

Vocabulary: least common multiple

Possible Trouble Spots: poor memory skills; needs more practice on basic multiplication facts.

Requisite Objectives:
Same as for Mini-lesson 5.1.4 (see p. 232).

Activity 1
To find the LCM of two or more numbers using prime factors, write the numbers 6 and 8 as products of their primes.

$$6 = 2 \cdot 3$$
$$8 = 2 \cdot 2 \cdot 2$$

Next, find the LCM using the successive multiples procedure.

6, 12, 18, (24)
8, 16, (24)

Then, find the prime factors of 24, using the factor tree procedure or the short division algorithm.

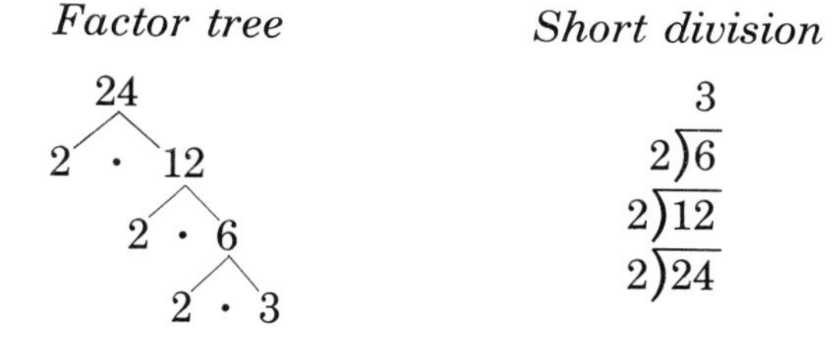

Point out that by the Fundamental Theorem of Arithmetic, regardless of the procedure used, the prime factors of 24 are $2 \cdot 2 \cdot 2 \cdot 3$. Compare the prime factors of 6, 8, and 24.

$$\begin{aligned} 6 &= 2 \cdot 3 \\ 8 &= 2 \cdot 2 \cdot 2 \\ 24 &= 2 \cdot 2 \cdot 2 \cdot 3 \end{aligned}$$

Notice that 2 was used as a factor only three times in the LCM, although 2 appears as a factor four times above (once in the factors for 6 and three times among the factors for 8).

Analyze other examples, such as 10 and 15, 6 and 27.

Additional considerations: Pupils are sometimes confused by the terminology *least common multiple* and *greatest common factor.* They seem to focus on the words *least* and *greatest* and so become confused. The least common multiple describes a number that indeed is *not* least but is greater than the factors involved or perhaps equal to one of them. In a sequence of multiples, the least common multiple is the first multiple common to all factors under consideration. In like manner, the greatest common factor is smaller than most of the composite numbers in question, or, at the most, equal to one of them. It often helps to underline the words *multiple* and *factor* in the phrases and to discuss with the pupils how these words are the key to understanding these two terms.

Note: Mini-lesson 5.1.2 is a prerequisite for Mini-lessons 5.1.3 and 5.1.4.

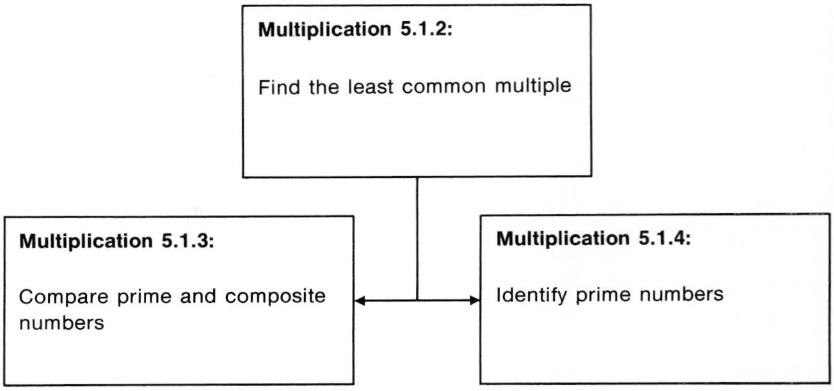

Multiplication 5.1.2:

Find the least common multiple

Multiplication 5.1.3:

Compare prime and composite numbers

Multiplication 5.1.4:

Identify prime numbers

EXERCISES

I. Find the GCF for the following sets of numbers:
1. any two prime numbers
2. 121, 125
3. 126, 28, 49
4. 13, 17, 31

II. Find the LCM for the following sets of numbers:
1. 2, 3, 5
2. 2, 5, 4

III. Use the prime factorization method for finding the LCM and GCF of 6 and 9.

IV. Answer true or false:
1. A composite number can be expressed as a product of primes in one and only one way if the order of factors is disregarded.
2. There are an infinite number of primes.

V. The least common multiple (LCM) of 6 and 8 is ______.

VI. Integers that are greater than 1 whose only divisors are 1 and the integer itself are ______.

VII. An integer that is greater than 1 and is not prime is ______.

VIII. Find the divisors or factors of 18.

IX. Can all of the factors of 18 be paired to yield 18 as a product?

X. Can all of the factors of 24 be paired to yield 24 as a product?

XI. The factors of 4 cannot be paired so that the products equal 4 unless 2 is thought to be paired with ______.

XII. Can all of the factors of the following numbers be paired to yield themselves as products using two different numbers for each pair?
1. 25
2. 36
3. 49
4. 64
5. 81
6. What do these numbers have in common?

XIII. What do you notice about the number of elements in the set of factors for numbers that are perfect squares?

XIV. Identify the name of the following theorem: A composite integer $n \geq 2$ can be expressed uniquely as a product of primes disregarding order.

XV. Express 372 as a product of primes.

XVI. Answer true or false to the following statements:
1. If n is an integer greater than 2, $(n^2 - 1)$ is never prime.
2. Every even number is the sum of two primes.

3. If n is an integer, then at least one prime lies between n and $2n$.
4. There is a multiple of 6 next to every prime.
5. Prime numbers whose difference is 2 are called "prime twins."
6. Two numbers are "relatively prime" if they have no common factor other than 1.

XVII. The largest common factor of two numbers is called ______ or ______.

XVIII. The greatest common factor (GCF) of 36 and 48 is ______.

SUGGESTED READINGS

Barnett, I. A. "The Fascination of Whole Numbers." *The Mathematics Teacher* 64 (February 1971): 103–108.

Beck, A.; Bleicher, M.; and Crowe, D. *Excursions into Mathematics.* New York: Worth, 1969.

Gullen, G. "The Smallest Prime Factor of a Natural Number." *The Mathematics Teacher* 67 (April 1974): 329–32.

Kennedy, Robert. "Divisibility by Integers Ending in 1, 3, 7, or 9." *The Mathematics Teacher* 65 (February 1971): 137–38.

Lappan, Glenda, and Winter, Mary Jean. "Prime Factorization." *The Arithmetic Teacher* 27 (March 1980): 24–27.

Morton, R. "Divisibility by 7, 11, 13 and Greater Primes." *The Mathematics Teacher* 61 (April 1968): 680–74.

Schatz, M. "Of the Infinitude of Primes." *The Mathematics Teacher* 68 (December 1975): 676–77.

Smith, Lehi. "A General Test of Divisibility." *The Mathematics Teacher* 71 (November 1978): 668–69.

Szetela, Walter. "A General Divisibility Test for Whole Numbers." *The Mathematics Teacher* 73 (March 1980): 223–25.

Yazab, Najiib. "Some Unusual Tests of Divisibility." *The Mathematics Teacher* 69 (December 1976): 667–68.

Rational Numbers and Fraction Concepts

This chapter and Chapters 15 to 18 are concerned with the set of numbers called rational numbers. *Fractions* are commonly used to represent rational numbers, as are *decimals, percents,* and *ratios.* Each is a notation that can be used to express the same rational number, as shown below.

Fraction	*Decimal*	*Percent*	*Ratio*
$\frac{1}{2}$	.5	50%	1:2

In this chapter, the focus is on rational numbers as represented by fractions.

RATIONAL NUMBERS

Recall that according to the closure principle a set of numbers is closed for a particular operation only if the operation yields a result (number) that is in the same set as the numbers used in the operation. This axiom implies that, for any number system, the system strives for closure. How, then, do we define the numbers that result from problems such as $3 \div 4 = \square$ and $4 \div 3 = \square$ (where the quotient of two integers is not an integer)? The set of numbers that will contain the results of *all* division operations is called the set of *rational numbers* (see Chart 14.1).

Chart 14.1. Set of Rationals

$$\left\{\begin{array}{c} \ldots \frac{^-3}{1}, \frac{^-2}{1}, \frac{^-1}{1}, \frac{0}{1}, \frac{1}{1}, \frac{2}{1}, \frac{3}{1}, \ldots \\ \ldots \frac{^-3}{2}, \frac{^-2}{2}, \frac{^-1}{2}, \frac{0}{2}, \frac{1}{2}, \frac{2}{2}, \frac{3}{2}, \ldots \\ \ldots \frac{^-3}{3}, \frac{^-2}{3}, \frac{^-1}{3}, \frac{0}{3}, \frac{1}{3}, \frac{2}{3}, \frac{3}{3}, \ldots \end{array}\right\}$$

The elements of the set of rational numbers as just defined are *fractions*. This term refers to the notation used to represent a division relationship (quotient) of two integers: For any integers a and b where $b \neq 0$,

$$a \div b = \frac{a}{b}$$

Notice in Chart 14.1 that rationals whose denominator is equal to 1 are integers:

$$\frac{0}{1} = 0 \qquad \frac{^-2}{1} = {^-2} \qquad \frac{3}{1} = 3$$

Notice also that when the numerator and the denominator are the same number, whether negative or positive, the rational number equals the integer $^{\pm}1$:

$$\frac{^-2}{2} = {^-1} \qquad \frac{3}{3} = 1 \qquad \frac{1}{^-1} = {^-1} \qquad \frac{2}{^-2} = {^-1}$$

TYPES OF FRACTIONS

Common fractions

The word *fraction* comes from the Latin *frangere,* meaning "to break." *Fraction* in early times meant "broken number." The term *common fraction* came from the Latin *fractiones vulgares,* which originally distinguished these from the sexagesimals used in astronomy. The term *common fraction* is used today (as compared with *decimal fraction* whose denominator is a power of 10). Common fractions are either proper or improper. Those with numerators less than their denominators are considered to be *proper fractions. Improper fractions* are those whose numerators are greater than or equal to their denominators. An improper fraction may be rewritten as a

Improper fractions

mixed number, which is comprised of a whole number and a proper fraction, or as a whole number:

$$(27/9 = 3) \qquad (28/9 = 3\tfrac{1}{9})$$

Simple fractions

Complex fractions

A *simple fraction* is a fraction whose numerator and denominator are both whole numbers. A *complex fraction* is a fraction whose numerator or denominator or both are fractional:

$$\frac{1}{\frac{2}{3}} \qquad \frac{\frac{2}{3}}{1} \qquad \frac{\frac{3}{4}}{\frac{2}{3}}$$

Numerator and denominator

The parts of a fraction are called *terms*. Latin writers called the upper term the *numerator* ("numberer") and the lower term the *denominator* ("namer").

Like and unlike fractions

When two fractions have the same denominator, they are called *like fractions; unlike fractions* have different denominators. Two unlike fractions may be renamed to have like denominators so they may be added or subtracted.

WHOLE NUMBERS AS FRACTIONS

The idea of a fraction as part of a group is shown below by the darkened circle. This implies 1 out of 3 equal parts—in this case, a part of a group rather than a part of a single unit or whole.

$\frac{1}{3}$

Now consider a whole group as many fractional parts. For example 3 pies, each divided into 6 servings.

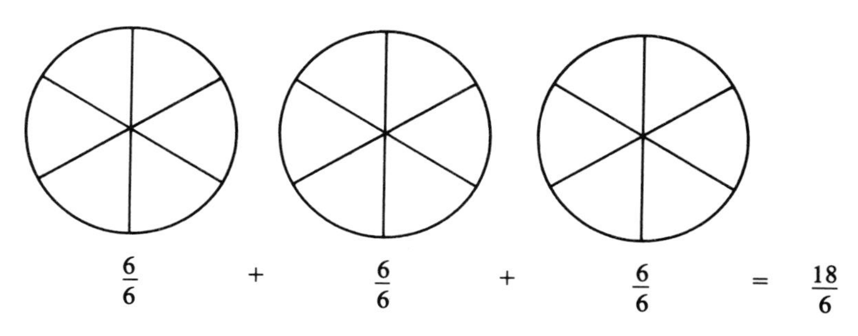

$$\frac{6}{6} + \frac{6}{6} + \frac{6}{6} = \frac{18}{6}$$

6 sixths + 6 sixths + 6 sixths = 18 sixths

$$(6 + 6 + 6)\left(\frac{1}{6}\right) = \frac{6 + 6 + 6}{6} = \frac{18}{6}$$

There are 18 sixths pies, but what is the simplest name? Do you order 18 sixths pies or 3 pies from a bakery?

$$\frac{18}{6} = \frac{6+6+6}{6} = \boxed{\frac{6}{6}} + \boxed{\frac{6}{6}} + \boxed{\frac{6}{6}} = 3$$

The ideas of proper fraction and whole numbers expressed as fractions are prerequisite to naming a mixed number as the sum of a whole number and a proper fraction.

$$\frac{19}{6} = \frac{18+1}{6} = \frac{18}{6} + \frac{1}{6} = 3 + \frac{1}{6} = 3\frac{1}{6}$$

NAMING MIXED NUMBERS

Suppose the owner of a restaurant has 3 1/6 pies left, each whole pie having been cut into sixths. How many servings of pie does he have? To answer this question, we translate the mixed numeral into the improper fraction notation:

$$3\frac{1}{6} = \frac{6}{6} + \frac{6}{6} + \frac{6}{6} + \frac{1}{6} = \frac{6+6+6+1}{6} = \frac{19}{6}$$

To begin, the whole numbers are named as improper fractions.

$$1 = \frac{\square}{6} = \frac{6}{6}$$

$$2 = \frac{6}{6} + \frac{6}{6} = \frac{6+6}{6} = \frac{12}{6}$$

$$3 = \frac{6}{6} + \frac{6}{6} + \frac{6}{6} = \frac{6+6+6}{6} = \frac{18}{6}$$

Then, the mixed numbers, or whole numbers, are renamed as improper fractions in the following manner:

$$3\frac{1}{6} = \frac{6}{6} + \frac{6}{6} + \frac{6}{6} + \frac{1}{6} = \frac{6 + 6 + 6 + 1}{6} = \frac{19}{6}$$

Mini-lesson 1.3.7 Identify the Fractional Part of a Whole

Vocabulary: whole, part, fraction, one-half, one-fourth, 1/4, 1/2, halves, fourths, figure

Possible Trouble Spots: does not notice that fractional parts of a whole must be equal in size; does not generalize to unfamiliar shapes

Requisite Objective:
Divide a symmetric three-dimensional object into parts of equal size

Activity 1
To provide a concrete model for the student, show a ping-pong ball that has been cut into 2 equal parts. Label each part "one-half." Allow students to manipulate and discuss the halves of the little ball. Other objects may be used. For example, commercially made cookies that are soft enough to slice in half and that are uniformly round, square, or rectangular are appropriate.

Activity 2
Have the child select from a group of parts the 2 equivalent parts that make up a figure:

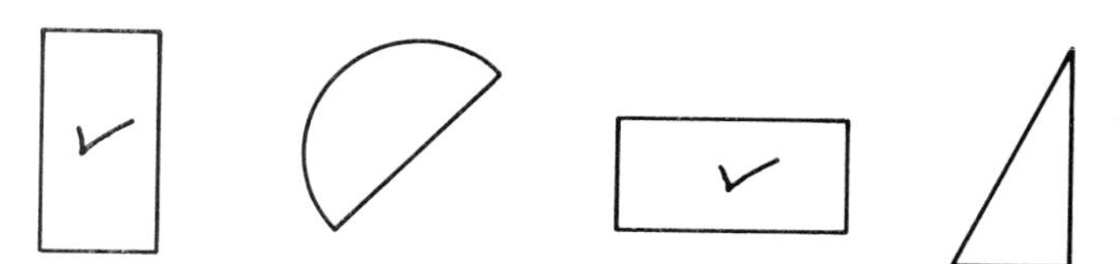

Then ask the child to construct the whole figure from the equivalent parts.

Activity 3
Draw various shapes and divide them into halves and quarters. Have students identify one-half and one-quarter by coloring the designated part(s). Follow this activity with more challenging tasks, such as the worksheet shown below, in order to assess the students' knowledge of a fractional part of a whole.

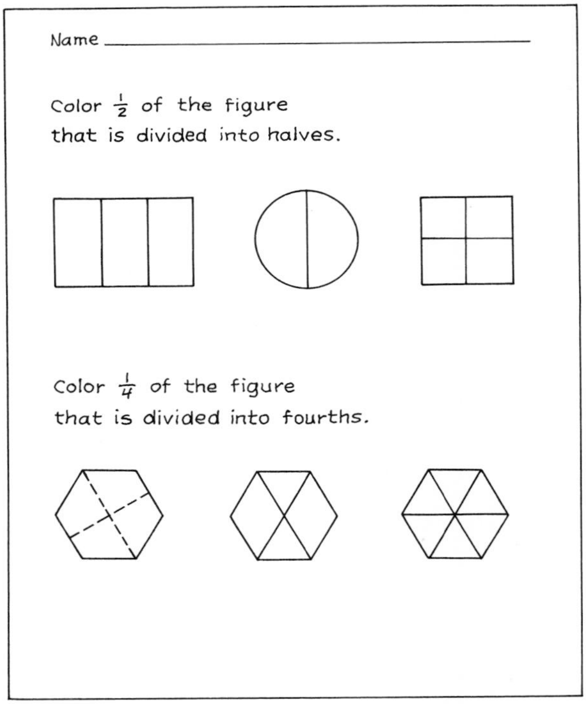

Name ______________________________

Color $\frac{1}{2}$ of the figure
that is divided into halves.

Color $\frac{1}{4}$ of the figure
that is divided into fourths.

Additional considerations: For those children who either *cannot* or for some reason *do not* attend to salient aspects of a situation you may have to provide cueing. For example, either outline or fill in one of the parts of the divided figure. Another cue is to label each of the fractional parts.

Fractions taught in the early grades should relate to meaningful activities in the children's daily lives. Some examples in outdated textbooks are pretty farfetched when applied to everyday life. For instance, one example had as an answer: "9 13/71 seconds after 6 o'clock."

Children come to school knowing a lot about fractions. They have been hearing and using language involving fractions in many situations. Here are some examples of preschoolers using fraction ideas:

"Sally, gimme *half* that candy bar."
"Daddy says he'll be home in a *half* hour."
"I need a *quarter* for lunch."

It is helpful for the teacher to draw out of pupils such experiences as listed above. When this is done, the children are made aware that they already know something about fractions and do not think of fractions as new and unrelated mathematics. However, teachers need to provide activities that will ensure that children do not connect particular configurations with particular fractional numbers. One child in a clinical interview stated that "one-fourth" was a part of a candy bar and "one-third" was part of a pie!

Mini-lesson 1.3.8 Identify the Fractional Part of a Set

Vocabulary: group, set, carton of eggs, collection

Possible Trouble Spots: fixates on fractional part of a whole and does not understand fractional part of a group; does not have good visual discrimination skills.

Requisite Objectives:
Sets 1.1.3: Recognize equivalent and nonequivalent sets.

Whole Numbers/Notation 1.1.5: Indicate the cardinal number property of sets.

Mathematical Applications 1.2.8: Recognize symmetry in nature and in own body.

Activity 1
Show the child sets that are divided in half and some that are not divided in half. Ask him to mark those that are divided in half. Help him notice that counting the number of objects on either side of the line will help him select the correct sets.

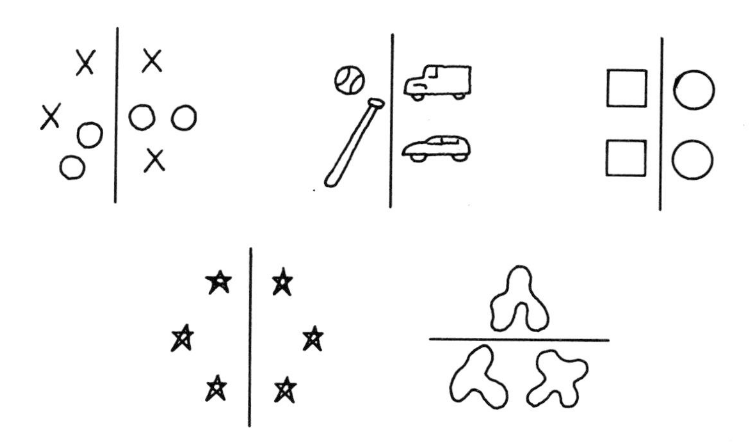

Activity 2
Have students complete worksheets such as the following.

Additional considerations: This mini-lesson is much more complex than the one on identifying a fractional part of a whole. First of all, the perceptual requirements are more complex. Secondly, the child needs to be able to assign objects in the original group to new groups in order to form fractional parts. He or she must be able to count objects to determine that the new groups are equivalent in number because it is the cardinality of the parts that constitutes fractional parts of a group. For example, if a student is to find one-half of a group of 8 marbles, he or she must realize that there must be 4 marbles in each of the new groups formed.

If students have difficulty finding the cardinal number property of a group, try using a balance. Give them 8 marbles and tell them to place one-half of the group of marbles in one pan on the balance, and the remaining half in the other pan. This procedure may also facilitate their counting.

Note: Mini-lesson 1.3.8 is a prerequisite for Mini-lessons 1.3.11 and 2.1.1.

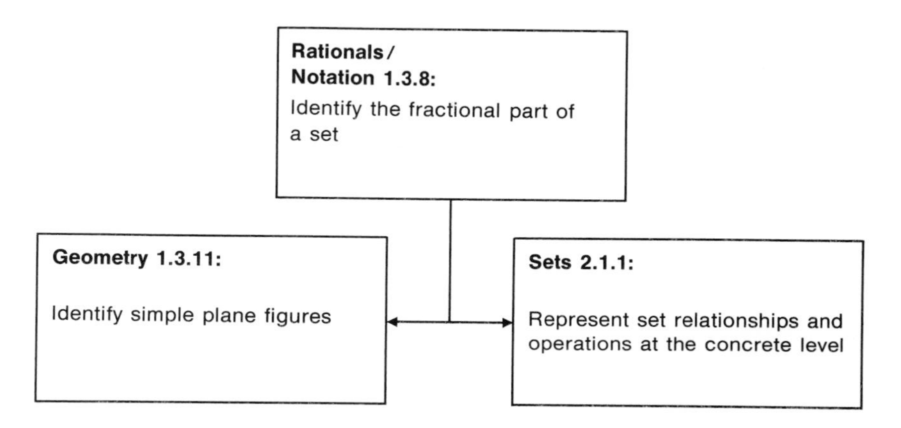

Mini-lesson 2.1.5 Select Concrete Representations of Equivalent Fractions, e.g., Halves and Fourths, Thirds and Sixths

Vocabulary: thirds, sixths

Possible Trouble Spots: visual perceptual problems may block a child from realizing that fractional pieces of a whole are equal in size; poor fine-motor skills may interfere with manipulative activities

Requisite Objectives:
Mathematical Applications 1.2.8: Recognize symmetry in nature and in own body.
Recognize fractional equivalents to a whole.

Activity 1
Have students identify symmetry as a basis for "one-half." For example, show hexagonal shapes divided into halves, thirds, and sixths; also show square or rectangular shapes divided or folded into halves and fourths.

Activity 2
Use Cuisenaire rods to compare equivalent fractions. In this case, the size of the unit changes according to selection of the bar that represents one whole.

Activity 3
Provide the child with paper and a pair of scissors and ask her to cut a unit strip and then strips of equal length to show thirds, fourths, fifths, and so on. The teacher may mark off fraction lengths on ditto

sheets as a guide for cutting the paper strips. Ask the child to use these strips to decide which fractional numbers—1/3 or 2/3, 3/4 or 2/3, 3/4 or 4/5—are larger.

1 unit		
$\frac{1}{3}$	$\frac{1}{3}$	$\frac{1}{3}$

Additional considerations: It is helpful for students to be allowed to discuss their observations in teams or small groups. Communication of this type serves as a self-correction device for those who are missing the point of what the term fractional part means.

Mini-lesson 3.2.3 Identify and Compare Equivalent Fractions

Vocabulary: equivalent fractions, sixths, eighths, twelfths

Possible Trouble Spots: does not notice salient aspects of situations; learns better via demonstration instead of manipulation of materials; needs much structure in what to do with the concrete materials; needs real world examples to become motivated to learn goal of lesson

Requisite Objectives:
Fractions/Notation 2.3.1: Select fractional equivalents to the number one at concrete and picture levels

Sets 3.2.1: Remove all disjoint equivalent sets from a given set

Separate a set into 6, 8, 10, or 12 equivalent sets

Activity 1
To develop awareness of fractional names for the number, one, give students puzzles of geometric shapes cut into equivalent fractional parts. As they remove parts have them say and record the size of each part (e.g., 1/6 from a figure divided into sixths, 1/8 from a puzzle comprised of eighths, etc.). In each case, when all of the parts of the puzzle are removed, ask the student to express this as a fraction (6/6, 8/8, and so on). Ask what he noticed in each case. (Both the numerator and denominator were the same number.)

Activity 2
Use sets to find equivalent fractions by engaging the child in the following activities.

A. Give the child 5 colored blocks and 5 plain blocks. Ask her to tell how many blocks are in the set and how many are colored. Guide her to notice that 5/10 of them are colored and that we can also say that 1/2 of them are colored (1/2 = 5/10).

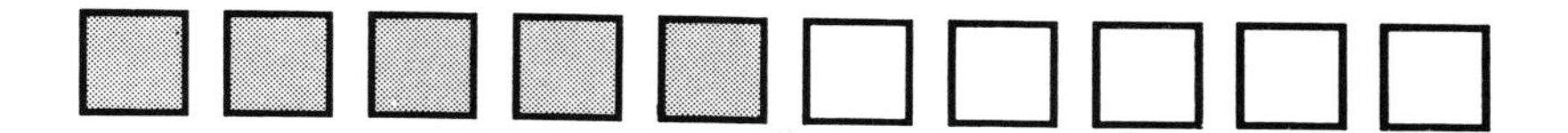

B. Give the child sets indicating various fractional parts and ask him to choose from the fractional notations the two notations that tell what part of each set is shaded.

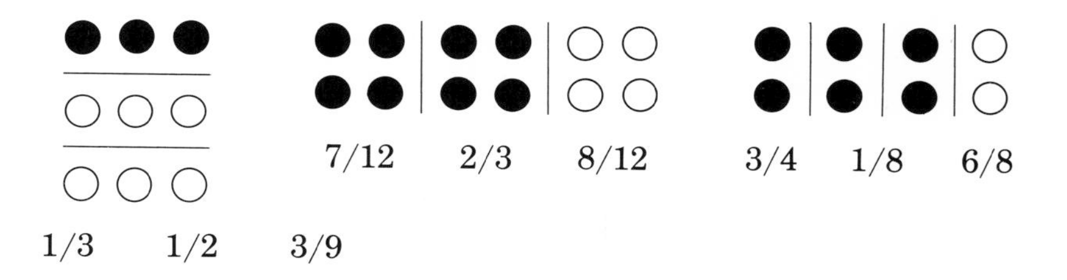

C. Show the child sets indicating various fractional parts and ask her to write two or more equivalent fractions to tell what part of each set is shaded.

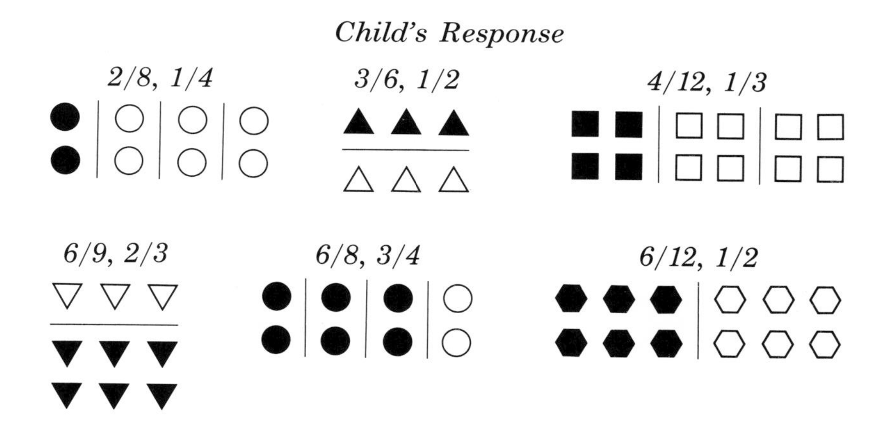

Activity 3
Develop a set of equivalent fractions using pictures.

A. Draw a picture to show 2/3. Then show each part cut into halves. Ask the child the following questions:

1. How many parts make up the whole? (6)
2. How many sixths are shaded? (4)

3. How many sixths are equivalent to 2/3? (4) Write an equation showing this equivalence. (2/3 = 4/6)

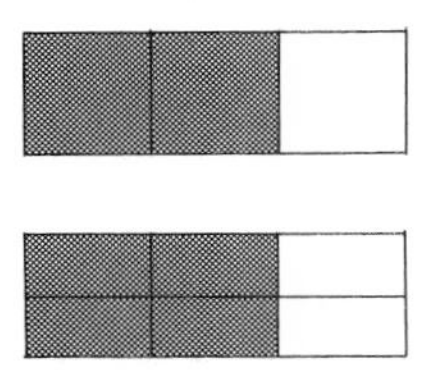

B. Then show the child pictures indicating 2/3, with each third cut into parts; 2/3, with each third cut into fourths; and 2/3, with each third cut into 5 parts. Ask the child questions like those in Activity 3A. You will, of course, need to alter the questions to fit the picture shown.

Activity 4
Students may use pictures to compare fractional values. They may use words to describe the comparisons. Some may use the signs for greater than (>) and less than (<).

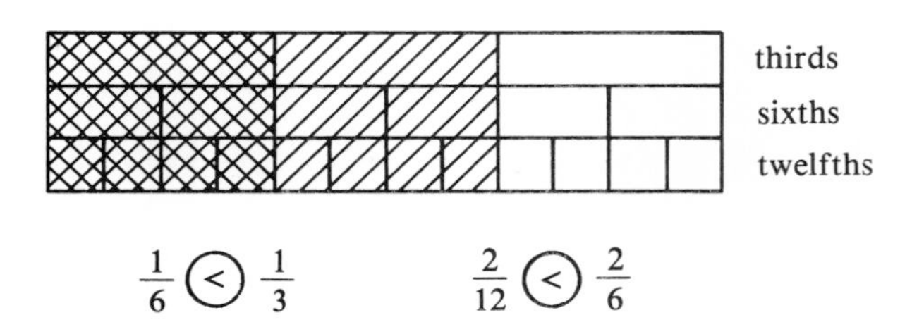

Activity 5
Students may then be ready to show equivalent fractions on a number line. Commence with something they are used to—a rectangular region—and compare this to using a number line.

A. Show the child a rectangle divided into 3 equivalent parts. Ask her the following questions:

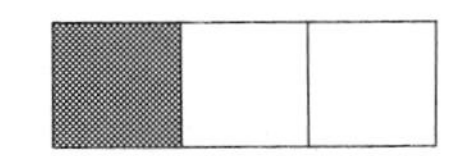

1. How many parts are shaded? (1)
2. How many equal parts are there in all? (3)
3. What fraction is shaded? (1/3)

B. Then show the child a number line from 0 to 1 divided into thirds and ask her the following questions:

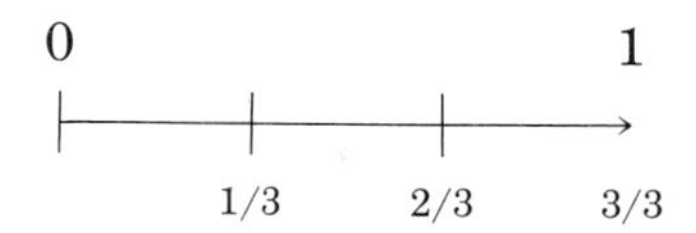

1. How many equal-sized parts is the number line divided into?
2. What is each part called? (1/3)

C. Show the child number lines indicating halves, fourths, and eighths. Remind the pupil that fractions naming the same number are equivalent. Ask him to find the fractions 1/2, 2/4, and 4/8 on the number line and to tell what he discovers about them. Then ask him to find two fractions that name zero.

D. Show the child number lines indicating thirds, sixths, and ninths and ask her to complete equivalent fractions.

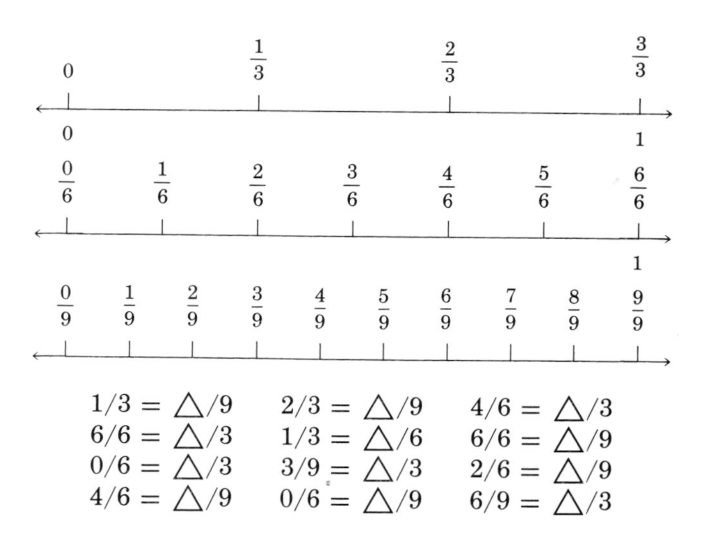

Note: A foot ruler is a number line on a stick and is a good device for aiding pupils in understanding fraction concepts.

Activity 6
Show the following table to the child. Ask him to look for patterns helpful in naming fractions equivalent to a given fraction. Guide him to notice that the fraction 2/3 is equivalent to 4/6, 6/9, 8/12, 10/15, 12/18 and that the fraction 2/5 is equivalent to 4/10, 6/15, 8/20, 10/25, 12/30.

×	1	2	3	4	5	6
1	1	2	3	4	5	6
2	2	4	6	8	10	12
3	3	6	9	12	15	18
4	4	8	12	16	20	24
5	5	10	15	20	25	30
6	6	12	18	24	30	36

(Arrows: $\frac{2}{5}$ points to rows 2 and 5; $\frac{2}{3}$ points to 12 and 18.)

Additional considerations: This is a crucial objective because skill in working with equivalent fractions underlies computation with fractions as well as estimation.

A review of "Possible Trouble Spots" points out the importance of teacher demonstration for many students. Demonstration can be done in a meaningful way, saves time, and helps those students who have a tendency to wander off task.

Note: Mini-lesson 3.2.3 is a prerequisite for Mini-lessons 3.2.4 and 3.3.4.

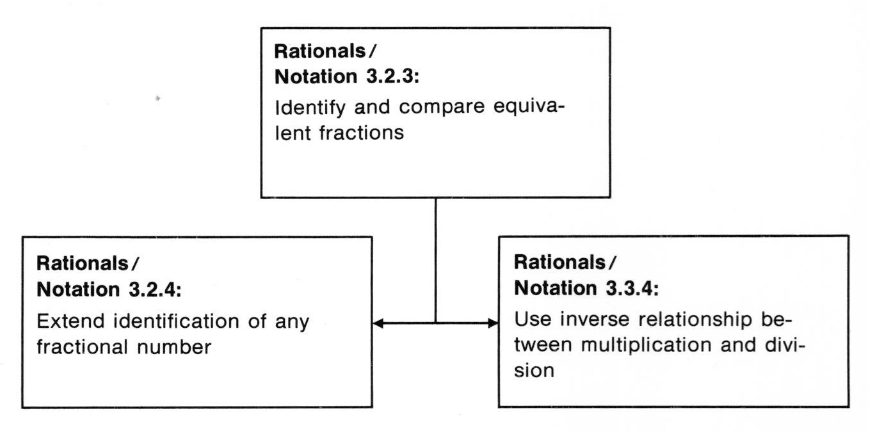

Mini-lesson 3.2.4 Extend Identification of Any Fractional Number

Vocabulary: numerator, denominator, shaded part

Possible Trouble Spots: visual discrimination difficulties inhibit work with pictures; auditory discrimination problems inhibit hearing the *th* used in naming the denominator number (one-fourth, two-sixths, etc.)

Requisite Objectives:
Select equivalent parts of a whole or of a set and correctly use the terms *half, third, fourth, sixth, eighth,* and so on.

Identify equivalent parts of a whole. Divide figures or sets into fractional parts.

To identify the numerator and denominator of fractions; to associate the numerator with a specific part of a whole or of a group; and to associate the denominator with the total number of equal parts of a whole or a group.

Read and write symbols for fractional numbers.

Activity 1
Fold a paper circle into 4 equivalent parts. Tell the child that the circle is divided into fourths, each part being 1/4, and that 1/4 is one of the 4 equal parts. Do the same with thirds. Fold rectangles into thirds, halves, fourths, eighths. Fold squares into thirds, halves, and so on. Then have the child divide a set of pictured objects into 2, 3, or 4 equivalent subsets that represent halves, thirds, or fourths.

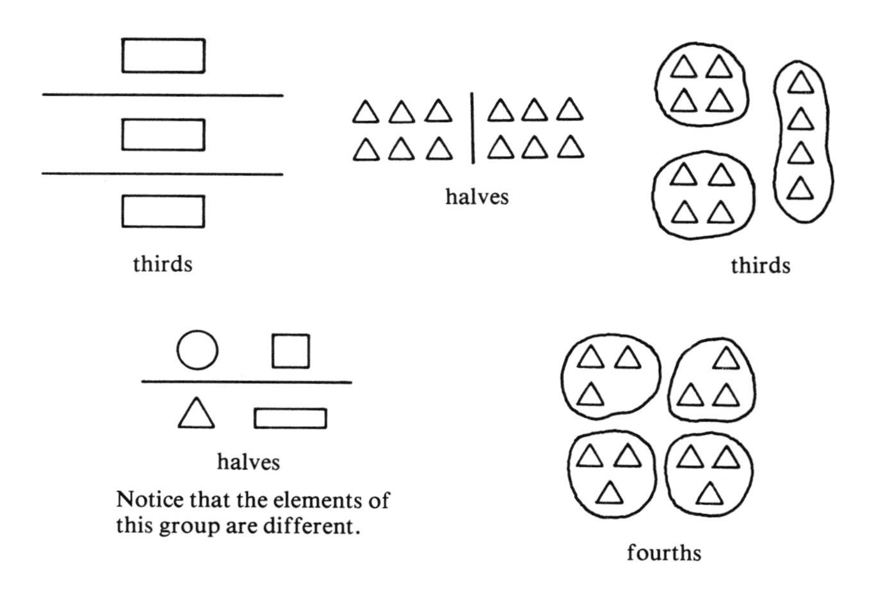

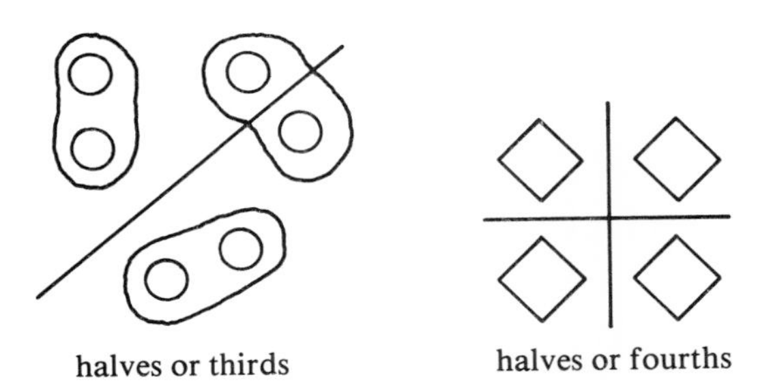

Activity 2
Have the child complete the following statement: "2/3 means _______ of _______ equal parts." Guide him to notice that 2 is the numerator and 3 is the denominator. Then show him figures divided into 3 equivalent parts and ask him to shade in 2/3 of the figures.

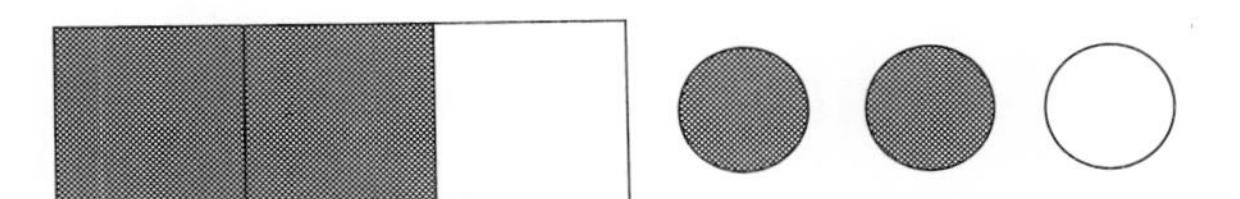

Activity 3
Give the child pictures with even numbers of objects and ask her to divide each set of objects into halves and label each part in as many ways as she can.

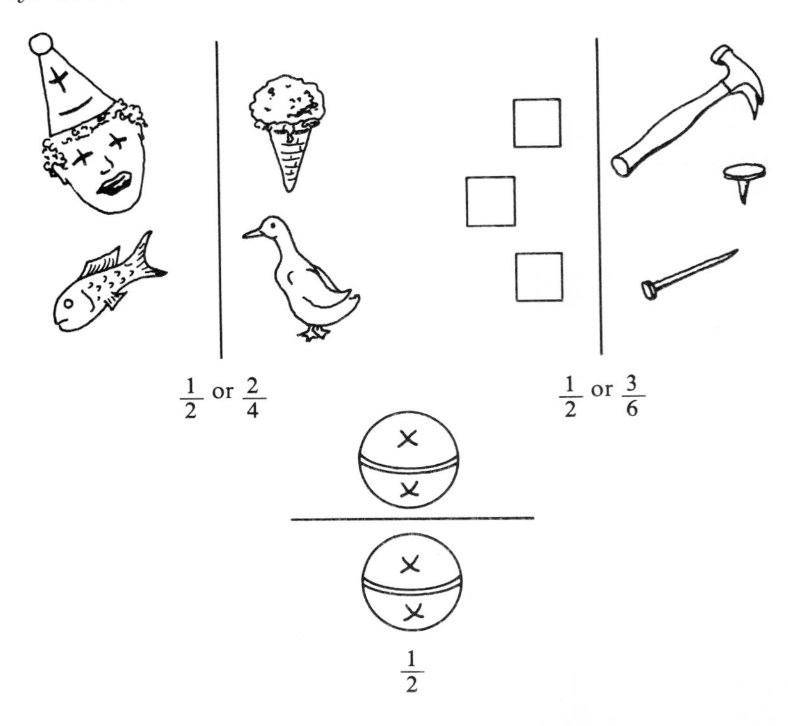

Additional considerations: The way of looking at fractional parts of a single unit and fractional parts of a discrete set of objects may be quite unrelated to a child. Furthermore, the elements of a set of objects may even be quite different as shown in Activities 1 and 3 above. The two ideas of fraction—part of a unit or part of a group—should, therefore, be presented to the pupil as two distinct ideas.

Note: Mini-lesson 3.2.4 is a prerequisite for Mini-lessons 3.2.10 and 3.2.20.

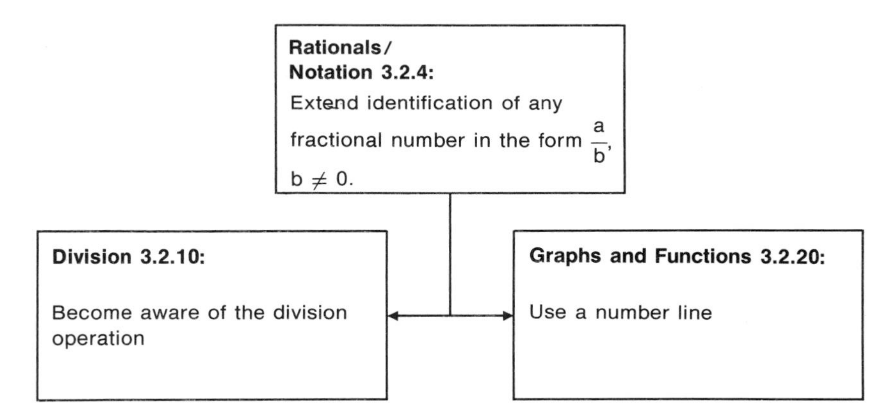

EXERCISES

I. Write the fractions 3/4 and 2/5 as decimals, as percents, and as ratios.

II. Answer true or false
1. Common fractions are either proper or improper.
2. Fractions whose numerators are greater than their denominators are called improper fractions.
3. Fractions whose numerators and denominators are equal are improper fractions.
4. A fraction whose numerator, denominator, or both are fractional are called complex fractions.

III. Write the following fractions as mixed numbers.

1. $\frac{7}{2}$ 2. $\frac{19}{8}$ 3. $\frac{25}{4}$ 4. $\frac{13}{5}$

IV. Write the following mixed numbers as fractions.

1. $1\frac{2}{3}$ 2. $6\frac{1}{5}$ 3. $2\frac{3}{8}$ 4. $1\frac{7}{8}$

SUGGESTED READINGS

Green, G. F. "A Model for Teaching Multiplication of Fractional Numbers." *The Arithmetic Teacher* 20 (January 1973): 5–9.

Linn, C. F. *The Golden Mean.* New York: Doubleday, 1974. Pp. 9–13.

Pittman, P. V. "Rapid Mental Squaring of Mixed Numbers." *The Mathematics Teacher* 70 (October 1977): 596–97.

Shokoohi, Gholam-Hossein. "Readiness of Eight-Year-Old Children To Understand the Division of Fractions." *The Arithmetic Teacher* 27 (March 1980): 40–43.

Renaming Fractions

Multiplying both terms of a fraction by the same number does not change its value. This principle is related to the *multiplicative identity,* which states that $n \times 1 = n$, and to the fact that the number 1 may be expressed in the form n/n. This relates to the principle that multiplying reciprocals yields the multiplicative identity, which in turn relies on the axiom that any nonzero integer (a) has a reciprocal ($1/a$). Thus, $a \times 1/a = a/a = 1$. Therefore, to rename 3/4 to twelfths, for example, both terms would be multiplied by 3:

$3/4 \times 3/3 = 9/12$

Another method is used next to rename 9/12 to fourths. Start with the principle that dividing both terms of a fraction by the same number does not change its value. Thus,

$$\frac{9}{12} \div \frac{3}{3} = \frac{3}{4}$$

Using the indirect open sentence $12 \div \square = 4$, the number that correctly completes the sentence is the new number name for the multiplicative identity.

Mini-lesson 4.1.3 Extend Expanded Notation to Include Fractions Through Thousandths

Vocabulary: tenth, hundredth, thousandth

Possible Trouble Spots: does not realize that place values to the right of units represent fractional parts of a unit; confuses place values to right of units with values of negative integers; does not understand the reciprocal idea that underlies the balance between tens and tenths, hundreds and hundredths, and so on

Requisite Objectives:
Whole Numbers/Notation 3.1.2: Extend place value through thousands

Activity 1
Place value extended to the right of the units can be used along with the counting boards to present a physical model of fractions with denominators of powers of ten.

A. Review whole number place values on the counting boards.

B. Show tenths, hundredths, and thousandths by extending counting boards to the right of the units board. Use this same procedure to represent 1/10, 1/100, and 1/1000 on an abacus. Have students record in expanded notation what is shown on the counting boards.

tens	units	tenths
9	9	9
8	8	8
7	7	7
6	6	6
5	5	5
4	4	4
3 ●	3	3
2	2 ●	2
1	1	1 ●

$(3 \times 10) + (2 \times 1) + (1 \times 1/10) = 32\ 1/10$

units	tenths	hundredths
9	9	9
8	8	8
7	7	7
6	6	6 ●
5	5	5
4	4	4
3	3	3
2 ●	2	2
1	1	1

$(2 \times 1) + (0 \times 1/10) + (6 \times 1/100) = 2\ 6/100$

tenths	hundredths	thousandths
9	9	9
8	8	8 ●
7	7	7
6	6	6
5	5	5
4	4	4
3	3	3
2	2	2
1	1	1

$(0 \times 1/10) + (0 \times 1/100) + (8 \times 1/1000) = 8/1000$

Activity 2
Have children show 1/10, 1/100, 1/1000 on counting boards and pictures of counting boards.

tens	units	tenths	hundredths	thousandths
9	9	9	9	9
8	8 ●	8	8	8
7	7	7	7	7
6	6	6	6	6
5	5	5	5	5 ●
4	4	4	4	4
3	3	3	3 ●	3
2 ●	2	2	2	2
1	1	1 ●	1	1

$$28\frac{135}{1000} = (2 \times 10) + (8 \times 1) + \left(1 \times \frac{1}{10}\right) + \left(3 \times \frac{1}{100}\right) + \left(5 \times \frac{1}{1000}\right)$$

Have them build to numbers in expanded notation through thousandths. See addition of fractions with unlike denominators in Chapter 16.

Additional considerations: Students in fourth grade have had several years' work with fractions. However, most of the activities involved small denominators (halves, thirds, sixteenths, and so forth).

The use of place values that deal with fractional parts based upon powers of ten is a helpful bridge to working with larger denominators for those students who have a good grasp of how place values are generated. Furthermore, an important reason for learning fractions with denominators of tenths, hundredths, thousandths, is that it serves to prepare students for work with the decimal-fraction part of our numeration system. Also, there are times when it is helpful to be able to translate from decimal fraction to common fraction, or vice versa, in order to make a computation easier. For example, I find it easier to change a problem such as .25 ÷ .05 to fractions before computing:

$$\frac{25}{100} \div \frac{5}{100} = \frac{25 \div 5}{100 \div 100} = \frac{5}{1} = 5$$

On the other hand, to add 1/10, 3/100, 17/1000, I change to decimals:

$$\begin{array}{r} .1 \\ .03 \\ \underline{.017} \\ .147 \end{array}$$

Note: Mini-lesson 4.1.3 is a prerequisite for Mini-lessons 4.1.11 and 4.2.2.

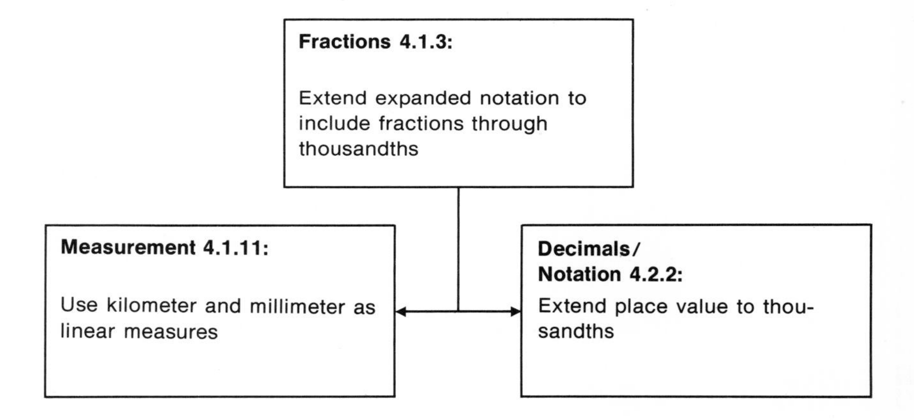

Mini-lesson 5.2.1 Compare the Least Common Multiple to the Least Common Denominator

Vocabulary: LCD, LCM

Possible Trouble Spots: Confused by the phrase *least common multiple* unless focus is on the word *multiple* (especially true when taught in conjunction with greatest common factor, where focus should be on the word *factor*).

Requisite Objectives:
Fractions/Notation 4.2.1: Write fractional numbers equivalent to one, from graphic representation

Fractions/Notation 4.2.2: Compare fractions and division

Rationals/Notation 4.3.1: Investigate graphic representation of multiplication of simple fractions using rectangular regions

Activity 1

To review use of rectangular regions to show simple multiplication of fractions, have students discuss the following graphic representations. Point out that the least common multiple of 2 and 3, namely, 6, is also the denominator of the product, $1/2 \times 1/3$. Also, have students compare the number of sixths in one-third and in one-half, thus relating the fact that fractions with unlike denominators can be translated to their equivalents where the denominators are the same. It is at this point that they should discuss the relationship of the lowest common multiple of two or more numbers to the lowest common denominator of two or more fractions.

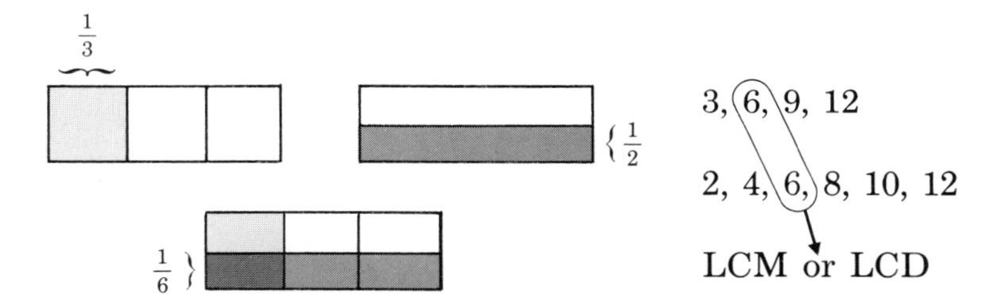

Activity 2

Provide students with sets of equivalent fractions that show that the lowest common multiple and lowest common denominator of two or more numbers is the same number.

$$\frac{1}{12}, \frac{1}{4}, \frac{1}{3}$$

12, 24
4, 8, 12
3, 6, 9, 12

By folding three sheets of paper for halves, thirds, and sixths, students can cut out the 1/2s and 1/3s and fit them over the 1/6s. This should be done with 1/2s, 1/4s, and 1/8s; 1/5s, 1/2s, 1/10s; 1/8s, 1/2s, 1/16s, etc., until students state that they are aware of the pattern that relates the LCD and LCM.

Activity 3
Have students list equivalent fractions as shown for 1/8 and 1/6. Then have them encircle the two fractions with the least common denominator. A completed example is presented below.

$$\frac{1}{8} = \frac{2}{16} = \frac{3}{24} = \frac{4}{32} = \frac{5}{40} = \frac{6}{48}$$

$$\frac{1}{6} = \frac{2}{12} = \frac{3}{18} = \frac{4}{24} = \frac{5}{30} = \frac{6}{36} = \frac{7}{42} = \frac{8}{48}$$

a. $\frac{2}{3}$ b. $\frac{3}{5}$

$\frac{5}{6}$ $\frac{1}{3}$

c. $\frac{3}{4}$ d. $\frac{5}{8}$

$\frac{2}{7}$ $\frac{2}{6}$

Additional considerations: Students often become confused with the terms least common multiple and greatest common factor. This confusion involves the logical discrepancy between the opposite ideas of something being both least and at the same time greatest. For example, the multiple of two numbers is as large as the smallest number, and usually greater than both. Looking at multiples for the numbers 3 and 4, we have:

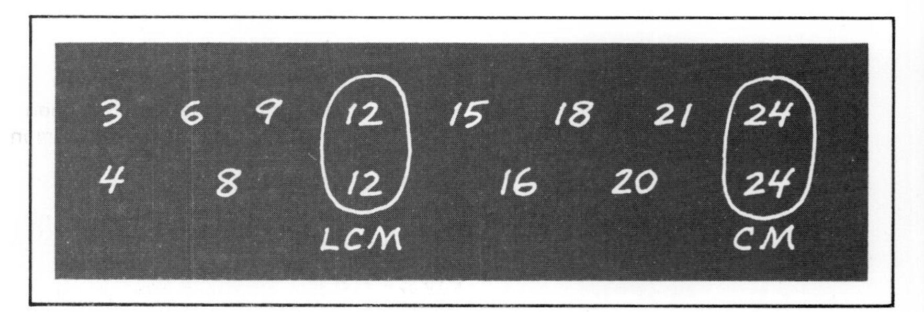

All of the multiples listed, except the numbers themselves, are greater than the given number. Shown are two common multiples, 12 and 24. However, since 12 is the first common multiple in the numerical sequence, it is the least common *multiple* (LCM), while 24 is a common multiple (CM); 36 is the next CM.

Greatest common factor should probably not be taught simultaneously with LCM. I have found it helpful to give students enough practice with LCM until they can create their own examples and explain the LCM to others. Then, I introduce the greatest common factor.

The opposite semantic illogic is inherent in the phrase greatest common factor (GCF). In this case, we are looking for a factor common to two or more numbers; the largest factor.

The confusion is due to the idea that factors are equal to or smaller than the numbers that they are a part of. Common factor is understood, but when the word greatest is introduced, the conflict in meaning emerges. Factors are shown below for the number 24 and 36. The GCF is 12; the factors in the sequence prior to the GCF may be common factors (CF), but they are not the *greatest* common factor. The next mini-lesson deals with GCF.

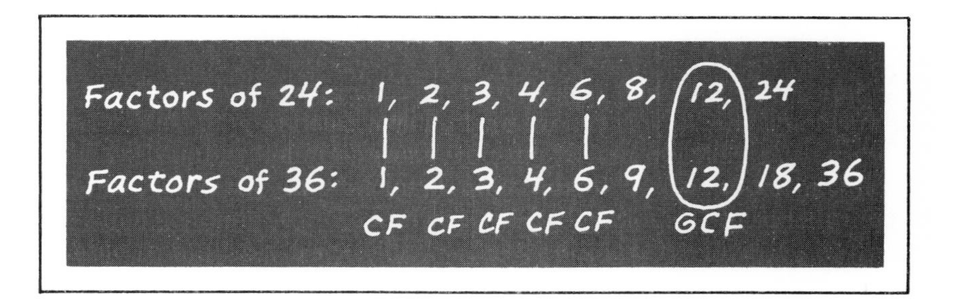

Note: Mini-lesson 5.2.1 is a prerequisite for Mini-lessons 5.2.2 and 5.3.3.

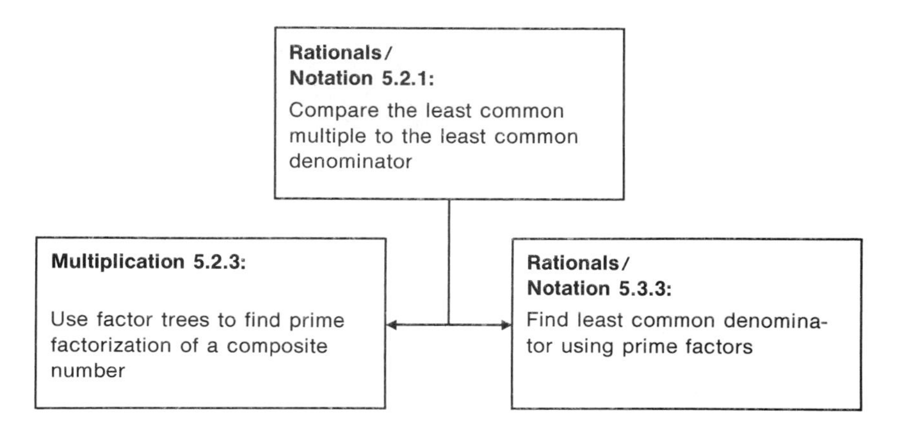

Mini-lesson 5.3.1 Find the Simplest Name for a Fraction Using the Greatest Common Factor as the Fractional Multiplicative Identity

Vocabulary: greatest common factor, GCF

Possible Trouble Spots: does not notice cause-effect relationships; does not conserve number; cannot rename a number as a product of factors; does not know basic division facts; does not know basic multiplication facts

Requisite Objectives:
Fractions/Notation 2.3.1: Select fractional equivalents to the number one at concrete and picture levels.
Multiplication 3.1.8: Multiply with 1, 2, 3, 4, 5 as factors; with 6, 7, 8, 9 as factors.

Division 3.2.12: Compute basic division facts.
Multiplication 5.2.4: Find the greatest common factor.
Division 5.2.7: Use divisibility rules for finding prime factors.

Activity 1
Have students use the prime factor method described in Chapter 13 to determine the greatest common factor.

A. They may use the following format:

$$\frac{6}{15} = \frac{2 \times 3}{5 \times 3} = \frac{2}{5} \times \frac{3}{3}$$

The GCF is 3.

B. Once the GCF is found, have students find the simplest name for fractions as follows:

$$\frac{36}{48} \div \frac{12}{12} = \frac{36 \div 12}{48 \div 12} = \frac{3}{4}$$

The GCF is 12.

C. Provide practice in finding equivalent fractions:

$\frac{8}{12} = \frac{2 \times 2 \times 2}{2 \times 2 \times 3}$	Factor the numerator and denominator into prime factors.
$= \frac{2 \times 2}{2 \times 2} \times \frac{2}{3}$	Indicate the fractional equivalents for 1.
$= \frac{4}{4} \times \frac{2}{3}$	Write $\frac{4}{4}$ as 1.
$= 1 \times \frac{2}{3}$	Use the multiplicative identity.
$= \frac{2}{3}$	

Activity 2
The divisibility rules are a helpful tool for finding factors of a number that lead to the GCF. Provide activities such as the following: "Test 6/15 for divisibility by 2, 3, 5. We find that 3 is the GCF."

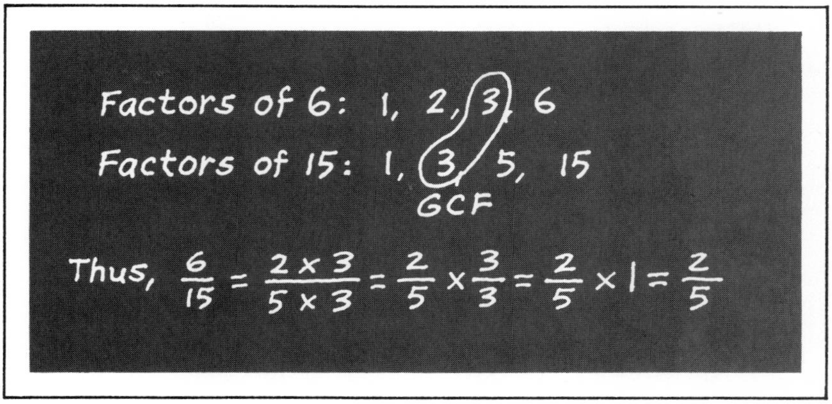

Factors of 6: 1, 2, 3, 6
Factors of 15: 1, 3, 5, 15
GCF

Thus, $\frac{6}{15} = \frac{2 \times 3}{5 \times 3} = \frac{2}{5} \times \frac{3}{3} = \frac{2}{5} \times 1 = \frac{2}{5}$

Additional considerations: To help students understand all that is involved in simplifying fractions, it is important to progress through step-by-step procedures as you explain simplifying fractions to the class.

Be aware that some students may not make use of cause-effect relationships. They will not intuitively see that by factoring the numerator and the denominator, the special name for 1 is made apparent. These students may not relate fractions with the principle that any number divided by 1 is that number ($n \div 1 = n$).

These students may also become confused by the renaming of 6/15 as 2/5. Renaming fractions involves a transformation in name, while the value of the fraction—the ratio between the numerator and the denominator—remains the same. This condition is related to the Piagetian task of conservation of number, which also involves noticing various transformations while number remains constant.

Mini-lesson 5.3.2 Find Equivalent Fractions, Using the Multiplicative Identity Property

Vocabulary: multiplicative identity

Possible Trouble Spots: does not relate solving indirect open sentences to finding equivalent fractions (see Activity 2); is distracted by graphic representations due to difficulty discriminating figure from ground

Requisite Objectives:
Rationals/Notation 4.2.1: Write fractional numbers equivalent to one, from graphic representation

Multiplication 4.3.5: Show use of indirect open sentence in finding equivalent fractions by multiplying

Activity 1
To develop skill in selecting special fractional names for 1, the multiplicative identity, have students use rectangular regions to identify fractions equivalent to 1.

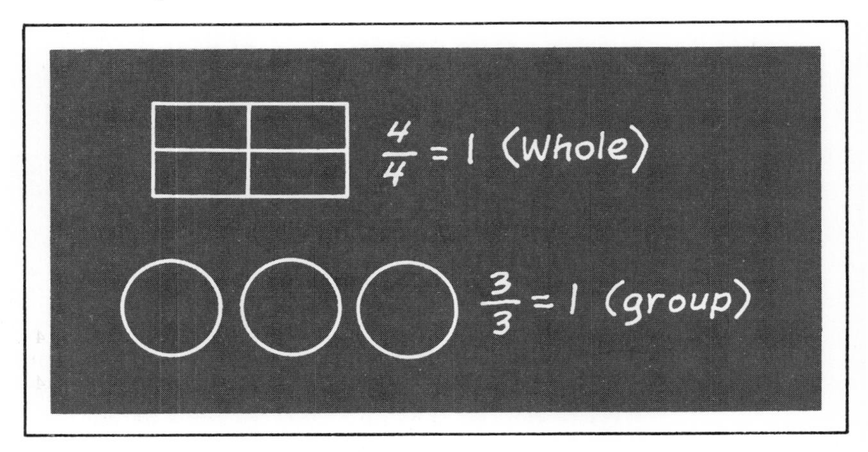

Activity 2
Given the problem

$$\frac{1}{3} = \frac{\square}{15}$$

have students find the number that will serve as the special name for 1 by using the following procedures:

$$\begin{aligned} 3 \times \square &= 15 \\ \square &= 5 \end{aligned}$$

The next step involves use of the fact that $\square = 5$ to name the multiplicative identity.

$$\frac{1}{3} \times \frac{\square}{\boxed{5}} = \frac{\quad}{15}$$

$$\frac{1}{3} \times \boxed{\frac{\boxed{5}}{\boxed{5}}} = \frac{\quad}{15}$$

$$\frac{1}{3} \times \frac{5}{5} = \frac{5}{15}$$

A. Find equal fractions by multiplying by the special name for 1:

$$\frac{2}{3} \times \boxed{\frac{2}{2}} = \frac{4}{6} \quad (2 \times 2,\ 3 \times 2)$$

$$\frac{2}{3} = \frac{6}{9} \quad (2 \times \boxed{3},\ 3 \times \boxed{3})$$

$$\frac{2}{3} = \frac{8}{12} \quad (2 \times \boxed{4},\ 3 \times \boxed{4})$$

B. Ask the child to write four equivalent fractions:

$$\frac{3}{4} = \frac{3 \times 2}{8} = \frac{3 \times 3}{12} = \frac{3 \times 4}{16} = \frac{3 \times 5}{20}$$

$$4 \times 2 \quad 4 \times 3 \quad 4 \times 4 \quad 4 \times 5$$

$$\frac{2}{5} = \frac{2 \times 2}{10} = \frac{2 \times 3}{15} = \frac{2 \times 4}{20} = \frac{2 \times 5}{25}$$

$$5 \times 2 \quad 5 \times 3 \quad 5 \times 4 \quad 5 \times 5$$

$$\frac{1}{6} = \frac{1 \times 2}{12} = \frac{1 \times 3}{\_\_} = \frac{1 \times 4}{\_\_} = \frac{1 \times 5}{\_\_}$$

$$6 \times 2 \quad 6 \times 3 \quad 6 \times 4 \quad 6 \times 5$$

Additional considerations: This mini-lesson emphasizes the importance of teaching multiplication of fractions prior to addition of fractions with unlike denominators. In order to add fractions with unlike denominators, the student must find equivalent fractions that have the same denominator. To find equivalent fractions, the student must be able to multiply a fraction by the multiplicative identity in fraction form:

$$\frac{a}{b} \times \frac{c}{c} = \frac{ac}{bc}$$

It is important to assess the student's ability to abstract from graphic representations to make sure that visual perception problems are not interfering. Simple tasks that involve identifying hidden figures are an appropriate assessment procedure.

Mini-lesson 5.3.3 Find Least Common Denominator Using Prime Factors

Vocabulary: least common denominator, LCD

Possible Trouble Spots: cannot rename a number as a product of primes; confuses least common multiple and greatest common factor; does not distinguish between the meaning of the terms *multiple* and *factor*

Requisite Objectives:
Multiplication 5.1.2: Find the least common multiple

Rationals/Notation 5.2.1: Compare the least common multiple to the least common denominator

Multiplication 5.2.3: Use factor trees to find prime factorization of a composite number

Activity 1
Find the LCM with color rods. Form two different one-color trains of the same length. For example, match a sequence (train) of yellow rods (3) to a sequence of red rods (4). (See color rods on p. 271.)

It will take 3 of the red rods and 4 of the yellow. The value of a train is found by either adding or multiplying the value of the parts: the value of the yellow train is $3 + 3 + 3 + 3$, or 4×3, or 12. The value of the red train is $4 + 4 + 4$, or 3×4, or 12.

The students should recognize that since there are no shorter 1-color trains comprised of yellow and red color rods that are equal in length, 12 is the smallest number that is a multiple of both 3 and 4. When using the 2-rod and the 3-rod, they should notice that the shortest 1-color trains have a value of 6 rather than 12. Thus, 6 is the least common multiple of 2 and 3.

Activity 2
Students should now review using factor trees to find prime factorization of a number. They may then use the LCM as the LCD for activities such as comparing the names of fifths and thirds as fractions with common denominators. Using the same procedure as in Activity 1, they find that the LCM of 3 and 5 is 15.

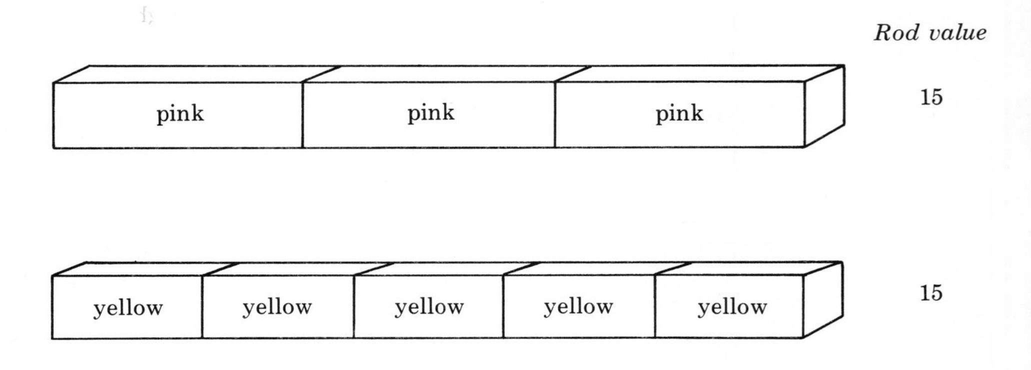

A. Once the prime factorization of two numbers is found, the LCD may also be found. For example, consider the numbers 35 and 15.

35: (7) × (5)   15: (3) × (5)

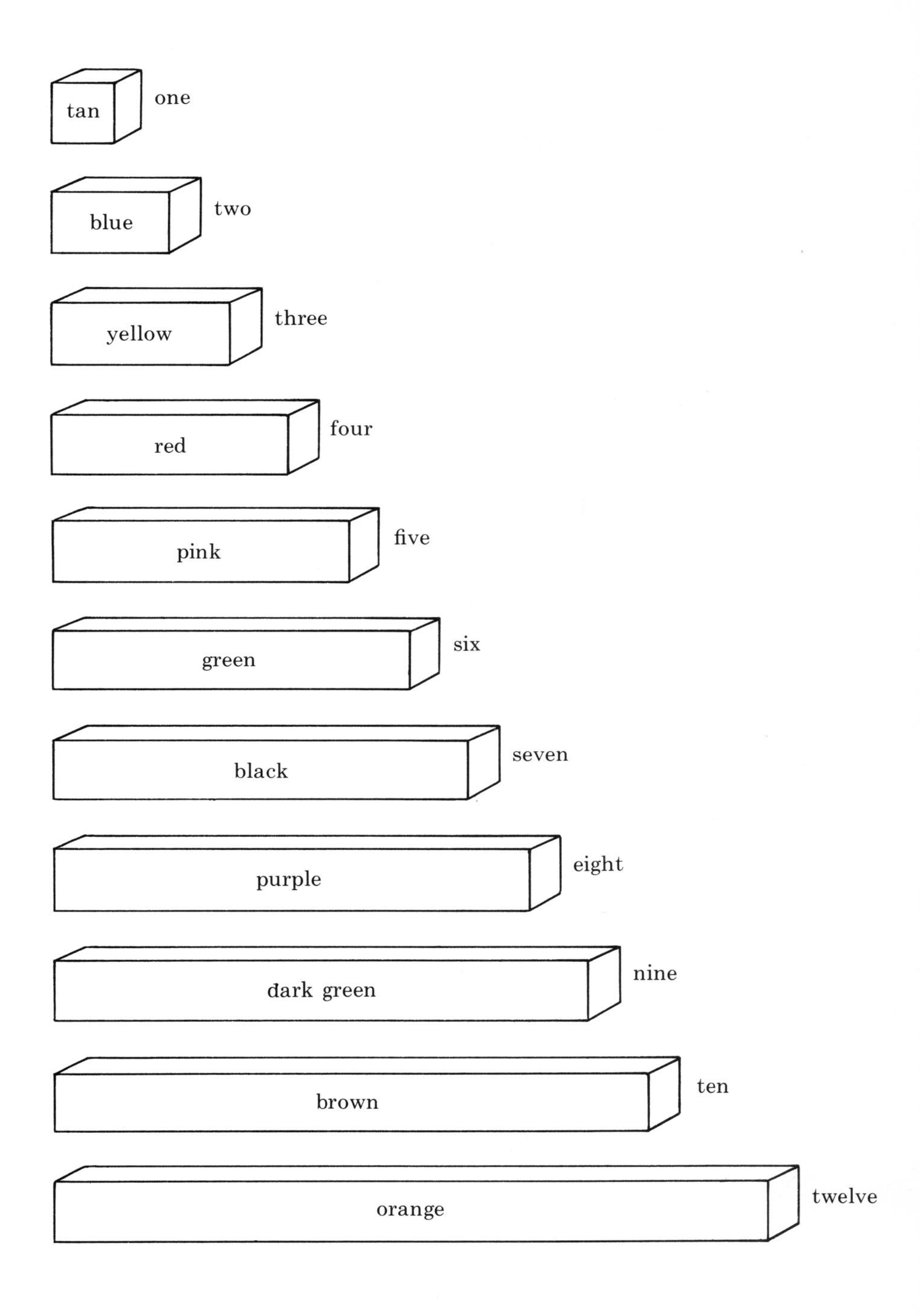
tan
one
blue
two
yellow
three
red
four
pink
five
green
six
black
seven
purple
eight
dark green
nine
brown
ten
orange
twelve

Since 5 appears as a factor in both numbers you may obtain the LCD directly as follows.

$$35 = 7 \times \textcircled{5}$$
$$15 = 3 \times \textcircled{5}$$
$$\text{LCD} = 5 \times 7 \times 3 = 105$$

B. The LCD for 35 and 75 is:

$$35 = \textcircled{5} \times 7$$
$$75 = \textcircled{5} \times 5 \times 3$$
$$\text{LCD} = 5 \times 5 \times 7 \times 3$$
$$\text{LCD} = 525$$

Activity 3

The following practice worksheet activities are suggested for consolidating the students' skills in using primes to find the lowest common denominator.

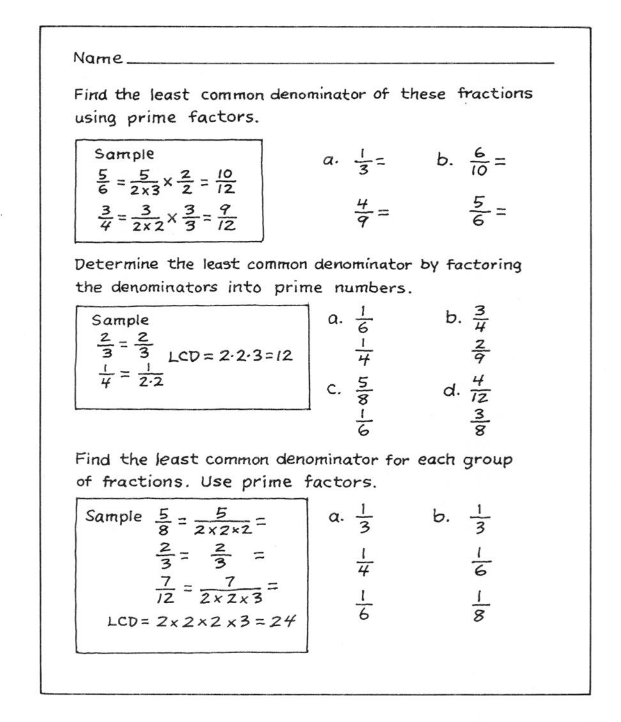

Name ____________________

Find the least common denominator of these fractions using prime factors.

Sample

$\frac{5}{6} = \frac{5}{2 \times 3} \times \frac{2}{2} = \frac{10}{12}$

$\frac{3}{4} = \frac{3}{2 \times 2} \times \frac{3}{3} = \frac{9}{12}$

a. $\frac{1}{3} =$ $\frac{4}{9} =$

b. $\frac{6}{10} =$ $\frac{5}{6} =$

Determine the least common denominator by factoring the denominators into prime numbers.

Sample

$\frac{2}{3} = \frac{2}{3}$

$\frac{1}{4} = \frac{1}{2 \cdot 2}$ LCD $= 2 \cdot 2 \cdot 3 = 12$

a. $\frac{1}{6}$ $\frac{1}{4}$

b. $\frac{3}{4}$ $\frac{2}{9}$

c. $\frac{5}{8}$ $\frac{1}{6}$

d. $\frac{4}{12}$ $\frac{3}{8}$

Find the least common denominator for each group of fractions. Use prime factors.

Sample $\frac{5}{8} = \frac{5}{2 \times 2 \times 2} =$

$\frac{2}{3} = \frac{2}{3} =$

$\frac{7}{12} = \frac{7}{2 \times 2 \times 3} =$

LCD $= 2 \times 2 \times 2 \times 3 = 24$

a. $\frac{1}{3}$ $\frac{1}{4}$ $\frac{1}{6}$

b. $\frac{1}{3}$ $\frac{1}{6}$ $\frac{1}{8}$

Additional considerations: Venn diagrams may be too abstract for some students.[1] However, if they are to be used as a graphic representation, it is helpful to simultaneously demonstrate the color rods as you record what the rods show with the Venn diagrams.

---

1. See Fredricka K. Reisman, *Diagnostic Teaching of Elementary School Mathematics* (Chicago: Rand McNally College Publishing Co.), p. 134, for a description of Venn diagrams.

Mini-lesson 5.3.4 Rename Improper Fractions as Mixed Numbers

Vocabulary: common fraction, improper fraction, mixed number

Possible Trouble Spots: does not understand that a fractional number can represent a value greater than 1

Requisite Objectives:
Categorize numbers as whole, mixed, improper fraction, proper fraction

Fractions/Notation 4.2.1: Write fractional numbers equivalent to one, from graphic representation

Activity 1
Students should be given concrete and picture level activities to represent the renaming of an improper fraction as a sum of a whole number and a common fraction. For example,

$$\frac{5}{2} = \frac{2}{2} + \frac{2}{2} + \frac{1}{2} = 1 + 1 + {}^{1}\!/_{2} = 2{}^{1}\!/_{2}$$

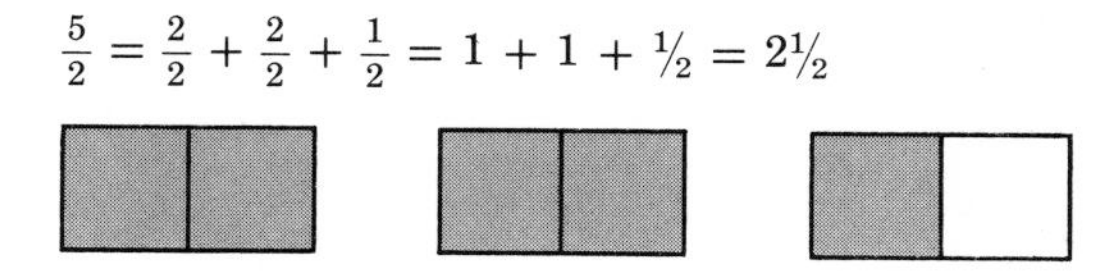

Activity 2
To assess whether students can identify different forms of equivalent numbers, provide tasks such as the following.

Name____

Place the following numbers in the appropriate column next to their equivalent number.

$1\frac{3}{8}$ $\quad 2\frac{2}{3}$ $\quad \frac{7}{8}$

7 $\quad 5\frac{3}{4}$ $\quad \frac{6}{3}$ $\quad 9\frac{2}{3}$ $\quad 8\frac{1}{7}$

$\frac{24}{4}$

Whole Numbers	Mixed Numerals	Improper Fractions	Proper Fractions
		$\frac{11}{8}$	
2			
		$\frac{29}{3}$	
		$\frac{57}{7}$	
		$\frac{23}{4}$	
		$\frac{21}{3}$	
			$\frac{14}{16}$
6			
		$\frac{8}{3}$	

Additional considerations: Students need concrete experiences upon which to construct an understanding of mixed numbers. However, these experiences must relate to their interests and level of understanding. Activities in textbooks concerning fractions, and mixed numbers in particular, often make use of the sale of stocks and bonds on the New York Stock Exchange or of mortgage payments. How many elementary school students have any understanding of such things? Interest paid on a savings account is abstract and concealed—even though many students may have savings accounts. If real life experiences are to be used as a vehicle for teaching fractions, then they must be real from the perspective of the student, not the adult. They must relate to economic situations that are realistic for the particular group of students. Such things as figuring the amount of material in meters needed to make drapes for a community project, following a recipe, using mixed numbers in shop, making a picture frame or a birdhouse are suggested. Employ creative problem solving with the class and have them help identify meaningful activities for learning about mixed numbers.

Mini-lesson 5.3.5 Translate Mixed Number to Equivalent Improper Fraction

Vocabulary: mixed number, improper fraction

Possible Trouble Spots: does not understand difference between density of fractions and spacing of whole numbers on a number line

Requisite Objectives:
Use of multiplicative identity

Addition of fractions with like denominators

Activity 1

In translating a mixed number to an improper fraction, the student must be aware that the final denominator is determined by the fractional part of the mixed number. This may be shown at all three levels of representation (concrete, picture, and written with fractions).

A. Using fraction bars, the student should be given an abundance of one set of color-rods to represent mixed numbers. For example, give 10 yellow bars arranged as 3 trains and 1 extra rod. This may be labeled:

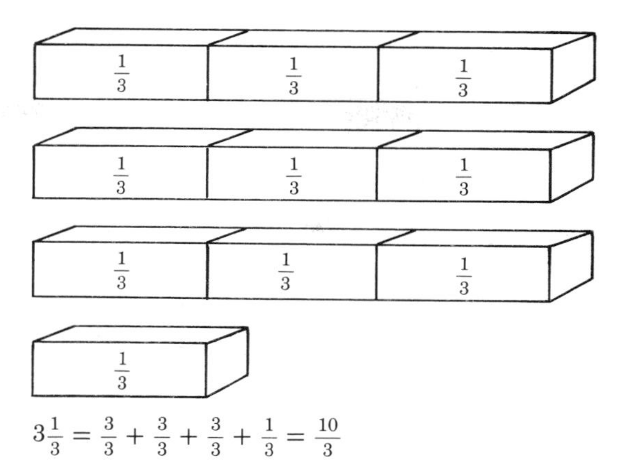

$$3\frac{1}{3} = \frac{3}{3} + \frac{3}{3} + \frac{3}{3} + \frac{1}{3} = \frac{10}{3}$$

B. Use pictures to represent translation of mixed number to equivalent improper fraction. For the mixed number 1 1/3, the following diagram may be used.

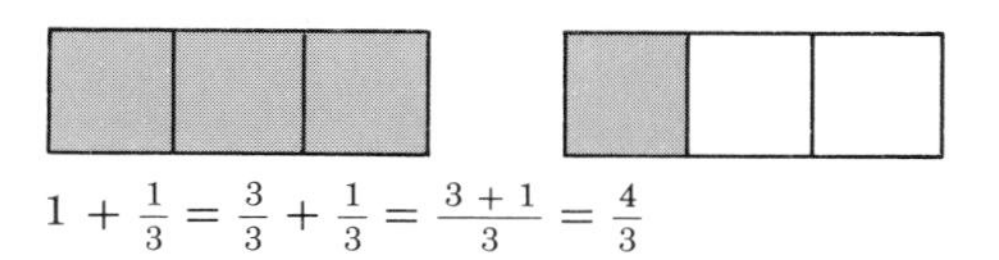

$$1 + \frac{1}{3} = \frac{3}{3} + \frac{1}{3} = \frac{3+1}{3} = \frac{4}{3}$$

C. After activities at the levels in *A* and *B* above, students may record translations to describe the concrete and picture representations. Tell them to find a pattern in their renaming that could serve as a shortcut. Some may have to be told about multiplying the whole number by the denominator and adding the numerator. The following sequence may serve as a model for making the shortcut pattern obvious.

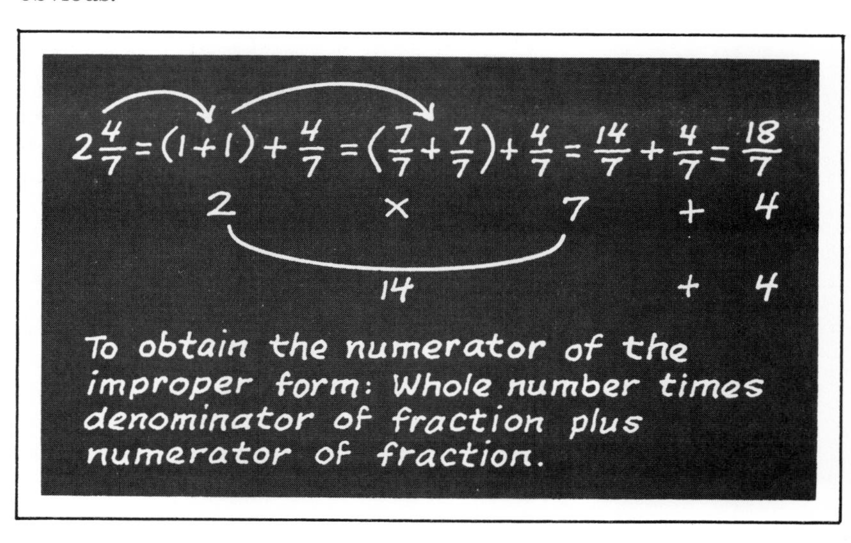

Unless a student understands the relationship between the whole-number part and the fraction part, it is best not to use the shortcut method.

Additional considerations: It is appropriate to inductively guide some pupils to this shortcut procedure after they understand components of the mixed form. For others, a step-by-step model that they may use as a cueing device is best because some students get lost in the "discovery-of-meaning" process.

If a number line is used as an instructional procedure, it is important that students see the spatial relationships among whole numbers and fractions. The characteristic of density of fractions is a difficult one for many students to grasp. That is the idea that between any two whole numbers, there are an infinite number of fractions. Also, that between any two fractions, there are an infinite number of fractions. When whole numbers and fractions are pictured on a number line and their relative positions discussed, the student is given a linear representation of mixed numbers.

Note: Mini-lesson 5.3.5 is a prerequisite for Mini-lessons 5.3.7 and 5.3.10.

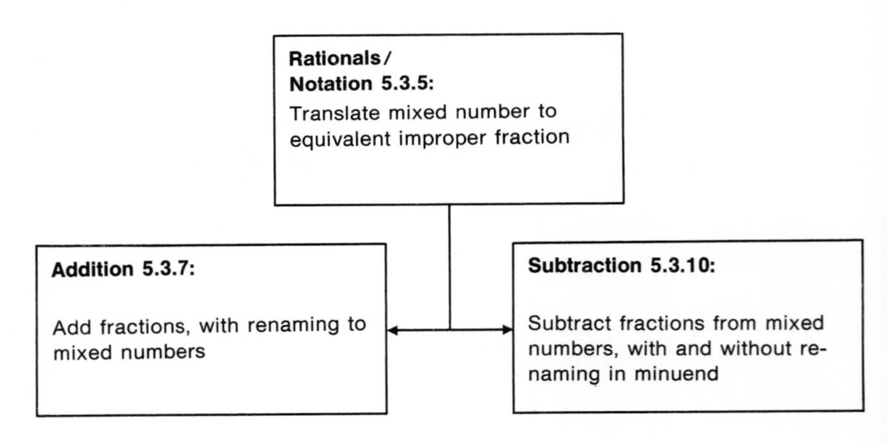

EXERCISES

I. $\frac{a}{b} + \frac{c}{d} = \frac{c}{d} + \frac{a}{b}$ shows that addition of fractions is ______.

II. Multiplication of fractions is associative. Thus, $\frac{x}{y} \cdot \left(\frac{z}{w} \cdot \frac{u}{v}\right) =$ ______.

III. The identity for addition of fractions is shown in the form ______, where ______ $\neq 0$.

IV. For fractions, $\frac{a}{b} = \frac{c}{d}$ only if _______ = _______ in whole numbers.

V. Two fractions $\frac{a}{b}$ and $\frac{b}{a}$ are related in that each is the _______ of the other.

VI. Order the fractions $\frac{40}{64}$, $\frac{14}{42}$, and $\frac{12}{24}$ in value from greatest to least.

VII. The product of a fraction and its multiplicative inverse is the _______.

VIII. The set of fractions has the property called _______, which is not present in the system of whole numbers.

IX. Since the quotient of two fractions is a fraction, the set of fractions is _______ for division.

X. Which of the following fractions is equivalent to $\frac{a \cdot m}{b \cdot m}$?

1. $\frac{m}{m}$ 3. $\frac{0}{m}$

2. $\frac{b}{a}$ 4. $\frac{a}{b}$

XI. Which of the following pairs of fractions represent the same number?

1. $\frac{7}{8}, \frac{14}{24}$ 3. $\frac{13}{12}, \frac{24}{26}$

2. $\frac{13}{24}, \frac{26}{52}$ 4. $\frac{7}{8}, \frac{21}{24}$

XII. Which of the following fractions is the simplest name for $\frac{72}{81}$?

1. $\frac{72}{81}$ 3. $\frac{16}{18}$

2. $\frac{18}{27}$ 4. $\frac{8}{9}$

XIII. Indicate what fractional part of each of the following shapes is colored using its simplest form.

1. 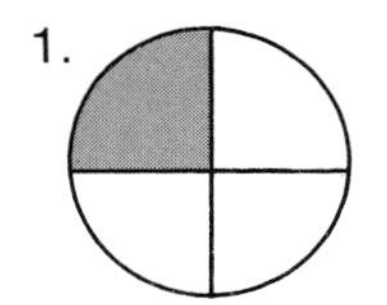

2. 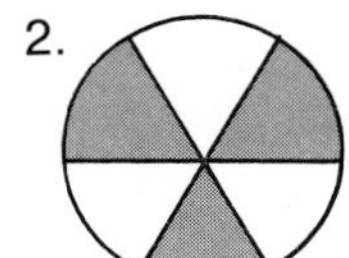

3. 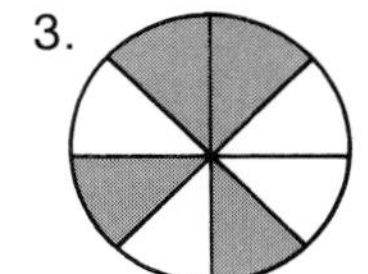

4. 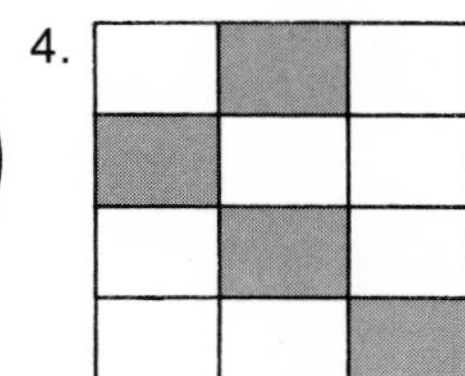

XIV. Solve the following problems:

1. $\frac{2}{3} = \frac{\square}{6}$

2. $\frac{3}{4} = \frac{\square}{8}$

3. $\frac{5}{10} = \frac{1}{\square}$

4. $\frac{6}{9} = \frac{2}{\square}$

XV. Explain why $\frac{-2}{1} = \frac{2}{-1} = -2$. (Hint: See chapter on division of integers.)

SUGGESTED READINGS

McKillip, William D.; Cooney, Thomas J.; Davis, Edward J.; and Wilson, James W. *Mathematics Instruction in the Elementary Grades.* Glenview, Ill.: Silver Burdett Co., 1978.

National Council of Teachers of Mathematics. *Rational Numbers.* Washington, D.C.: National Council of Teachers of Mathematics, 1972.

Ockenga, Earl, and Duea, Joan. "Ideas." *The Arithmetic Teacher* 25 (January 1978): 28–32.

Smith, Seaton E., Jr.; and Backman, Carl A. *Teacher-Made Aids for Elementary School Mathematics.* Reston, Va.: National Council of Teachers of Mathematics, 1974.

Stern, Catherine, and Stern, Margaret B. *Children Discover Arithmetic.* New York: Harper and Row, 1971.

Computation with Fractions

MULTIPLICATION OF FRACTIONS

For years, elementary schools have taught addition and subtraction of fractions with unlike denominators prior to multiplication and division of fractions. But multiplication of fractions is a prerequisite for finding equivalent fractions, and finding equivalent fractions is involved in addition and subtraction of fractions with unlike denominators.

Mini-lesson 4.3.1 Investigate Graphic Representation of Multiplication of Simple Fractions Using Rectangular Regions

Vocabulary: of to indicate multiplication

Possible Trouble Spots: visual-perception difficulties; difficulty going from the vertical overhead projection on a screen to the horizontal plane at the desk; poor eye-hand coordination

Requisite Objectives: Graphs and Functions 3.2.20: Use a number line

Activity 1
Illustrate $2 \times 2/3$ on a number line.

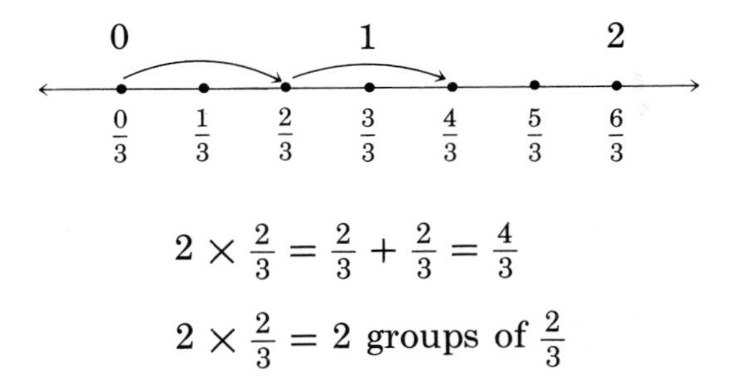

$$2 \times \frac{2}{3} = \frac{2}{3} + \frac{2}{3} = \frac{4}{3}$$

$$2 \times \frac{2}{3} = 2 \text{ groups of } \frac{2}{3}$$

Activity 2
To illustrate $2/3 \times 3/4$ with a unit region, four pieces of equal-sized paper are needed. Fold these as shown in illustrations *a*, *b*, and *c*.

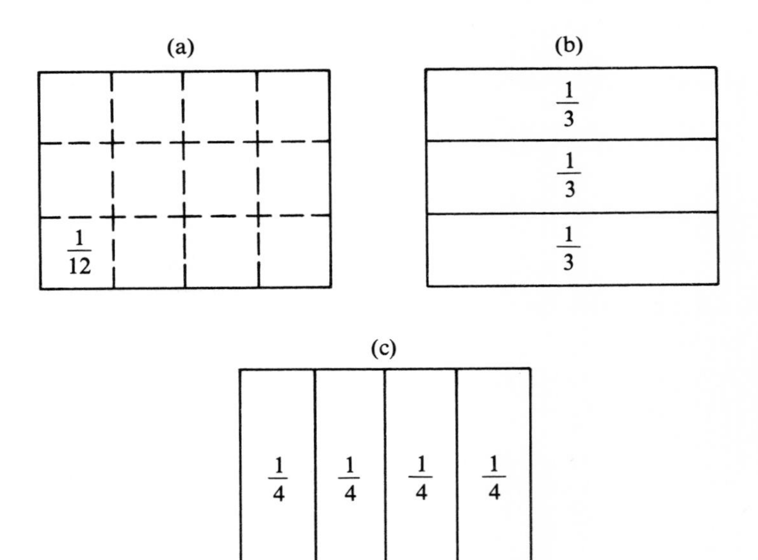

Paper *a* becomes the model for the new denominator (twelfths). Cut sheets *b* and *c* along their folds after discussing the idea that each strip is 1/3 and 1/4 of the whole respectively. Fold the fourth sheet like sheet *a* but cut this one into twelfths. These will be labeled "twelfth measures." Since the problem asks for 2/3 of 3/4, fit 3 of the 1/4 strips onto sheet *a* in their vertical positions, as shown.

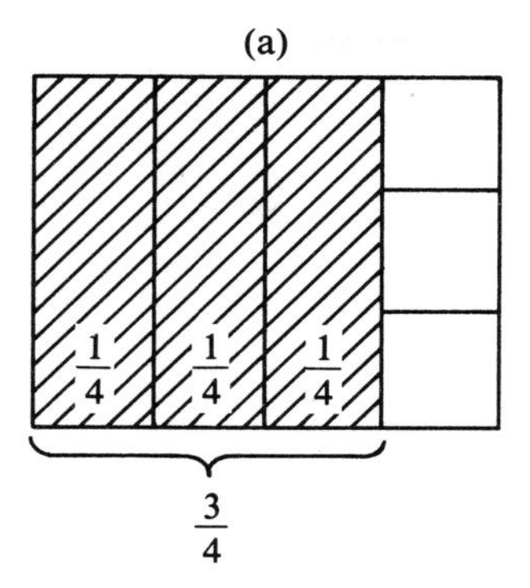

Then ask him to place 2 of the 1/3 strips onto sheet *a* in their horizontal positions, as shown. This action provides a model for renaming 1/3 as 4/12 and 1/4 as 3/12. Next, tell the child to place a twelfth measure

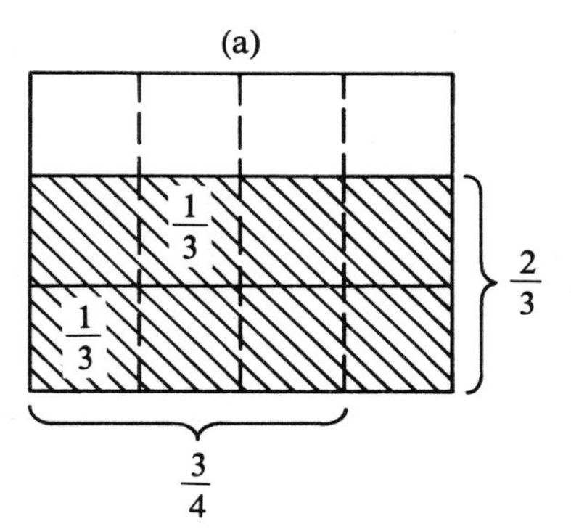

on each overlapping portion of the 1/4 strips and the 1/3 strips. Notice that since sheet *a* is folded into 12 units, each twelfth measure indicates 1 out of 12 equal parts, or 1/12. Tell the child that 2/3 of 3/4 will be represented by those squares on sheet *a* that have a twelfth measure on a square.

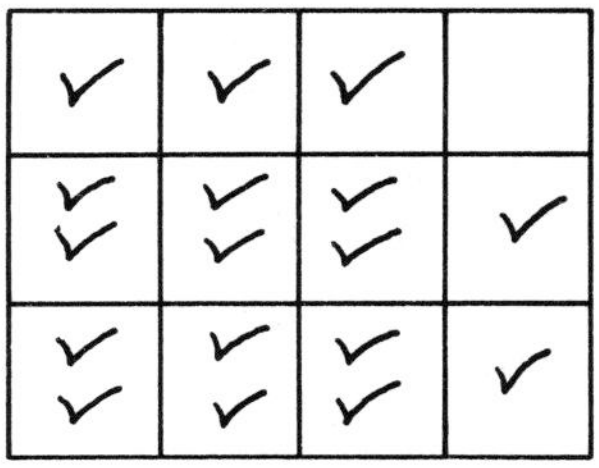

The squares with 2 ✓'s indicate 2/3 of 3/4. There are 6 of these, or 6 twelfths (6/12).

Activity 3
Present the problem $2/3 \times 3/4 = \square$ using the illustrations as shown. The checkerboard squares in diagram *c* represent 2/3 of 3/4, or 6/12. Notice that this is 1/2 of the total, so $6/12 = 1/2$.

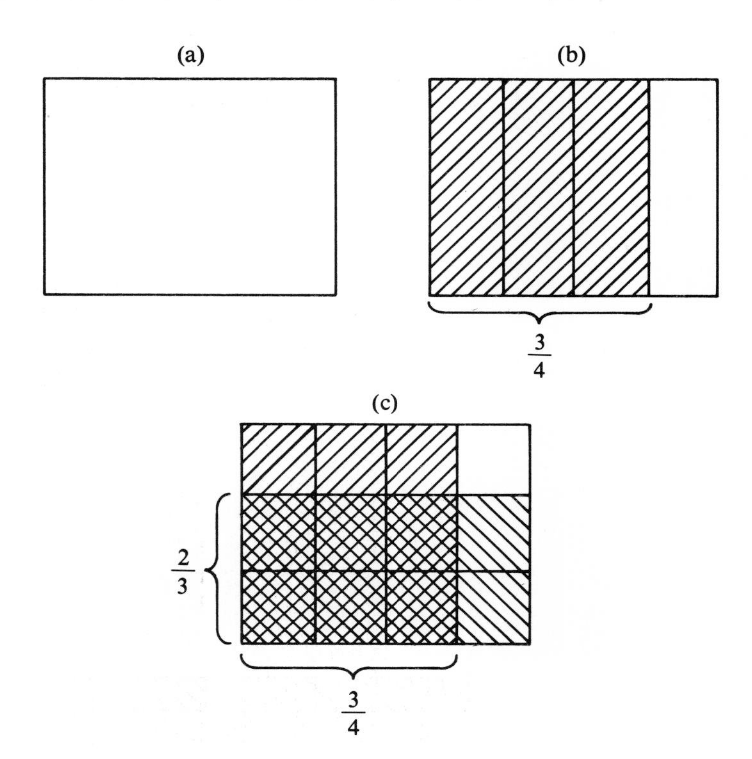

> *Note:* Overlays on an overhead projector work quite nicely for these sequences of operations with fractions.

Additional considerations: Children gain a greater understanding of the meaning of the word *of* in multiplying fractions when they are introduced to practical examples of arithmetic ideas, such as the following:

1. "Fold this paper to find a half of a half."
2. "Give me a quarter of your candy bar."

3. "Mary ate 2/3 of the box of 7 chocolates. How many pieces of candy did she eat?"

To associate the word *of* with multiplication, use the following procedures:

A. Say to the child, "Fold a sheet of paper to show that one-half of one-half equals one-fourth." Guide him to notice that the folded sheet also shows that a quarter of one whole is a quarter.

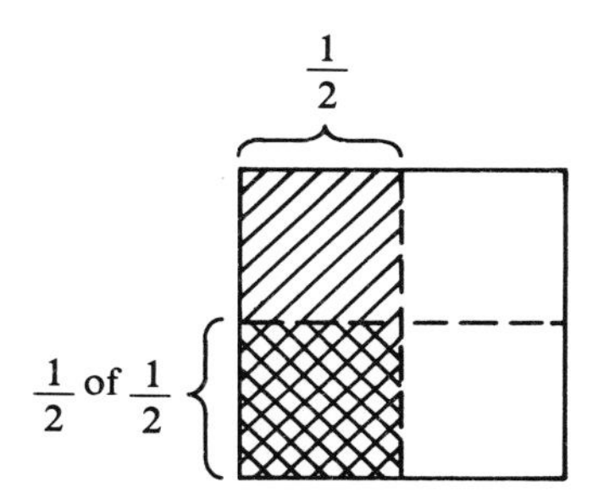

B. Say to the child, "Cut each of seven pieces of paper into thirds and mark two-thirds of each piece of paper." The child can then either count the thirds or rearrange the thirds to form wholes again. The pupil finds she has 4 2/3 pieces of paper.

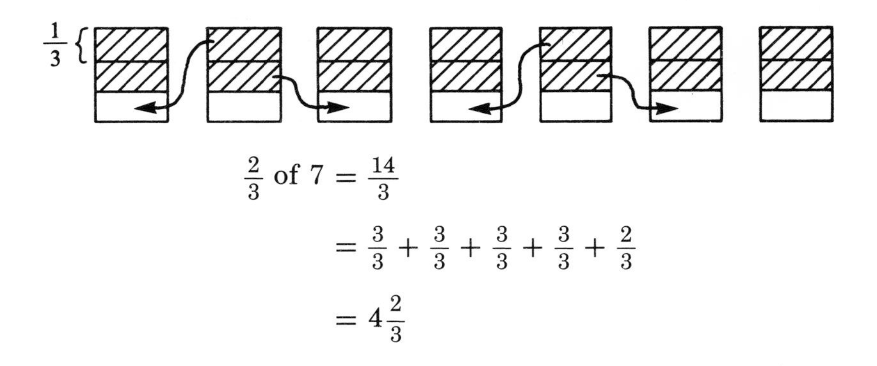

$$\frac{2}{3} \text{ of } 7 = \frac{14}{3}$$

$$= \frac{3}{3} + \frac{3}{3} + \frac{3}{3} + \frac{3}{3} + \frac{2}{3}$$

$$= 4\frac{2}{3}$$

Then show the child figures shaded in different fractional parts and ask her to record various problems using the word *of* to indicate multiplication.

Children may be encouraged to look for a pattern for finding the answers in Activities 1 and 2. Guide the children to notice that the pattern involves multiplying the two numerators together to get the numerator of the product and multiplying the two denominators together to get the denominator of the product.

$$\frac{1}{2} \times \frac{1}{2} = \frac{1 \times 1}{2 \times 2} = \frac{1}{4}$$

$$\frac{3}{4} \times \frac{2}{3} = \frac{3 \times 2}{4 \times 3} = \frac{6}{12}$$

$$\frac{5}{6} \times \frac{2}{3} = \frac{5 \times 2}{6 \times 3} = \frac{10}{18}$$

C. Show the child several phrases containing a multiplication idea expressed by the word *of*. Ask him to replace the word *of* with the multiplication sign. For example: two bags of 5 apples (2×5)

Note: Mini-lesson 4.3.1 is a prerequisite for Mini-lessons 4.3.10 and 4.3.13.

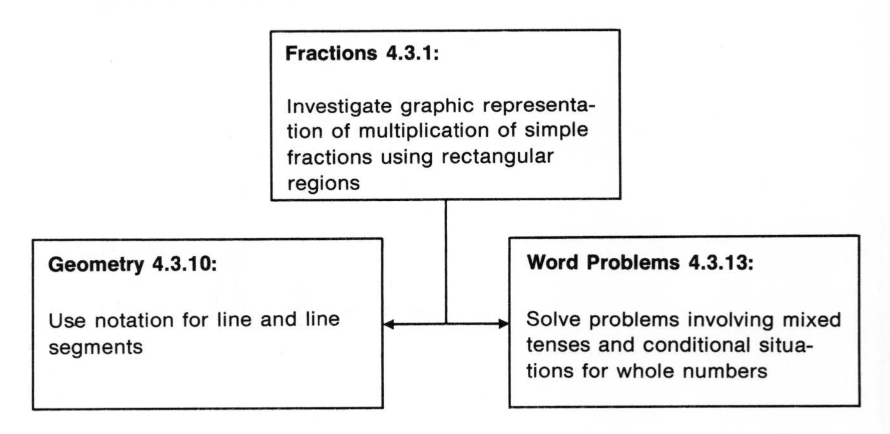

Simplifying Fractions Before Multiplying

Simplifying the names of fractions before obtaining the product is a helpful aid not only for elementary school mathematics but also for later computations.

Mini-Lesson 5.2.2 Simplify a Fraction Before Multiplying

Vocabulary: simplify, prime factor, equivalent notation, rename

Possible Trouble Spot: does not conserve number

Requisite Objectives:
State many names for a number

Activity 1
Guide the child in simplifying the fraction 6/12 by use of primes.

$$\frac{6}{12} = \frac{2 \cdot 3 \cdot 1}{2 \cdot 2 \cdot 3} = \boxed{\frac{2}{2}} \cdot \boxed{\frac{3}{3}} \cdot \frac{1}{2} = \frac{1}{2}$$

Review that prime factor product expressions may be used to represent numbers and that the numerator and denominator of a fraction may be divided by the same number without changing the value of the fraction.

$$\frac{6}{12} \div \frac{6}{6} = \frac{1}{2}$$

Activity 2
Solve the problem $2/3 \times 5/6 = \square$ using various properties of the operations. Simplify first and then multiply.

$\frac{2}{3} \times \frac{5}{6} = \frac{2 \times 5}{3 \times 6}$	Equivalent notation
$= \frac{5 \times 2}{3 \times 6}$	Commutative property for multiplication
$= \frac{5}{3} \times \left(\frac{2}{6} \div \frac{2}{2}\right)$	Right identity for division
$= \frac{5}{3} \times \frac{1}{3}$	Simplified $\frac{2}{6}$ by dividing with the multiplicative identity renamed as $\frac{2}{2}$
$= \frac{5}{9}$	Basic multiplication facts

Divide both the numerator and the denominator of the product expression

$$\frac{2 \times 5}{3 \times 6}$$

by 2 and, therefore, simplification takes place before further multiplication occurs.

Simplification may occur by dividing both numerator and denominator by the greatest common factor.

$$\frac{2}{3}\times\frac{5}{6}=\frac{2\times 5}{3\times 6}=\frac{\overset{1}{\cancel{2}}\times 5}{3\times\underset{1}{\cancel{2}}\times 3}=\frac{5}{9}$$

or

$$\frac{\overset{1}{\cancel{2}}}{3}\times\frac{5}{\underset{3}{\cancel{6}}}=\frac{5}{9}$$

Additional considerations: The term *reduce* as used historically meant to "rename" a fraction as its simplest name—maintaining the same fractional number with the smallest denominator. Some early writers also referred to "reducing one-fourth to twentieths," implying that the new part size becomes smaller.

The word *cancel* means to "annul or destroy." For example, we cancel, or strike out, the 2 in $\cancel{2}/\cancel{4} = 1/2$ or in

$$\frac{\cancel{2}\times 1}{\cancel{2}\times 2}=1/2$$

The terms *reduce* and *cancel* have come into disfavor since overweight people "reduce" and checks and stamps are "cancelled." We say, rather, that fractions are "renamed." However, since children still may encounter these terms, the teacher may wish to relate to the children the historical interpretation of these words to facilitate meaning.

In some cases of cancelling and reducing, division of the numerators and denominators may occur more than once. For example:

$$\frac{4}{5}\times\frac{15}{20}=\frac{\overset{1}{\cancel{2}}\cdot\overset{1}{\cancel{2}}\cdot 3\cdot\overset{1}{\cancel{5}}}{5\cdot\underset{1}{\cancel{2}}\cdot\underset{1}{\cancel{2}}\cdot\underset{1}{\cancel{5}}}=\frac{3}{5}$$

or

$$\frac{4}{5}\times\frac{15}{20}=\frac{\overset{1}{\cancel{4}}\times\overset{3}{\cancel{15}}}{\underset{1}{\cancel{5}}\times\underset{5}{\cancel{20}}}=\frac{3}{5}$$

or

$$\frac{\overset{1}{\cancel{4}}}{\underset{1}{\cancel{5}}}\times\frac{\overset{3}{\cancel{15}}}{\underset{5}{\cancel{20}}}=\frac{3}{5}$$

Sometimes, in multiplying mixed numbers, it is more convenient to rename them as improper fractions to see whether simplification before multiplication is possible.

$$2\frac{1}{4} \times 4\frac{2}{3} = \frac{\overset{3}{\cancel{9}}}{\underset{2}{\cancel{4}}} \times \frac{\overset{7}{\cancel{14}}}{\underset{1}{\cancel{3}}} = \frac{21}{2} = 10\frac{1}{2}$$

The teacher should be alert for those who simplify expressions as

$$\frac{20 + 1}{2}$$

to

$$\frac{\overset{10}{\cancel{20}} + 1}{\underset{1}{\cancel{2}}} = 11$$

Such errors indicate a lack of some prerequisites. The teacher who notices a child making this error should diagnose whether the pupil knows that $21/2 = 20/2 + 1/2$. A helpful aid is to have the child fold 11 pieces of paper in half. Another suggestion is to ask the child to rename 21/2 as wholes.

$\frac{2}{2} + \frac{2}{2} + \frac{2}{2} + \frac{2}{2} + \frac{2}{2}$	"Ten halves"
$\frac{2}{2} + \frac{2}{2} + \frac{2}{2} + \frac{2}{2} + \frac{2}{2}$	"Twenty halves"
$\frac{1}{2}$	"Twenty-one halves"

By counting the wholes, the child sees that there are 10 wholes and 1/2 extra, or 10 1/2.

Note: Mini-lesson 5.2.2 is a prerequisite for Mini-lessons 5.3.2 and 5.3.6.

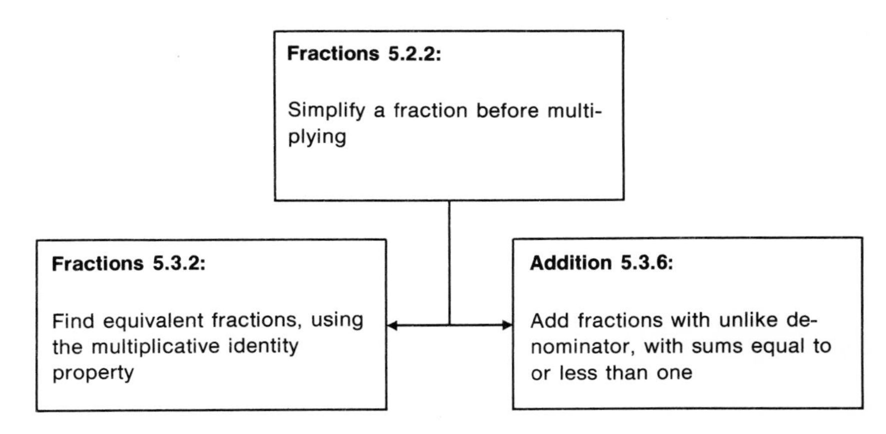

MULTIPLICATION OF MIXED NUMBERS

The distributive property is the basis of the algorithm for the multiplication of mixed numbers just as it is for the algorithm for the multiplication of whole numbers. There are essentially three types of algorithms, as shown below. How many properties can you identify?

1. A mixed number times a whole number

$$\begin{array}{r l} 36 & \\ \times\ 2\frac{1}{4} & \\ \hline 72 & (2 \times 36) \\ 9 & \left(\frac{1}{4} \times 36\right) \\ \hline 81 & \end{array}$$

$$\begin{aligned} 2\tfrac{1}{4} \times 36 &= \left(2 + \tfrac{1}{4}\right) \times 36 \\ &= (2 \times 36) + \left(\tfrac{1}{4} \times 36\right) \\ &= 72 + 9 \\ &= 81 \end{aligned}$$

2. A whole number times a mixed number

$$\begin{array}{r l} 36\frac{1}{4} & \\ \times\ 5 & \\ \hline 30 & (5 \times 6) \\ 150 & (5 \times 30) \\ \hline 180 & \\ +\quad \frac{5}{4} & \left(5 \times \frac{1}{4}\right) \\ \hline 180\frac{5}{4} & \end{array}$$

$$\begin{aligned} 5 \times 36\tfrac{1}{4} &= 5 \times \left(36 + \tfrac{1}{4}\right) \\ &= (5 \times 36) + \left(5 \times \tfrac{1}{4}\right) \\ &= 180 + \tfrac{5}{4} \\ &= 180 + \tfrac{4}{4} + \tfrac{1}{4} \\ &= 180 + 1 + \tfrac{1}{4} \\ &= 181\tfrac{1}{4} \end{aligned}$$

3. A mixed number times a mixed number

$$\begin{array}{r l} 15\frac{1}{2} & \\ \times\ 8\frac{1}{5} & \\ \hline 40 & (8 \times 5) \\ 80 & (8 \times 10) \\ 4 & \left(8 \times \frac{1}{2}\right) \\ 3 & \left(\frac{1}{5} \times 15\right) \\ \hline 127 & \\ \frac{1}{10} & \left(\frac{1}{5} \times \frac{1}{2}\right) \\ \hline 127\frac{1}{10} & \end{array}$$

$$\begin{aligned} 8\tfrac{1}{5} \times 15\tfrac{1}{2} &= \left(8 + \tfrac{1}{5}\right) \times \left(15 + \tfrac{1}{2}\right) \\ &= (8 \times 15) + \left(8 \times \tfrac{1}{2}\right) + \\ &\quad \left(\tfrac{1}{5} \times 15\right) + \left(\tfrac{1}{5} \times \tfrac{1}{2}\right) \\ &= 120 + 4 + 3 + \tfrac{1}{10} \\ &= 127\tfrac{1}{10} \end{aligned}$$

If pupils have displayed the necessary prerequisites for performing the operations in the first and second algorithms, they should be able to compute the third algorithm by applying the Distributive Property of Multiplication over Addition at the symbolic level.

To solve the problem 5 × 1 1/3, show as (5 × 1) + (5 × 1/3) using a number line.

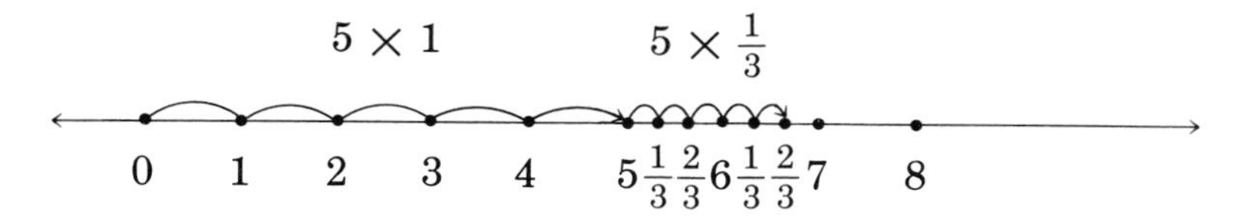

Record the following representation, as shown on the number line.

$$\begin{aligned}5 \times 1\tfrac{1}{3} &= (5 \times 1) + \left(5 \times \tfrac{1}{3}\right)\\ &= \quad 5 \quad + \quad \tfrac{5}{3}\\ &= 5 + \tfrac{3}{3} + \tfrac{2}{3}\\ &= (5 + 1) + \tfrac{2}{3}\\ &= 6\tfrac{2}{3}\end{aligned}$$

DIVISION OF FRACTIONS

What is involved in computing 3/4 ÷ 2/5? At this point, the child should have the necessary prerequisite skills, concepts, and principles to perform this division. These prerequisite concepts and an example for each are listed in the following task analysis structure.

Prerequisites for division of fractions

1. There are different notations for division.

$$6 \div 3 \quad \text{or} \quad 3\overline{)6} \quad \text{or} \quad \frac{6}{3}$$

Therefore

$$\frac{3}{4} \div \frac{2}{5} = \frac{\frac{3}{4}}{\frac{2}{5}}$$

2. Any number divided by 1 is that number ($n/1 = n$). If the denominator (2/5) of the complex fraction above could be eliminated, the problem would be easier to solve. This elimination can occur if 2/5 is renamed to 1. Since a number times its reciprocal is another name

for 1 ($a \times 1/a = a/a = 1$), 2/5 is multiplied by its reciprocal as follows:

$$\frac{2}{5} \times \frac{5}{2} = \frac{10}{10} = 1$$

3. If both the numerator and the denominator are multiplied (or divided) by the same number, the value of the original fraction is not changed, because the resulting multiplier b/b is another name for 1, the multiplicative identity. This is called the "golden rule for fractions." If the denominator of the complex fraction is multiplied by 5/2, then its numerator must also be multiplied by 5/2. Notice that the following complex fraction equals 1.

$$\frac{\frac{5}{2}}{\frac{5}{2}} = 1$$

Thus, the original complex fraction

$$\frac{\frac{3}{4}}{\frac{2}{5}}$$

is multiplied by 1, changing its name but not its original number value.

$$\frac{\frac{3}{4}}{\frac{2}{5}} \times \boxed{\frac{\frac{5}{2}}{\frac{5}{2}}}$$

Notice that there are two separate uses of the multiplicative identity concept in steps 2 and 3 below. Thus, the step-by-step rationale is shown for inverting the divisor and multiplying in division of fractions.

$$\frac{\frac{3}{4} \times \frac{5}{2}}{\frac{2}{5} \times \frac{5}{2}} = \frac{\frac{3}{4} \times \frac{5}{2}}{\frac{10}{10}} = \frac{\frac{3}{4} \times \frac{5}{2}}{1} = \frac{3}{4} \times \frac{5}{2}$$

Examples in which inversion is unnecessary are as follows:

(a) $\dfrac{3}{8} \div \dfrac{3}{4} = \dfrac{3 \div 3}{8 \div 4} = \dfrac{1}{2}$

(b) $\dfrac{25}{36} \div \dfrac{5}{9} = \dfrac{25 \div 5}{36 \div 9} = \dfrac{5}{4} = \dfrac{4 + 1}{4}$

$$= \frac{4}{4} + \frac{1}{4} = 1 + \frac{1}{4} = 1\frac{1}{4}$$

ADDITION OF FRACTIONS

Like Fractions

Have children read the problem aloud

When adding like fractions, reading the problem aloud is helpful. Children have been using essentially the same language pattern when they add "*one* candy bar and *two* candy bars to obtain *three* candy bars." In an example described in an earlier chapter involving the youngster leaving his doctor's office saying "One lollipop for me, one lollipop for sister, and one lollipop for brother; I have three lollipops," the child could have been labeling the countable idea in terms of fractions or place values instead of lollipops.

(a) 1 (lollipop) + 1 (lollipop) + 1 (lollipop) = 3 (lollipops)
(b) 1 (ten) + 1 (ten) + 1 (ten) = 3 (tens) = 30
(c) 1 (fourth) + 1 (fourth) + 1 (fourth) = 3 (fourths)

$$\frac{1}{4} + \frac{1}{4} + \frac{1}{4} = \frac{3}{4}$$

Using labels for the countable ideas

The denominator may be considered the label for the phenomenon being counted and the numerator as the counter.

The above sum may also be expressed as (1 + 1 + 1) lollipops = 3 lollipops. Examples *b* and *c* above could also be written as (1 + 1 + 1) ten = 3 tens and (1 + 1 + 1) fourths $= \frac{1+1+1}{4} = \frac{3}{4}$.

Then

$$\frac{3+2}{7} = \frac{5}{7}$$

and

$$\frac{8}{10} + \frac{1}{10} = \frac{8+1}{10} = \frac{9}{10}$$

This demonstrates that the numerators are added, not the denominators. Activities involving examples like these will help prevent the following computational errors.

$$\begin{array}{r} \frac{3}{8} \\ -\frac{1}{3} \\ \hline \frac{2}{5} \end{array} \quad \text{or} \quad \begin{array}{r} \frac{1}{3} \\ +\frac{2}{5} \\ \hline \frac{3}{8} \end{array}$$

After children know how to add fractions having the same name, such as $1/4 + 1/4 + 1/4 = 3/4$ or $1/8 + 1/8 + 1/8 + 1/8 + 1/8 = 5/8$, they may be introduced to examples such as the following:

(a) $\frac{1}{8} + \frac{2}{8} = \frac{3}{8}$

(b) $\frac{2}{5} + \frac{2}{5} = \frac{4}{5}$

(c) $\frac{1}{6} + \frac{5}{6} = \frac{6}{6} = 1$

(d) $\frac{3}{4} + \frac{3}{4} = \frac{6}{4} = \frac{4}{4} + \frac{2}{4} = 1 + \frac{2}{4} = 1\frac{2}{4} \left(\text{or} = \frac{3}{2} = 1\frac{1}{2}\right)$

(e) $1\frac{2}{4} + 2\frac{1}{4} = 3\frac{3}{4}$

Note: The associative and commutative properties are helpful when adding fractions expressed as mixed numbers. Note the steps below that show these properties.

$$1\frac{3}{4} + 2\frac{2}{4} = 1 + \frac{3}{4} + 2 + \frac{2}{4}$$

$$= 1 + \left(\frac{3}{4} + 2\right) + \frac{2}{4}$$ Associative Property for Addition

$$= 1 + \left(2 + \frac{3}{4}\right) + \frac{2}{4}$$ Commutative Property for Addition

$$= (1 + 2) + \left(\frac{3}{4} + \frac{2}{4}\right)$$ Associative Property for Addition

$$= 3 + \frac{5}{4}$$

$$= 3 + \frac{4 + 1}{4}$$

$$= 3 + \frac{4}{4} + \frac{1}{4}$$

$$= 3 + 1 + \frac{1}{4}$$

$$= 4 + \frac{1}{4}$$

$$= 4\frac{1}{4}$$

Addition of rational numbers

Following are some examples of addition of rational numbers. (Recall that a *rational number* is the quotient of two integers with the divisor not equal to zero.) Rational numbers may be named by fractions in the form a/b $(b \neq 0)$, where a and b are integers.

1. A whole number plus a rational number expressed as a proper fraction

$$\begin{array}{r} 4 \\ +\ \frac{3}{8} \\ \hline 4\frac{3}{8} \end{array}$$

2. A mixed number and a proper fraction, with no renaming

$$\begin{array}{r} 4\frac{2}{8} \\ +\ \frac{3}{8} \\ \hline 4\frac{5}{8} \end{array}$$

3. A mixed number and a proper fraction, with renaming

$$\begin{array}{r} 3\frac{5}{8} \\ +\ \frac{3}{8} \\ \hline 3\frac{8}{8} \end{array} = 3 + 1 = 4$$

4. Mixed numbers, with no renaming

$$\begin{array}{r} 2\frac{1}{4} \\ +1\frac{2}{4} \\ \hline 3\frac{3}{4} \end{array}$$

5. Mixed numbers, with renaming to whole number

$$\begin{array}{r} 2\frac{1}{4} \\ +1\frac{3}{4} \\ \hline 3\frac{4}{4} \end{array} = 4$$

6. Mixed numbers, with renaming to mixed number

$$\begin{array}{r} 2\frac{3}{4} \\ +1\frac{2}{4} \\ \hline 3\frac{5}{4} \end{array} = 3 + 1 + \frac{1}{4} = 4\frac{1}{4}$$

Unlike Fractions

What is necessary to add unlike fractions, such as 3/4 + 5/6?

1. A common denominator must be found. We do this by finding the least common multiple (LCM) of the denominators (as discussed in Chapter 13).

$$\begin{array}{rl} \frac{3}{4} & 4, 8, \textcircled{12} \\ +\frac{5}{6} & 6, \textcircled{12} \\ \hline & \text{LCM} = 12 \end{array}$$

2. The LCM is used as the new denominator for both fractions.

$$\begin{array}{r} \frac{3}{4} = \frac{}{12} \\ +\frac{5}{6} = \frac{}{12} \\ \hline \end{array}$$

3. The original fractions are renamed to their equivalents, and the denominator will be the LCM. Open, indirect sentences are helpful in this renaming process.

$$4 \times \square = 12$$
$$4 \times \boxed{3} = 12$$
$$6 \times \triangle = 12$$
$$6 \times \triangle = 12$$

4. The fractions as renamed are added.

$$\frac{3}{4} \times \frac{\boxed{3}}{\boxed{3}} = \frac{9}{12}$$

$$+\frac{5}{6} \times \frac{\triangle 2}{\triangle 2} = \frac{10}{12}$$

$$\frac{19}{12} = \frac{12}{12} + \frac{7}{12} = 1\frac{7}{12}$$

SUBTRACTION OF FRACTIONS

Like Fractions

There are several types of subtraction problems with like denominators.

1. Rationals expressed as proper fractions, with no renaming

$$\begin{array}{r} \frac{2}{4} \\ -\frac{1}{4} \\ \hline \frac{1}{4} \end{array}$$

2. Rationals expressed as proper fractions, with renaming

$$\frac{3}{4} - \frac{1}{4} = \frac{2}{4} = \frac{\overset{1}{\cancel{2}}}{\underset{1}{\cancel{2}} \times 2} = \frac{1}{2}$$

> *Note:* In the second problem above, the teacher may guide the child to think "twos out of two, one; twos out of two, one; the numerator is one; the denominator is one times two, or two; the answer is one-half." Dividing both numerator and denominator by 2 is the same as dividing by 2/2 or 1.
>
> $$\frac{2}{4} \div \frac{2}{2} = \frac{2 \div 2}{4 \div 2} = \frac{1}{2}$$

The cancelling notation was probably popular in previous eras because of the belief that children could not divide a fraction by a fraction in the early stages of learning about fractional operations. The form

$$\frac{2 \div 2}{4 \div 2} = \frac{1}{2}$$

however, is nothing more than a double division of whole numbers. A difficulty with the cancellation method is that sometimes a pupil will say zero is the numerator, as a result of cancelling the numerator number. For example:

$$\frac{2}{4} = \frac{\not{2}}{\not{2} \times 2} = \frac{0}{2}$$

or

$$\frac{6}{8} = \frac{\not{2} \times 3}{\not{2} \times 2 \times 2} = \frac{0}{4}$$

The child thinks that $\not{2} \times 3$ implies 0×3, which is indeed zero. The 2×2 in the denominator somehow offsets this same logic, so he or she obtains 4 in the denominator instead of zero. Placement of 1 as in the following example avoids this confusion.

$$\frac{6}{9} = \frac{2 \times \overset{1}{\not{3}}}{3 \times \underset{1}{\not{3}}} = \frac{2 \times 1}{3 \times 1} = \frac{2}{3}$$

3. Rationals expressed as improper and proper fractions, with no renaming

$$\begin{array}{r} \frac{9}{8} \\ -\frac{7}{8} \\ \hline \frac{2}{8} \end{array}$$

4. A whole number minus a rational number expressed as a proper fraction, with renaming of the whole number as a mixed number and using the denominator of the subtrahend as a special name for 1

$$\begin{array}{rcr} 4 & = & 3\frac{8}{8} \\ -\frac{3}{8} & = & \frac{3}{8} \\ \hline & & 3\frac{5}{8} \end{array}$$

5. Rationals expressed as mixed and proper fractions, with renaming in the minuend

$$\begin{array}{rr} 3\frac{2}{8} = 2 + 1\frac{2}{8} = 2 + \frac{8}{8} + \frac{2}{8} = & 2\frac{10}{8} \\ -\ \frac{3}{8} = \qquad\qquad\qquad & -\ \frac{3}{8} \\ \hline & 2\frac{7}{8} \end{array}$$

6. Rationals expressed as mixed numbers, with no renaming in the minuend

$$\begin{array}{r} 3\frac{2}{8} \\ -1\frac{1}{8} \\ \hline 2\frac{1}{8} \end{array}$$

7. Rationals expressed as mixed numbers, with renaming in the minuend

$$\begin{array}{rr} 3\frac{2}{8} = 2 + \frac{8}{8} + \frac{2}{8} = & 2\frac{10}{8} \\ -1\frac{5}{8} \qquad\qquad & 1\frac{5}{8} \\ \hline & 1\frac{5}{8} \end{array}$$

Unlike Fractions

Subtraction of fractions with unlike denominators include the following:

1. Rationals expressed as proper fractions

$$\begin{array}{r} \frac{3}{4} \times \frac{3}{3} = \frac{9}{12} \\ -\frac{1}{6} \times \frac{2}{2} = \frac{2}{12} \\ \hline \frac{7}{12} \end{array}$$

2. Improper fractions, with renaming in the remainder

$$\begin{array}{l} \frac{9}{4} \times \frac{3}{3} = \frac{27}{12} \\ -\frac{1}{6} \times \frac{2}{2} = \frac{2}{12} \\ \hline \qquad\qquad\ \frac{25}{12} = \frac{12}{12} + \frac{12}{12} + \frac{1}{12} = 2\frac{1}{12} \end{array}$$

3. A mixed number and a proper fraction, with no renaming in the minuend

$$\begin{array}{r} 3\frac{4}{5} \times \frac{8}{8} = 3\frac{32}{40} \\ -\ \frac{3}{8} \times \frac{5}{5} = \ \ \frac{15}{40} \\ \hline 3\frac{17}{40} \end{array}$$

LCM = 40
5, 10, 15, 20, 25, 30, 35, (40)
8, 16, 24, 32, (40)
Note that children often forget the whole number in the renaming process.

4. Mixed and proper fractions, renaming within the minuend

$$\begin{array}{r} 3\frac{1}{5} \times \frac{8}{8} = 3\frac{8}{40} = 2 + 1 + \frac{8}{40} = 2 + \frac{40}{40} + \frac{8}{40} = 2\frac{48}{40} \\ -\ \frac{3}{8} \times \frac{5}{5} = \frac{15}{40} \qquad\qquad\qquad\qquad\qquad\qquad\qquad \frac{15}{40} \\ \hline 2\frac{33}{40} \end{array}$$

5. Rationals expressed as mixed numbers, with no renaming in the minuend

$$\begin{array}{r} 3\frac{4}{5} \times \frac{8}{8} = 3\frac{32}{40} \\ -1\frac{3}{8} \qquad = 1\frac{15}{40} \\ \hline 2\frac{17}{40} \end{array}$$

6. Rationals expressed as mixed numbers, with renaming in the minuend

$$\begin{array}{r} 3\frac{1}{5} \times \frac{8}{8} = 3\frac{8}{40} = 2 + \frac{40}{40} + \frac{8}{40} = 2\frac{48}{40} \\ -1\frac{3}{8} \times \frac{5}{5} = 1\frac{15}{40} = \qquad\qquad\qquad 1\frac{15}{40} \\ \hline 1\frac{33}{40} \end{array}$$

Mini-lesson 6.1.1 Apply Number Properties to Operations on Fractions

Vocabulary: commutative, associative, distributive, closure, identity

Possible Trouble Spots: does not notice salient aspects of a situation; poor memory skills; has difficulty generalizing to new situations

Requisite Objectives:
Addition 2.2.3: Use additive identity

Subtraction 2.2.7: Investigate the addition properties as they apply to subtraction

Multiplication 3.1.1: Use multiplication properties

Division 4.1.9: Investigate properties of multiplication in regard to division

Activity 1

After reviewing number properties applied to whole numbers, assess the students' ability to transfer to naming properties applied to fractions. This may take the form of recall tasks, matching columns, or multiple choice items.

Activity 2

Show the child various fraction operations and ask him to name the property.

(a) $\frac{3}{4} + \frac{7}{9} = \frac{7}{9} + \frac{3}{4}$

(b) $\frac{3}{6} \times \frac{1}{2} = \frac{1}{2} \times \frac{3}{6}$

(c) $\frac{4}{8} \div \frac{2}{2} = \frac{2}{4}$

(d) $3 \times \frac{47}{7} = \left(3 \times \frac{40}{7}\right) + \left(3 \times \frac{7}{7}\right)$

$= \frac{120}{7} + \frac{21}{7}$

$= \frac{141}{7}$

(e) $\left(\frac{1}{2} + \frac{1}{3}\right) + \frac{1}{4} = \frac{1}{2} + \left(\frac{1}{3} + \frac{1}{4}\right)$

Key: (a) Commutative Property for Addition

(b) Commutative Property for Multiplication

(c) Identity related to division

(d) Distributive Property for Multiplication over Addition

(e) Associative Property for Addition

Additional considerations: Encourage students to use language that is comfortable for them. For example, use "order" instead of "commutative," use "grouping" for the associative property, and "law for multiplying by 1" instead of "multiplicative identity."

Keep in mind that an understanding of the properties is most useful to those who need help in computing. For example, students may use their knowledge of the Distributive Property of Multiplication over Addition in multiplying mixed numbers, as shown in the following example.

$$\textit{Error:}\quad \begin{array}{r} 3\frac{1}{4} \\ \times 2\frac{1}{2} \\ \hline 6\frac{1}{8} \end{array}$$

Using Distributive Property of Multiplication over Addition:

$$\begin{aligned}
3\tfrac{1}{4} \times 2\tfrac{1}{2} &= \left(3 + \tfrac{1}{4}\right)\left(2 + \tfrac{1}{2}\right) \\
&= (3 \times 2) + \left(3 \times \tfrac{1}{2}\right) + \left(\tfrac{1}{4} \times 2\right) + \left(\tfrac{1}{4} \times \tfrac{1}{2}\right) \\
&= 6 + \tfrac{3}{2} + \tfrac{2}{4} + \tfrac{1}{8} \\
&= 6 + \left(\tfrac{2}{2} + \tfrac{1}{2}\right) + \tfrac{2}{4} + \tfrac{1}{8} \\
&= (6 + 1) + \tfrac{1}{2} + \tfrac{2}{4} + \tfrac{1}{8} \\
&= 7 + \left(\tfrac{4}{8} + \tfrac{4}{8}\right) + \tfrac{1}{8} \\
&= 7 + 1 + \tfrac{1}{8} \\
&= 8\tfrac{1}{8}
\end{aligned}$$

Note: Mini-lesson 6.1.1 is a prerequisite for Mini-lessons 6.1.2 and 6.1.6.

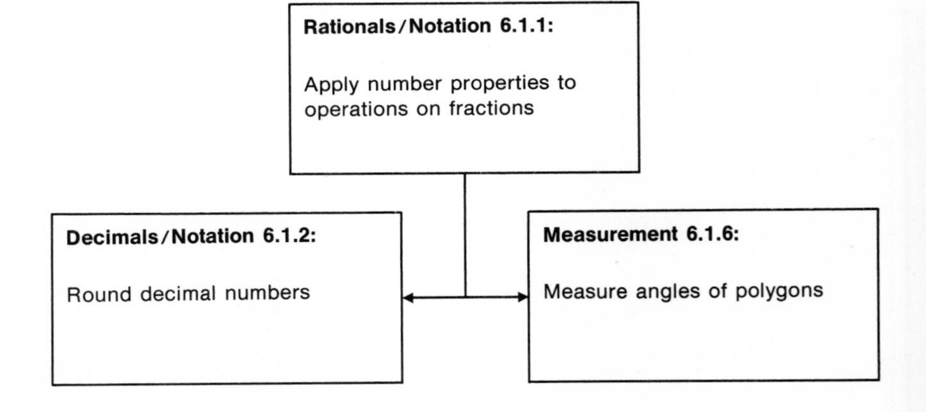

EXERCISES

I. Perform the following operations and write the answers in simplest form:

1. $\frac{1}{4} + \frac{1}{4} + \frac{1}{4} + \frac{1}{4} + \frac{1}{4}$
2. $\frac{2}{3} + \frac{2}{3} + \frac{2}{3} + \frac{2}{3}$
3. $\frac{1}{2} + \frac{1}{4}$
4. $\frac{1}{3} + \frac{1}{6}$
5. $\frac{5}{8} - \frac{3}{8}$
6. $\frac{3}{4} - \frac{1}{2}$
7. $\frac{4}{5} - \frac{1}{2}$

II. Use the chart below to perform the following computations:

1. $\frac{1}{2} + \frac{3}{18}$
2. $\frac{3}{4} - \frac{3}{16}$
3. $\frac{1}{10} + \frac{1}{2} + \frac{1}{5}$

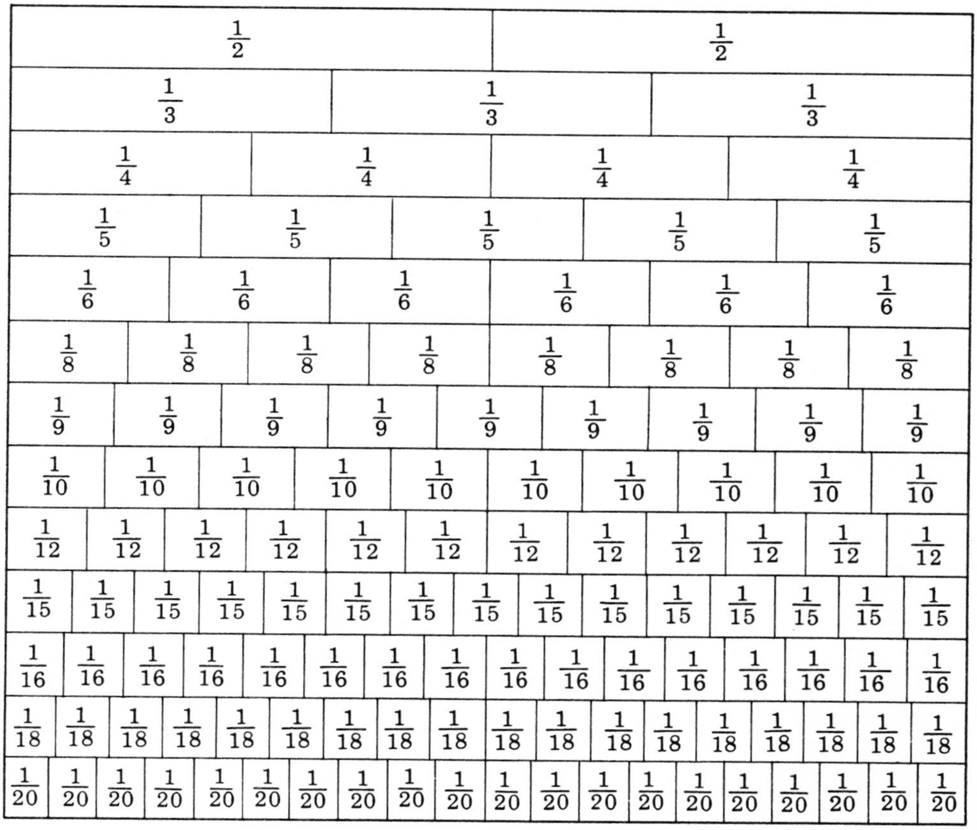

$\frac{1}{2}$	$\frac{1}{2}$

$\frac{1}{3}$	$\frac{1}{3}$	$\frac{1}{3}$

$\frac{1}{4}$	$\frac{1}{4}$	$\frac{1}{4}$	$\frac{1}{4}$

$\frac{1}{5}$	$\frac{1}{5}$	$\frac{1}{5}$	$\frac{1}{5}$	$\frac{1}{5}$

$\frac{1}{6}$	$\frac{1}{6}$	$\frac{1}{6}$	$\frac{1}{6}$	$\frac{1}{6}$	$\frac{1}{6}$

$\frac{1}{8}$	$\frac{1}{8}$	$\frac{1}{8}$	$\frac{1}{8}$	$\frac{1}{8}$	$\frac{1}{8}$	$\frac{1}{8}$	$\frac{1}{8}$

$\frac{1}{9}$	$\frac{1}{9}$	$\frac{1}{9}$	$\frac{1}{9}$	$\frac{1}{9}$	$\frac{1}{9}$	$\frac{1}{9}$	$\frac{1}{9}$	$\frac{1}{9}$

$\frac{1}{10}$	$\frac{1}{10}$	$\frac{1}{10}$	$\frac{1}{10}$	$\frac{1}{10}$	$\frac{1}{10}$	$\frac{1}{10}$	$\frac{1}{10}$	$\frac{1}{10}$	$\frac{1}{10}$

$\frac{1}{12}$	$\frac{1}{12}$	$\frac{1}{12}$	$\frac{1}{12}$	$\frac{1}{12}$	$\frac{1}{12}$	$\frac{1}{12}$	$\frac{1}{12}$	$\frac{1}{12}$	$\frac{1}{12}$	$\frac{1}{12}$	$\frac{1}{12}$

$\frac{1}{15}$	$\frac{1}{15}$	$\frac{1}{15}$	$\frac{1}{15}$	$\frac{1}{15}$	$\frac{1}{15}$	$\frac{1}{15}$	$\frac{1}{15}$	$\frac{1}{15}$	$\frac{1}{15}$	$\frac{1}{15}$	$\frac{1}{15}$	$\frac{1}{15}$	$\frac{1}{15}$	$\frac{1}{15}$

$\frac{1}{16}$	$\frac{1}{16}$	$\frac{1}{16}$	$\frac{1}{16}$	$\frac{1}{16}$	$\frac{1}{16}$	$\frac{1}{16}$	$\frac{1}{16}$	$\frac{1}{16}$	$\frac{1}{16}$	$\frac{1}{16}$	$\frac{1}{16}$	$\frac{1}{16}$	$\frac{1}{16}$	$\frac{1}{16}$	$\frac{1}{16}$

$\frac{1}{18}$	$\frac{1}{18}$	$\frac{1}{18}$	$\frac{1}{18}$	$\frac{1}{18}$	$\frac{1}{18}$	$\frac{1}{18}$	$\frac{1}{18}$	$\frac{1}{18}$	$\frac{1}{18}$	$\frac{1}{18}$	$\frac{1}{18}$	$\frac{1}{18}$	$\frac{1}{18}$	$\frac{1}{18}$	$\frac{1}{18}$	$\frac{1}{18}$	$\frac{1}{18}$

$\frac{1}{20}$	$\frac{1}{20}$	$\frac{1}{20}$	$\frac{1}{20}$	$\frac{1}{20}$	$\frac{1}{20}$	$\frac{1}{20}$	$\frac{1}{20}$	$\frac{1}{20}$	$\frac{1}{20}$	$\frac{1}{20}$	$\frac{1}{20}$	$\frac{1}{20}$	$\frac{1}{20}$	$\frac{1}{20}$	$\frac{1}{20}$	$\frac{1}{20}$	$\frac{1}{20}$	$\frac{1}{20}$	$\frac{1}{20}$

III. Change the following mixed fractions to improper fractions:

1. $1\frac{3}{4}$
3. $2\frac{1}{5}$
3. $4\frac{2}{3}$
4. $2\frac{13}{16}$

IV. Perform the following operations:

1. $3\frac{1}{2} \times 1\frac{1}{5}$
2. $2\frac{2}{3} \times 3\frac{3}{4}$
3. $4\frac{1}{2} \times 3\frac{1}{3}$
4. $2\frac{1}{4} \div 1\frac{1}{2}$
5. $7\frac{1}{2} \div 6\frac{1}{4}$
6. $6\frac{2}{3} \div 2\frac{2}{3}$

V. $\frac{2}{6} (= / \neq) \frac{1+1}{6}$

VI. The sum $\frac{a}{b} + \frac{c}{d}$ can be represented as $\frac{a}{b} \cdot \frac{d}{d} + \frac{b}{b} \cdot \frac{c}{d}$ since b and d are ________.

VII. $\frac{3}{5} + \left(\frac{2}{3} + \frac{1}{2}\right) = \frac{3}{5} + \left(\frac{1}{2} + \frac{2}{3}\right)$ shows that addition of fractions is ________.

VIII. $\frac{2}{3} + \left(\frac{1}{5} + \frac{3}{2}\right) = \left(\frac{2}{3} + \frac{1}{5}\right) + \frac{3}{2}$ shows that addition of fractions is ________.

IX. $\frac{0}{1} = \frac{0}{1} \cdot \frac{k}{k} =$ ________ if k is a nonzero whole number.

X. State the basic property for each of the following:

1. $\frac{a}{b} + \left(\frac{c}{d} + \frac{e}{f}\right) = \left(\frac{a}{b} + \frac{c}{d}\right) + \frac{e}{f}$
2. $\frac{a}{b} + \frac{e}{f} = \frac{e}{f} + \frac{a}{b}$
3. $\frac{a}{b} + \frac{0}{k} = \frac{0}{k} + \frac{a}{b} = \frac{a}{b}$
4. If $\frac{a}{b}$ and $\frac{c}{d}$ are fractions, then $\frac{a}{b} \div \frac{c}{d}$ is a fraction.

XI. Which of the following statements is valid for fractions?

1. Examples of fractions are $\frac{1}{4}, \frac{-3}{-1}, \frac{0}{5}, \frac{0}{-8}, \frac{1}{-4}$.
2. Subtraction is commutative.
3. Subtraction is associative.
4. Subtraction has the identity $\frac{0}{k}$, $k \neq 0$ for all cases.
5. Examples of fractions are $\frac{1}{0}, \frac{\pi}{3}, \frac{\sqrt{2}}{1}$.

SUGGESTED READINGS

Green, G. F. "A Model for Teaching Multiplication of Fractional Numbers." *The Arithmetic Teacher* 20 (January 1973): 5–9.

Linn, C. F. *The Golden Mean.* New York: Doubleday, 1974.

Pittman, P. "Rapid Mental Squaring of Mixed Numbers." *The Mathematics Teacher* 70 (October 1977): 596–97.

Shokoohi, Gholam-Hassein. "Readiness of Eight-Year-Old Children to Understand the Division of Fractions." *The Arithmetic Teacher* 27 (March 1980): 40–43.

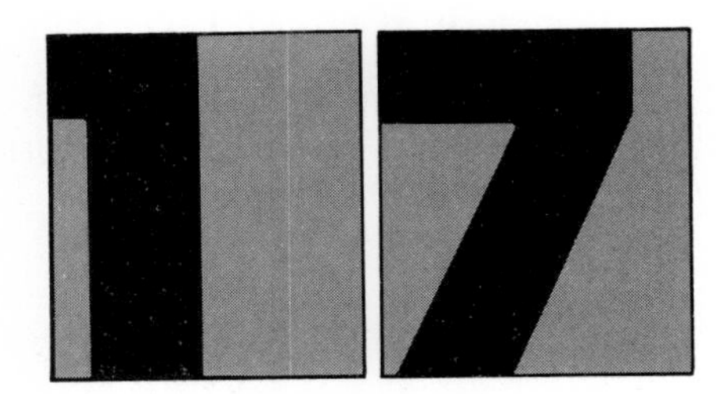

Decimals and How To Write Them

WHAT IS A DECIMAL?

The terms *decimal* and *decimal fraction* denote a fraction whose denominator is a power of ten. The forms 5/10, 65/100, and so on, are commonly called decimal fractions, while the word decimal names their equivalents in the form .5 and .65. *Mixed decimals* are written in the form 23.06. A decimal, therefore, has no written denominator; instead, its denominator is expressed by a dot written before the numerator. If "three thousandths" is read aloud, the person hearing this could not tell if the written form were 3/1000 or .003.

Introduce decimals early in the curriculum

It is suggested that the topic of decimals be introduced earlier than the traditional sequencing has allowed. Since the writing of decimals is merely an extension to the right of the decimal place-value system, there is no need to veil half of the numeration system until fifth or sixth grade. This does not imply that young children master the topic of decimals but rather that they investigate the ideas associated with place values to the right of the decimal point.

AIDS FOR TEACHING DECIMAL NOTATION

Many instructional aids used for teaching whole numbers are appropriate for decimals. For example, place-value charts may be extended to include decimals. The heavy vertical line separating units from tenths (shown below) represents the decimal point.

hundreds	tens	units	tenths	hundredths

The units place on an abacus may be moved to the left to allow for decimal places to its right.

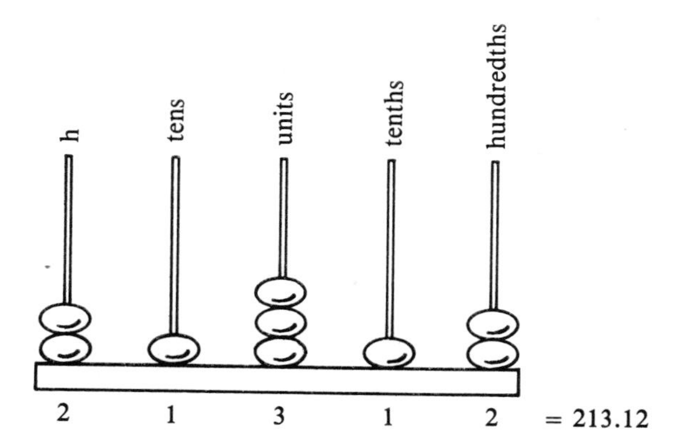

Counting boards placed to the right of the units board may be used to represent tenths, hundredths, thousandths, and so on.

The teacher may use pennies and dimes to represent decimal extensions in relation to a dollar; and meter, decimeter, centimeter, and millimeter strips may be used to provide various experiences with decimals, including thousandths.

Two relationships that underlie teaching decimal notation will be presented: (1) an interpretation of decimals as an extension of place value to the right of the units obtained by following the left-to-right pattern of divide-by-ten, and (2) the reciprocal relationship between whole numbers and fractions (for example, the reciprocal of 10 is 1/10, and the reciprocal of 1/10 is 10/1 or 10; this second relationship is particularly appropriate for older students who may also study negative exponents as pointed out below). The reciprocal relationship has been involved as students renamed the number 1 in the form a/a, $a \neq 0$. But they may not have understood that the number expressed as a/a $(a \neq 0)$ was the standard name obtained when reciprocals were multiplied:

$$\frac{a}{1} \times \frac{1}{a} = \frac{a \times 1}{1 \times a} = \frac{a \times 1}{a \times 1} = \frac{a}{a} \times \frac{1}{1} = \frac{a}{a} \times 1 = \frac{a}{a}$$

Negative exponents are also related to the reciprocal relationship. At this time you may wish to review the section in Chapter 19 concerning the expression of decimal place values in exponential notation.

INTERPRETING DECIMAL FRACTIONS

Terminology

The teacher should point out to the children that the *th* on the end of decimal names is related to the names of denominators, such as one-fourth for 1/4, one-fifth for 1/5, one-sixth for 1/6, and so on. The word *and* is often used to read the decimal point; "three and twenty-five thousandths" is written $3\frac{25}{1000}$ or 3.025. A zero is used to show "no tenths." In the mixed decimal 200.203, the zeros indicate that there are no tens, no units, and no hundredths.

Mixed decimals

A *mixed decimal* (7.32) is composed of a whole number and a fractional number. This is similar to a mixed fraction $\left(7\frac{32}{100}\right)$ or an improper fraction (732/100).

Simple decimals

Simple, or *pure,* decimals have only a fractional part (.72). The proper fraction 3/4 indicates a fractional part of the place value to its left. The numeral $.00\frac{1}{3}$ is read "one-third hundredth," and its value may be expressed in fractional form as $1/3 \times 1/100$, or 1/300.

Equivalent decimals

The method of reading and writing decimals may include such renamings as $.4 \longrightarrow .40 \longrightarrow .400$ (four-tenths has the same value as forty hundredths, which in turn has the same value as four hundred thousandths). (Four hundred-thousandths represents .00004.) The concept of equivalent fractions is basic to the renaming process. This may be shown at the concrete and picture levels. A cube comprising 100 blocks can be used to show that 1 of its unit cubes is .01 of the large cube. Graph paper may be introduced to show, for example, that 1 strip of 1000 squares is one-tenth (.1) of a whole grid of 10,000 (as shown in the illustration).

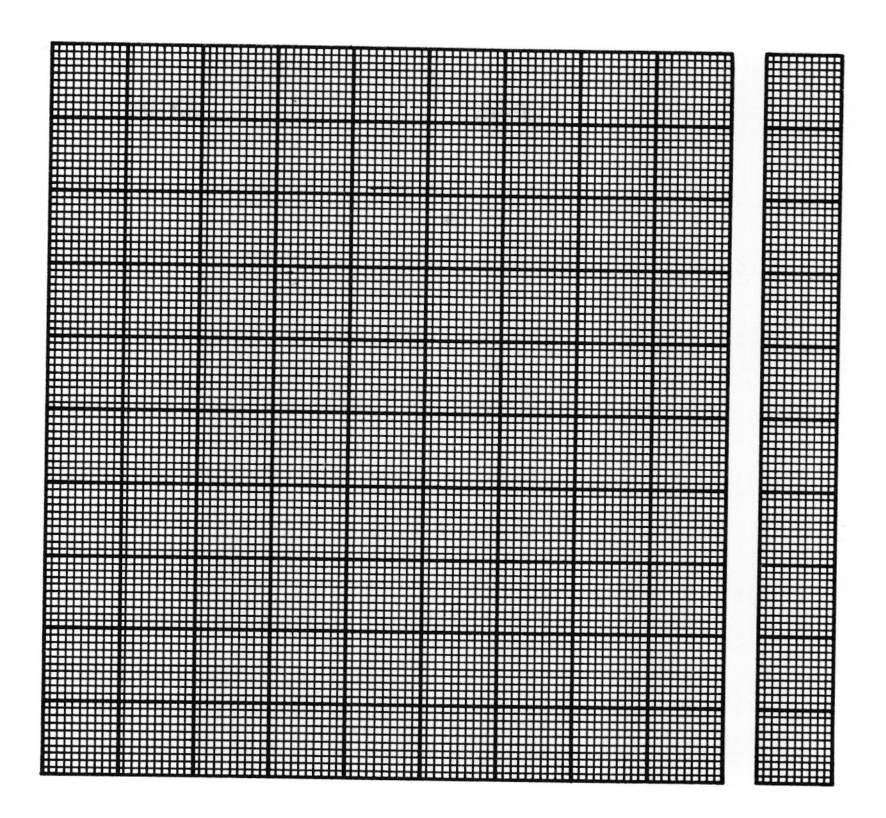

The individual squares in the strip comprise ten-hundredths (.10) of the grid, but this is the same as the strip, thus .1 and .10 are different names for the same strip.

This type of renaming underlies the addition 5/10 + 6/100 + 7/1000 = .567. Finding the greatest common multiple of the denominators is a prerequisite.

$$\frac{5}{10} \times \frac{100}{100} = \frac{500}{1000}$$

$$\frac{6}{100} \times \frac{10}{10} = \frac{60}{1000}$$

$$+\frac{7}{1000} \qquad = \frac{7}{1000}$$

$$\frac{567}{1000} = .567$$

Mini-lesson 4.2.2 Extend Place Value to Thousandths

Vocabulary: decimal point, tenths, hundredths, thousandths, decimals

Possible Trouble Spots: does not understand place-value principle; does not realize that the decimal point is a cue that numbers to its right designate fractional parts of one unit; does not relate reciprocal relationship to decimal notation; does not see the balance of the place-value system, i.e., tens/tenths; hundreds/hundredths; thousands/thousandths, and so on

Requisite Objectives:
Know purpose of decimal point

Divide by ten

Rationals/Notation 3.3.4: Use inverse relationship between multiplication and division

Rationals/Notation 3.3.5: Identify multiplicative inverse as the reciprocal of a number

Rationals/Notation 3.3.6: Multiply reciprocals to obtain the identity element for multiplication

Rationals/Notation 4.1.3: Extend expanded notation to include fractions through thousandths

Activity 1
Review notation of whole numbers by asking students to explain how they obtained the place values to the left of units.

A. They should express awareness of place values as products of powers of ten as shown below.

Suggested board work

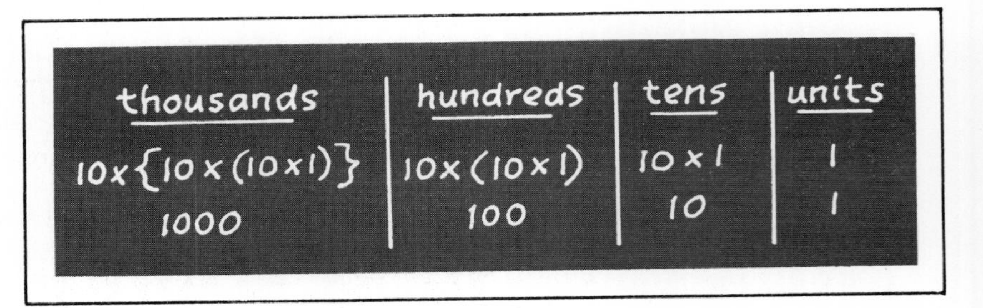

B. Then ask them to begin the pattern at the thousands place and show the pattern for obtaining place values that name smaller numbers. The following visual aid may be used to begin the pattern of names of the decimal place values. Students in grade 3 should be able to express tenths.

Suggested board work

thousands	hundreds	tens	units	tenths
1000	$1000 \div 10$	$100 \div 10$	$10 \div 10$	$1 \div 10$
	$\frac{1000}{10}$	$\frac{100}{10}$	$\frac{10}{10}$	$\frac{1}{10}$
	100	10	1	.1

The move from units to tenths is crucial for showing that we are leaving numbers greater than the number 1 and now are considering numbers that name a fractional part of the number 1.

Since students at this level cannot yet divide a fraction by a fraction, change your instructional strategy at this point. Simply tell the students that another way of writing one-tenth is *.1* and insert it in the place value chart under *tenths.*

C. Now focus should move to the names of the next two place values to the right. It is often helpful to point out the symmetry around the units place. A balance idea to show the reciprocal relationship may be used.

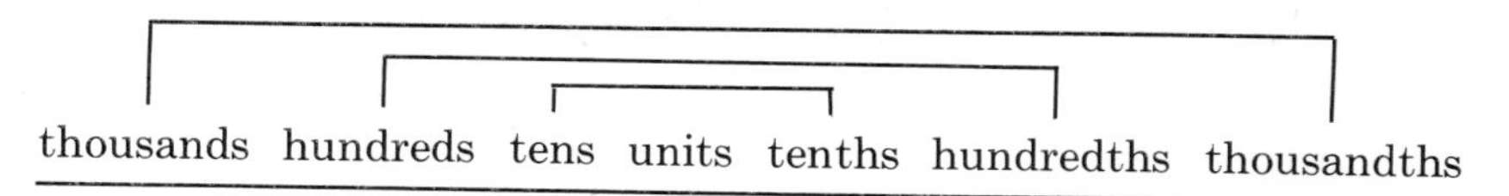

Activity 2

Show the child arrow graphs involving division by 10 to introduce the idea of extending the decimal notation to tenths, hundredths, thousandths, and so on. Also, review multiplication by 10 to obtain larger numbers.

$$3000 \xrightarrow{\div 10} 300 \xrightarrow{\div 10} 30 \xrightarrow{\div 10} 3 \xrightarrow{\div 10} .3$$

$$5000 \xleftarrow{\times 10} 500 \xleftarrow{\times 10} 50 \xleftarrow{\times 10} 5 \xleftarrow{\times 10} .5$$

Activity 3
Engage the child in mapping activities. Ask him to describe the patterns of writing numbers.

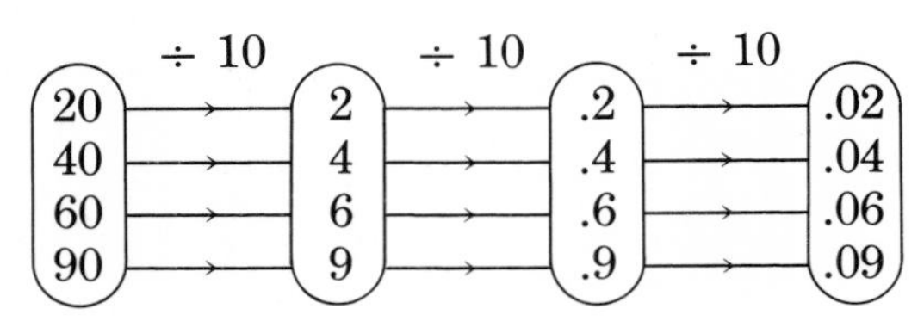

Additional Considerations: Being able to generate decimal place value involves ability to generalize the *divide-by-10* relationship. This is difficult for some students, especially for those developing at a slower rate intellectually than the majority of others of the same age, or for those who do not notice salient aspects of situations, such as patterns that underlie decimal place value notation. Rather than taking long periods of time while attempting to get such students to see the decimal place value generalization, it may be more feasible to spend that time teaching them how to use a hand calculator for finding decimal notation. To generate decimal place values, the student merely presses the whole number 1, then presses the + key, then the keys 1, 10, then the ÷ key, to obtain .1. On some calculators, the sequence is 1, ÷, 10, =. By successive dividing of .1 by 10, the display will show .01, .001, and so on. The decimal point does not have to be pressed as it will appear independently. Use of the hand calculator makes teaching decimal notation prior to grade six possible. When students arrive at grade six, they can then focus on computations and applications of decimals.

Note: Mini-lesson 4.2.2 is a prerequisite for Mini-lessons 4.2.10 and 4.3.2.

Decimals/Notation 4.2.2:

Extend place value to thousandths

Mathematical Applications 4.2.10:
Record simple equivalent metric measures in daily activities, for example, .7 m or .01 m

Decimals/Notation 4.3.2:
Translate fractions to decimals

TRANSLATING FRACTIONS TO DECIMALS

Common fractions

A common fraction may be converted to a simple decimal, which is either finite or repeating. Common fractions include proper fractions (numerator $<$ denominator) and improper fractions (numerator $\geq$ denominator).

Finite decimals

Finite decimals (also called *terminating decimals*) contain a specific number of digits. Such decimals as 1.5, .25, and .75 are finite decimals.

Repeating decimals

Repeating decimals, which are *infinite,* or *nonterminating,* have the same digit repeated, as represented by three dots (.666 . . .), or have patterns of digits repeated (.142857142857 . . .). The notation used to show the repeating digit or group of repeating digits (called the *repetend*) is a bar (*vinculum*), which is placed over the repetend. Dots, which are placed over the first and last digits of the repeated pattern, are also used.

$$.285714285714\ldots = .\overline{285714} = .\dot{2}8571\dot{4}$$
$$.666\ldots = .\overline{6} = .\dot{6}$$
$$.83333\ldots = .8\overline{3} = .8\dot{3}$$

A terminating decimal may be expressed as a repeating decimal:

$$.75 = .75\overline{0}$$
$$.184 = .184\overline{0}$$

There are also infinite decimals that do not show a repeating pattern; they are called *nonrepeating decimals* and represent irrational numbers such as π (3.14159 . . .) or $\sqrt{2}$ (1.41428 . . .).

The concept of a fraction as an expressed quotient underlies the conversion of a common fraction to a decimal. The fraction a/b may be expressed as $a \div b$, $b \neq 0$, where a is the dividend and b the divisor. The division algorithm allows that $a = bq + r$, where $0 \leq r < b$ (q = quotient, r = remainder). If the remainder is zero, the decimal is said to terminate; it is finite.

Mini-lesson 4.3.2 Translate Fractions to Decimals

Vocabulary: decimal equivalent

Possible Trouble Spots: does not understand equivalence generalization; does not understand that a number may be expressed using different notations

Requisite Objectives:
Division 4.1.10: Divide any whole number by a 1-digit number
Decimals/Notation 4.2.2: Extend place value to thousandths

Activity 1
Ask the child to translate fractions to decimals, with no renaming, such as 3/10 = .3; 75/100 = .75; 3/100 = .03; 3/1000 = .003.

Activity 2
Ask the child to convert 1/4 to a decimal. Show her two sets of open sentences like the following and guide her to take the simplest path.

(a) $\frac{1}{4} = \frac{\square}{10} \qquad 4 \times \square = 10$

(b) $\frac{1}{4} = \frac{\square}{100} \qquad 4 \times \square = 100$

She should notice that example *b* involves naming a whole number to make the sentence true, whereas the solution to example *a* is more complicated. Therefore, she should select hundredths as the desired decimal fraction place value.

$$\frac{1}{4} \times \frac{\boxed{25}}{\boxed{25}} = \frac{\boxed{25}}{100} = .25$$

Activity 3

Method for converting fractions to decimals

The following method can be used to convert a fraction to a decimal.[1]
To convert 3/8 to a decimal, recall that

$$1 = 1.0 = 1.00, \text{ so } 1/4 = \frac{1.0}{4} = \frac{1.00}{4}$$

As 1/4 may be written in the form $\frac{1.00}{4}$, so can $3/8 = \frac{3.00}{8} = \frac{3.000}{8}$.

$$\begin{array}{r} .25 \\ 4\overline{)1.00} \\ 8 \\ \hline 20 \\ 20 \\ \hline \end{array} \qquad \begin{array}{r} .375 \\ 8\overline{)3.000} \\ 2\,4 \\ \hline 60 \\ 56 \\ \hline 40 \\ 40 \\ \hline 0 \end{array}$$

Zeros may be added to the right of a decimal to obtain an equivalent decimal.

1. For alternate methods of converting fractions to decimals, see Fredricka K. Reisman, *Diagnostic Teaching of Elementary School Mathematics: Methods and Content* (Rand McNally College Publishing Co., 1977), pp. 380–81.

Additional Considerations: The NAEP (August 1979, p. 1) reported that skill in naming decimal fractions was very different for 9-year-olds than for those 13 and 17 years of age.[2] One test item on naming decimals and the percentage response for each option is shown in Table 17.1.

Table 17.1 Which of the Following Is One and Twenty-Four Hundredths?

	Percent Responding		
	Age 9	*Age 13*	*Age 17*
1.024	11	14	15
*1.24	19	65	76
2401	6	1	0
12400	57	18	7
I don't know	6**	1**	1**

*Correct response.
**Figures do not total 100% because of rounding and/or nonresponse.

The NAEP report offers an explanation for the particularly poor response as follows:

> These results are understandable, since many 9-year-olds have had little or no exposure to decimal fractions in school. Although the performance of 13-year-olds and 17-year-olds on this and a similar exercise was considerably higher than that of 9-year-olds, one-third of 13-year-olds and a quarter of 17-year-olds did *not* correctly identify the numerical form of the decimal number.

Teachers should be aware of these NAEP results and use them to diagnose areas of the mathematics curriculum in their own classes that need more effective instruction. Use of hand calculators may be a key to open the topic of decimal notation to third and fourth graders in school. They are exposed to decimal notation in many real life situations (money, baseball statistics, newspaper advertisements, newscasts on television). It is in the school life that students have to wait for decimals until grade six.

2. National Assessment of Educational Progress (NAEP). *Mathematical Knowledge and Skills,* Report No. 09-MA-02 (August 1979): 1–2.

Note: Mini-lesson 4.3.2 is a prerequisite for Mini-lessons 4.3.3 and 4.3.4.

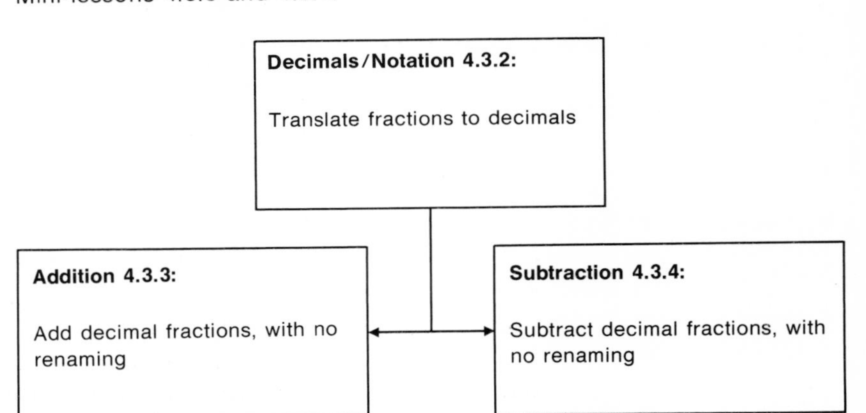

CONVERTING DECIMALS TO COMMON FRACTIONS

Converting a terminating decimal to a common fraction merely involves translating from the decimal notation to the fraction notation.[3]

(a) $.5 = \frac{5}{10} = \frac{\overset{1}{\cancel{5}}}{2 \times \underset{1}{\cancel{5}}} = \frac{1}{2}$ Renaming the fraction in lowest terms.

(b) $.003 = \frac{3}{1000}$

(c) $1.33 = 1 + \frac{33}{100} = 1\frac{33}{100}$ or $1.33 = \frac{133}{100} = 1\frac{33}{100}$

(d) $$.13\frac{1}{3} = \frac{13\frac{1}{3}}{100} = \frac{\frac{40}{3}}{100} = \frac{\frac{40}{3} \times \boxed{\frac{3}{1}}}{100 \times \boxed{\frac{3}{1}}} = \frac{\frac{40 \times 3}{3 \times 1}}{100 \times \frac{3}{1}} = \frac{\frac{3 \times 40}{3 \times 1}}{100 \times \frac{3}{1}}$$

$$= \frac{\frac{\overset{1}{\cancel{3}}}{\underset{1}{\cancel{3}}} \times \frac{40}{1}}{100 \times \frac{3}{1}} = \frac{\frac{40}{1}}{\frac{300}{1}} = \frac{40}{300}$$

3. For examples showing how to convert a repeating decimal to a common fraction, see Reisman, *Diagnostic Teaching of Elementary School Mathematics,* pp. 383–84.

Note: $\frac{3}{1} / \frac{3}{1}$ is the special name for 1, which underlies dividing out the 3s as shown. Also $40/1 = 40$; $300/1 = 300$ to obtain 40/300.

Mini-lesson 5.1.1 Extend Expanded Notation to Thousandths

Vocabulary: expanded notation

Possible Trouble Spots: does not understand that the value of a numeral is the sum of the products obtained by multiplying face value times place value; does not see the relationship between dividing by ten and reciprocals of whole number place values ($1 \div 10 = 1/10$)

Requisite Objectives:
Whole Numbers/Notation 3.1.3: Write numerals in expanded notation.

Multiply by powers of ten.

Activity 1
Review writing whole numbers in expanded notation as follows:

$$135 = 100 + 30 + 5$$

or

$$135 = (1 \times 100) + (3 \times 10) + (5 \times 1)$$

Activity 2
Extend expanded notation to decimals using written practice activities as follows:

A. Fill in the missing numbers. (Shown in boldface.)

$$35.64 = (3 \times 10) + (5 \times 1) + \left(6 \times \frac{1}{\mathbf{10}}\right) + \left(4 \times \frac{1}{\mathbf{100}}\right)$$

$$35.64 = (3 \times 10) + (5 \times 1) + (6 \times \mathbf{.1}) + (4 \times \mathbf{.01})$$

B. Identify the place value of each 6.
Example: 1.67 6 tenths

1. 6.54
2. 3.76
3. 4.006
4. 5.607
5. 2.658

C. Write as decimals.

1. seven tenths
2. three hundredths
3. sixty-five hundredths
4. eight thousandths
5. four hundred ninety-six thousandths
6. three tens, one unit, and four thousandths

Activity 3
Some students will be able to translate the expanded form to the exponential form:

$$335.64 = (3 \times 10^2) + (3 \times 10^1) + (5 \times 10^0) + (6 \times 10^{-1}) + (4 \times 10^{-2})$$

This activity should follow work with negative exponents and their use in expressing the reciprocal of a whole number. For example, using exponential notation the whole number 100 is expressed as 10^2; the reciprocal of this number (1/100) is expressed as $1/10^2$ or 10^{-2}. Review the section on exponential notation in Chapter 19.

Additional Considerations: The counting boards, abacus, and place-value chart are appropriate models for showing expanded notation.

Note: Mini-lesson 5.1.1 is a prerequisite for Mini-lessons 5.1.8 and 5.1.12.

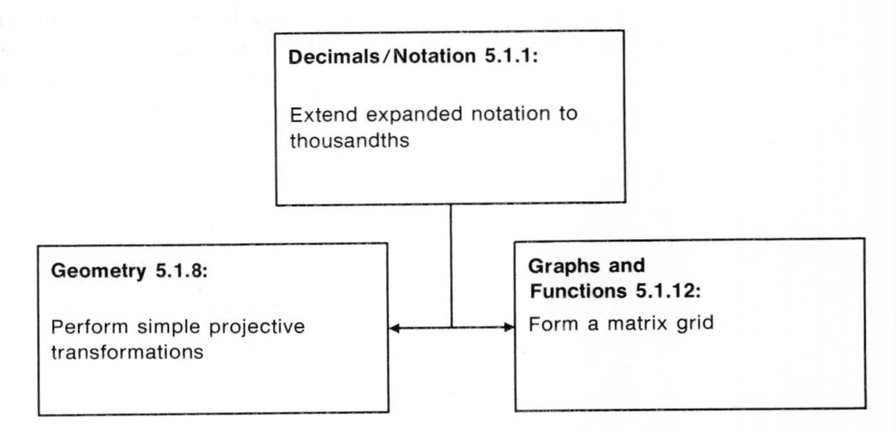

Mini-lesson 6.1.2 Round Decimal Numbers

Vocabulary: round

Possible Trouble Spots: does not understand rounding of whole numbers; has difficulty sequencing numbers

Requisite Objectives:
Whole Numbers/Notation 4.1.2: Round whole numbers
Placing numbers in order

Activity 1
Review Mini-lesson 4.1.2 on rounding whole numbers.

Activity 2
Point out to students that the pattern involved in rounding decimals is related to that for rounding whole numbers. The following steps should be followed in rounding decimals. The example used is the number 2.8117.

Procedure A: Decide on a specific number of decimal places for rounding; for example, 2 decimal places, or hundredths.

Procedure B: Locate and mark the digit in the desired place (hundredths).

$$2.8\overset{\downarrow}{1}17$$

Procedure C: Check the digit to the right of the arrow (the thousandths position for this example). Students may call this the *determining digit,* because it determines if the number to be rounded remains the same or increases by one.

Procedure D: If the determining digit is less than 5, then leave the digit marked by the arrow as is. If the determining digit is 5 or greater, increase the marked digit by one. In either case, after the rounding has occurred, all digits to the right of the marked number are dropped. Thus, for our example, 2.8117 rounded to two places is 2.81.

Activity 3
Have children answer the following:

1. Round to the nearest tenth:

.79 12.600 .237

2. Round to the nearest hundredth:

3.760 1.6105 8.4143

3. Round to the nearest thousandth:

8.2686 24.3260814 7.481761

4. Round to the nearest ten-thousandth:

.4761351 .12345 .532109

Key: 1. .8, 12.6, .2
2. 3.76, 1.61, 8.41
3. 8.269, 24.326, 7.482
4. .4761, .1235, .5321

Activity 4
Supply each missing decimal. Then round the circled number to units.

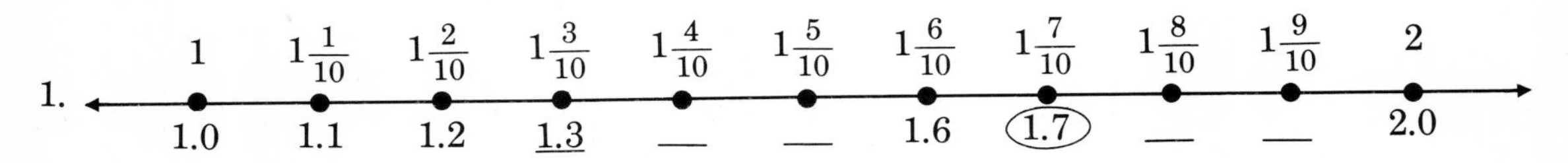

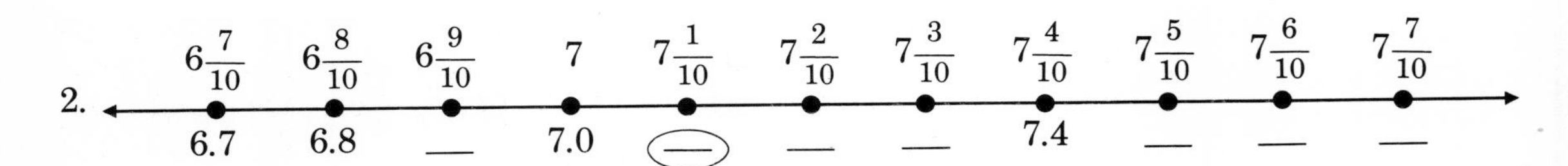

Additional Considerations: Rounding is an arbitrary convention. It is therefore appropriate to teach this topic in a direct and structured manner.

Note: Mini-lesson 6.1.2 is a prerequisite for Mini-lessons 6.1.3 and 6.1.10.

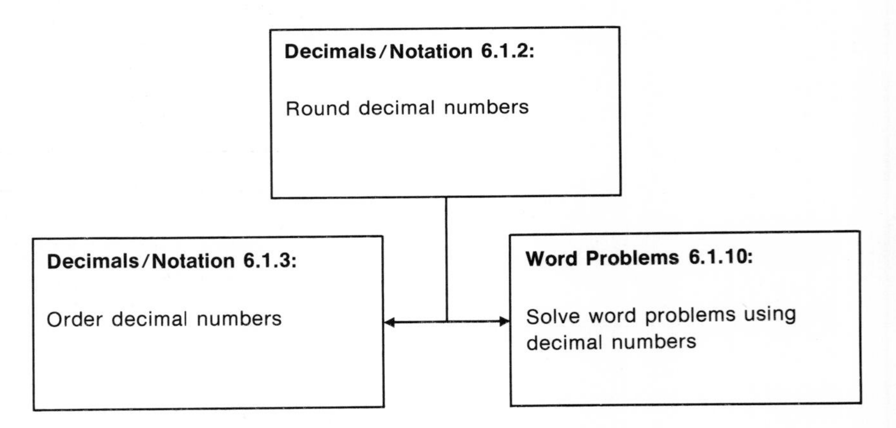

Mini-lesson 6.1.3 Order Decimal Numbers

Vocabulary: place in order, greatest, smallest

Possible Trouble Spots: cannot order numbers; does not understand inequality relations

Requisite Objectives:
Graphs and Functions 3.2.20: Use a number line
Compare value of numbers

Activity 1
This mini-lesson involves inequality and greater/less-than principles. If you are given two decimals, and there is a positive number that may be added to the first decimal to obtain a sum equal to the second decimal, then (1) the two decimals are unequal, and (2) the second decimal is greater. For example, in the case of the decimals .3 and .2, we may add .1 to .2 to obtain .3. Thus .3 is greater than .2 and comes next if ordering by tenths. This principle may be shown on a number line.

Additional considerations: Notice the use of inflectional endings that express comparatives and superlatives in task directions. The inflectional endings *-er* and *-est* are often either ignored or misused, especially by students with language-related learning disabilities. It is sometimes necessary to highlight these inflectional endings by underlining them, writing them in color, writing them in larger letters, or separating them from the stem of the word.

Note: Mini-lesson 6.1.3 is a prerequisite for Mini-lessons 6.1.6 and 6.1.10.

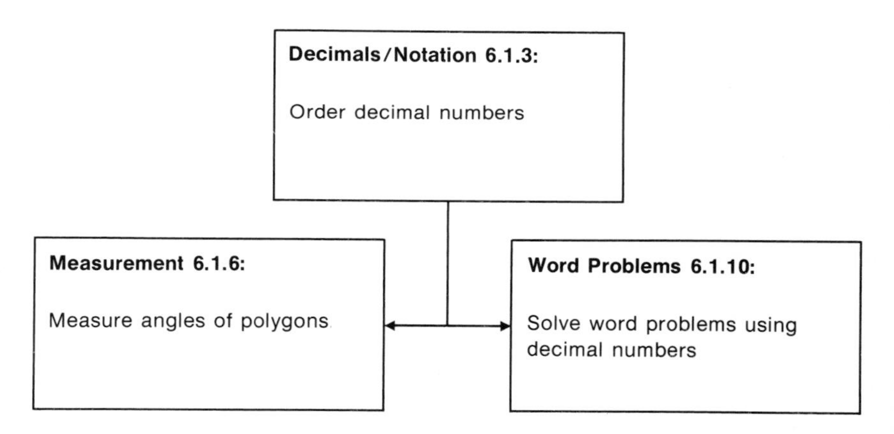

Mini-lesson 6.2.2 Complete Mathematical Sentences Showing Patterns Related to Decimal Notation

Vocabulary: arrow graph, mapping, pattern

Possible Trouble Spots: does not notice salient aspects of situations; does not identify underlying patterns

Requisite Objectives:
Decimal/Notation 6.2.1: Identify patterns of decimal place value using mappings

Perform successive operations

Activity 1
Review mapping patterns for successive dividing-by-ten in Minilesson 4.2.2 above.

Activity 2
Use indirect open mathematical sentences to show patterns for decimal notation. Present sentences such as the following:

$$2 \div \square = .02$$
$$.2 \div \square = .02$$

Review with the child the fact that in multiplication the sum of the number of zeros in the factors is equal to or greater than the sum of the number of zeros in the product. This principle relates to the placement of the decimal point in the examples cited above. In $2 \div 10 = .2$, the divisor has one zero and the quotient fills one decimal place and that in $.2 \div 10 = .02$, the divisor has one zero and the quotient fills one place more than does the dividend (.2).

$$\begin{array}{rrrr}
2 \times \square = 2 & 2 \times \square = 2 & 2 \div \square = 2 & 200 \div \square = 20 \\
2 \times \square = 20 & 2 \times \square = 20 & 20 \div \square = 2 & 20 \div \square = 2 \\
20 \times \square = 200 & 2 \times \square = 200 & 200 \div \square = 20 & 2 \div \square = \underline{.2} \\
200 \times \square = 2000 & 2 \times \square = 2000 & 2000 \div \square = 200 & .2 \div \square = \underline{.02}
\end{array}$$

Activity 3
Present the following division problems to the child and ask him to describe the pattern of determining place values for decimals.

$$100 \div 10 = 10$$
$$10 \div 10 = 1$$

$1 \div 10 = .\underline{1}$ "One place to the right of the units must be filled."

$.1 \div 10 = .\underline{0}\underline{1}$ "Two places to the right of the units must be filled."

$.01 \div 10 = .\underline{0}\underline{0}\underline{1}$ "Three places to the right of the units must be filled."

Activity 4
Present a set of integers such as 11, 38, 9, 71, 3, 132, 701, 300, 492 and ask the child to divide each integer by 10, using the decimal point correctly. Then ask her either to read the answers aloud or to write them.

Additional considerations: Indirect open sentences are an abstract and difficult topic for a majority of students; thus, Activity 2 will not be appropriate for all students. Use of the hand calculator serves as an alternative procedure for analyzing patterns like those in Activity 2 by transforming the indirect sentences to direct open sentences as follows:

$$\begin{array}{llll} 2 \div 2 = \square & & 2 \div .2 = \square \\ 20 \div 2 = \square & & .2 \div .02 = \square \\ 200 \div 20 = \square & & \\ 2000 \div 200 = \square & & \end{array}$$

Note: Mini-lesson 6.2.2 is a prerequisite for Mini-lessons 6.2.3 and 6.2.4.

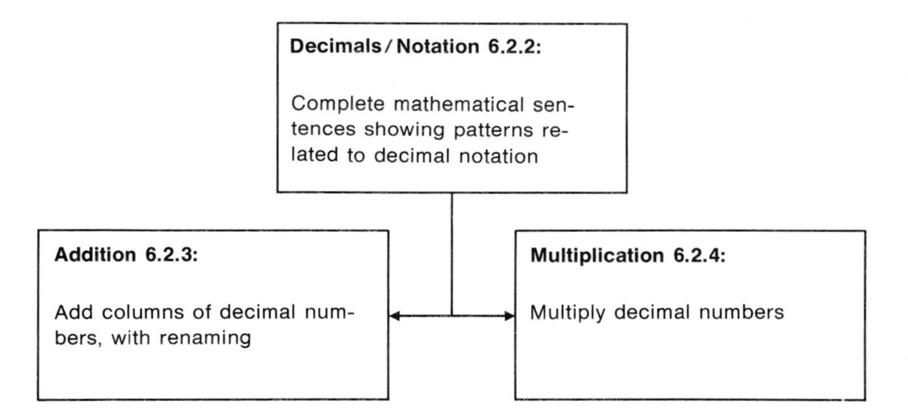

EXERCISES

I. Write the following numbers in expanded form and then in exponential notation:

1. .3 2. .15 3. .0004 4. 908.6

II. Write equivalent fractions for the following decimals and mixed decimals:

1. 0.316 2. 0.61 3. 0.01
4. 11.093 5. 8.008

III. Write decimals for the following fractions:

1. $\frac{34}{10^2}$ 2. $\frac{683}{10^3}$ 3. $\frac{683}{10^1}$ 4. $\frac{35}{10^6}$ 5. $\frac{9}{5}$ 6. $\frac{3}{1}$ 7. $\frac{14}{10}$ 8. $\frac{-31}{100}$

IV. Convert 13/15 to a decimal. Will it repeat? Why?

V. Convert .5 to a fraction.

VI. Match the ideas in the columns below:

Column I	*Column II*
1. fraction to decimal notation	a. $.75 = \frac{75}{100}$
2. terminating decimal to fraction notation	b. 1.41421 . . .
3. irrational number	c. $\frac{-4}{5} = -0.8$

VII. Express the following numbers as repeating decimals:

1. $\frac{2}{11}$
2. $\frac{1}{19}$
3. $6\frac{2}{3}$
4. $\frac{-11}{15}$

VIII. Is the decimal .101001000100001 . . . representative of a rational number?

IX. Numbers that can be expressed as repeating decimals are called _____.

X. Numbers that can be represented by nonrepeating decimals are called _____.

XI. Give an example of an irrational number.

XII. Draw a place-value chart to show the following numbers:
1. 66.34 2. 19.04 3. 900.01

XIII. Show the following numbers on counting boards.
1. 66.34 2. 19.04 3. 900.01

XIV. Write as decimals.
1. thirteen hundredths
2. four tenths
3. twenty-four thousandths
4. one hundred fifteen thousandths
5. two hundredths

XV. Round each to the nearest (1) hundredth (2) tenth, and (3) whole number.
1. .167 2. .869 3. 1.208 4. 47.326

XVI. Supply each missing decimal.

1. 4.5 4.6 4.7 ___ ___ 5.1 ___ ___ ___

2. 63.2, 63.3, 63.4, _____, _____, _____, _____, _____, _____

XVII. Write the decimals in order from least to greatest.
1. .5 .56 .556
2. .070 .03 .2
3. .300 .7 .40

SUGGESTED READINGS

Hobbs, B. F., and C. H. Burris. "Minicalculators and Repeating Decimals." *The Arithmetic Teacher* 25 (April 1978): 18–20.

National Assessment of Educational Progress (NAEP). *Mathematical Knowledge and Skills*. Report No. 09-MA-02. (August 1979): 1–2.

Sgroi, J. T. "Patterns of Repeating Decimals: A Subject Worth Repeating." *The Mathematics Teacher* 70 (October 1977): 604–605.

Wilson, Guy. *Teaching the New Arithmetic*. New York: McGraw-Hill Book Co., 1951. Pp. 238–57.

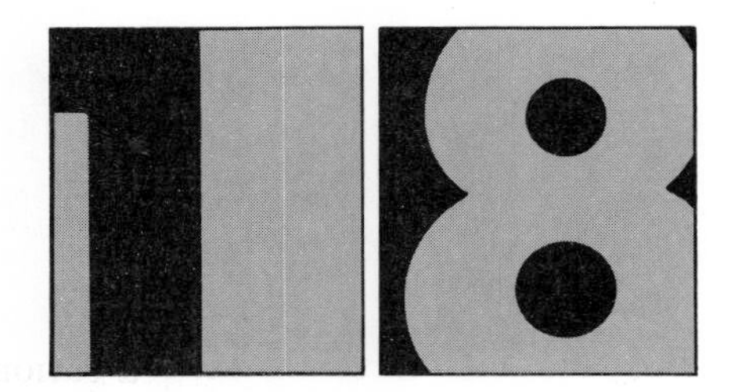

Computations with Decimals

ADDITION AND SUBTRACTION OF DECIMALS

The same properties apply to the addition and subtraction of decimals as to that of integers. As with whole numbers, place values are lined up and a vertical algorithm is used. Placing the decimal points one under another or using graph paper helps the learner to line up the place values.

$$\begin{array}{r} {\scriptstyle 1}\\ .4 \\ 1.04 \\ {\scriptstyle 1} \\ +19.803 \\ \hline 21.243 \end{array} \qquad \begin{array}{r} {\scriptstyle 6\ \ 12} \\ \not{7}.\not{2}1 \\ -\ .30 \\ \hline 6.91 \end{array}$$

MULTIPLICATION AND DIVISION OF DECIMALS

Placing the decimal point

The processes of multiplication and division are the same for integers and decimals, except for placing the decimal point in the product or quotient. The rationale for placing the decimal point is dependent in particular upon the place-value concept and the distributive property. These two ideas are used for solving the following multiplication problems.

Example: 3×1.5

3×1.5	
$= 3 \times (1 + .5)$ or $3 \times \left(1 + \frac{5}{10}\right)$	Renaming
$= (3 \times 1) + \left(3 \times \frac{5}{10}\right)$	Distributive Property for Multiplication over Addition
$= 3 + \frac{15}{10}$	Multiplication of rationals (note the product 3×1 is considered a rational since $3 = \frac{3}{1}$)
$= 3 + \frac{10}{10} + \frac{5}{10}$	Renaming improper fractions
$= 3 + 1 + \frac{5}{10}$	Renaming multiplicative identity
$= (3 + 1) + \frac{5}{10}$	Associative Property for Addition
$= 4 + \frac{5}{10}$	Addition of whole numbers
$= 4.5$	Decimal notation

Example: $.3 \times 1.5$

Vertical algorithm

$$\begin{array}{r} 1.5 \\ .3 \\ \hline 1\,5 \\ 3\,0 \\ \hline .4\,5 \end{array}$$

$.3 \times 1.5 = \frac{3}{10} \times \left(1 + \frac{5}{10}\right)$	Renaming
$= \left(\frac{3}{10} \times 1\right) + \left(\frac{3}{10} \times \frac{5}{10}\right)$	Distributive Property for Multiplication over Addition
$= \frac{3}{10} + \frac{15}{100}$	Multiplication of rationals
$= \frac{30}{100} + \frac{15}{100}$	Renaming equivalent fractions
$= \frac{45}{100}$	Addition of rationals
$= .45$	Decimal notation

Note that .45 results from the following thought processes. $3 \times 15 = 45$; tenths $\times$ tenths equals hundredths.

Notice that there is a pattern for the relationship between the number of places used to the right of the units place in the product and the number of places in the factors to the right of the units place. The places to the right of the units place are called "decimal places."

The teacher may introduce the pupils to an estimation procedure for pointing off the decimal point in the product. For the example 2.35×35.8, 2.35 is close to 2 and 35.8 is close to 36; $2 \times 36 = 72$, which tells us there will be two places to the left of the decimal point. The product, in fact, is 84.130, with two places to the left of the decimal point.

This strategy, however, does not work when the products are close to 10, 100, 1000, and so on. For example 1.76×5.1, $2 \times 5 = 10$ suggests two places to the left of the decimal point. But, in fact, $1.76 \times 5.1 = 8.967$, with only one digit to the left of the decimal point. However, with an estimate of 10 for the product, the estimation method can still be helpful, since pupils can ask themselves which of the products .8976, 8.976, 89.76, and so on, is closest to 10, thereby determining the placement of the decimal point.

PLACING THE DECIMAL POINT

Present various multiplication problems involving decimals and point out the pattern with respect to placement of the decimal point in the product.

$$.\underline{1} \times .\underline{1} = .\underline{01}$$

① ① ②

$$.\underline{01} \times .\underline{1} = .\underline{001}$$

② ① ③

$$.\underline{001} \times .\underline{01} = .\underline{00001}$$

③ ② ⑤

$$3.\underline{6} \times 2 = 7.\underline{2}$$

① ①

$$360 \times .\underline{002} = .\underline{720} \text{ or } .72$$

③ ③

Notice that the decimal point in the product is placed so that there will be as many decimal places in the product as there are decimal places in the factors. The rationale for this statement may be shown with fractions:

$$2.35 \times 35.8 = \frac{235}{100} \times \frac{358}{10} = \frac{235 \times 358}{100 \times 10} = \frac{84130}{1000} =$$

$$84.\underline{1}\,\underline{3}\,\underline{0} = 84.13$$

> *Note:* The circled numerals and the lines drawn under each factor and product (as on p. 326) point up to the pupil the relations among the number of decimal places. For example, in $.006 \times .02$, the pupil may think "six times two equals twelve" and "one one-thousandth times one one-hundredth equals one one-hundred-thousandth." Therefore, there will be five decimal places in the product. He or she may then indicate five decimal places as follows: . _ _ _ _ _ . The pupil then places the product 12 at the farthest position, _ _ _ 12, and completes the product by filling in the number of zeros needed to reach the decimal point: . 00012 .

Division of decimals akin to division of fractions

Division of decimals may be done as with division of fractions.

Example: $.45 \div .3$

$$\frac{45}{100} \div \frac{3}{10}$$

$$= \frac{45 \div 3}{100 \div 10} = \frac{15}{10}$$

Notice that the placement of the decimal point in the product depends upon the number of decimal places as indicated in the denominator of the answer, 15/10. Since the denominator indicates that the tenth place is used, the equivalent decimal for 15/10 is 1.5.

This procedure is useful for examples such as $64 \div .008$. As a prerequisite, the child may rename 64 to its rational form, 64/1. The teacher may tell the child to use the "invert the divisor" method and then to translate to a decimal fraction.

$$64 \div .008$$

$$= \frac{64}{1} \div \frac{8}{1000}$$

$$= \frac{\overset{8}{\cancel{64}}}{1} \times \frac{1000}{\underset{1}{\cancel{8}}}$$

$$= 8000$$

The algorithm for division of fractions may also take the following forms:

(a) $$.45 \div .3 = \frac{.45}{.3} \times \frac{10}{10} = \frac{4.5}{3} = 3\overline{)4.5}\ \ \text{(quotient } 1.5\text{)}$$

(b) $$.0045 \div .03 = \frac{.0045}{.03} \times \frac{100}{100} = \frac{.45}{3} = 3\overline{).45}\ \ \text{(quotient } .15\text{)}$$

(c) $$98 \div .14 = \frac{98}{.14} \times \frac{100}{100} = \frac{9800}{14} = 14\overline{)9800}$$

9800	
1400	100
8400	
1400	100
7000	
2800	200
4200	
2800	200
1400	
1400	100
	700

By multiplying both the divisor and the dividend by the power of 10 necessary to convert the divisor into a whole number, the problem is simplified to dividing a decimal by a whole number. The procedure used when dividing a decimal by a whole number is to place the decimal point in the quotient directly over the decimal point of the dividend.

Decimal division algorithm

This use of the multiplicative identity is shown in various ways in the decimal division algorithm. Using the same examples, carets or arrows may be used to indicate the placement of the decimal point in the quotient.

$$.3_{\wedge}\overline{).4_{\wedge}5}\ \ (1.5) \quad \text{or} \quad .3\overline{).4\ 5}\ \ (1.5)$$

$$.03_{\wedge}\overline{).00_{\wedge}45}\ \ (.15) \quad \text{or} \quad .03\overline{).00\ 45}\ \ (.15)$$

$$.14_{\wedge}\overline{)98.00_{\wedge}}\ \ (7\ 00.) \quad \text{or} \quad .14\overline{)98.00}\ \ (7\ 00.)$$

The emphasis in the following algorithm is on the difference obtained when subtracting the number of decimal places in the divisor from the number of decimal places in the dividend.

```
.3).45 |
    30 |  10
    -- |
    15 |
    15 |   5
    -- |  ---
     0 | 1.5
```

Recognizing patterns as aid to placing the decimal point

A decimal multiplication-division grid (as shown) yields decimal place values. It is suggested that construction of a fraction grid precede that of a decimal grid.

Fraction Grid

×/÷	1	$\frac{1}{10}$	$\frac{1}{100}$	$\frac{1}{1000}$
1	1	$\frac{1}{10}$ $\frac{1}{10^1}$	$\frac{1}{100}$ $\frac{1}{10^2}$	$\frac{1}{1000}$ $\frac{1}{10^3}$
$\frac{1}{10}$	$\frac{1}{10}$ $\frac{1}{10^1}$	$\frac{1}{100}$ $\frac{1}{10^2}$	$\frac{1}{1000}$ $\frac{1}{10^3}$	$\frac{1}{10{,}000}$ $\frac{1}{10^4}$
$\frac{1}{100}$	$\frac{1}{100}$ $\frac{1}{10^2}$	$\frac{1}{1{,}000}$ $\frac{1}{10^3}$	$\frac{1}{10{,}000}$ $\frac{1}{10^4}$	$\frac{1}{100{,}000}$ $\frac{1}{10^5}$
$\frac{1}{1000}$	$\frac{1}{1{,}000}$ $\frac{1}{10^3}$	$\frac{1}{10{,}000}$ $\frac{1}{10^4}$	$\frac{1}{100{,}000}$ $\frac{1}{10^5}$	$\frac{1}{1{,}000{,}000}$ $\frac{1}{10^6}$

Decimal Grid

×/÷	1 (10^0)	.1 (10^{-1})	.01 (10^{-2})	.001 (10^{-3})
1 (10^0)	1 (10^0)	.1 (10^{-1})	.01 (10^{-2})	.001 (10^{-3})
.1 (10^{-1})	.1 (10^{-1})	.01 (10^{-2})	.001 (10^{-3})	.0001 (10^{-4})
.01 (10^{-2})	.01 (10^{-2})	.001 (10^{-3})	.0001 (10^{-4})	.00001 (10^{-5})
.001 (10^{-3})	.001 (10^{-3})	.0001 (10^{-4})	.00001 (10^{-5})	.000001 (10^{-6})

In multiplying decimals, the shortcut for placing the decimal point in the product is adding the decimal places in the factors. What pattern do you see for placing the decimal point in division of decimals?

Notice the patterns in the decimal and fraction grids. It appears from the decimal grid that

$$10^{-6} \div 10^{-3} = 10^{-3}$$
$$10^{-5} \div 10^{-3} = 10^{-2}$$
$$10^{-4} \div 10^{-3} = 10^{-1}$$
$$10^{-3} \div 10^{-3} = 10^{0}$$
$$10^{-2} \div 10^{-1} = 10^{-1}$$

The dividend exponent minus the divisor exponent equals the quotient exponent.

Thus, hundredths ÷ tenths = tenths, or a dividend pointed off to two places (.34) divided by a divisor pointed off to one place (.2) yields a quotient pointed off to one place (1.7).

$.34 \div .2 = 1.7$

② ① ①

The number of places in the dividend minus the number of places in the divisor equals the number of places in the quotient.

Knowledge of exponential notation allows gifted pupils to see the relationships among the exponent and the number of zeros in the denominator and the number of decimal fraction places used. This will be helpful later when using scientific notation (see Chapter 22).

EXERCISES

I. Compute the following examples and check by showing the indicated operations on counting boards.

1. $\begin{array}{r} 2.4 \\ +6.3 \\ \hline \end{array}$ 2. $\begin{array}{r} .8 \\ +.4 \\ \hline \end{array}$

3. $\begin{array}{r} 8.5 \\ -4.1 \\ \hline \end{array}$ 4. $\begin{array}{r} 16.2 \\ -15.7 \\ \hline \end{array}$

II. Complete the magic square. The sum of the numbers in each row, column, and diagonal must be 2.1.

	.3	.8
.5	.7	
.6		.4

III. Arrange the following numbers vertically and add.
1. 5.2 + 93 + .16 + .127
2. .97 + 6 + 1.3 + .276
3. .89 + 268 + .4 + .10

IV. Arrange in order from simplest to most complex.

1.	2.	3.	4.	5.
$\begin{array}{r} 6.45 \\ +7.93 \\ \hline \end{array}$	$\begin{array}{r} .9 \\ +.4 \\ \hline \end{array}$	$\begin{array}{r} 42.675 \\ -65.377 \\ \hline \end{array}$	$\begin{array}{r} .8 \\ +.1 \\ \hline \end{array}$	$\begin{array}{r} .9 \\ -.2 \\ \hline \end{array}$

V. Compute and check by multiplying.
$.02\overline{)3.6}$

SUGGESTED READING

Bright, George W. "Ideas." *The Arithmetic Teacher* 24 (March 1977): 217–18.

Whole Numbers Extended: Rounding, Simplifying Expressions, and Exponents

ROUNDING

Judging and estimating are activities engaged in by children even before they enter kindergarten. The early choices made regarding which bow to wear to a friend's birthday party or which toy to play with are important experiences that underlie ability to estimate. The important role parents can play in this regard is to allow their child to make these choices. Also, it is important that children be *aware* that they are engaging in decision making. This relates to metacognition and cognitive monitoring discussed in Chapter 1.

Judging can range from a simple binary choice (for example, "Is a crayon red or not red?") to complicated topics such as rounding numbers and using scientific notation. Related to judging, such as if three or more sets are equivalent, are estimation skills. Estimation is usually not taught until fourth grade, when rounding two-digit divisors is introduced. This is unnecessarily late.

As indicated in the August 1979 report of the *National Assessment of Educational Progress,* 9-, 13-, and 17-year-olds did poorly on various types of rounding and estimating problems. This is thought to reflect lack of exposure to estimating activities, especially for the 9-year-olds.

Estimating is particularly important in everyday usage, e.g., deciding if one has enough money to make various purchases. Even when using a hand calculator, estimating is necessary because so often a wrong key is pressed or a battery runs out.

Learning to round numbers is quite difficult for many children. It requires a knowledge of place value and also of the basic ground rules for rounding up and rounding down. Bennett and Nelson have presented a precise description:

> To *round off* a number to a given place, locate the place value and then check the digit to its right. If the digit to the right is 5 or greater, then all digits to the right are replaced by 0s and the given place is increased by 1. If the digit to the right is 4 or less, then all digits to the right of the given place value are replaced by 0s.[1]

Rounding off to various places is shown below for the number 25,124,925. The place value to which the number is rounded is underscored. The number is first rounded to ten millions, then to millions, hundred thousands, and so on, until, in the last example, it is rounded to tens.

$$\underline{2}5{,}124{,}925 \longrightarrow 30{,}000{,}000$$
$$2\underline{5}{,}124{,}925 \longrightarrow 25{,}000{,}000$$
$$25{,}\underline{1}24{,}925 \longrightarrow 25{,}100{,}000$$
$$25{,}1\underline{2}4{,}925 \longrightarrow 25{,}120{,}000$$
$$25{,}12\underline{4}{,}925 \longrightarrow 25{,}125{,}000$$
$$25{,}124{,}\underline{9}25 \longrightarrow 25{,}124{,}900$$
$$25{,}124{,}9\underline{2}5 \longrightarrow 25{,}124{,}930$$

Mini-lesson 4.1.2 Round Whole Numbers

Vocabulary: round up, round down, round numbers, approximate

Possible Trouble Spots: Needs practice in making simple judgments; does not understand place value; cannot rename numbers from a given place to its adjoining place (24 tens = 240 units); does not sequence large numbers

---

1. Bennett, Albert B. Jr., and Nelson, Leonard T., *Mathematics: An Informal Approach* (Boston: Allyn and Bacon, 1979), p. 61.

Requisite Objectives:
Whole Numbers/Notation 4.1.1: Extend place value through millions

Activity 1
Review place value by providing activities such as shown below.

1. Write 2,143 in expanded form:
2. Write these numerals in the short form: 1,000 + 700 + 60 + 5
3. Which digit is in the thousands place in each of the following numerals? 1,621; 6,083
4. Which digit is in the hundreds place in each of the following numerals? 8,471; 4,035
5. Write the number that comes before and after 5,872.

Activity 2
Round to the nearest tens. Using a vertical number line, box the distance between the two tens that surround the number to be rounded. Then draw a line at the midpoint between these two boundaries as shown.

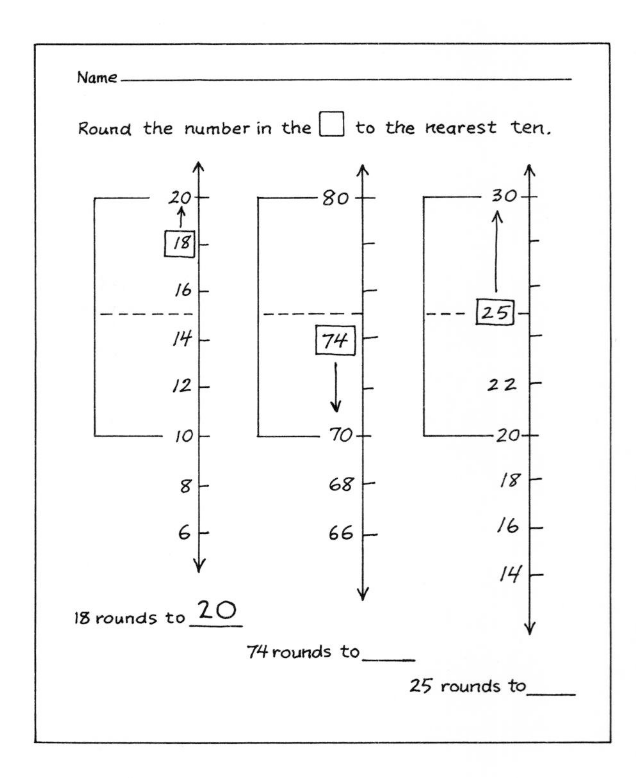

This procedure presents a visual picture of the physical distance between the number to be rounded and its boundaries. Obviously the boxed number is closer to 20.

Activity 3
Have students use number lines for rounding activities.

A. The vertical number line lends itself to the term *round up* as 18 is rounded up to 20. In like manner 74 is *rounded down* to 70. The ground rule for a number at the midpoint, that is, with the digit *5* in the units place, is to round up. Indicating the tens boundaries is important lest students become confused as to what the boundaries are.

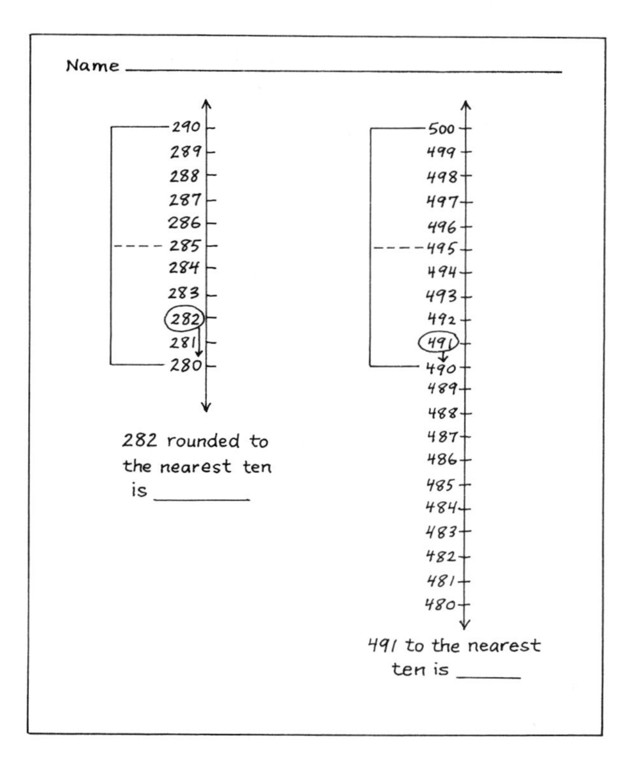

B. Have students round to the nearest hundred using the vertical number line technique.

C. Have students round to thousands.

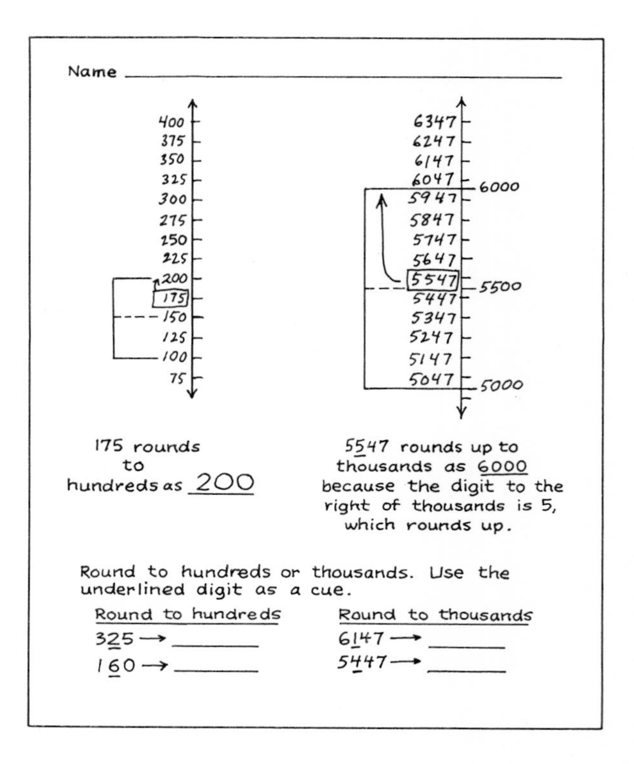

Additional considerations: Estimation skills are one of the least practiced areas in school, yet one of the most used in life situations. Rounding numbers is only one aspect of estimation. Students should estimate time needed to complete tasks, money needed to buy something, money to be earned through working, and so on. Estimation is a topic that is particularly adaptable to problem-solving activities. Students should identify the salient aspects of a given circumstance. For example, if numbers are to be rounded, then digits in the place-value position relevant to the rounding need to be identified. If money is needed for purchases, then estimates must be made of cash on hand. If a task takes a certain amount of time to complete, then amount of time available will dictate successful completion of the task. Next, criteria for comparing should be decided upon. Finally, judgments should be made and the validity of the estimation verified.

Note: Mini-lesson 4.1.2 is a prerequisite for Mini-lessons 4.1.4 and 4.3.7.

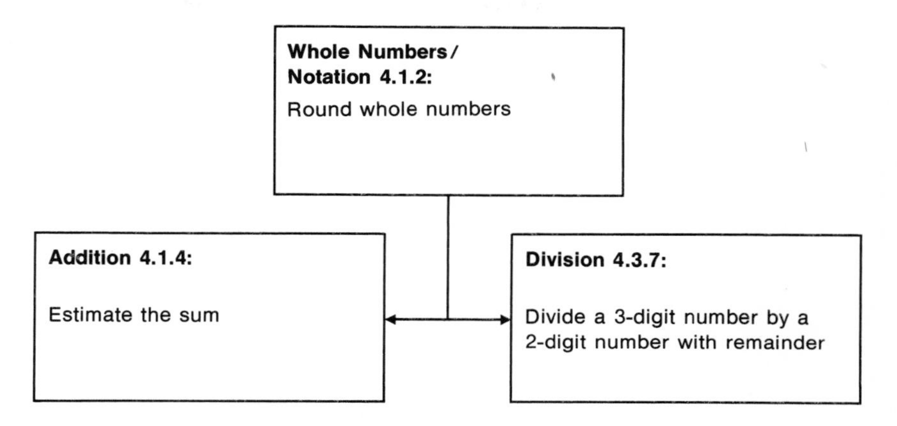

Mini-lesson 7.2.1 Simplify Expressions Involving Order of Operations

Vocabulary: brackets, braces, parentheses

Possible Trouble Spots: may make errors in computation; may forget sequence convention; topic may be too complex for student

Requisite Objectives:
Perform the binary operations of addition, subtraction, multiplication, and division

Activity 1
Tell students that there is an agreed-upon order of simplifying expressions with different operations. When no (parentheses), {braces}, or [brackets] are given, the following is the convention:

1. Simplify powers.
2. Perform multiplication and division in order from left to right.
3. Perform addition and subtraction in order from left to right.

Example:

$$\begin{aligned} 24 - 8 \div 2 \times 2 + 1 &= 24 - \underbrace{8 \div 2} \times 2 + 1 \\ &= 24 - \underbrace{4 \times 2} + 1 \\ &= \underbrace{24 - 8} + 1 \\ &= \underbrace{16 + 1} \\ &= 17 \end{aligned}$$

4. Have students simplify the following expressions.

1. $16 + 8 \div 4 \times 2 - 4 = \square$
2. $24 - 4 \div 2 \times 2 = \square$
3. $60 - 10 \div 2 + 3 = \square$
4. $63 \div 7 - 6 \div 3 = \square$
5. $4 \times 5 - 36 \div 9 \div 4 = \square$

Activity 2

Tell students when grouping symbols such as (), {}, or [] are given, they should perform the operation within the innermost grouping first and work toward the outermost grouping. Provide time for two or more students to team as they try some examples such as the following:

$$\begin{aligned}[\{(9 - 3) \times 6\} + 2] &= [\{6 \times 6\} + 2] \\ &= [36 + 2] \\ &= 38\end{aligned}$$

Then have students simplify the following expressions.

1. $(16 + 4) \div [(8 \times 2) \div 4] = \square$
2. $[(5 \times 20) + (6 \div 3)] - 51 = \square$
3. $[(4 \times 2) + (6 \times 10) - (3 \times 20) + 56] \div 8 = \square$
4. $[(8 \times 5) - \{30 - (5 \times 2)\}] \div 2 = \square$
5. $2 \times \left\{\dfrac{[5\,(2 \times 2)] + (8 \times 3)}{[(24 \div 6) + 4] - 3}\right\} = \square$

Additional Considerations: This topic is an important prerequisite for algebra. Therefore, spaced practice will serve as a review as well as for consolidation of the skill. A systematic outline of steps such as shown in the activities is a useful model to help the student solve problems of this type.

EXPONENTS

Some people describe an exponent in this manner: "The exponent tells how many times the base is multiplied by itself." Let's see what that *really* says. Take the example 8^2. If 8 is multiplied by itself twice we have

$$8 \times 8 \text{ (eight multiplied by itself once)} = 8^2$$
$$8 \times 8 \times 8 \text{ (eight multiplied by itself twice)} = 8^3$$

It is therefore mathematically incorrect to say that an exponent indicates how many times a base number is multiplied by itself. *The exponent tells how many times the base is used as a factor.* Thus, $8^2 = 8 \times 8$.

In exponential language, a base is not thought of as a model group but as the factor that is to be repeated in an exponential expression. The products are called *powers* of the base. For example, the fourth power of 3 (written 3^4) is 81; the third power of 5 (written 5^3) is 125.

The power of the base is often read using the forms "first," "second," "third," "fourth," "fifth," and so on. These are referred to as *ordinal forms*. Often b^2 is called "base square," and b^3 is "base to the cube power" or "the cube of the base."

Four ideas about exponents are presented below. The first idea has already been described above and is summarized here.

1. *For any positive whole number (the set of positive whole numbers* $= \{1, 2, 3 \ldots\}$*) used as a base* (b), $b^n = \{b \times b \times b \times \ldots \times b\}$ *(n times).*

The exponent is represented by n and the base by b. In the following example, numbers are assigned to b and n.

base (b)	*exponent (n)*			*number expressed as a base-ten numeral*
3	4	$3^4 = 3 \times 3 \times 3 \times 3$	=	81
4	2	$4^2 = 4 \times 4$	=	16
10	3	$10^3 = 10 \times 10 \times 10$	=	1000
2	8	$2^8 = 2 \times 2 \times 2 \times 2 \times 2 \times 2 \times 2 \times 2$	=	256
8	2	$8^2 = 8 \times 8$	=	64

Example a:

$$2^2 \cdot 2^3$$

$$2^2 \cdot 2^3 = 2^{2+3} \text{ or } 2^5$$

$$2 \cdot 2 \times 2 \cdot 2 \cdot 2 = 2 \cdot 2 \cdot 2 \cdot 2 \cdot 2$$

$$4 \times 8 = 32$$

Example b:

$$b^7 \cdot b^3$$

$$b^7 = b \cdot b \cdot b \cdot b \cdot b \cdot b \cdot b$$

$$b^3 = b \cdot b \cdot b$$

So

$$b^7 \cdot b^3 = b^{7+3} = b^{10}$$

2. *Numbers expressed in exponential form having the same base may be multiplied by adding the exponents. For all positive whole numbers m, n, and a,* $a^m \cdot a^n = a^{m+n}$.

3. What happens in division? The following exercises involve dividing exponential number expressions. What pattern do you see? Do the exponents sum or is some other operation employed?

Example a:

$$2^5 \div 2^2$$

$$2^5 = 2 \cdot 2 \cdot 2 \cdot 2 \cdot 2 \qquad = 32$$

and

$$2^2 = 2 \cdot 2 \qquad = 4$$

$$32 \div 4 = 8 = 2 \cdot 2 \cdot 2 = 2^3$$

So

$$2^5 \div 2^2 = 2^{5-2} = 2^3$$

Example b:

$$3^4 \div 3^2$$

$$\frac{3^4}{3^2} = \frac{3 \cdot 3 \cdot 3 \cdot 3}{3 \cdot 3} = \frac{3}{3} \cdot \frac{3}{3} \cdot \frac{3}{1} \cdot \frac{3}{1}$$

But

$$\frac{3}{3} = 1$$

and

$$\frac{3}{1} = 3$$

So

$$\frac{3}{3} \cdot \frac{3}{3} \cdot \frac{3}{1} \cdot \frac{3}{1} = 1 \cdot 1 \cdot 3 \cdot 3 = 3^2$$

So

$$3^4 \div 3^2 = 3^{4-2} = 3^2$$

Example c:

$$b^3 \div b^3,\ b \neq 0$$

$$\frac{b^3}{b^3} = \frac{b \cdot b \cdot b}{b \cdot b \cdot b} = \frac{b}{b} \cdot \frac{b}{b} \cdot \frac{b}{b} = 1 \cdot 1 \cdot 1 = 1$$

So

$$b^3 \div b^3 = b^{3-3} = b^0 = 1$$

Example d:

$$b^2 \div b^5$$

$$b^2 \div b^5 = \frac{b^2}{b^5} = \frac{b \cdot b}{b \cdot b \cdot b \cdot b \cdot b} = \frac{b}{b} \cdot \frac{b}{b} \cdot \frac{1}{b} \cdot \frac{1}{b} \cdot \frac{1}{b} = \frac{1}{b^3}$$

So

$$\frac{b^2}{b^5} = \frac{1}{b^{5-2}} = \frac{1}{b^3}$$

A prerequisite agreement for the notation

$$\frac{b \cdot b}{b \cdot b \cdot b \cdot b \cdot b} = \frac{b}{b} \cdot \frac{b}{b} \cdot \frac{1}{b} \cdot \frac{1}{b} \cdot \frac{1}{b}$$

is that for $\frac{a}{b}$, *if a and b are real numbers* (*see Chapter 2*) *and b does not equal zero, then*

$$\frac{a}{b} = a \cdot \frac{1}{b}$$

So

$$\frac{b^2}{b^{2+3}} = \frac{b^2}{b^2 \cdot b^3} = \boxed{\frac{b^2}{b^2}} \cdot \frac{1}{b^3} = \frac{1}{b^3}$$

> *Note:* Example *c* illustrates why b^0 ($b \neq 0$) is defined equal to 1. 0^0 is undefined.

4. If we remove the restriction that the exponents m and n are positive whole numbers, we can investigate the meaning of negative exponents through the property $a^m \cdot a^n = a^{m+n}$. We found that for $b^3 \div b^3$ ($b \neq 0$), we could also write the following:

$$\frac{b^3}{b^3} = 1 \qquad \frac{b^3}{b^3} = b^{3-3} = b^0 = 1 \qquad (b \neq 0)$$

Also

$$\frac{b^5}{b^5} = 1 \qquad \frac{b^5}{b^5} = b^{5-5} = b^0 = 1 \qquad (b \neq 0)$$

$$\frac{b^n}{b^n} = 1 \qquad \frac{b^n}{b^n} = b^{n-n} = b^0 = 1 \qquad (b \neq 0)$$

We then defined the zero exponent as $a^0 = 1 (a \neq 0)$. It appears that in b^m/b^n, the n in the denominator is the negative exponent in b^{m-n}.

If $b^{m-n} = b^{m+(-n)}$, then

$$\frac{b^m}{b^n} = b^{m+(-n)} \; (b \neq 0)$$

Examine the following illustrations:

$$b^3 \cdot b^{-3} = b^{3+(-3)} = b^{3-3} = b^0 = 1 \qquad (b \neq 0)$$

$$b^7 \cdot b^{-7} = b^{7+(-7)} = b^{7-7} = b^0 = 1 \qquad (b \neq 0)$$

Since the product is always 1, we know that the two given numbers in each example are multiplicative inverses (see Chapter 9). For example, if $b \neq 0$, then the inverse of b^2 is $1/b^2$. When two multiplicative inverses are multiplied, what is the product?

$$b^2 \cdot \frac{1}{b^2} = \frac{b^2}{b^2} = 1 \qquad (b \neq 0)$$

If we assume that $b^2 \cdot b^{-2} = 1$, then we may conclude that b^{-2} is the multiplicative inverse of b^2. A number has one and only one multiplicative inverse. Thus, b^{-2} is just another name for $1/b^2$. A negative exponent may be described as follows, where $b \neq 0$ and n is any integer.

$$b^{-n} = \frac{1}{b^n}$$

The notation of a negative exponent implies that its base number is a divisor. Extending place value systems to the right of the units place, or b^0, the form b^{-1} implies $1/b$.

So

$$10^{-1} = \frac{1}{10}$$

$$10^{-2} = \frac{1}{100}$$

$$5^{-2} = \frac{1}{25}$$

Mini-lesson 7.2.2 Use Nonnegative Exponential Notation

Vocabulary: exponent, exponential notation, power, base

Possible Trouble Spots: Visual perceptual difficulty may inhibit noticing the exponent; lacks understanding of multiplication operation

Requisite Objectives:
Perform multiplication and division

Name the base-ten place values

Activity 1
Write equations such as the following to show the function of the exponent:

$$2 \times 2 \times 2 \times 2 \times 2 = 2^5$$
$$2 \times 2 \times 2 \times 2 = 2^4$$
$$2 \times 2 \times 2 = 2^3$$
$$2 \times 2 = 2^2$$

Tell students that the smaller digit to the upper right is called an exponent and that the number that serves as the factor is called the base. Model the language to be used as follows in relationship to the first example given above. The 5 is the exponent. It tells how many times the base, 2, is used as a factor. "Two is used as a factor five times," read "Two to the fifth power." 2^2 is read: "Two squared"; "2 is used as a factor twice"; or "Two to the second power."

Activity 2
After practice of this nature, present an example in the form $27 \div 3$. Instruct the students to find how many times 3 is used as a factor in 27 by division:

$$27 \div ③ = 9$$
$$9 \div ③ = 3$$
$$3 \div ③ = 1 \qquad \square = 3$$

Next, have students find the following:

$$8 = 2^{\square}$$
$$8 \div ② = 4$$
$$4 \div ② = 2$$
$$2 \div ② = 1 \qquad \square = 3$$
$$16 = 4^{\square}$$
$$16 \div ④ = 4$$
$$4 \div ④ = 1 \qquad \square = 2$$
$$25 = 5^{\square}$$
$$25 \div ⑤ = 5$$
$$5 \div ⑤ = 1 \qquad \square = 2$$

Activity 3
The following paper-pencil exercises are suggested for practice.

Name ____________

State how each of these is read.

1. 4^5 ____________
2. 5^4 ____________
3. 10^2 ____________
4. n^7 ____________

Use exponents to express each product as a power.

5. $7 \times 7 \times 7 \times 7 =$ ______ 6. $5 \times 5 \times 5 =$ ______
7. $6 \times 6 \times 6 \times 6 \times 6 =$ ______ 8. $3 \times 3 \times 3 \times 3 =$ ______
9. $10 \times 10 \times 10 =$ ______ 10. $1 \times 1 \times 1 \times 1 \times 1 =$ ______

Find the number named by each of the following.

11. $3^4 =$ ____ 12. $7^2 =$ ____ 13. $2^4 =$ ____ 14. $5^3 =$ ____
15. $1^5 =$ ____ 16. $5^1 =$ ____ 17. $10^5 =$ ____ 18. $12^3 =$ ____
19. $4^2 \times 2^3 =$ ______ 20. $3^2 \times 3^4 =$ ______
21. $6^3 \times 1^4 =$ ______ 22. $10^2 \times 5^3 =$ ______
23. $14 \times 8^3 =$ ______ 24. $7 \times 9^2 \times 3^4 =$ ______

Name the exponent that should replace □.

25. $3^{□} = 81$ □ = ____ 26. $2^{□} = 32$ □ = ____ 27. $8^{□} = 8$ □ = ____
28. $10^{□} = 10{,}000$ □ = ____ 29. $10^{□} = 100{,}000$ □ = ____ 30. $7^{□} = 2401$ □ = ____

Name ____________

Complete the table.

Number	Product of tens	Power of ten
100		
	10 x 10 x 10	
		10^4
	10 x 10 x 10 x 10 x 10	
1,000,000		
		10^7
	10x10x10x10x10x10x10x10	

Name ______________________________

Write a power of ten for the place value of each circled digit.

1. (1)46,369 ____ 2. 63(7) ____ 3. 95(0)46 ____

4. (8)3,125 ____ 5. 4(8)8 ____ 6. (2)507 ____

7. 7(5)1,920,864 ____ 8. 6,954,(2)30,817 ____

9. (3)47,269,015 ____ 10. (9),716,245,830 ____

11. 64(0),261,378 ____ 12. (1)8,952,076,643 ____

Write the number for each expanded notation.

13. $(8 \times 10^3) + (0 \times 10^2) + (3 \times 10^1) + (9 \times 10^0) =$ ____

14. $(3 \times 10^2) + (0 \times 10^1) + (9 \times 10^0) =$ ____

15. $(1 \times 10^2) + (2 \times 10^1) + (5 \times 10^0) =$ ____

16. $(2 \times 10^3) + (8 \times 10^2) + (5 \times 10^1) + (0 \times 10^0) =$ ____

17. $(7 \times 10^3) + (3 \times 10^2) + (4 \times 10^1) + (6 \times 10^0) =$ ____

18. $(2 \times 10^3) + (0 \times 10^2) + (0 \times 10^1) + (2 \times 10^0) =$ ____

19. $(4 \times 10^4) + (8 \times 10^3) + (1 \times 10^2) + (6 \times 10^1) + (7 \times 10^0) =$ ____

20. $(9 \times 10^4) + (5 \times 10^3) + (0 \times 10^2) + (3 \times 10^1) + (2 \times 10^0) =$ ____

Additional considerations: Exponential notation is an extremely helpful shortcut. However, time spent in initial learning experiences should not be in shortcut form. Showing the students all of the multiplication that underlies the meaning of an exponent provides meaning to the notation. For example, show the meaning of an exponent by expressing products in the form $2^4 = 2 \times 2 \times 2 \times 2$.

Note: Mini-lesson 7.2.2 is a prerequisite for Mini-lessons 7.2.3 and 7.2.5.

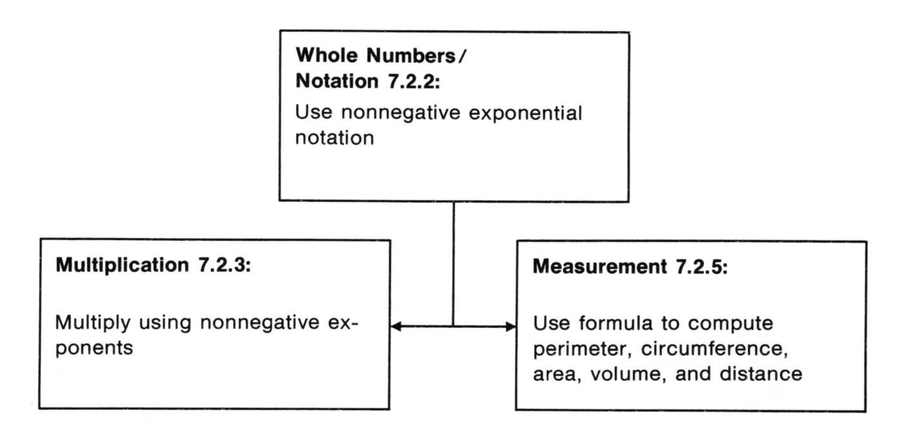

EXERCISES

I. Write each of the following in words.
1. 3^4
2. 5^3

II. Write the following as numerals.
1. nine cubed
2. four to the fourth power

III. What is the exponent in the following?
1. 4^5
2. 2^4
3. 5^3
4. 3^2

IV. Rewrite the following using exponentials:
1. $2 \times 2 \times 2 \times 2$
2. $3 \times 3 \times 3$
3. $6 \times 6 \times 6 \times 6 \times 6 \times 6$
4. 9×9

V. Express each of the following as a power of ten.
1. 1,000
2. 100,000
3. 100
4. 1,000,000

VI. Express each of the following as a product of a digit and a power of ten.
1. 80,000
2. 200,000

VII. Use exponents to name each product as a power.
1. $3 \times 3 \times 3 \times 3 \times 3 =$
2. $7 \times 7 \times 7 \times 7 =$
3. $11 \times 11 \times 11 =$
4. $2 \times 2 \times 2 \times 2 \times 2 \times 2 =$
5. $5 \times 5 \times 5 \times 5 \times 5 =$
6. $13 \times 13 \times 13 \times 13 =$

VIII. Round the following numbers as indicated.
1. 403 to tens
2. 7685 to tens
3. 12,345 to ten thousands
4. 210,555 to hundreds

Nonmetric Geometry, Figurate Numbers, Relations, and Graphs

POINT, LINE, AND PLANE

Point

Just as the idea of number is accepted intuitively, so it is for the idea of a point. A number is recorded by a symbol called a "numeral"; this symbol represents the number. Similarly, a *point* may be represented by a dot (•), sometimes with a letter attached (A •). A point has no length, no depth, no width—only position; and the dot that represents it indicates its position. For example, this dot (•) indicates a different position on the page from that of the dot shown above. The set of all points is called *space*. The configurations of geometry are subsets of points in space.

Line and ray

A *line* is a set of points. There are *straight lines,* which may be represented by a continuous drawing not changing direction, and *curved lines,* no part of which is straight but continuously changes direction.

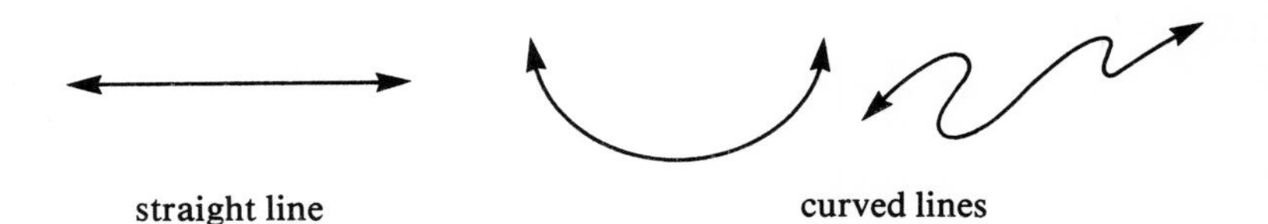

The notation $\overleftrightarrow{AB}$, read "line AB," represents a line that extends indefinitely in either direction, as indicated by an arrow on either end $\overset{A}{\bullet}\!\!\longleftrightarrow\!\!\overset{B}{\bullet}$. $\overrightarrow{AB}$, read "ray AB," is the *ray* whose end point is A and extends indefinitely in one direction from A $\overset{A}{\bullet}\!\!\longrightarrow$.

Intersecting lines

For any two distinct points in space, there exists one straight line that contains these two points. *Intersecting lines* have at least one point in common.

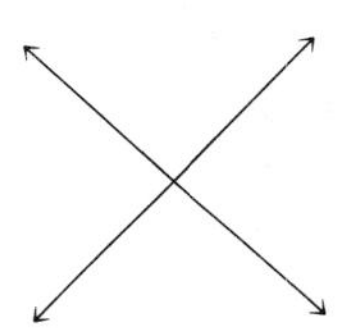

Collinear set of points

A set of points is said to be "collinear" if all points of the set are contained in the same straight line. Otherwise, they are considered to be "noncollinear."

Line segment

The notation for a *line segment,* that part of a line whose endpoints are A and B, is $\overline{AB}$, read "line segment AB." The length of a line segment is the distance between its two endpoints. To this distance is assigned a unique positive real number, which is the measure of the length of the line segment joining the two points.

Congruent line segments

Two line segments of the same length are said to be *congruent.* If the point R is located between the points P and Q so that $\overline{PR} = \overline{RQ}$, then R is the *midpoint* of the line segment $\overline{PQ}$. The point R is said to *bisect* the line segment $\overline{PQ}$. Zero distance is allowed for. If the points P and Q are graphs of the same number, then these are the same point, and the measure of $\overline{PQ} = 0$.

Union of two rays

The set operations of union (see number 2, Chapter 4, For Further Study) and intersection provide a helpful language for describing geometric conditions. For example, the *union of two rays* is the set of all points that are in one ray, the other ray, or both rays. The union of ray AB ($\overrightarrow{AB}$) and of ray AC ($\overrightarrow{AC}$) is the line named by any two of its points

$$\overset{C}{\bullet}\longleftrightarrow\overset{A}{\bullet}\longleftrightarrow\overset{B}{\bullet}\ ;\ \overrightarrow{AB} \cup \overrightarrow{AC} = \overleftrightarrow{AC} = \overleftrightarrow{CA} = \overleftrightarrow{AB} = \overleftrightarrow{BA}.$$

Naming a ray

Note that the order of letters in naming a ray *does* make a difference; the first letter named identifies the endpoint and the second letter named tells the direction. Thus, on $\overleftrightarrow{AB}$, the rays are named as follows:

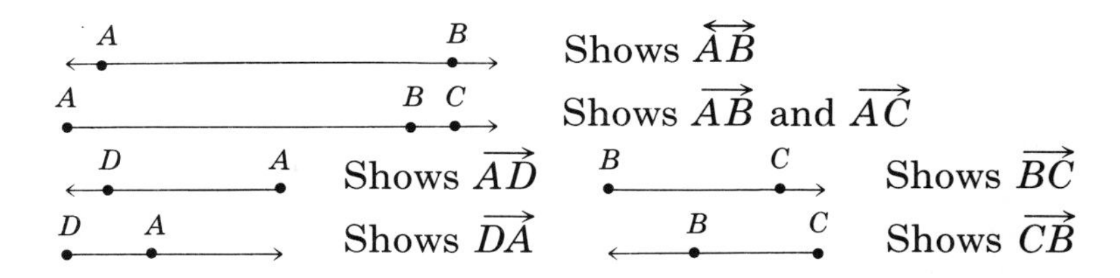

Plane

A *plane* is a set of points represented in two dimensions. Any three points that are not in a straight line determine a plane. The endpoints of the legs of a three-legged table or a photographer's tripod determine a plane.

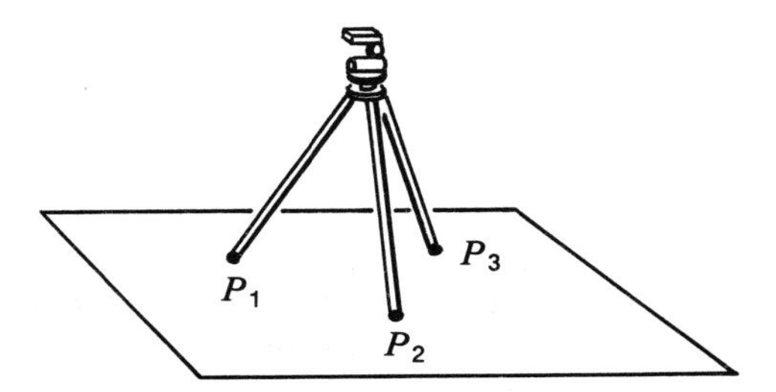

If a straight line contains two points of a plane, then the line lies entirely within the plane. For any given line and a point not on the line, there is one and only one plane that contains both the point and the line. Any three noncollinear points lie in one and only one plane. Two intersecting lines lie in one and only one plane. Lines or points that lie in the same plane are said to be "coplanar."

Parallel lines and skew lines

Given a line and a point, one and only one line can be drawn through the point so that both lines are parallel. *Parallel lines* (in Euclidean space) are coplanar lines that do not intersect. Nonparallel lines within a plane are said to be "nonplanar." *Skew lines* are lines in different planes that do not intersect. In the illustration below, parallel lines are represented by the railroad tracks labeled t_1 and t_2; the part of the bridge labeled *SR* represents a line skew to both t_1 and t_2.

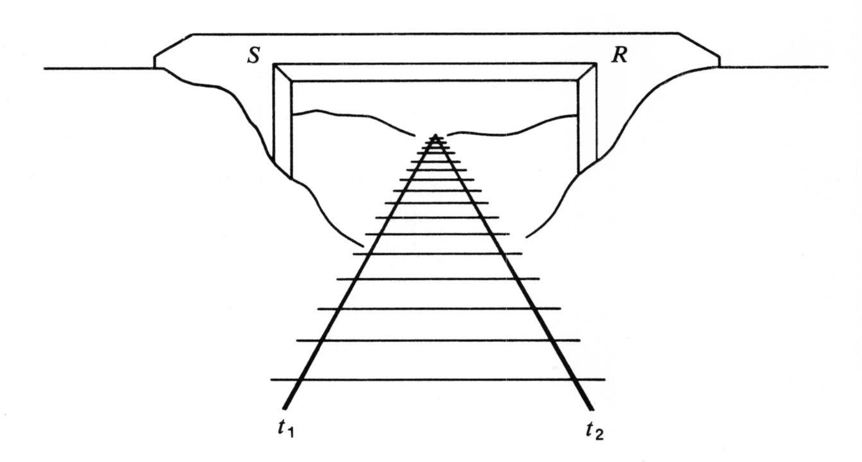

Parallel planes

Parallel planes are planes that do not intersect. The intersection of two planes is a straight line. Planes M and N intersect in line $\overleftrightarrow{AB}$.

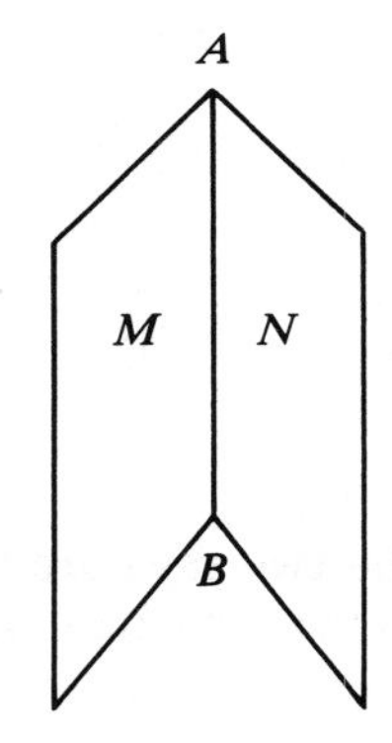

The intersection of a line and a plane may be either a point or a line.

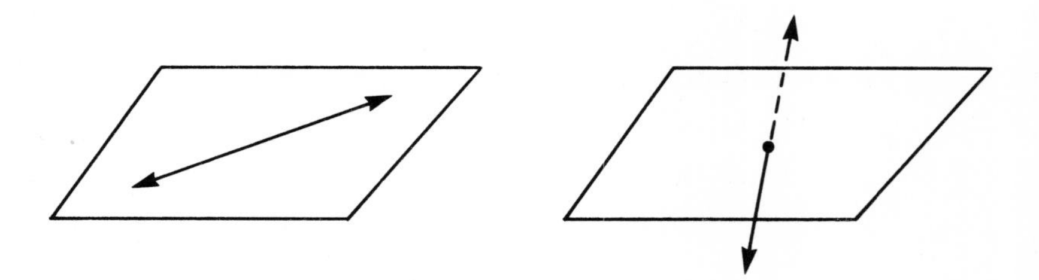

It is also possible that the line and the plane have no points in common; in other words, the line would be parallel to the plane.

Half lines and half planes

A point on a line is the boundary of each of the two *half lines* it creates, forming three disjoint sets of points, the point itself and the two resulting sets of points. Similarly, a line will separate a plane into two *half planes.* This line is not included in either of the half planes but is the boundary between them. The two half planes have no points in common. Again, three disjoint sets of points are formed.

Convex set of points

A half plane is an example of a *convex* set of points. A set of points is a convex set if it is such that for any two points P and Q located anywhere in the set, the segment PQ lies entirely within the set.

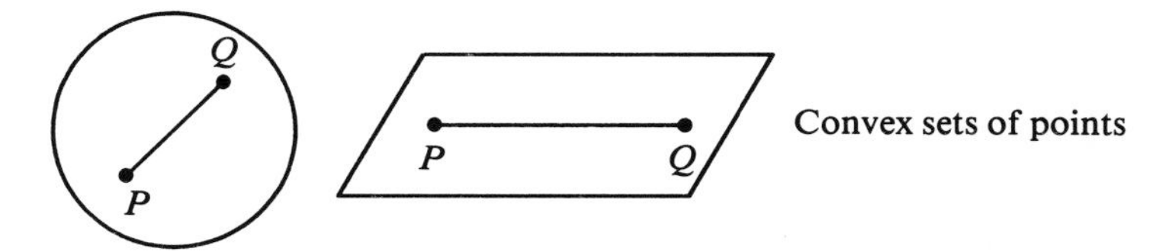

Convex sets of points

Concave set of points

If it is *not* true that the line segment connecting any two points lies completely within the plane, then the set of points determining the plane is said to be "concave."

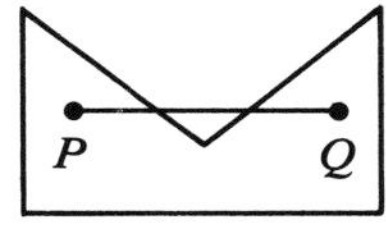

Concave sets of points

ANGLES

An *angle* is the union of two unique rays with a common end point. The two rays are the sides of the angle; the common endpoint is its *vertex*. The symbol $\angle$ represents one angle; $\angle$s represents more than one angle. Angles are indicated by the letters of their sides and the letter of the common point.

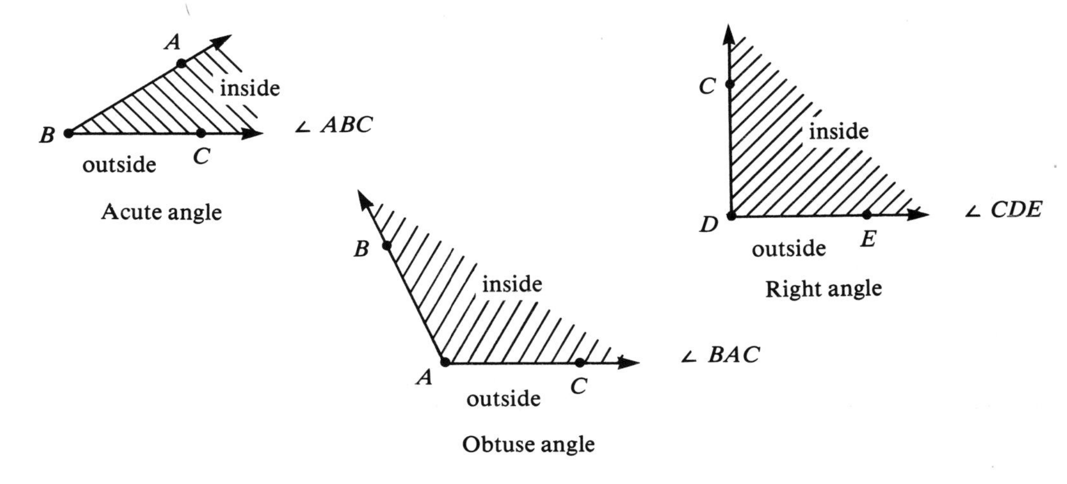

Naming angles

The vertex is always named in the middle. For example, in the acute angle above, $\overrightarrow{BA} \cup \overrightarrow{BC} = \angle ABC$ or $\angle CBA$. Angles may also be named by a single letter representing its vertex $\angle B$ or by a Greek letter, alpha $\angle \alpha$ or beta $\angle \beta$. Lower-case letters ($a, b, c, \ldots$) or numbers (1, 2, 3, . . .) may also be used for notation.

An angle separates a plane into three disjoint sets: the angle itself, the interior of the angle, and the exterior of the angle.

The size of angle BAC

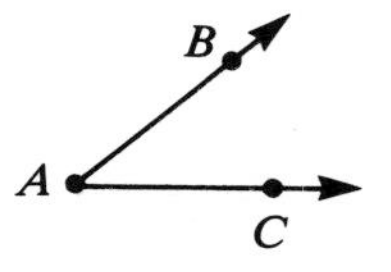

may be interpreted as the amount of rotation necessary to make AC coincide with (fall on top of) $\overrightarrow{AB}$.

POLYGONS

A *polygon* is a figure formed by the union of at least three line segments in the same plane where each line segment intersects two other line segments at their endpoints. Each line segment is a side of the figure, and the corners of the figure are vertices.

A *triangle* is a polygon with three sides and three vertices; a *quadrilateral* has four sides and four vertices; a *pentagon,* five, an *octagon,* eight, a *decagon,* ten. All of the polygons shown below are "convex polygons" because the set of their interior points is a convex set (as described in this chapter). A "regular polygon" has all of its sides and all of its angles congruent.

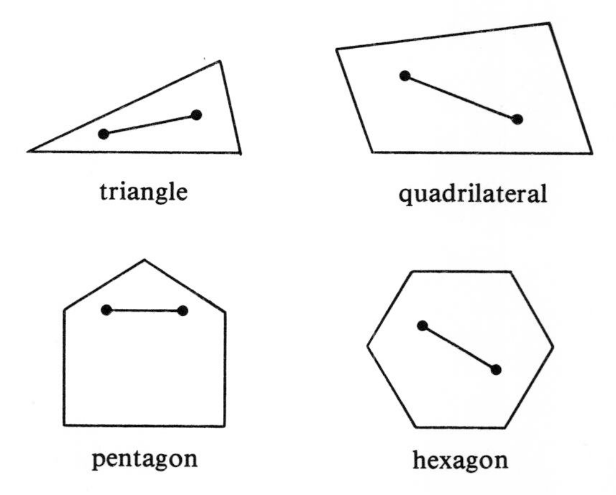

The following figure is *not* a convex polygon; it is a concave polygon.

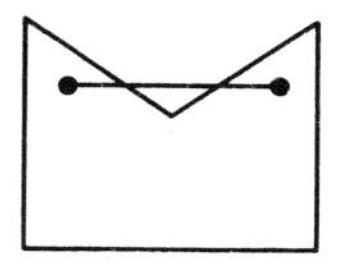

CIRCLES

A *circle* is a simple closed curve in a plane, all points of which are at a fixed distance from a fixed point. The fixed distance, a line segment, is called the *radius* and the fixed point the *center* of the circle. A line segment whose end points are points of a circle is a *chord* of the circle. A *diameter* of a circle is a chord that contains the center of the circle, and its length is twice the length of the radius. A *central angle* ($\angle APB$) is an angle whose vertex is the center of the circle. If the sides of the central angle intersect the circle in points A and B, then these two points and all points on the circle form a *minor arc* ($\widehat{AB}$); the two points and all points on the circle exterior to the angle form a *major arc* ($\widehat{ACB}$). A *semicircle* is that measure of an arc equal to half the measure of the circumference of the circle.

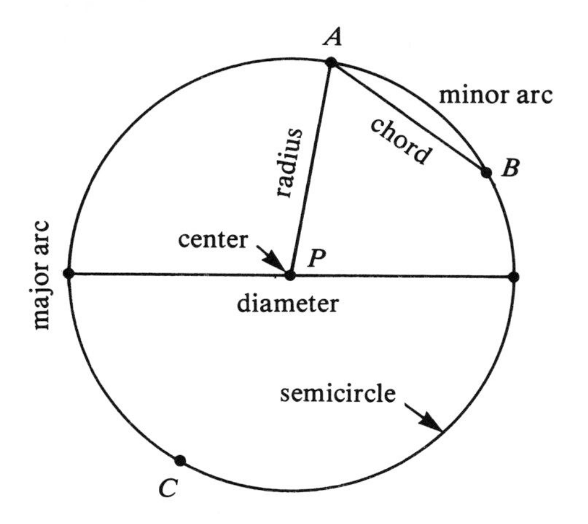

Two or more circles with the same center are *concentric* circles.

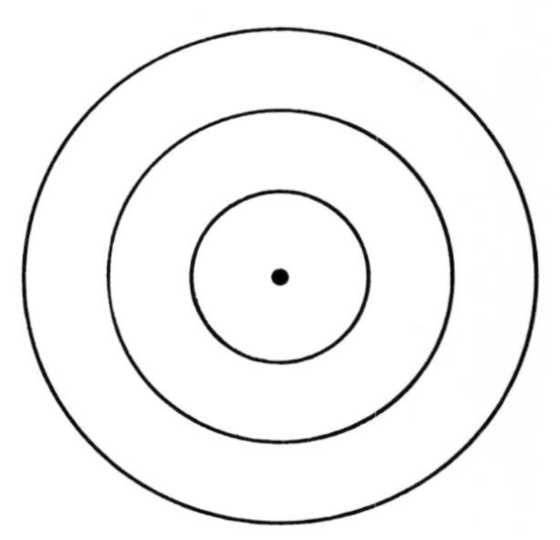

> *Note:* A circle and a circle with its interior are often confused. The teacher should make it clear to pupils that the circle is the curved line and does not include the interior region.

As with any simple closed curve in a plane, a circle separates the plane into three sets of points:

1. Set of points in the interior of the curve
2. Set of points of the curve
3. Set of points in the exterior of the curve

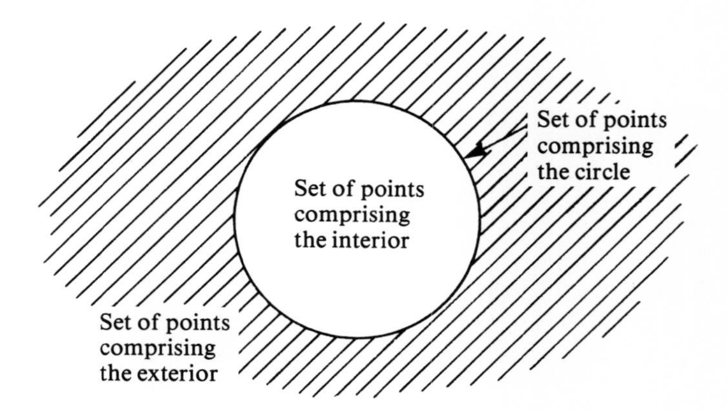

CONSTRUCTIONS

The following mini-lesson for the intermediate grades involves constructing geometric figures using a straightedge (unmarked ruler) for drawing a line segment to represent a straight line and a compass for drawing a circle.

The basic constructions described below depend upon the intersection properties of two lines, a line and a circle, and two circles.

Mini-lesson 5.1.10 Construct Geometric Figures

Vocabulary: vertex, endpoint, angle, ray, acute, obtuse, right angle, congruent, parallel, symmetric, quadrilaterals, trapezoid, parallelogram, rhombus, polygon, construct, radius, arc, intersect, segment, triangle, bisect, midpoint

Possible Trouble Spots: poor eye-hand coordination; visual perception problems; poor fine motor skills; inability to attend to salient aspects of a situation; distractibility and/or hyperactivity; difficulty following directions

Requisite Objectives:
Geometry 1.1.8: Show topological relationships: proximity, enclosure, separateness, order

Measurement 1.1.9: Compare sizes of objects

Geometry 1.2.6: Recognize topologically equivalent shapes

Measurement 1.2.7: Arrange geometric figures by number of sides and size

Geometry 1.3.11: Identify simple plane figures

Geometry 2.3.8: Identify line, line segment, and parallel lines

Geometry 3.2.15: Identify angles

Geometry 4.2.5: Identify 3-dimensional geometric shapes

Activity 1
To construct at a point on a given line an angle congruent to a given angle, ask the child to draw a line l and an acute angle $\angle ABC$. Ask her to mark a point P on the line l and to select $\overrightarrow{PT}$ of the line. Tell her to use the vertex B of $\angle ABC$ as a center and any convenient radius to draw the arc of a circle cutting the sides of the angle at points M and N. She then uses the same radius and P as a center to draw the arc of a circle, cutting $\overrightarrow{PT}$ in Q. With Q as a center and $\overline{MN}$ as a radius, the child draws the arc of a circle, intersecting the original arc in point R. The $\overrightarrow{PR}$ completes the angle QPR, which is congruent to angle ABC ($\angle QPR \cong \angle ABC$).

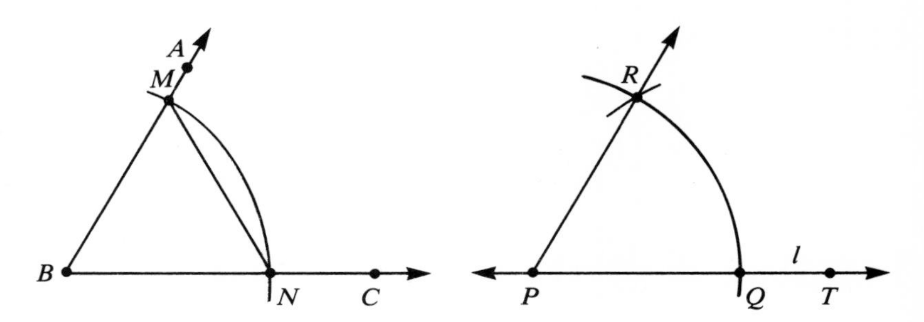

Activity 2
To construct a triangle whose sides are congruent to three given line segments, ask the child to draw three line segments of varying length, a, b, c. Tell him to draw a line l and on it mark the point A. Ask him to use A as a center and c as a radius to draw the arc of a circle cutting l in B; A as the center and b as radius to draw arc $\widehat{PQ}$; B as a center and a as a radius to draw arc $\widehat{RS}$ to intersect $\widehat{PQ}$. The intersection point is C. Ask the child to draw $\overline{AC}$, $\overline{BC}$. ΔABC is the required triangle.

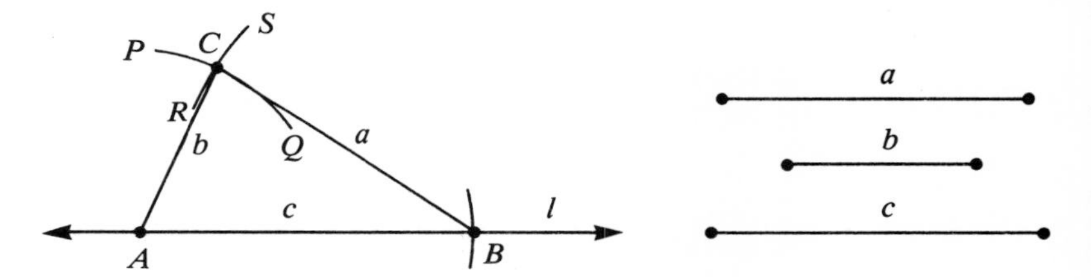

Activity 3
To bisect a given angle, ask the child to draw an acute angle ABC. Tell her to use B as a center and any radius to draw the arc of a circle cutting $\overrightarrow{BA}$ and $\overrightarrow{BC}$ in points D and E. Then ask her to use D as a center and a radius greater than $\overline{DE}$ to draw a circle. Tell her to do the same with E as a center. The circles will intersect at point F. The ray $\overrightarrow{BF}$ bisects $\angle ABC$.

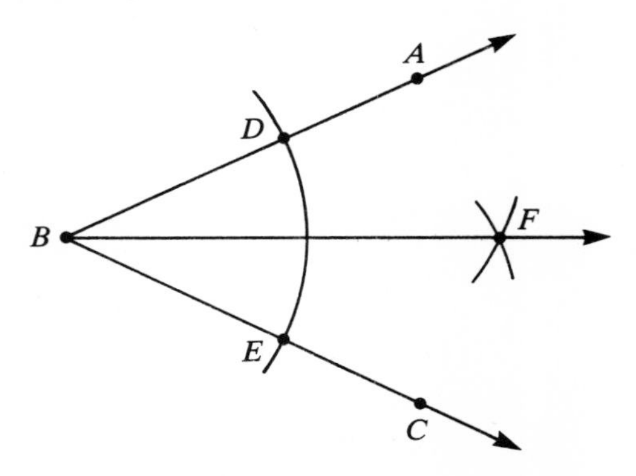

Activity 4
To bisect a given line segment, ask the child to draw $\overline{PQ}$. Then ask him to use P as a center and a radius greater than $1/2\overline{PQ}$ to draw a circle. Tell him to do the same with Q as a center. These two circles intersect in points M and N. The point O on $\overline{MN}$ is the midpoint of $\overline{PQ}$.

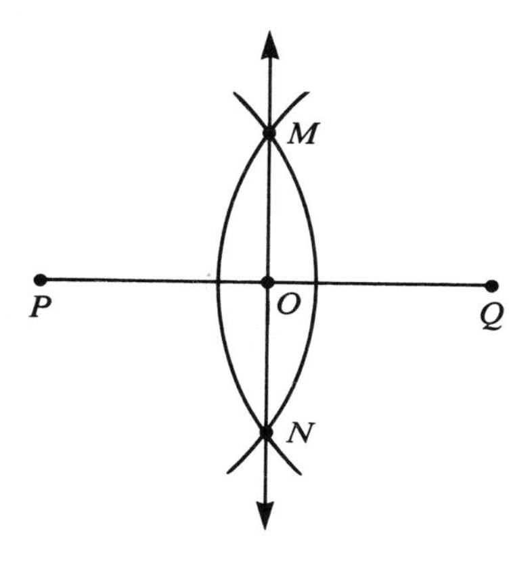

Activity 5
To construct a line perpendicular to a given line, ask the child to draw $\overleftrightarrow{PQ}$ and to select an arbitrary segment. Then ask her to proceed as in Activity 4. $\overline{MN}$ determines a line perpendicular to the line determined by $\overline{PQ}$.

Additional considerations: Following are suggestions for helping students name shapes they draw (rectangle, rhombus, hexagon, trapezium, right triangle, and so on). Two aids in helping children to recognize and say the names of the common shapes are

A. Building a bulletin board. Each time a new shape is introduced and named, the child draws the shape on the wall chart and writes its name. The bulletin board then becomes a reference.

B. Providing a variety of books showing shapes and names. Allow the children to show and discuss these shapes.

Suggestions for helping recognize specific shapes include discussing the rhombus as a subset of parallelograms and pointing out that a trapezoid has one pair of opposite sides parallel. Also discuss with the children how smaller shapes are put together. Each small shape touches at least one other small shape along a complete edge. Many children with visual perception problems are not able to bridge from concrete to pictorial representations of mathematical concepts. Therefore, specific activities involving concrete to picture translations are helpful. For matching solid forms to their picture representation, obtain the following materials: solid forms of three or more geometric shapes, for example, square, circle, triangle; pictures of each of these in three-dimensional perspective; and pictures of each in two-dimensional representation:

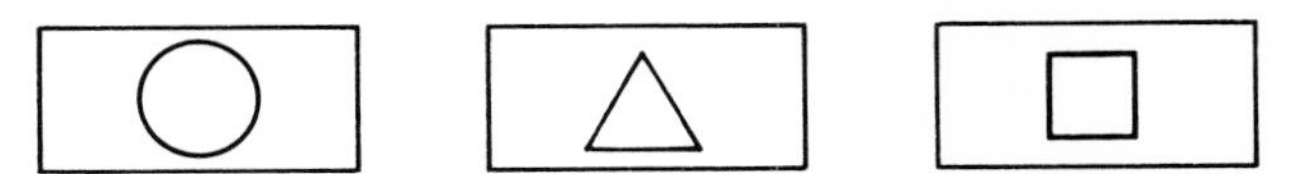

It may be necessary to allow the child to trace with his or her index finger the outline of each of the solid forms and of the pictures to facilitate matching them. The teacher should also consider the number of choices involved in this type of matching task in order to distinguish between the child's inability to discriminate forms and poor performance due to having to choose from too large a set of stimuli. It may be necessary to reduce the number of choices to facilitate successful performance.

You may need to simplify the task even more by using assorted colors of only one geometric shape.

Provide six pairs of a geometric shape (for example, squares) in various colors: red, yellow, blue, green, orange, and purple. Show the squares randomly and ask the child to select the square that is the same color as the square held up. If the child has difficulty with this task, try reducing the number of choices. For example, use only four colors for a child who appears to have difficulty making choices. At first, accentuate differences among the colors and then gradually minimize them as the child begins to discriminate. This procedure also helps the child learn to attend to salient aspects of a situation.

Note: Mini-lesson 5.1.10 is a prerequisite for Mini-lessons 5.3.15 and 6.1.6.

Geometry 5.1.10:

Construct geometric figures

Graphs and Functions 5.3.15:

Interpret coordinates of a point on a grid

Measurement 6.1.6:

Measure angles of polygons

FIGURATE NUMBERS

A series of numbers that can be expressed as geometric forms is called *figurate numbers.* This type of series is derived from an arithmetical progression of numbers in which the terms have a common difference that is an integer. This is described in Brooks:

> Take the series of natural numbers in which the common difference is 1, as represented by line A in the example [below]; then, the series B will be figurate numbers; series C derived from series B, and series D derived from series C will be figurate numbers. Other series could be obtained by beginning with any other arithmetical series whose first term is 1, and common difference is an integer. Thus, the series derived from the progression 1, 3, 5, 7, 9, etc., is 1, 4, 9, 16, 25, etc.
>
> A, 1-2-3-4-5-6-7
> B, 1-3-6-10-15-21-28
> C, 1-4-10-20-35-56-84
> D, 1-5-15-35-70-126-210[1]

Notice how series A yields the figurate series B, C, and D by addition. For example, $1 + 2$ (series A) $= 3$ (series B); $1 + 2 + 3$ (series A) $= 6$ (series B); $1 + 2 + 3 + 4$ (series A) $= 10$ (series B); and so on.

Triangular Numbers

The figurate numbers in series 1, 3, 4, 10, . . . , are called *triangular numbers* because the number of units they express can be arranged in the form of triangles.

X	X	X	X	X
	XX	XX	XX	XX
		XXX	XXX	XXX
			XXXX	XXXX
				XXXXX
1	3	6	10	15

Square Numbers

The figurate series derived from 1, 3, 5, 7, 9, . . . , in which the common difference is 2, can be arranged in squares. This figurate series is 1, 4, 9, 16, 25, . . . , and is pictured as shown:

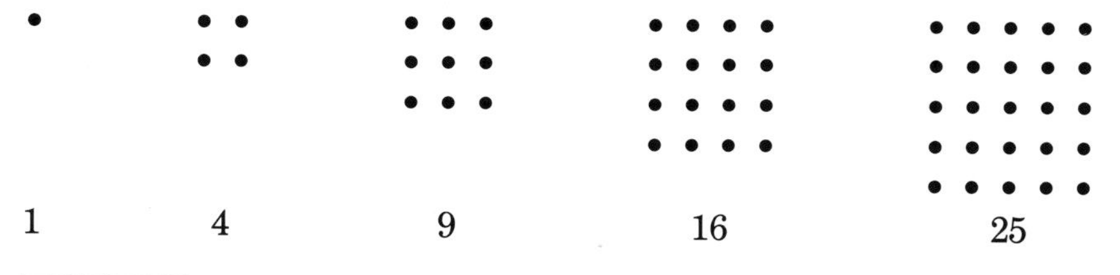

1. Edward Brooks, *Philosophy of Arithmetic* (Lancaster, Pa.: Normal Publishing Co., 1880), p. 385.

Polygon Numbers

The series 1, 4, 7, 10, . . . , in which the common difference is 3, yields the figurate numbers in the series 1, 5, 12, 22, . . . , which are called pentagonal numbers because they may be arranged in pentagons. In the same manner, we can obtain figurate numbers that may be expressed as many other shapes, such as hexagons (six-sided), heptagons (seven-sided), octagons (eight-sided). Brooks points out that

> the number of the sides of the polygon which the figurate number series represents is always two greater than the common difference of the series from which they were derived. When the common difference of the series in arithmetical progression is 1, the sums of the terms give the triangular numbers; when the common difference is 2, the sums of the terms are square numbers; when the difference is 3, the sums are the pentagonal numbers, and so on.[2]

RELATIONS

Relation as a set of ordered pairs

A *relation* can be defined as a set of ordered pairs. Relations may be shown in several ways: by formulae, tables, graphing, and pairing.

Another common method of designating a relation is to use the letter R. For example, the relation "is a sister of" for "Mary (m) is a sister of John (j)" is written mRj. R stands for "is a sister of." To show that Mary (m) is the sister of not only John (j) but also of Lora (l) and Betsy (b), we would write mRj, mRl, mRb. This relation may be designated by the following set of ordered pairs:

R = {(Mary, John), (Mary, Lora), (Mary, Betsy)}

Other examples, where the members of pairs are represented by x and y, are shown below:

x is shorter than y, xRy
x is a child of y, xRy
x is heavier than y, xRy
x is less than y, xRy

Notice how the order of the elements is important in these relations. For example, x cannot be both taller than and shorter than y at the same time. By the same reasoning, x cannot be a child of y and also y be a child of x. Thus, a relation is a set of *ordered* pairs.

Equivalence Relations

Reflexivity

The equals relation symbolized by = may be described in the following phrases: "x is as heavy as x"; "x is as full as x"; "x is as young

2. Brooks, *Philosophy of Arithmetic,* p. 387.

as x"; "x is as tall as x." If x is replaced by various nouns in the previous examples, a special characteristic called the "reflexive property" is shown: "John is as heavy as John"; "sugar bowl is as full as sugar bowl"; "Lisa is as young as Lisa"; "Harry is as tall as Harry." Thus, if xRx for all elements x, then R is said to be reflexive; $x = x$ is true for all substitutions. The reflexive property is one of three properties of equivalence relations. The other two are symmetry and transitivity.

Symmetry

Symmetry involves the idea that if $x = y$, then $y = x$. Therefore, in the example "is a sister of," mRb and bRm. In the example of Mary and John, we may use the relation "is a sibling of," mRj and jRm. This property underlies the number idea that if $12 = 7 + 5$, then $5 + 7 = 12$. If xRy and yRx for all x elements of R and for all y elements of R, then the relation is said to be symmetric; if $x = y$ then $y = x$.

Transivity

Equivalence relations have a transitive property described by "is less than," "is greater than," "is shorter than," "is heavier than." A specific example showing transitivity might be "If John is shorter than Bill, and Bill is shorter than Greg, then John is shorter than Greg." (A relation where transitivity does not apply might be "If John loves Mary, and Mary loves Greg, then John loves Greg.") The transitive relation states: If xRy and yRz, then xRz, and R is a transitive relation. Referring again to Mary (m), Lora (l), and Betsy (b), the transitive property designates that if Mary is a sister of Lora, and Lora is a sister of Betsy, then Mary is a sister of Betsy: If mRl and lRb, then mRb.

Reflexivity, Symmetry, and Transitivity as Related to Number Concepts

If a child does not realize that 6 raisins are 6 raisins regardless of their physical arrangement in space, the child is not grasping the reflexive idea that $6 = 6$.

In fact, all three relations underlie renaming equivalent fractions, as illustrated in the following three-step sequence.

Example: To show $\frac{3}{4} = \frac{9}{12}$ $\left(\frac{a}{b} = \frac{c}{d} \text{ if and only if } ad = bc\right)$

Step 1: $\frac{3}{4} = \frac{3}{4}$ (Reflexivity)

Step 2: If $\frac{3}{4} = \frac{6}{8}$, then $\frac{6}{8} = \frac{3}{4}$ (Symmetry)

Step 3: If $\frac{3}{4} = \frac{6}{8}$, and $\frac{6}{8} = \frac{9}{12}$, then $\frac{3}{4} = \frac{9}{12}$ (Transitivity)

This shows that equals is an equivalence relation over the set of fractions and is an important concept in addition and subtraction of numbers in fraction form with unlike denominators.

Additional implications of these three relations in the early grades involve a child's knowing that $5 = 3 + 2$ may be substituted for $2 + 3 = 5$. This example of symmetry may not be intuitively grasped by youngsters.

We often use different names for the same number. Underlying this notion is the transitive property exemplified as follows: If $2 + 3 = 5$ and $5 = 4 + 1$, then $2 + 3 = 4 + 1$.

Inequality Relations

Sets may serve as models for number comparisons. Notice that there is a similarity between the subset symbol ($\subseteq$) and the "less than or equal to" symbol ($\leq$). Also notice the similarity between the proper subset symbol ($\subset$) and the symbol for "less than" ($<$).

Given

$$A = \{1, 2, 3, 4\}$$

If

$$B = \{1, 2, 3\}$$

and

$$C = \{1, 2, 3, 4\}$$

then

$B \subset A$ and $n(B) < n(A)$ or *3 is less than 4*

$C \subseteq A$ and $n(C) \leq n(A)$

CARTESIAN COORDINATES

We have used ordered pairs in connection with cross product of sets. An ordered pair may also determine the position of a point. For example, the ordered pair (3, 1) describes the position of a point. Pairs of numbers used like this are called the *Cartesian coordinates* of the

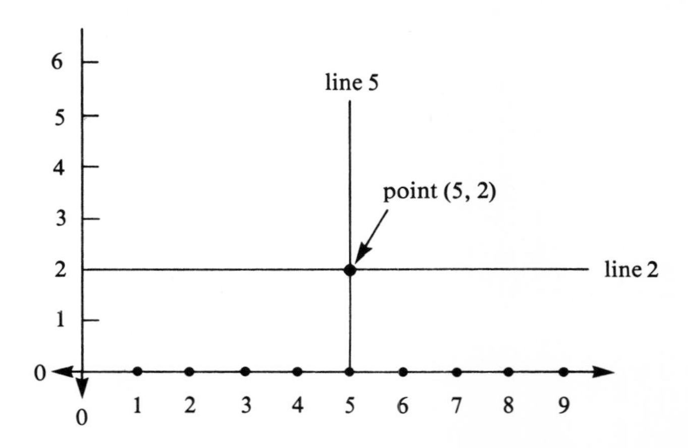

point. The point whose coordinates are (0, 0) is called the *origin*. The two number lines forming the grid are the *axis*. The location of a point may be found by the intersection of two lines. For example, the point whose coordinates are (5, 2) is found at the intersection of line 5 and line 2.

The position of line 5 is determined by moving along the horizontal axis to point 5. The position of line 2 is determined by moving up the vertical axis to point 2.

Mini-lesson 5.3.15 Interpret Coordinates of a Point on a Grid

Vocabulary: Cartesian coordinates, grid, graph, ordered pair

Possible Trouble Spots: difficulty with spatial relationships; inability to attend to salient aspects of situations; visual perception difficulty; counts incorrectly on a number line

Requisite Objectives:
Graphs and Functions 3.2.20: Use a number line

Graphs and Functions 3.3.9: Extend use of a number line to set of integers

Graphs and Functions 4.1.17: Interpret and use a grid

Activity 1
To draw geometric figures on a grid when given the Cartesian coordinates, ask the student to locate on graph paper the coordinates (3, 1), (5, 1), (7, 3), (5, 6), (3, 6), and (1, 3). Then ask him to connect the points in the order given and to connect the last point located with the first one. Ask him to name the figure drawn.

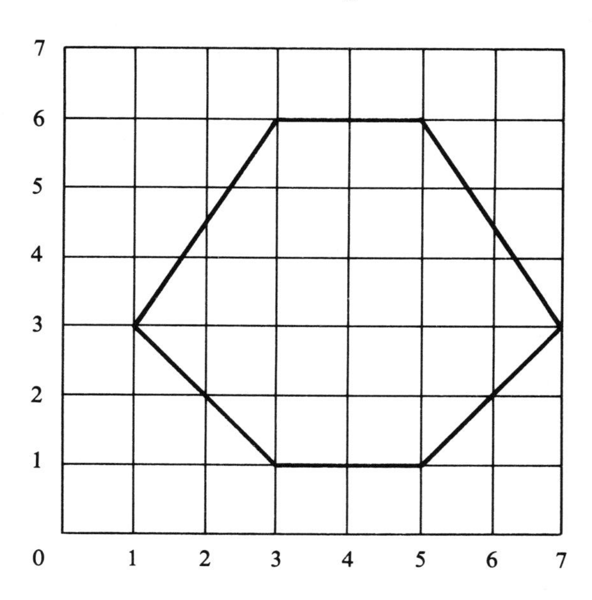

Activity 2
To determine the Cartesian coordinates of a figure from its position on a grid, ask the children to draw polygons on grids and then determine the Cartesian coordinates of the vertices. Then have the children exchange their sets of ordered pairs and compare the figures.

Activity 3
To use a geoboard as a grid, ask the child to show a polygon on a geoboard and then to determine the coordinates of the vertices. Some geoboards are made of plexiglass and may be used on an overhead projector. This activity is similar to activities 1 and 2 but is appropriate for younger children because it is at the concrete level.

Activity 4
To identify the positions of ordered pairs on an x, y axis, divide the class into two teams—one team using dots to graph its positions and the other team using X's. The teams take turns calling out ordered pairs as you or a pupil graph the positions. The object of the game is for a team to get four dots (or four X's, depending on the team's symbol) in a row, either vertically, horizontally, or diagonally. The teacher can change the conditions of the game by changing the numbers on the x, y axis or by changing the quadrant used. Only the numbered positions can be used to name the ordered pairs.

A sample game is played next. Remember, ^-x is negative only if x is positive. In like manner, ^-y is negative only if y is positive.

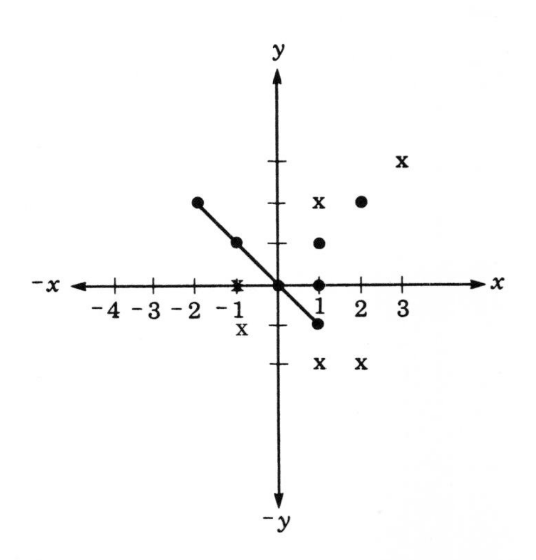

Below are listed the ordered pairs called out by each team.

	Team (•)	Team (X)
Starts	1, 1	
		1, 2
	0, 0	
		$^{-}1$, $^{-}1$ (to block a diagonal)
	2, 2	
		3, 3 (to block)
	1, 0	
		$^{-}1$, 0
	1, $^{-}1$	
		1, $^{-}2$ (to block)
	$^{-}1$, 1	
		2, $^{-}2$ (to block)
	$^{-}2$, 2 (Won!)	

The set of ordered pairs that the winning team called out is {(1, 1), (0, 0), (2, 2), (1, 0), (1, $^{-}1$), ($^{-}1$, 1), ($^{-}2$, 2)}. In the relation called out by the winning team, the domain (the set of all first elements of the ordered pairs of a relation) is {1, 0, 2, 1, 1, $^{-}1$, $^{-}2$} and its range (the set of all second elements of the ordered pairs of a relation) is {1, 0, 2, 0, $^{-}1$, 1, 2}.

Additional considerations: The teacher should be aware that the coordinate on the horizontal axis is named before that on the vertical axis on an x/y graph, whereas the coordinates are named in the opposite order for longitude and latitude. This difference in the order of naming the coordinates is sometimes confusing to the student who is introduced to the two topics simultaneously.

Note: Mini-lesson 5.3.15 is a prerequisite for Mini-lessons 6.2.7 and 6.2.8.

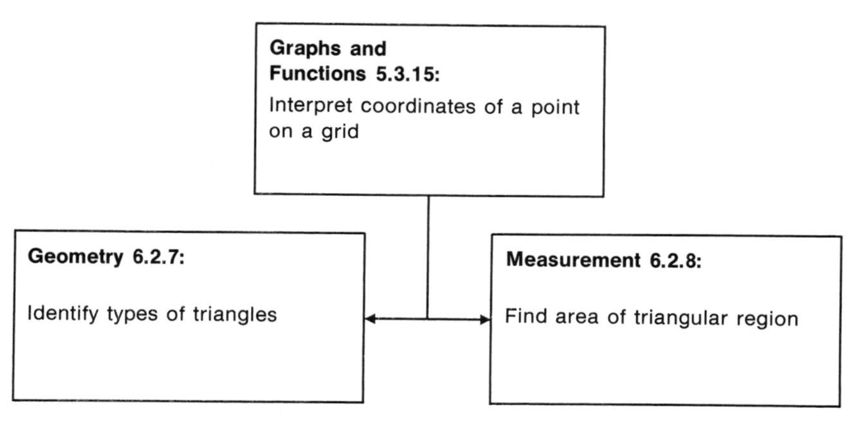

RELATIONS AND FUNCTIONS

As stated previously, a *relation* is a set of ordered pairs of elements that are in a certain sequence. The drawing below shows why the ordered pair (2, 3) is different from the ordered pair (3, 2).

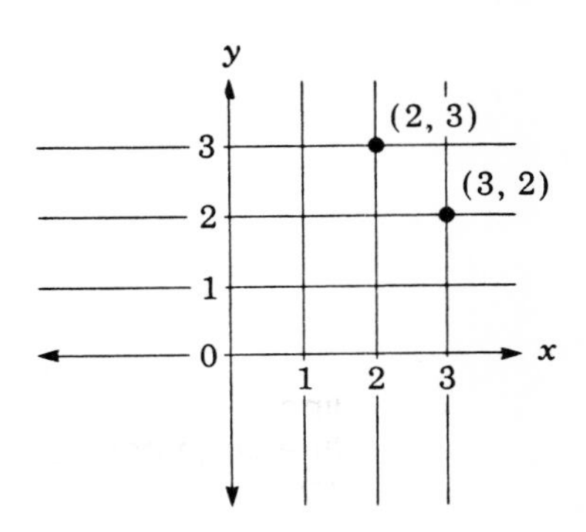

In the ordered pair (x, y), the first number gives a position on the x axis, the second number gives a position on the y axis. Thus, for the ordered pair (2, 3), we go 2 jumps to the right (positive direction) on the x axis and then go 3 jumps up (positive direction) on the y axis.

A *function* is a set of ordered pairs in which no two of the ordered pairs have the same *first* element. Thus, the set of ordered pairs $\{(1, 3), (2, 5), (3, 7)\}$ is not only a relation but also a function. The equation $y = x^2$, where x is any real number, defines a function and can be expressed $f(x) = x^2$. A partial listing of the set is $\{(0, 0), (^-1, 1), (1, 1), (5, 25), (1/2, 1/4), (\sqrt{2}, 2), \ldots\}$.

The letters f, g usually denote functions. For example, $f = \{(1, 2), (7, 14)\}$. If (x, y) is a member of f, write "$y = f(x)$," read "y equals f of x."

EXERCISES

Set A

I. Look around the classroom and outside and discuss the various shapes. See if any shapes are made up of a set of smaller shapes. For example, a rectangular window frame is made up of a set of squares, and some fences made of wire comprise a set of rhombus shapes.

II. 1. Given a set of congruent shapes, such as squares, make a variety of shapes with them.

2. Given sets of right-angled triangles and of equilateral triangles, make various shapes.

III. 1. Given dittoed copies of grids, color as many different shapes as possible on each grid. Draw one shape on each to get started. First do a grid of equilateral triangles. Then complete grids of squares, right triangles, and parallelograms.

2. Given a mixed set of shapes, form various figures with each shape touching at least one other shape along a complete edge as shown below.

3. Given strips of cardboard punched at each end with paper fasteners holding them together, experiment with making shapes from 3 or more strips. The illustration below shows two kinds of polygons made from 4 strips.

IV. Match the following:

line _____	1. •————•
line segment _____	2. ←————•
ray _____	3. ←————→

V. What geometric figure is formed by the $\cap$ of two planes?

VI. Draw lines to connect three noncollinear points.

1. Are $\angle ABC$ and $\angle CBA$ the same angle?
2. Are $\angle ABC$ and $\angle BCA$ the same angle?
3. What polygon is formed?

Set B

I. 1. Make dot patterns for 6 dots and for 15 dots in the form of triangles so that there is always the same number of dots along the edges of each triangle.

2. Such a number is called a _____.

II. 1. Draw and order a series of triangular numbers to include the two patterns above.

2. How many dots must be added to each triangular number to obtain the dot pattern for the next larger triangular number?

3. What triangular number has 10 dots in its longest row?

III. 1. Complete the following pattern.

$$\begin{aligned} 1 &= 1 \\ 2 + 1 &= 3 \\ 3 + 2 + 1 &= \square \\ 4 + 3 + 2 + 1 &= \square \\ 5 + 4 + 3 + 2 + 1 &= \square \end{aligned}$$

2. What do you notice?
3. Is it reasonable to think of 1 as a triangular number?

IV. Find some other pairs of numbers, which, when added together, are square numbers.

Note: If you add the third and fourth triangular numbers in the series on p. 367, you obtain a square number.

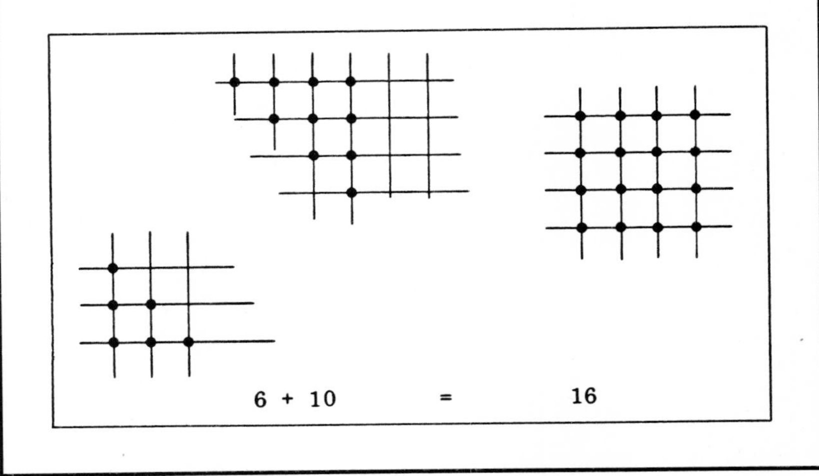

V. Complete the following table and state the relationship between the position and the double of each triangular number.

Position of triangular number	1	2	3	4	5	6
Triangular number	1	3	6			
(Triangular number) × 2	2		12			

VI. Diagram all square numbers up to 49.

VII. State the sums of the following:

1. the first two odd numbers
2. the first three odd numbers
3. the first four odd numbers

VIII. Make a geoboard with 9 nails arranged in 3 rows and 3 columns. Make 8 different triangles on the board.

1. How many of the triangles in number VIII above have a right angle? These are called _____.
2. How many of the triangles in number VIII above have two of their edges the same length? These are called _____.

IX. A triangle with no edges of the same length is called a _____.

X. How many different straight lines can be drawn joining each of the 6 points below with each of the other points?
(*Hint:* See answer for number XI.)

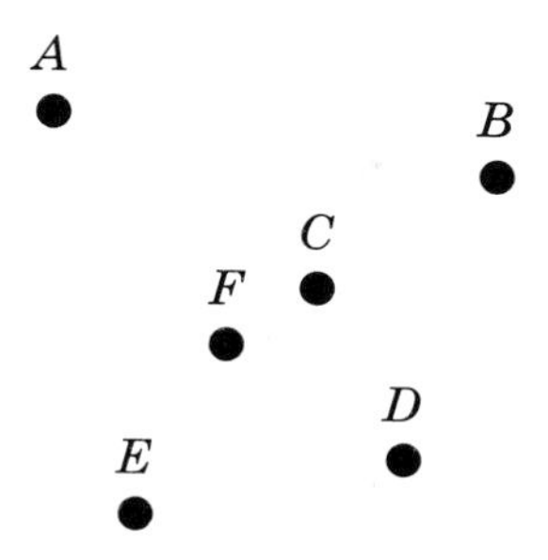

XI. How many lines can be drawn between the following:

1. 4 points
2. 7 points
3. 9 points
4. 20 points
5. 100 points

XII. 1. Complete the following table of points and lines.

	Points	Lines
a.	1	0
b.	2	1
c.	3	3
d.	4	6
e.	5	
f.	6	
g.	7	
h.	8	
i.	9	
j.	10	

2. What do you notice about the numbers in the lines column?

XIII. Use checkers or discs to make the following triangular pattern:

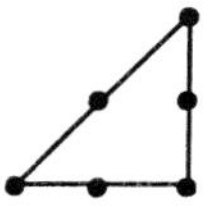

Moving one disc changes this to a rectangular pattern:

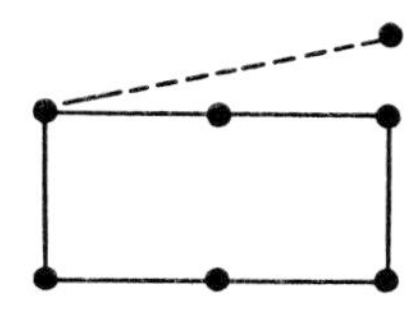

Tell which triangular patterns can be changed to rectangular patterns.

XIV. Draw a circle and mark off 24 equidistant points around the circle. Join each of the 24 points to every other point. The resulting pattern is called a mystic rose. How many lines are there in the pattern?

Mystic rose pattern

XV. Draw a grid and show the following positions:

1. (10, 3)
2. (3, ⁻11)
3. (11, ⁻11)
4. (2, ⁻11)
5. (11, 2)

XVI. When straight lines intersect, the opposite angles are called _____.

XVII. Graph the points (1, 1), (11, 1), (11, 11), (1, 11) and join them in order; draw a line between (1, 1) and (11, 1); between (11, 1) and (11, 11), and so on. What is the resulting figure?

XVIII. Multiply each number in number XVII above by 2. What is the resulting figure when these points are joined in order?

XIX. Rearrange the pieces in the figure below to form a square without using the unlettered piece in the middle.

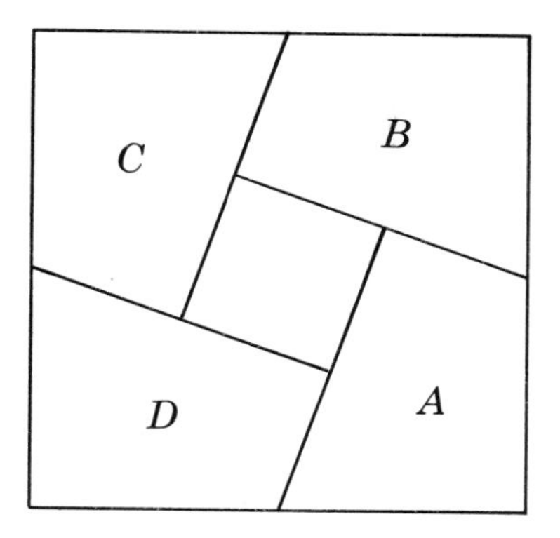

XX. Rearrange the four pieces in the rectangle below to form a parallelogram.

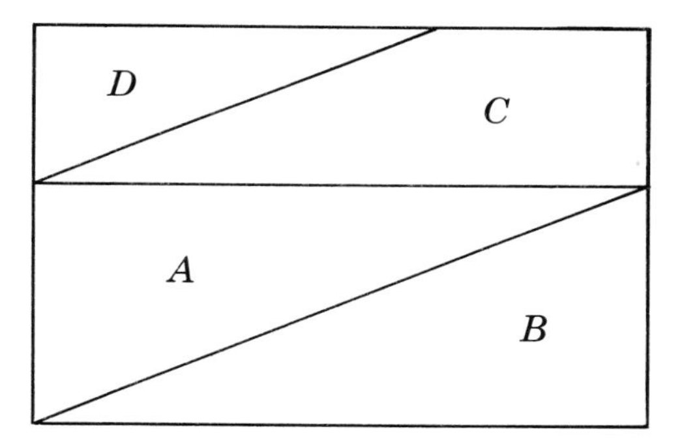

Set C

I. Draw arrow graphs to show "is the sister of" for the children in the families represented below (*B* represents a boy; *G* represents a girl):

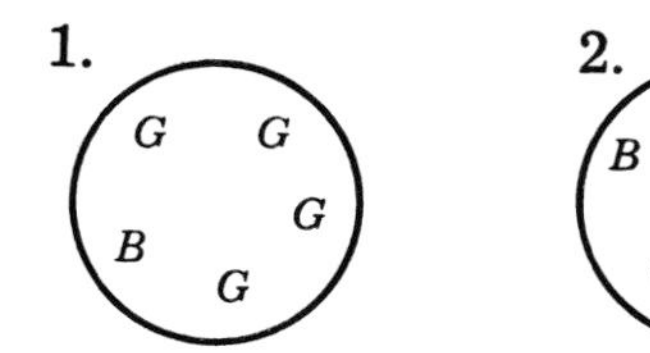

II. Draw an arrow graph to show "is 3 more than" for this set of numbers.

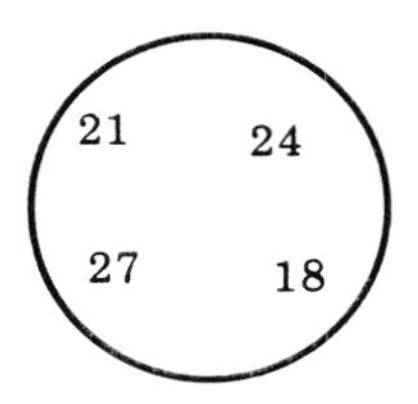

III. Write the indicated relation.

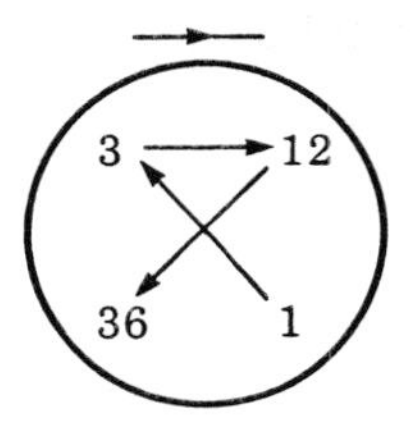

IV. Show the relationships between two sets.

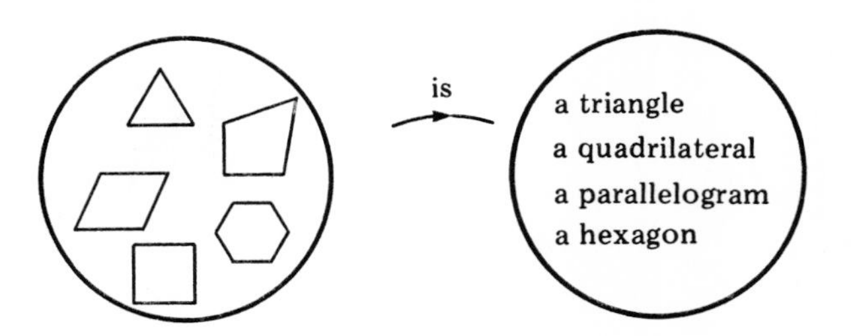

V. Write the relationship between two sets.

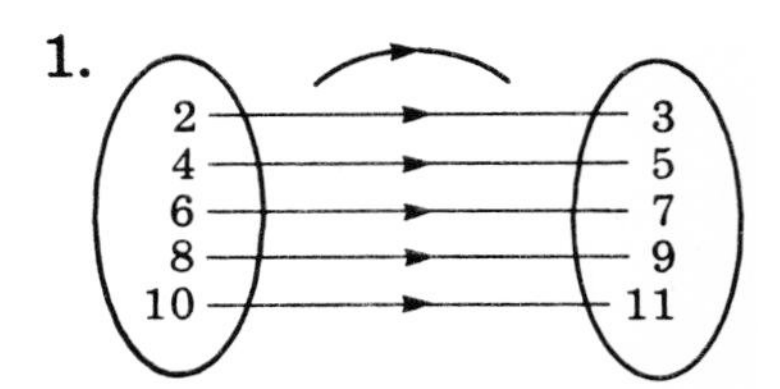

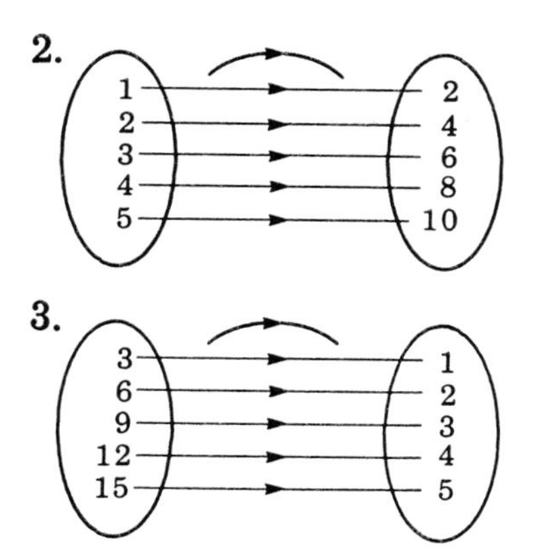

VI. Write the set of ordered pairs for the information in the arrow graphs.

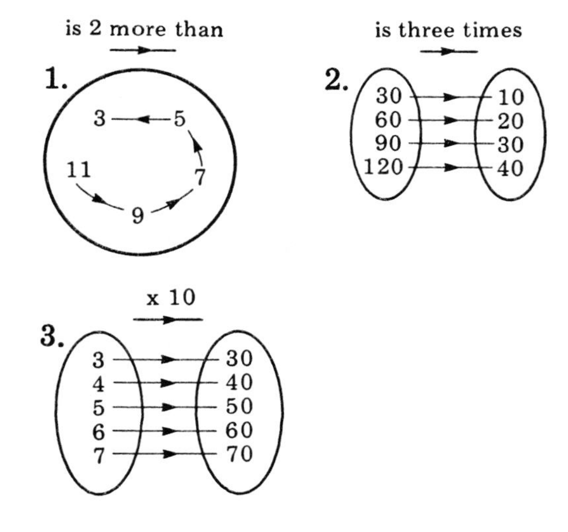

VII. The set of first members is called the "domain." The set of second members is called the "range."

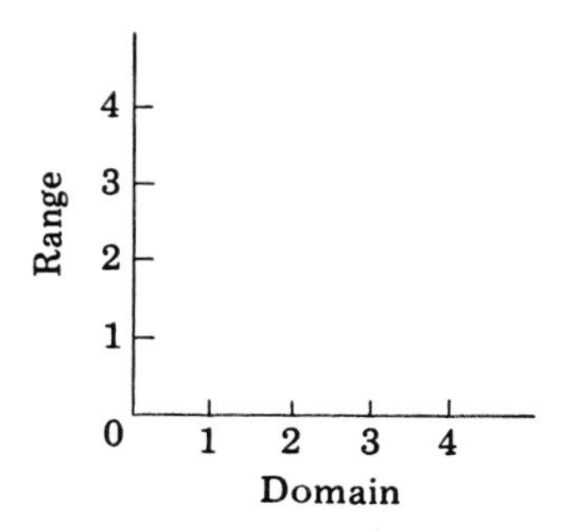

Graph the following ordered pairs and then state the domain and the range for each set.

1. {(0, 0), (3, 1), (6, 2), (9, 3)}
2. {(0, 0), (1, 3), (2, 6), (3, 9)}
3. {(1, 1), (2, 2), (3, 3), (4, 4), (5, 5)}

Connections between members within and between sets have been shown by statements ("is a sister of"), by arrow graphs (→—), by graphing points on a grid, and as a set of ordered pairs. Regardless of the form showing this connection, it is called a "relation."

VIII. Copy and complete the arrow graph below. Then complete the following ordered pairs:

1. (4,) 2. (6,) 3. (16,) 4. (Δ,)
5. Is the relation a function?

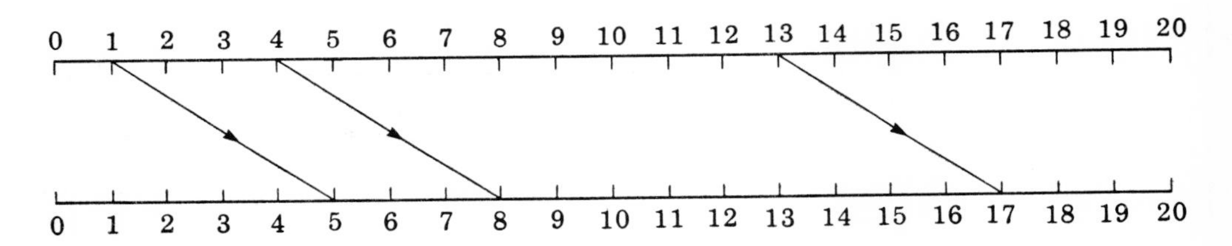

IX. The relation "is half of" is shown below with arrows pointing from a number to another twice as large. Draw a diagram for "is twice."

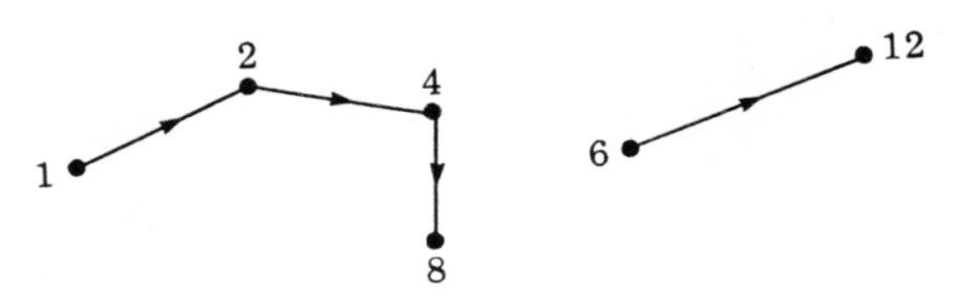

X. 1. What is the relation shown in the figure below?

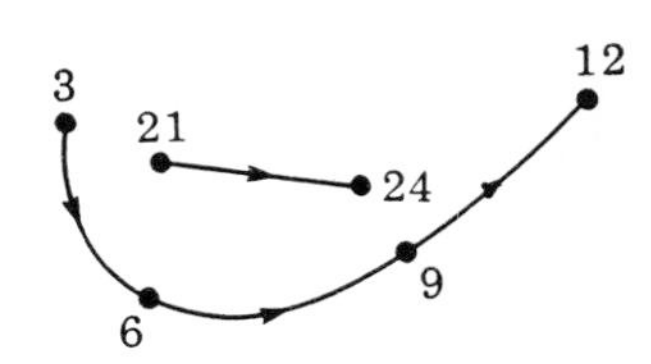

2. What relation would the diagram show if the arrows pointed in the opposite direction?

XI. Draw a diagram to illustrate the relation "is greater than" for the set {1, 2, 3, 4}. (A relationship that shows the link between two elements is called a "binary relation.")

XII. The relations "is less than" and "is greater than" are inverse relations. State the inverse for each of the following:

1. is twice
2. is older than
3. is the same as
4. is the father of
5. lives in the same house as

XIII. The inverse of Lisa "is as tall as" her mother is her mother "is as tall as" Lisa. The relation "is as tall as" is its own inverse. This is called a "symmetric relation." It is shown by lines with arrows in both directions.

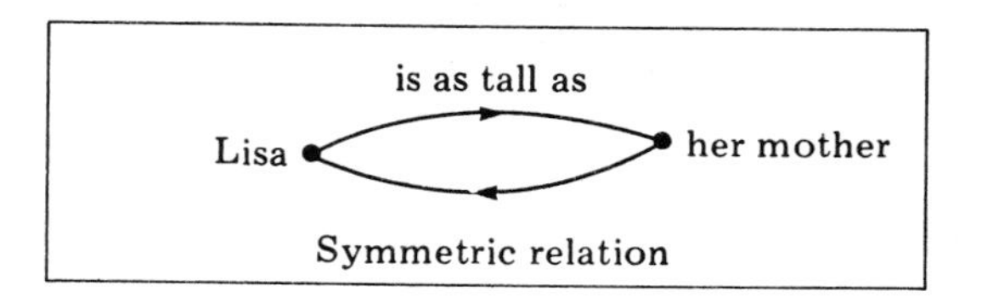

Symmetric relation

Which of the following are symmetric relations?

1. is as heavy as
2. is less than
3. is equal to
4. is the mother of
5. is parallel to
6. is the same age as

XIV. Following are listed the names of children in three separate families. Complete the diagram to show the relation "is the sister of."

Andrew Smith (A)
Betsy Smith (B)
Connie Smith (C)

Dottie Campbell (D)
Edie Campbell (E)

Fred Jones (F)
Greg Jones (G)
Helen Jones (H)

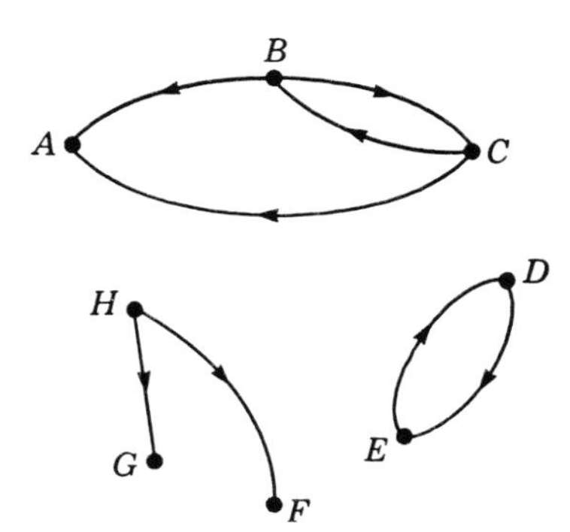

XV. The points A to H represent a set of children. The relation is "is the brother of."

1. Describe each relation shown; for example, C "is the brother of" B.

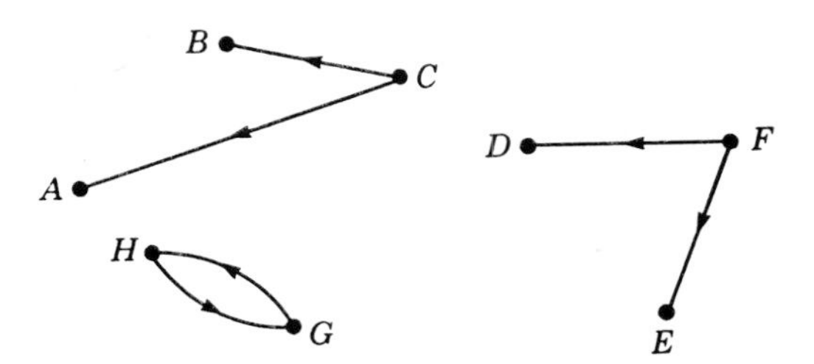

2. C must be a boy since he is the brother of A and of B. What is the sex of each of the children represented in the diagram?

XVI. The following diagrams show the relations "is less than" ($<$), "is greater than" ($>$), and "is less than or equal to" ($\leq$).

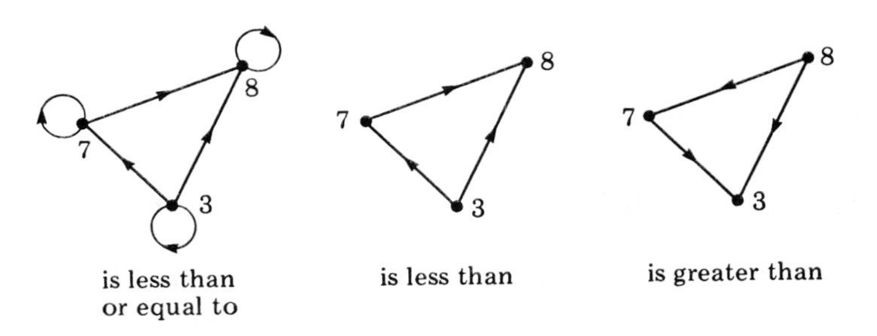

To show "or equal to," the rings at each dot are drawn representing the idea that each number is equal to itself. The arrow here is drawn from the point back to the same point. This represents a "reflexive relation."

Reflexive

Show the relation "is greater than or equal to" for the set {2, 7, 11}. The arrows leading from one dot to another show the relation "is greater than."

XVII. Designate whether the relation "is a factor of" is symmetric, reflexive, or both.

XVIII. Draw diagrams for the following relations:

1. $5 < 7$; $7 < 11$
2. $x < y$; $y < z$. Is $x < z$?
3. *A* is heavier than *B*; *B* is heavier than *C*. Is *A* heavier than *C*?
4. Lisa is taller than Linda; Linda is taller than Joan. Is Lisa taller than Joan?

These are called "transitive relations."

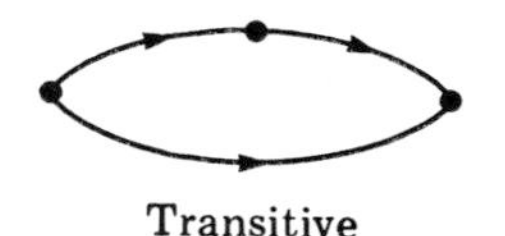

Transitive

XIX. A relation that is symmetric, reflexive, and transitive is called an "equivalence relation." Which of the following are equivalence relations?

1. is similar to
2. is a factor of
3. lives in the same city as
4. is the aunt of
5. is the same age as
6. loves

SUGGESTED READINGS

Backman, Carl A., and Cromie, Robert G. *Introduction to Concepts of Geometry*. Englewood Cliffs, N.J., Prentice-Hall, 1971.

Backman, Carl A., and Smith, Seaton E., Jr. "Activities with Easy-To-Make Triangle Models." *The Arithmetic Teacher* 19 (February 1972): 156–57.

Bush, Mary T. "Seeking Little Eulers." *The Arithmetic Teacher* 19 (February 1972): 105–107.

Clancy, Jean C. "An Adventure in Topology." *The Arithmetic Teacher* 6 (November 1959): 278–79.

Geometry in the Mathematics Curriculum. Thirty-sixth Yearbook of the National Council of Teachers of Mathematics, edited by Kenneth B. Henderson. National Council of Teachers of Mathematics, 1906 Association Drive, Reston, Va. 22091 (1973).

Piaget, J.; Inhelder, B.; and Szeminska, A. *The Child's Conception of Geometry*. New York: Basic Books, 1960.

Robinson, Edith. "The Role of Geometry in Elementary School Mathematics." *The Arithmetic Teacher* 13 (December 1966): 3–10.

Transformational Geometry and Topological Relationships

RIGID TRANSFORMATIONS

Invariance, the idea that certain features of an object or a situation remain unchanged, underlies many aspects of elementary school mathematics. Piaget's discussions on conservation involve invariance.

The concept of invariance is basic to geometry as well as to arithmetic. Invariance in geometry may be studied by observing selected transformations that are fun for children. These include rigid transformations, projective transformations, and topological transformations.

Translations, rotations, and *reflections* are called *rigid transformations* because they involve changes in location only and do not involve changes in size or shape.

A *transformation of translation,* sometimes called a "slide," may be described as "moving along a line"; *rotation,* also referred to as a "turn," may be exemplified by "turning through an angle." Children who reverse the figures d and b, p and q, 3 and Ɛ, 7 and ٢, 6 and ∂ are performing still another transformation called a *reflection,* or *flip.*

Mini-lesson 2.1.9 Show Rigid Transformations, (e.g., Slides, Turns, Flips)

Vocabulary: turn, rotation; slide, translation; flip, reflection

Possible Trouble Spots: inability to recognize the relationship between an invariant and an accompanying transformation; does not attend to salient aspects of a situation

Requisite Objectives:
Geometry 1.2.6: Recognize topologically equivalent shapes

Mathematical Applications 1.2.8: Recognize symmetry in nature and in own body

Mathematical Applications 1.3.15: Make simple binary judgments

Activity 1
To perform transformations of translations, have the child move a book along the chalkboard. Guide him to notice that no change occurs in the shape or size of the book but that a change occurs only in the placement of the book on the chalkboard.

> *Note:* Translations occur in one plane.

Activity 2
A. To perform transformations of rotation, have the child describe situations in which something can be made to move around a fixed point, such as the hands of a clock face, the handle on a pepper mill, or the pivot of a pinwheel.

B. To identify shapes having rotational symmetry, show the child the shapes in the illustration. Tell her to picture each of the five shapes in a frame having the same shape (at the concrete level, you might make a receptacle from clay or cardboard into which the various shapes will fit). Ask her to tell how far a shape removed from the "frame" has to be turned to fit into its frame again. When a shape can be rotated into its frame again before it has made a complete turn, it is said to have "rotational symmetry."

C. Make a square and mark a dot in one of the corners. Position the square so that the dot appears in the upper left-hand corner. Ask the child to rotate the square through a 1/4-turn clockwise $\frac{1}{4}$. Repeat this process but have the dots in different starting positions (as shown in the illustrations). Assess the child's knowledge of fractions for this activity.

The child should either show the result by a drawing or by actually repositioning a cardboard square. A pin pushed through the point where the two diagonals cross serves as a vertex.

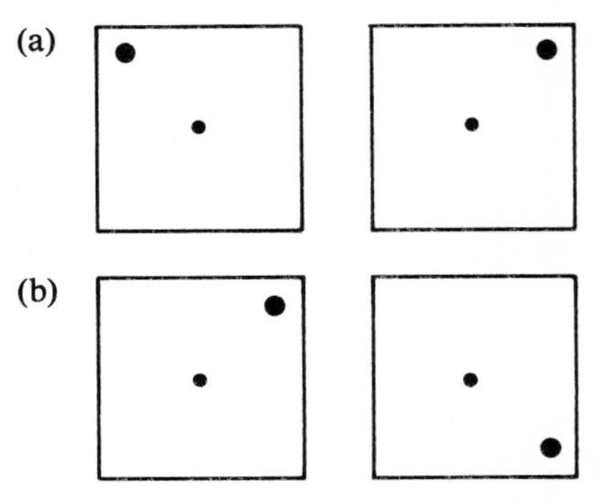

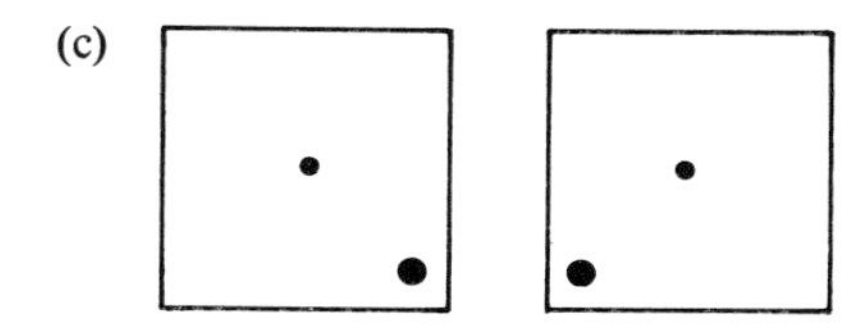

Then ask the child to rotate the squares through a 1/2 turn clockwise, a 3/4 turn clockwise, and one full rotation, either in a clockwise or a counterclockwise direction.

D. To combine rotations, use squares as described previously. Ask the child to show the result when a 1/4 turn is followed by a 1/2 turn in a clockwise direction. Have the dots in various starting positions (as shown). Ask her if she gets the same results when the order of rotation is reversed; that is, when a 1/2 turn is followed by a 1/4 turn. She should recognize that the results are the same.

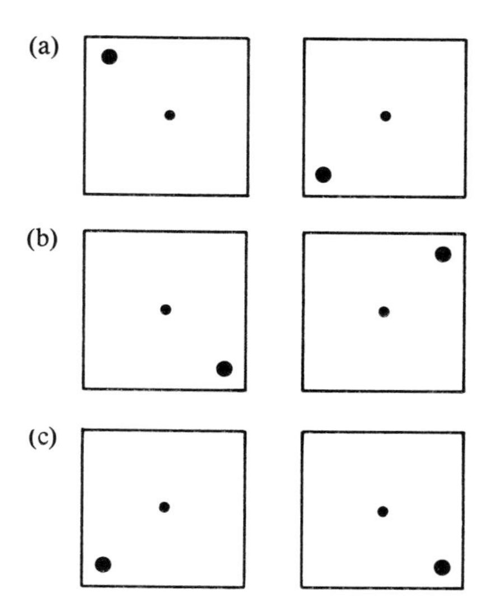

E. To rotate triangles about their centers, allow the child to experiment with triangles of different shapes, using the same procedures as in the previous rotation activity.

Activity 3

A. To perform transformations of reflection, give the child flannel cutouts of letters of the alphabet. Have him fold these along their central axes (both vertically and horizontally). Discuss the concept of reflection (sometimes called "flips" or "mirror images") in regard to the letters.

B. Ask the child to fold a piece of paper once and to draw and color on the "once-folded" paper one-half of the letter *A*, as shown. Ask her to cut out the shape. She should have a capital letter *A*.

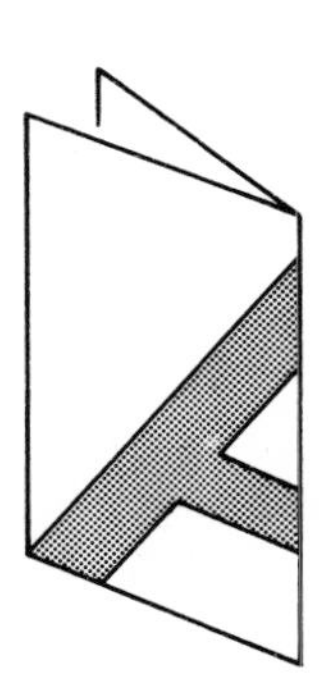

Direct the child to cut out other letters in this manner. Help her to name letters that can be made by cutting a "twice-folded" paper and a "more-than-twice-folded" paper. The fold is called the figure's "line of symmetry." (Cutting out paper dolls from folded paper is a comparable activity.)

Activity 4
To classify letters according to their symmetry, discuss with children observations related to this type of symmetry. The following generalizations concerning the symmetry of classes of letters should follow activities dealing with transformations of reflection.

Type of symmetry	*Letters of alphabet*
Line symmetry only	A, B, C, D, E, K, M, T, U, V, W, Y
Rotational symmetry only	N, S, Z
Both line and rotational symmetry	H, I, O, X
No symmetry	F, G, J, L, P, Q, R

Additional considerations: Rigid transformations that are presented in this mini-lesson may be investigated in our environment. For example, reflections in a mirror or as one peers into a pool of water, rotations in the shapes of jellyfish such as the aurelia, the rotational symmetry of snowflakes, and the vertical symmetry of the butterfly or of the beetle. Symmetry may also be pointed out in many pieces of furniture and in architecture. The Taj Mahal in India and the Alhambra in Spain are beautiful examples of symmetry.

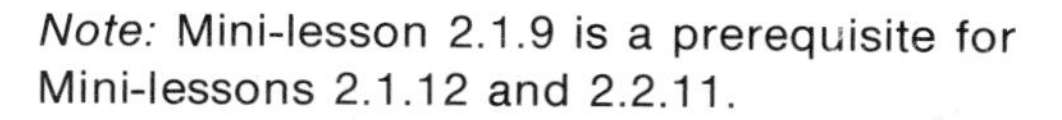

Note: Mini-lesson 2.1.9 is a prerequisite for Mini-lessons 2.1.12 and 2.2.11.

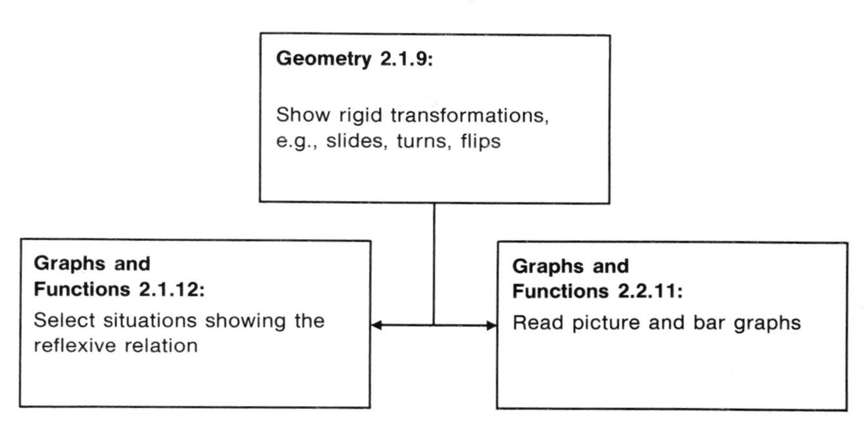

PROJECTIVE TRANSFORMATIONS

Projective transformations involve changes of visual perception as in the shape and/or size of an object's shadow differing from the object's shape and/or size. In some cases, a person's shadow may be much longer than the person is tall, or it may be shorter.

Projective transformations involve the comparisons of ratios and may be shown in two ways of casting shadows. One method uses point-source light rays (for example, light from a pen light or flashlight) and another way involves casting shadows with parallel light rays (for example, from the sun or a slide projector from about six feet away). The latter method is called "affine transformations." In projective transformations, a straight object casts a straight shadow and certain ratios are preserved from the object to its shadow. Viewing an object from a different perspective also results in projective transformations.

Mini-lesson 5.1.8 Perform Simple Projective Transformations

Vocabulary: ray, perspective, projection

Possible Trouble Spots: Does not organize visual information; difficulty with spatial relationships

Requisite Objectives:
Ability to take another's view
Geometry 1.2.6: Recognize topologically equivalent shapes

Geometry 1.3.11: Identify simple plane figures

Geometry 2.3.8: Identify line, line segment, and parallel lines

Activity 1

To make shapes larger and smaller, have children participate in the following activities.

A. Show the child an illustration like the one shown and discuss with him how the smaller lion's head has been made larger. Direct the child to compare the eyes, noses, and mouths of the two lions. Point out how the straight line (rays) meet at the point A. Have him measure the distance from A to the tip of the nose on the smaller head and then measure the distance from A to the tip of the nose on the larger head. Question the child as to what he notices. Ask him to repeat this process for other parts of the heads. Then ask him how much larger is the large head.

B. Tell the child to draw on the left side of his paper a shape like the one pictured in step 1. Tell her to position a point as shown. Then

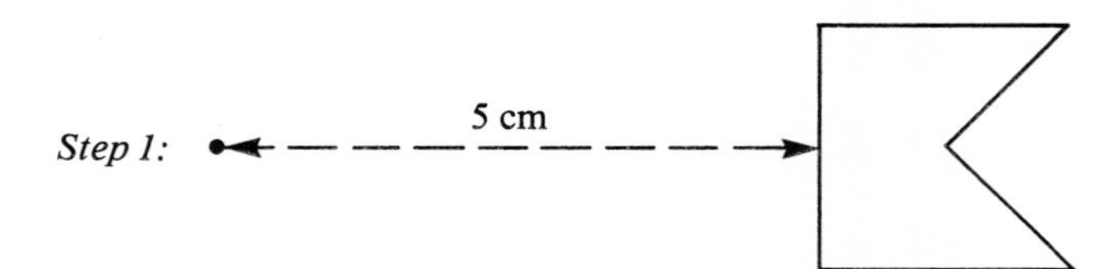

have her draw rays from the point through the vertices of the shape as shown in step 2. Tell the child to use these lines to draw the shape larger and then smaller. Discuss with her what characteristics remain the same when the shape is made larger or smaller.

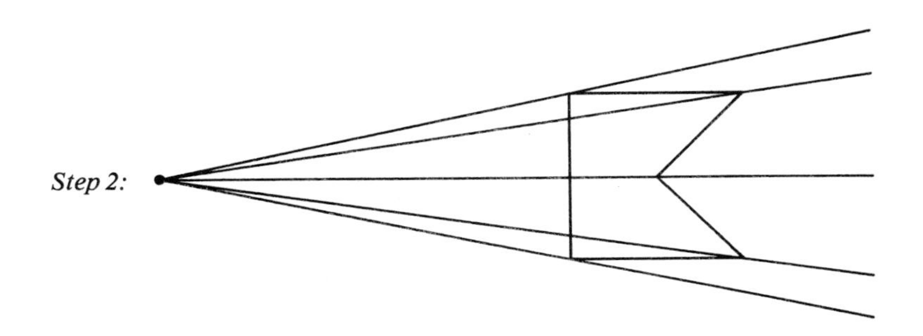

C. Ask the child to construct a triangle with sides equalling 3 cm, 4.5 cm, and 6 cm. Tell him to use the ray method (described in Activity 1B) to make larger and smaller triangles.

> *Note:* Inches may be used as the measure if a metric stick is not available (see Chapter 22 for a discussion of the metric system of measure). Strings may also be used by young children to measure the sides of a triangle and to represent the rays.

Activity 2
To match shapes when viewed from different perspectives, ask the child to look at the five shapes as shown in the illustrations. Tell him that each row of drawings below shows three views of one of the shapes (view from above, front, and side). Ask him to match each row of drawings with the corresponding shape.

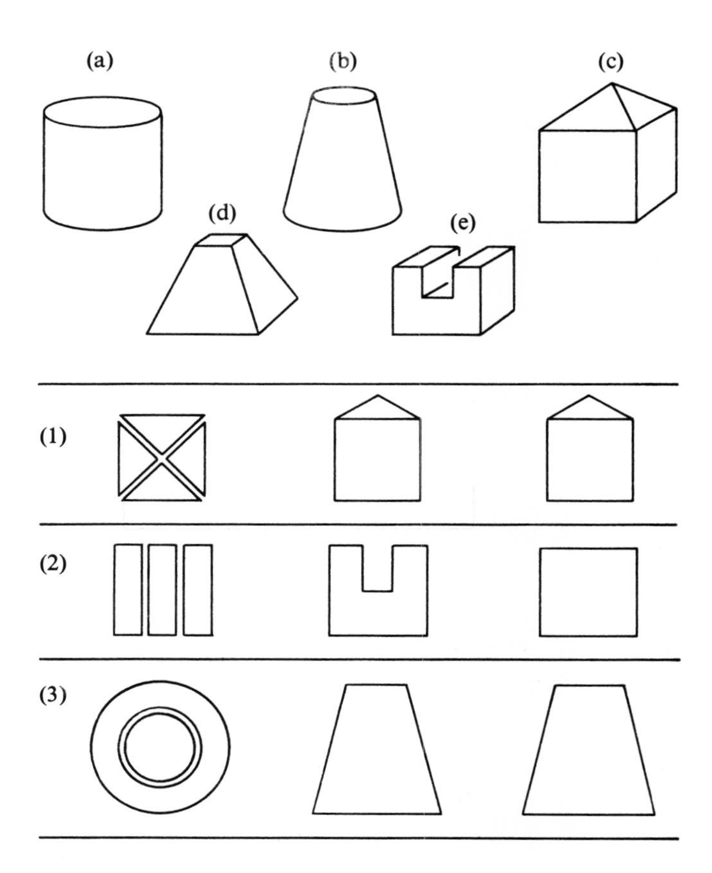

Additional considerations: Many perceptual tasks go beyond mere recognition of configurations; they require that the student receive and organize visual information.

Perceptual organization tasks are more complex than the simple recognizing and choosing of a form that matches another form. Perceptual organization requires the student to actively construct an organizing structure both in regard to the instructional materials and the goal of the activity. Therefore, it is important that the student keep the task in mind and know its purpose. It is here that cognitive monitoring discussed in Chapter 1 may be applied as a cueing technique to keep the students on task. Cognitive monitoring may also help sustain the students' motivation to complete the task and keep them from being distracted. Students may self-monitor their behavior, thus providing simple self-checks on their performance to keep on task and approach as closely as possible the required goal.

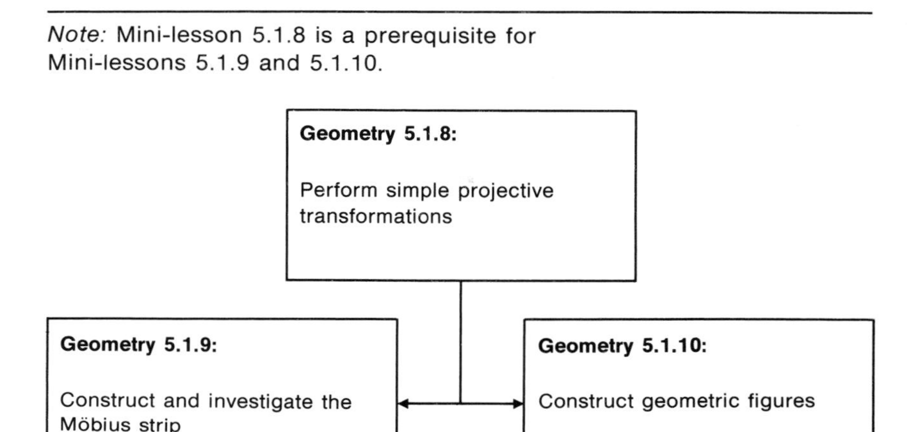

TOPOLOGICAL TRANSFORMATIONS

Topology is that branch of geometry dealing with the idea that shapes may be stretched out or squeezed or may assume a variety of different appearances. *Topological transformations* involve changes in size and shape. For example, a rubber band may take on the shape of a circle, a triangle, a square, a rectangle, an ellipse, or even an octagon. Have you ever looked in the carnival mirrors that distort your appearance? Topological changes are occurring as you become fat in one mirror and a bean pole in another. The Piagetian task of flattening a ball of clay involves a topological transformation. Topological transformations do not involve disconnecting connected attributes (as in cutting apart a rubber band) nor connecting disconnected attributes (the endpoints of a rope cannot be connected). Therefore, *breaking, tearing, folding,* or *connecting are not involved in topological transformations.* Straightness, number of sides, and ratios need not be preserved in topological transformations.

Topological equivalence

Objects that can be changed from one to another are considered to be topologically equivalent. *Topological equivalence* is shown when a young child is asked to copy on paper the shape of a triangle or a square, and the drawing resembles a circle.

The elements or objects in topology are such geometric figures as points, lines, curves, surfaces, triangles, circles, and spheres, but the relations used to compare them are different from Euclidean relations of congruence and similarity. In Euclidean geometry, two figures are congruent if one can be superimposed on the other with corresponding sides equal. Two figures are similar if their shapes are the same with corresponding angles equal. Topological equivalence implies that two geometric figures can be twisted or stretched into each others' shapes.

Activities involving topological transformations

Children may explore two-dimensional topological transformations by stretching and twisting shapes drawn on a balloon, rubber sheet, or elastic tape. Three-dimensional topological transformations may be analyzed by changing the shape of a ball of clay without breaking, tearing, folding, or connecting. In fact, preschool and kindergarten environments are conducive to exploring many geometric or topological ideas. Some of these ideas involve the closeness of things (proximity), shapes (configurations or separation of parts of objects), orders of things (order), and boundaries of bodies and objects (enclosure).

Young children make topological discriminations

It has been found that topological relations are developed before Euclidean relations. Piaget observed that 3- or 4-year-old children could make topological discriminations (both manually and pictorially) such as a

> closed from an open figure, an object with a hole in it from one without a hole, and a closed loop with something inside from one with something outside or on the loop's boundary. But the ability to discriminate between rectilinear and curvilinear figures . . . does not develop until several years later (about age 9–10). Thus, the same child who can readily distinguish an open from a close circle may be quite unable to discriminate between the closed circle and rectilinear closed figures such as squares or diamonds."[1]

Topological relationships

There are four basic topological relationships: *proximity, separation, order,* and *enclosure.*

1. *Proximity.* This refers to the nearness of objects. Children may explore the idea of proximity by pointing to the objects or placing an object near another object. They may also describe a picture, telling what is near a particular item in the picture.

2. *Separation.* Experiences involving the notion of separation should help children differentiate one object from another object or the parts of an object from one another. Children may be given models of a face with moveable eyes, nose, mouth, and ears. They should try to put the face together without any of the parts touching. If some of the parts do touch one another, the teacher should ask the children to make the face look even more realistic. Children may exchange models and comment on one another's attempts. This will help develop their observation skills.

---

1. John H. Flavell, *The Developmental Psychology of Jean Piaget* (Princeton, N. J.: D. Van Nostrand Co., 1963), p. 329.

3. *Order*. Understanding order helps to determine the sequence of objects and events. Proximity and separation activities must precede order experiences because children need to show the relative closeness of objects and, at the same time, be able to discriminate between them before they can put objects in some order. Ordering objects such as placing beads according to size of a color pattern is helpful.

Children may also order objects according to function: putting the cars of a train (an engine, some cars, and a caboose) in order; dressing a doll (the undergarments will necessarily have to be put on first); putting on socks and shoes; opening a closed door before walking through the doorway.

4. *Enclosure*. This involves the idea that something is either enclosed by some kind of boundary or it is not. A figure that is enclosed by two others can move freely only between these two endpoints. To give children an understanding of enclosure, the teacher may have one child stand between two others on a line. The pupil then moves on the line but does not pass either of the children at the endpoints. The child is enclosed by the two children.

Another activity that illustrates enclosure involves having several children form a ring around a child. The child in the center is enclosed by the ring.

Piaget and Inhelder emphasize that spatial notions are developed from actions performed on objects in space; at first, these are motor actions, but then they become internalized or imagined actions.[2] For example, adults see objects as being together or separated in space as a result of past actions of placing objects together and separating them.

Mini-lesson 1.2.6 Recognize Topologically Equivalent Shapes

Vocabulary: equivalent, topological, Möbius strip, Klein bottle

Possible Trouble Spots: does not recognize the relationship between an invariant and an accompanying transformation; does not see cause-effect relationships; difficulty with spatial relationships; difficulty constructing mental images

Requisite Objective:
Geometry 1.1.8: Show topological relationships: proximity, enclosure, separateness, order

Activity 1
To create topologically equivalent shapes, have children engage in the following activities.

---

2. Jean Piaget and B. Inhelder, *The Child's Conception of Space* (London: Routledge & Kegan Paul, 1956).

A. Give the child a large rubber band and ask her to make a series of shapes, such as a facial profile, a ghost, a dog, and so forth. Guide her to notice that there are some basic features of the rubber band that do not change; it remains a simple closed curve.

Note: The shapes in Activity 1A are simple closed curves because the line in each figure does not pass through the same point more than once, but it does return to its starting point.

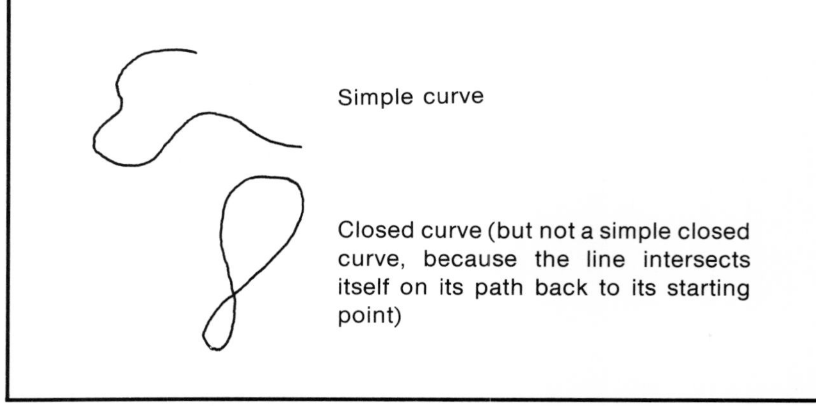

B. Give the child a segment of rope and ask him to form the figures as shown in the illustration. Ask him how these figures differ from those in Activity 1A. The figures shown here are simple open curves.

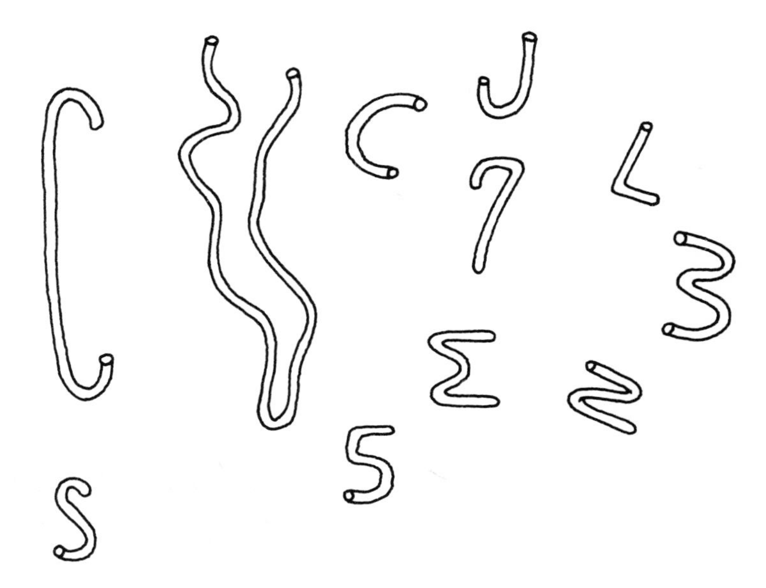

C. Then discuss with the child how an elastic doughnut could be bent or stretched into a coffee cup so that they are topologically equivalent.

Source: Mario F. Triola, *Mathematics and the Modern World* (Menlo Park, Calif.: Cummings Co., 1973), p. 302. Reprinted with permission.

D. Show the child other groups of topologically equivalent figures, as shown.

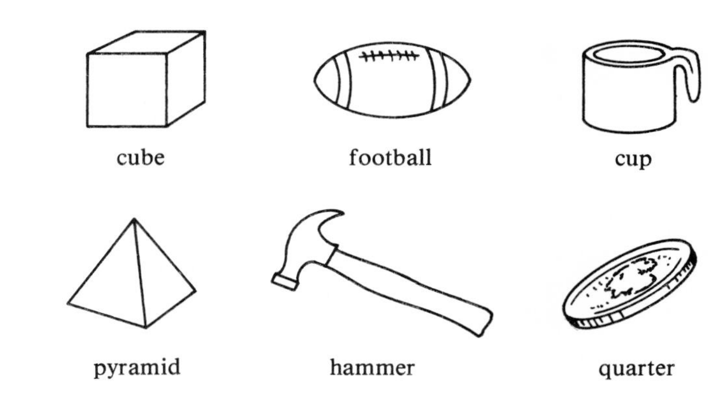

Source: Triola, *Mathematics and the Modern World,* p. 303. Reprinted with permission.

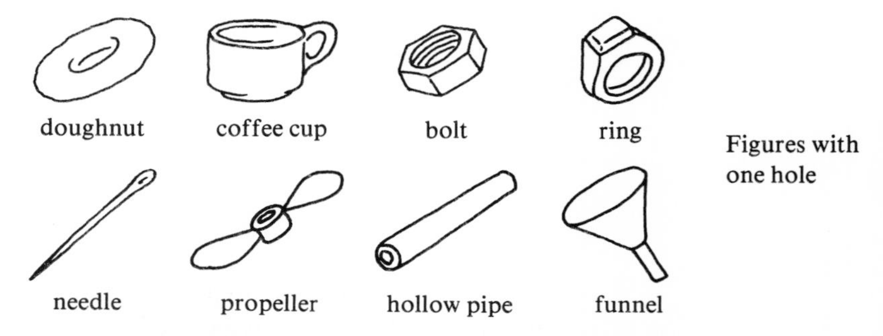

Source: Triola, *Mathematics and the Modern World,* p. 303. Reprinted with permission.

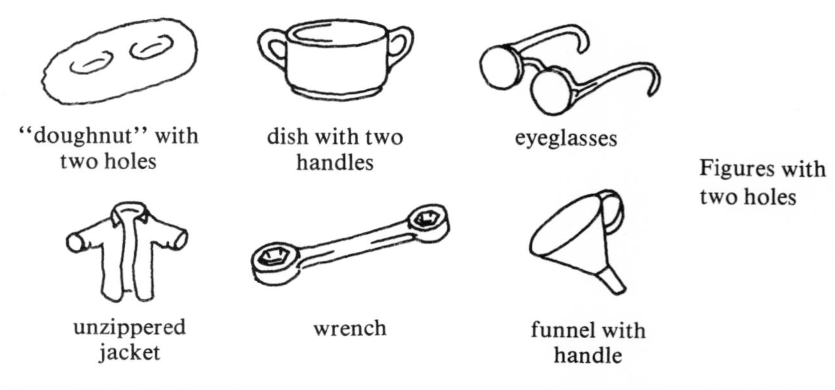

Source: Triola, *Mathematics and the Modern World,* p. 303. Reprinted with permission.

Activity 2

To investigate the Möbius strip, have children participate in the following activities.

A. Explain to the child that the Möbius strip has only one side; it is a one-sided surface. (Ferdinand Möbius discovered the Möbius strip in the nineteenth century.)

Give the child a strip of paper measuring 25 cm by 3 cm; 50 cm by 4 cm; or 11 by 1/2 inches. Have her twist the strip once and stick the ends together. Ask if she can find the inside of the strip.

Then tell the child to mark a point anywhere on the twisted strip and color the strip starting from this point and continuing until this point is reached again without raising the crayon. Ask her to tell which part of the strip is colored.

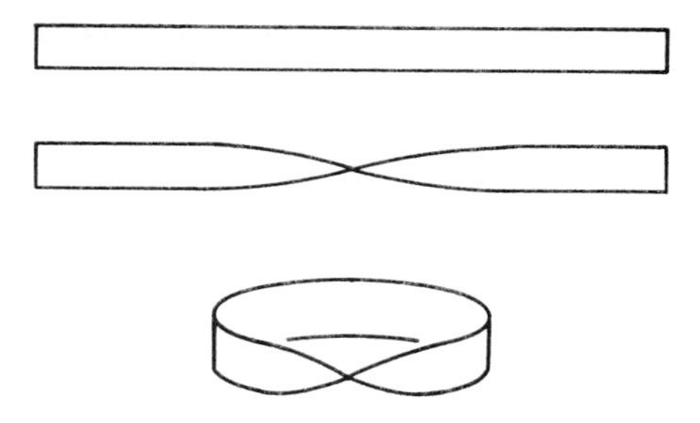

B. Give the child a strip of paper as above. Have him twist the strip once and stick the ends together. Then have him draw a line along the middle of each side of the strip and cut along the line. Next have him cut along two lines as shown below.

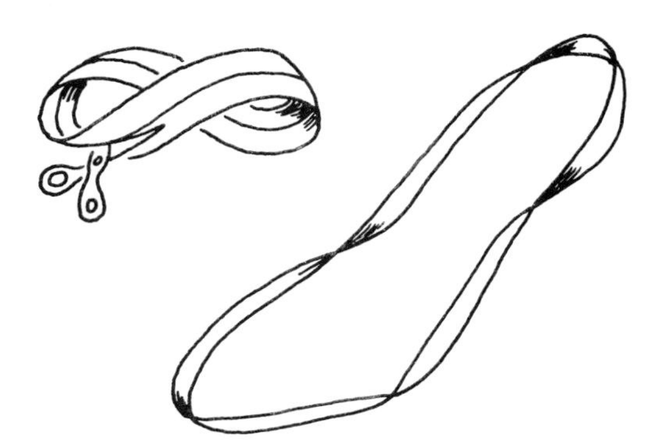

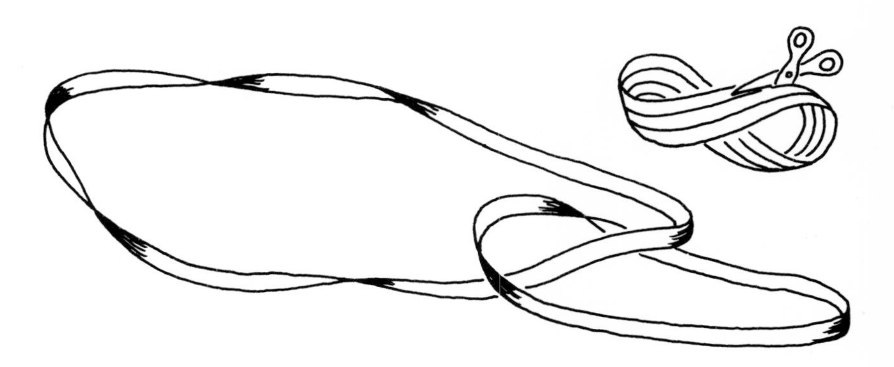

The child will see that the result of the two cuts is two intertwined bands. One is a new Möbius strip, the other a two-sided band with two full twists. A conveyor belt made with a twist is a Möbius strip. It wears longer and more uniformly because the entire surface touches the rotating wheels and carries the desired objects.

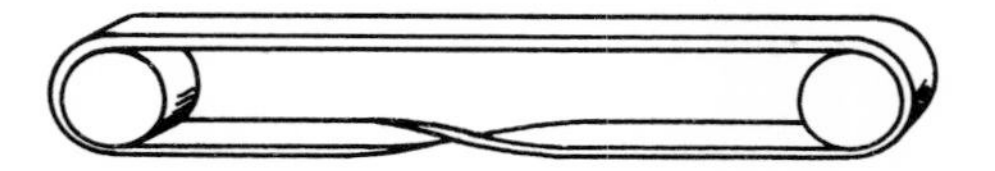

Klein Bottle

Another one-sided surface is the Klein bottle, originated by Felix Klein (1882). A Klein bottle cut in half becomes two Möbius strips.

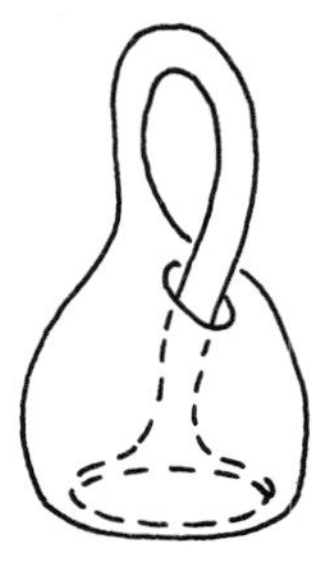

Additional considerations: Students who have difficulties such as those listed under "Possible Trouble Spots" for this mini-lesson need help in attending to relevant aspects of the various changes in form that are described. Cues such as isolating the various topological relationships help such students notice the relevant relationships.

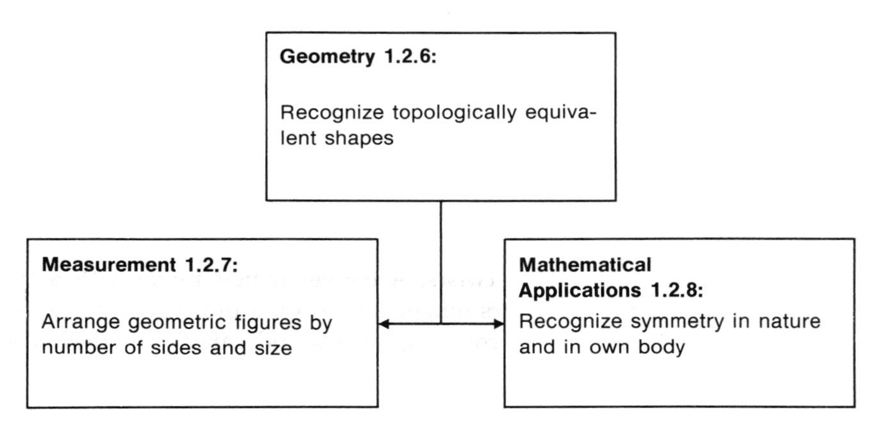

NETWORKS

Network theory, another aspect of topology, involves arcs and vertices. It was initiated by the Swiss mathematician Leonard Euler (pronounced oil' er) (1707–1783) when he analyzed an interesting question involving a group of bridges in the town of Königsberg (then in Germany). In Königsberg, there were two islands connected by seven bridges crossing the Pregel River (see Figure 21.1). A question arose: "Is it possible to take a walk so as to cross each bridge (representing an arc) once and only once?" The starting point may be on either of the islands or on either of the banks. The finishing point need not be the same as the starting point.

Figure 21.1 illustrates each piece of land marked with a point. Lines connect these points so that there is a line over each bridge.

Path

At this point, the reader should try to find a path that will answer the question. "A *path* in a network is simply a set of one or more arcs and vertices that one can trace without lifting the pencil off the network and without tracing an arc of the network."[3]

The problem of the Königsberg bridges is similar to drawing the network of dashed lines without lifting the pencil from the paper and without retracing any lines.

3. Carl A. Backman and Robert G. Cromie, *Introduction to Concepts of Geometry* (Englewood Cliffs, N. J.: Prentice-Hall, 1971), p. 243.

Figure 21.1 Network of Bridges in Königsberg

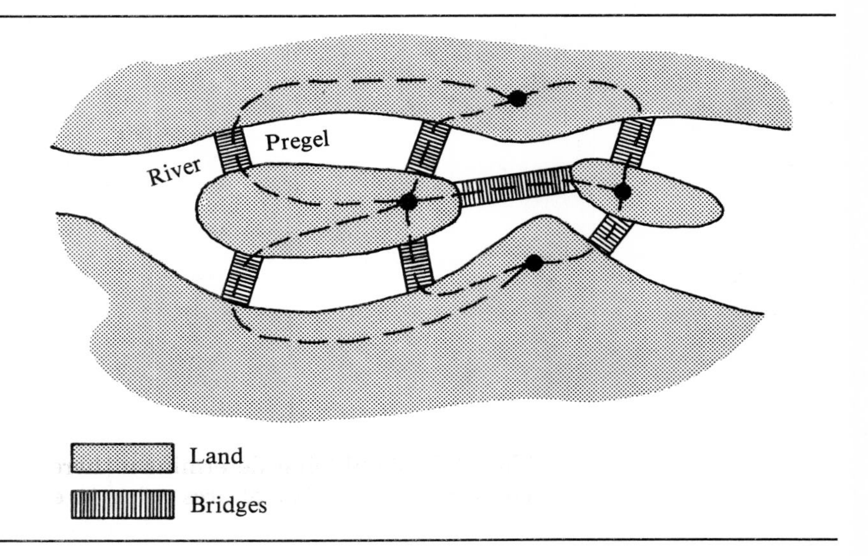

Arcs and junctions

To answer the question, Euler counted the number of lines (*arcs*) going to each point (called a *junction* or *vertex*). For each junction, he said whether the number of lines was odd or even. Notice that all four vertices are odd, since they all have an odd number of lines or arcs coming to them.

Traversability of networks

Euler observed that when a point or vertex has an odd number of arcs connected to it, that vertex must be either a starting point or a terminal point. If there are three or more vertices at which an odd number of arcs converge, some retracing is necessary. Since there are four odd vertices in the bridges problem, it is said to have no solution. This network is not traversable because the following two conditions do not apply: (a) the pencil may not be lifted from the paper and (b) no arc may be retraced. It should be noted that there is no limitation to the number of times a vertex may be traversed.

Euler discovered, then, that a network can be drawn only if either (a) it has no odd vertices or (b) it has two odd vertices. In this latter case, it can be drawn only by starting at an odd vertex.

Mini-lesson 7.3.1 Tell If a Network Is Traversable Using Euler's Procedure

Possible Trouble Spots: poor problem-solving skills; difficulty making inferences; difficulty abstracting and dealing with complexity; distractibility

Requisite Objective:
Geometry 7.3.2: Find relationships between the number of spaces, vertices, and arcs in any network

Activity 1
To decide if networks are traversable using Euler's procedure, ask the child to trace a shape like the one shown without lifting his pencil or retracing a line.

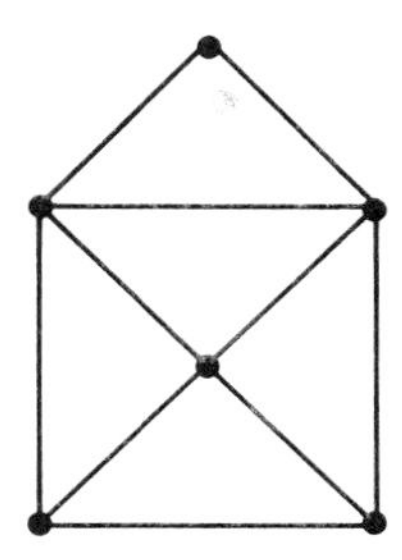

The child should first determine if there are only *two* odd vertices by counting the number of arcs going to each vertex (as shown).

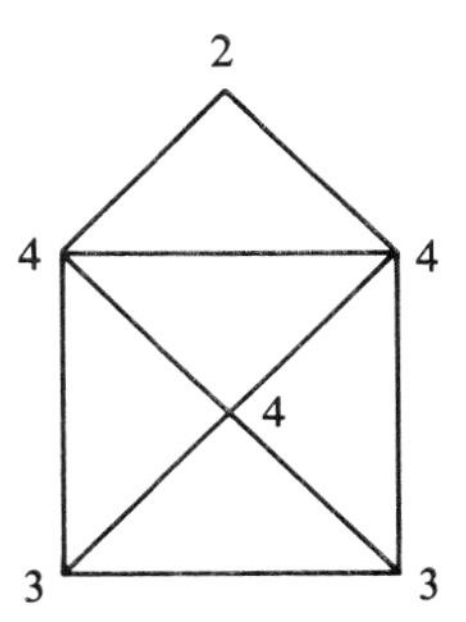

When the child has determined that there are only two odd vertices, she might traverse the path shown in the illustration.

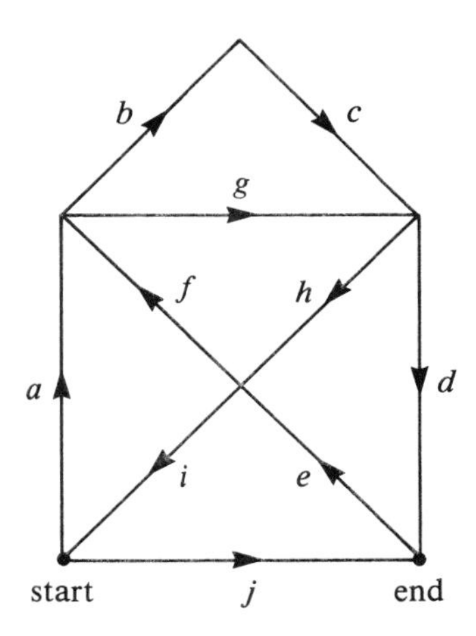

Activity 2
To complete networks by drawing arcs when given odd and even vertices, give the child dittoed sheets of sets of vertices, with each vertex marked as odd or even (as shown). Have her draw arcs and then tell which figures contain networks that are traversable. The child could conclude that networks *c* and *d* are traversable because they contain only even vertices; networks *b* and *e* are not traversable because they contain more than two odd vertices; and network *a* is traversable only by starting at one of its two odd vertices.

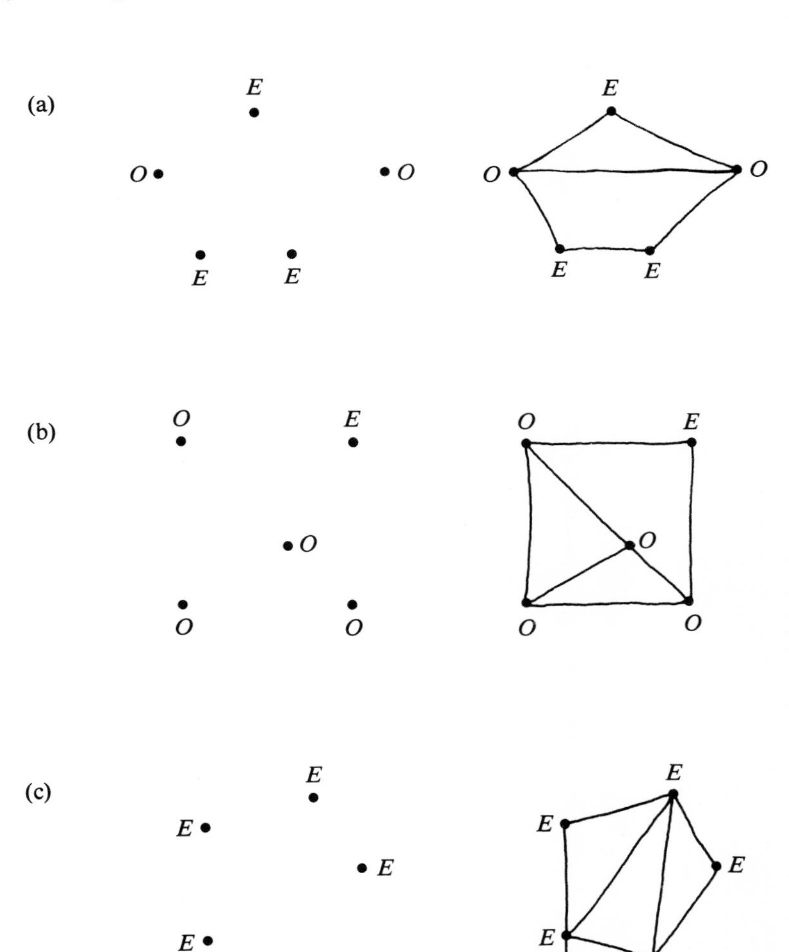

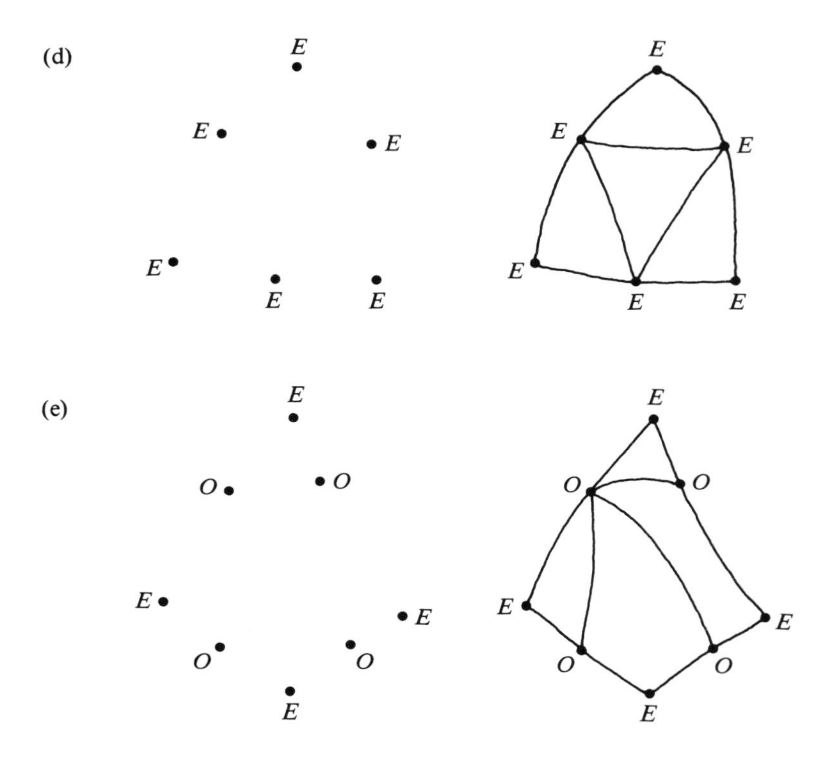

Activity 3

To find relationships between the number of spaces, junctions (vertices), and lines (arcs) in any network, show the child networks on which the spaces have been colored. Ask him to count the number of nonoverlapping spaces, junctions, and lines in each of the networks, including the space surrounding the figure as an additional region (for example, in illustration *a*, there are 6 spaces).

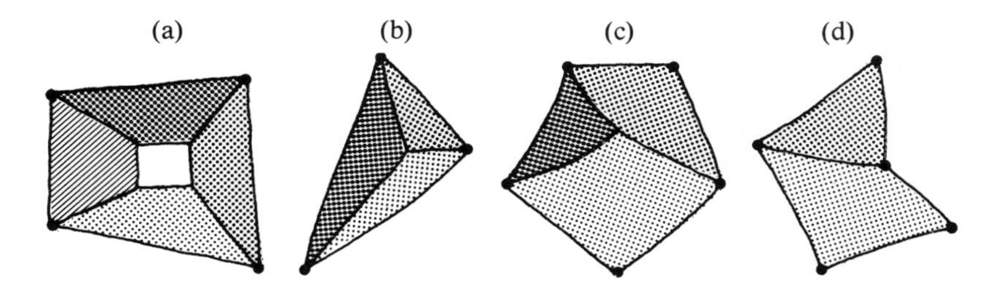

Leonard Euler discovered a relationship among the number of spaces, the number of junctions, and the number of lines formed by the network. Guide the pupil to find this relationship by analyzing the networks shown. He may be helped to see the pattern by recording the information about each network in a table as shown below. The child, of course, should fill in the answers.

Network	*Number of spaces* (S)	*Number of junctions* (J)	*Number of lines* (L)
(a)	6	8	12
(b)	4	4	6
(c)	4	6	8
(d)	3	5	6

If the child is unable to see the relationship, ask him to try adding the number of spaces to the number of junctions and compare this sum to the number of lines for each network. Could you draw a network with (a) 2 spaces, 4 junctions, and 5 lines? (b) 3 spaces, 4 junctions, and 5 lines? or (c) 2 spaces, 3 junctions, and 5 lines?

Activity 4

To find relations among components of polyhedra, explain to the child that *polyhedra* are many-sided objects that may be described as networks of points and lines in three dimensions. Point out that Euler found a relationship among the number of faces (spaces), vertices (junctions), and edges (lines or arcs) in polyhedra. Help the child construct a table like the one shown.

Shape	*Number of faces* (F)	*Number of vertices* (V)	*Number of edges* (E)
Cube	6	8	12
pyramid (square base)	5	5	8
tetrahedron	4	4	6
prism	5	6	9
octahedron	8	6	12
hexagonal prism	8	12	18
dodecahedron	12	20	30

After helping the child complete the table, ask her to write a mathematical sentence to describe the relationship among the number of faces, the vertices, and the edges. The child should write "$F + V = E + 2$."

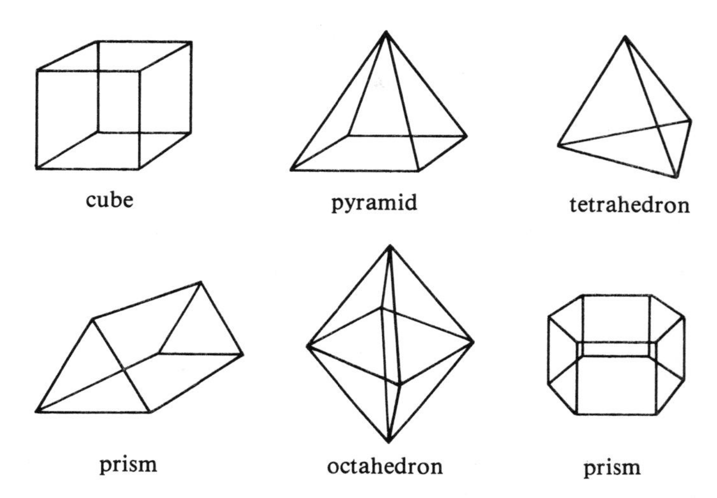

Additional considerations: For students who have difficulty constructing the underlying relationships for understanding network theory, it may be helpful to emphasize the relevant dimensions by use of cues such as color, tactile diagrams, and discussions where salient features are pointed out.

Note: Mini-lesson 7.3.1 is a prerequisite for Mini-lessons 7.3.4 and 7.3.6.

Geometry 7.3.1:

Tell if a network is traversable using Euler's procedure

Graphs and Functions 7.3.4:
Find a formula to describe a set of ordered pairs

Mathematical Applications 7.3.6:
Solve simple geometric problems

EXERCISES

I. Show the line of symmetry by drawing over it.

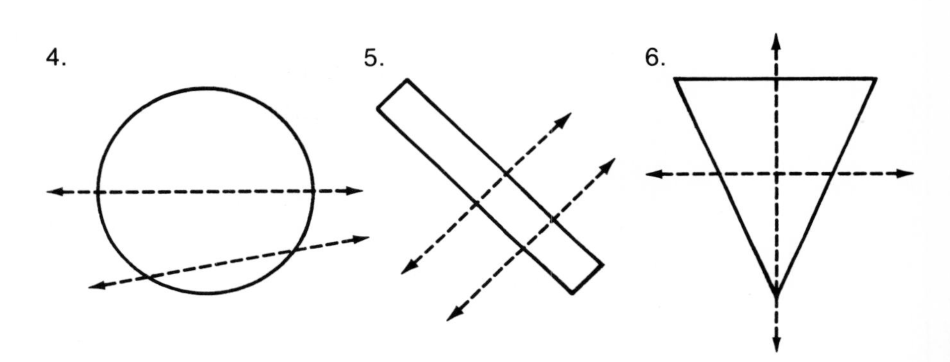

II. Classify the letters of the alphabet according to those with
1. vertical line of symmetry
2. rotational symmetry
3. horizontal line of symmetry
4. no symmetry

III. Tell which of the following networks are traversable.

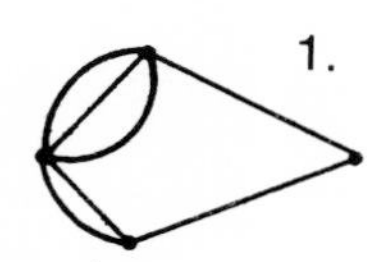
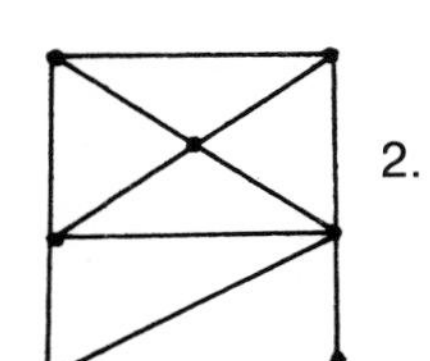
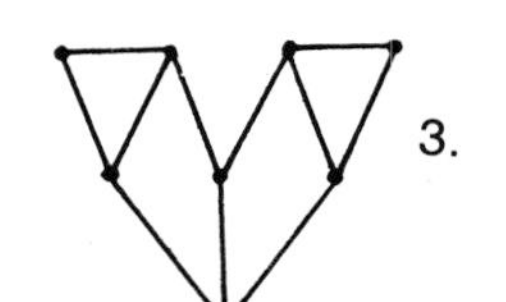
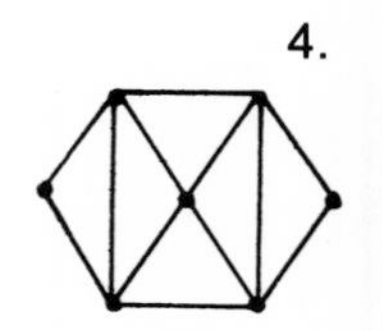

IV. Draw two networks such that each has five vertices but with the condition that only one of the networks is traversable.

V. Match the following rigid transformations with the relevant example.

1. mirror image	_____ slide
2. water skiing	_____ turn
3. backward writing of letters	_____ flip
4. windmill	
5. propeller	
6. shadows	

SUGGESTED READINGS

Bergamini, D. "The Mathematics of Distortion." *Mathematics.* New York: Time-Life Books, 1963. Pp. 176–91.

Euler, L. "The Seven Bridges of Königsberg." In J. Newman, *The World of Mathematics, 1.* New York: Simon and Schuster, 1956. Pp. 571–80.

Glenn, W. H., and Johnson, D. "Topology: the Rubber-Sheet Geometry." *Exploring Mathematics on Your Own.* New York: Doubleday, 1961. Pp. 197–234.

Newman, J., and Kasner, E. "Rubber-Sheet Geometry." *Mathematics and the Imagination.* New York: Simon and Schuster, 1940. Pp. 265–98.

Measurement Including Metric Geometry

STANDARD AND NONSTANDARD MEASURES

Measurements are made with standard and nonstandard instruments. A *standard measure* is one that is generally acceptable. A standard is needed to avoid confusion when measuring different attributes of objects, such as length, area, mass, capacity, time, velocity, humidity, temperature. A system of measures is one in which the various standard measures are related to one another. The particular standard measure must be suitable to the task that is being undertaken. The set of standards must be appropriate to the range of measurement.

A *unit* is a standard measure of an attribute. For example, a minute is a unit of time duration; a meter, centimeter, inch, yard are all units of length; a gram, ounce, pound are units of weight.

To compare measures of objects using nonstandard procedures, compare the lengths, heights, and widths of various objects. String, strips of paper, wood, hands, the foot, a book may be used as measures. Compare results to see that different-sized objects yield different measures.

The differences in measure result from different-sized measuring units rather than from inaccurate measurement. These are examples of nonstandard measures.

THE METRIC SYSTEM

One of the standard measures used to measure lengths is called the *meter*. The notation for meter is m; 3 meters is written "3 m." A *decimeter* is 1/10 of a meter.

> *Note:* The m is not an abbreviation for meter; it is a symbol for meter and, therefore, is not followed by a period.

The notation symbol for decimeter is dm; 7 decimeters is written "7 dm." The *centimeter* (cm), 1/100 of a meter, is needed to measure small things and to measure more precisely. A centimeter is further divided into 10 parts called *millimeters* (mm).

Ten meters is called a *dekameter;* 100 meters is a *hectometer;* and 1,000 meters is a *kilometer* (pronounced in Europe and England as kilom' eter and in the United States as kil' o meter).

The metric system was first suggested by the Flemish mathematician Simon Stevin in 1585. It was adopted in France in the latter eighteenth century and has been legalized for many years in Great Britain and recently in the United States. The system of weights and measures includes measures of *length* ("meter" is the unit measure), of *surface* (unit is the "area"), of *capacity* (unit is the "liter"), and of *weights* (unit is the "gram").

Mini-lesson 1.3.12 Measure Linear Distance Using Nonstandard and Standard Units

Vocabulary: standard measure, metric system, unit, meter, decimeter, centimeter, gram

Possible Trouble Spots: does not conserve quantity; does not understand how to estimate

Requisite Objective:
Measurement 1.1.9: Compare sizes of objects

Activity 1
To measure objects using standard procedures, have children participate in the following activities.

A. Have the children use a meter stick or string, 1 m long, to measure lengths. Ask them to measure heights of friends and to record and compare their results. Since most objects will not be exactly 1 m long, suggest that the children record their measures as "a little over 3 m" or "just under 12 m."

B. Many things measured are smaller than 1 m in length. Ask the child to list objects less than 1 m long.

C. Have the child make a meter marked off into decimeters. Ask him to use this measure and list objects that have lengths of 2 dm, 5 dm, and 7 dm.

Activity 2
To record standard measures, have the child measure heights of friends and record the measures in meters and decimeters. The measure shown is 1 dm in length.

1 decimeter

The metric system is built on a tens system in a manner similar to our decimal numeration system. This comparison is helpful for children in a transitional period of having to "think metric." For older children, learning to use the metric system will probably involve some comparisons with the system of measures already familiar to them.

To relate metric measures to powers of ten, the teacher may review the procedures for generating place values in our decimal notation system. Children may be guided to see that the ascending values for each unit are obtained by multiplying the measure by 10. These ascending values have Greek prefixes indicating their values (as "dekameter" for 10 meters). The descending values are obtained by multiplying by 1/10 (as in decimal notation). These have Latin prefixes (as "decimeter" for 1/10 meter).

Activity 3
To discover patterns of metric notation, engage the children in the following activities.

A. Ask the child to use a strip of posterboard or paper to make a ruler and to mark off 15 cm. Then have the child complete the following table after measuring objects in decimeters and centimeters and ask her to tell what patterns she notices.

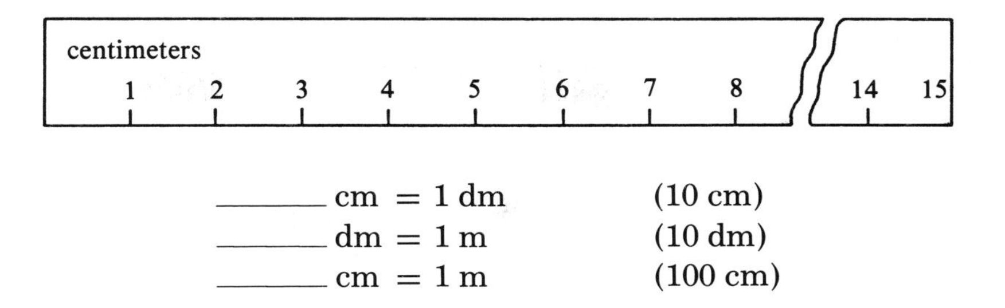

_______ cm = 1 dm (10 cm)
_______ dm = 1 m (10 dm)
_______ cm = 1 m (100 cm)

B. Ask the children to make a chart showing metric measurements, such as Chart 22.1, while discussing the values of the prefixes. Help them notice patterns such as the occurrence of powers of ten and the use of the same prefixes across measures. This chart may be used for reference while the children are becoming accustomed to using the metric system terms. Use of a horizontal arrangement of entries may help children to make associations with the place-value chart.

Chart 22.1. Metric Measurements

			Length			
kilo-meter (km)	*hecto-meter (hm)*	*deka-meter (dkm)*	*meter (m)*	*deci-meter (dm)*	*centi-meter (cm)*	*milli-meter (mm)*
1000 m	100 m	10 m	1 m	$\frac{1}{10}$ m	$\frac{1}{100}$ m	$\frac{1}{1000}$ m
			Weight			
kilo-gram (kg)	*hecto-gram (hg)*	*deka-gram (dkg)*	*gram (g)*	*deci-gram (dg)*	*centi-gram (cg)*	*milli-gram (mg)*
1000 g	100 g	10 g	1 g	$\frac{1}{10}$ g	$\frac{1}{100}$ g	$\frac{1}{1000}$ g
			Capacity			
kilo-liter (kl)	*hecto-liter (hl)*	*deka-liter (dkl)*	*liter (l)*	*deci-liter (dl)*	*centi-liter (cl)*	*milli-liter (ml)*
1000 *l*	100 *l*	10 *l*	1 *l*	$\frac{1}{10}$ *l*	$\frac{1}{100}$ *l*	$\frac{1}{1000}$ *l*
			Surface			
	hectare 10000 m^2		*are* 100 m^2		*centiare* 1 m^2	

Activities involving estimation in the metric system will help bridge to thinking metric and to proficiency in applying metric measures realistically.

C. Ask the child to:

1. Measure the width of the fingernail on his fourth finger or the width of a paper clip (both are about 1 cm long)

2. Weigh a paper clip (about 1 g)
3. Use a micrometer gauge to find the diameter of the wire of a paper clip (about 1 mm)
4. Weigh sticks of butter (about 9 sticks weigh 1 kg)
5. Measure his hand span (about 15 to 20 cm)
6. Measure his pace (about 1/2 m to 1 m)
7. Weigh himself on a bathroom scale graduated in kilograms
8. Measure his height, using a metric linear scale taped to the wall

Additional considerations: It should be pointed out that all measurement is approximate. Measurement instruments are sources of error in themselves. For example, rulers may not all have the same measurements due to the manufacturing process, changes in temperature, or the varying thicknesses of their marks.

PERIMETER AND CIRCUMFERENCE

Perimeter is the linear measure around a two-dimensional shape (a plane figure). The perimeter of a circle is called the *circumference* of the circle. The perimeter of a polygon is the sum of the linear measures of the line segments making up the figure.

Mini-lesson 4.1.12 Find Perimeter of Polygons

Vocabulary: perimeter, base, height, length, width, side

Possible Trouble Spots: does not conserve length; poor eye-hand coordination; visual discrimination problems; difficulty with spatial relationships

Requisite Objectives:
Measurement 2.1.11: Use appropriate measuring device, for example, ruler or straight edge for length

Graphs and Functions 2.1.12: Select situations showing the reflexive relation

Measurement 2.2.9: Use centimeter for linear measure

Measurement 2.3.9: Estimate measures, including meter, decimeter, and centimeter

Graphs and Functions 3.1.14: Performs transformations involving conservation of length, time, discontinuous quantity, continuous quantity (liquid)

Activity 1
To find the perimeter of objects, have the children participate in the following activities.

A. Ask the child to use a piece of string or thread to find perimeters of irregular shapes and then of regular shapes (such as squares, rectangles, and parallelograms).

B. Draw polygons on graph paper or construct them on geoboards and ask the child to "count" the linear units around the figure. (This activity is geared toward children who cannot yet add.)

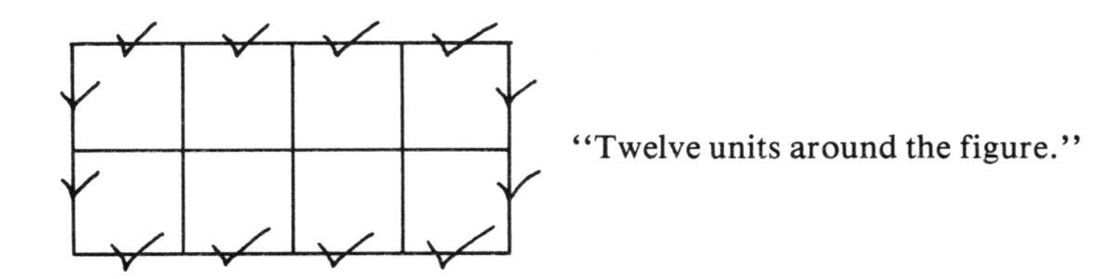

C. Present to the child various figures with some of the faces marked with letters and individual drawings of the lettered faces with their measurements (as shown). Help the child measure the perimeter of each of the faces. (Perimeter of $P = 18$ cm; $Q = 36$ cm; $R = 20$ cm; $S = 24$ cm; $T = 18$ cm; $U = 26$ cm.)

Guide her to notice that perimeters can be found by addition or sometimes by multiplication. For example, the perimeter of shape P is $6 + 6 + 6 = 18$, or $3 \times 6 = 18$, or 18 cm.

For shape Q, the addition is $6 + 12 + 6 + 12 = 18 + 18 = 36$.

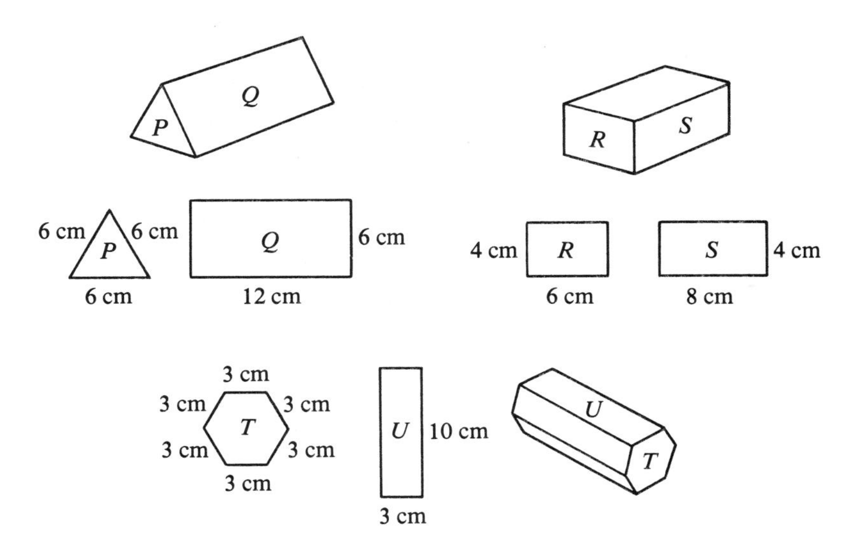

Activity 2

To measure the circumference of a circle, have the child participate in the following activities.

A. Before asking the child to find the circumference of a circle, help her measure the diameter of a circle. Draw a circle and ask the child to find the diameter by moving a ruler across the circle and noting the maximum length obtained or by placing the circular region between two parallel boundaries (pencils will do) and measuring the distance between.

B. Give the child a circular object, such as the bottom of a plastic bottle, a coin, or a bicycle wheel. Ask him to find the circumference by marking a point of the edge of the circular region, placing the mark at the beginning of a ruler or at point A on a line, and rolling the object along the ruler or the line until the mark reaches the ruler again (point B). The distance $\overline{AB}$ indicates the circumference.

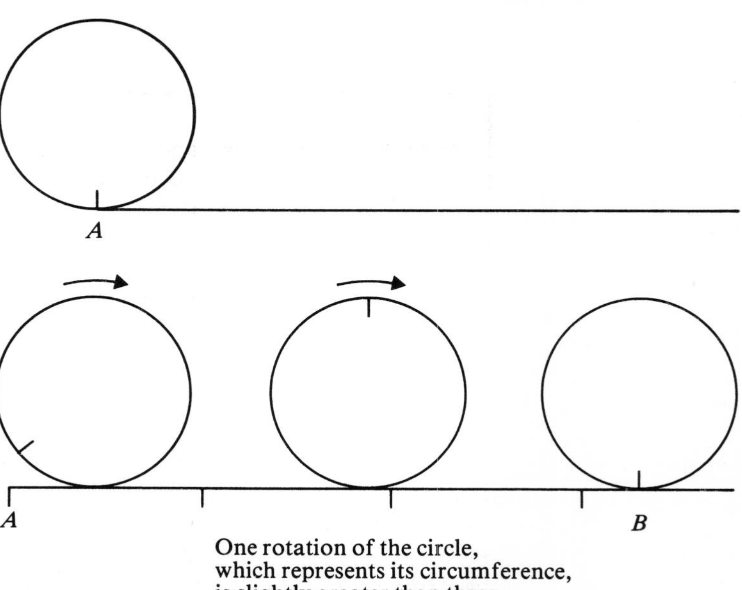

One rotation of the circle, which represents its circumference, is slightly greater than three times the measure of its diameter.

C. Ask the child to wrap a piece of string or strip of paper around a circular model and to cut the string or paper so that the strip is equal in length to the circumference of the object. Then ask her to cut additional strips equal in length to the diameter. Tell her to place the strip that represents the measure of the circumference on a flat surface and by its side the diameter strips end to end.

Ask her to repeat this process with various sizes of circles so that she will notice that for a given circle, the measure of its diameter is about 1/3 its circumference. The circumference of a circle is about three times the diameter.

Additional considerations: The perimeters of shapes that children can handle may be found by direct measurement or by wrapping string around the shapes and then measuring the lengths of the string.

AREA

Area is the amount of surface covered. This surface may be measured by considering how many congruent shapes would be needed to cover it entirely, with *no* gaps between the shapes.

> *Note:* If the square is decided upon as the easiest unit to use, care should be taken to avoid using the formula used to find the area (length multiplied by breadth or bottom times height) before the child understands the concept of area.

Tessellations

Such patterns made up of shapes fitting together so there are no gaps and no overlappng places are called *tessellations*. Tessellations may be formed from sets of two, three or more different shapes as well as from repeating rectangular, hexagonal, square, rhombus, or triangular regions. Some examples of tessellations are shown below.

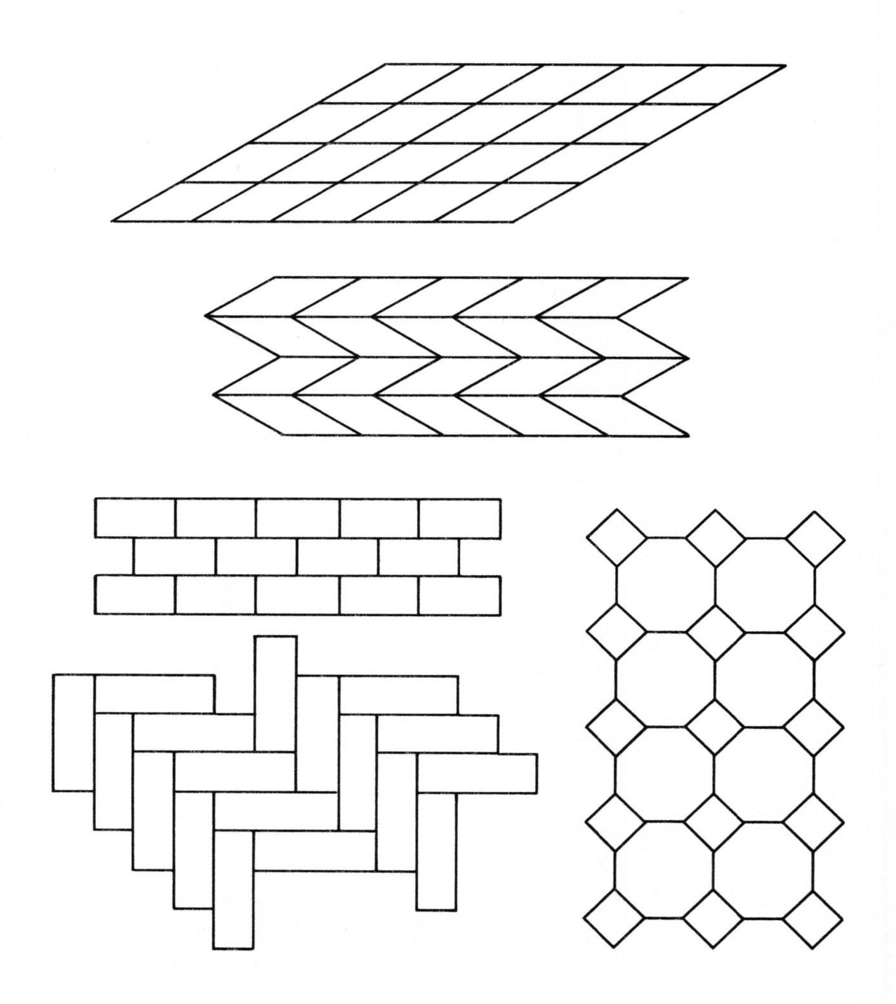

Mini-lesson 4.1.13 Find Area of Polygon Regions

Vocabulary: area, square measure, m^2, cm^2, dm^2, pi

Possible Trouble Spots: does not conserve area; visual spatial difficulty; difficulty estimating and making judgments

Requisite Objectives:
Measurement 2.3.9: Estimate measures, including meter, decimeter, and centimeter

Graphs and Functions 3.1.14: Performs transformations involving conservation of length, time, discontinuous quantity, continuous quantity (liquid)

Activity 1
To use multiplication to find the area of a rectangular region, show the child various drawings of shapes, using squares joined edge to edge. Guide him to see that several shapes can be made whose areas are all equal and that shapes with a given perimeter can have different areas.

Activity 2
To find the areas of regular and irregular shapes, duplicate the shapes shown in the illustration or have the child draw them on graph paper. Ask her to find the area of each shape a, b, c, e, g ($4\ \text{cm}^2$). Ask the child to color shapes each having an area of $16\ \text{cm}^2$ (shapes h, i, j). Then ask her to color some shapes each having a perimeter of 12 cm. Ask her to give the area of each shape (area of $d = 5\ \text{cm}^2$; area of $f = 9\ \text{cm}^2$).

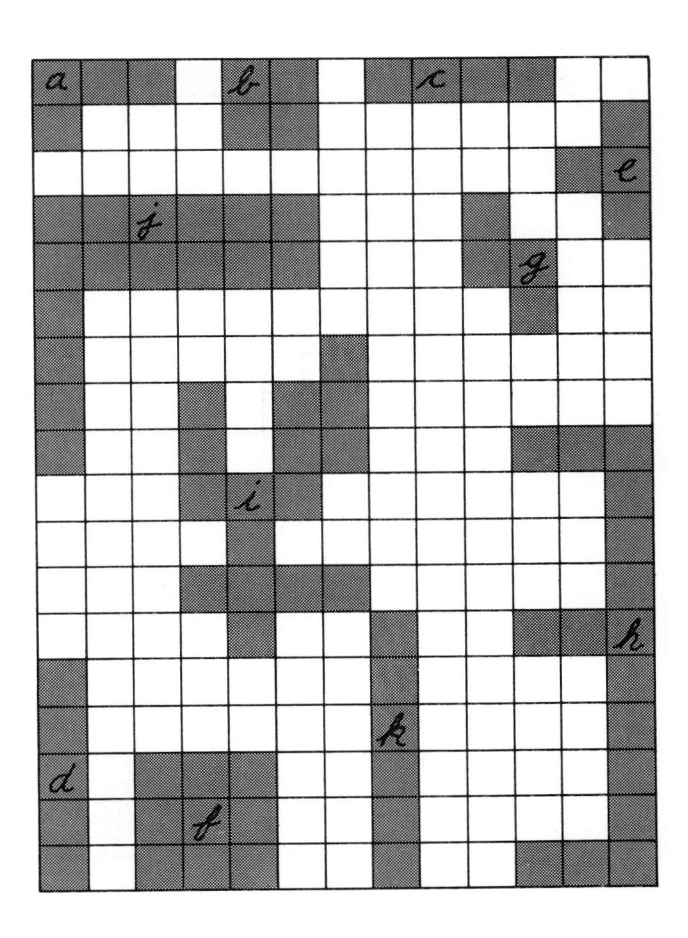

The child should obtain area by first counting squares in a shape. Then guide her to use multiplication of rows times columns for shapes *b*, *c*, *d*, *f*, and *k*. For example, she should see the patterns:

$$\begin{array}{lllll} b & \longrightarrow & 4\text{ cm}^2 & \longrightarrow & 2 \times 2 \\ c & \longrightarrow & 4\text{ cm}^2 & \longrightarrow & 1 \times 4 \\ d & \longrightarrow & 5\text{ cm}^2 & \longrightarrow & 5 \times 1 \\ f & \longrightarrow & 9\text{ cm}^2 & \longrightarrow & 3 \times 3 \\ k & \longrightarrow & 6\text{ cm}^2 & \longrightarrow & 6 \times 1 \end{array}$$

Activity 3

To find the area of a triangular region, have children participate in the following activities.

A. Find the area of a 3-by-2 rectangle. Have the child use the terms *bottom* (b) and *high* (h) along with (or in place of) *length* and *width* (as shown in the diagram).

Area of a rectangular region

Guide him to notice that a diagonal drawn on the rectangular region yields two triangles of equal measure and, therefore, the area of a triangular region is 1/2 the area of a rectangular region.

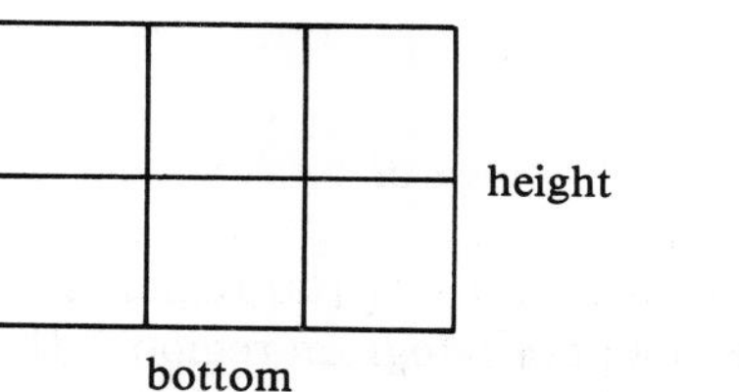

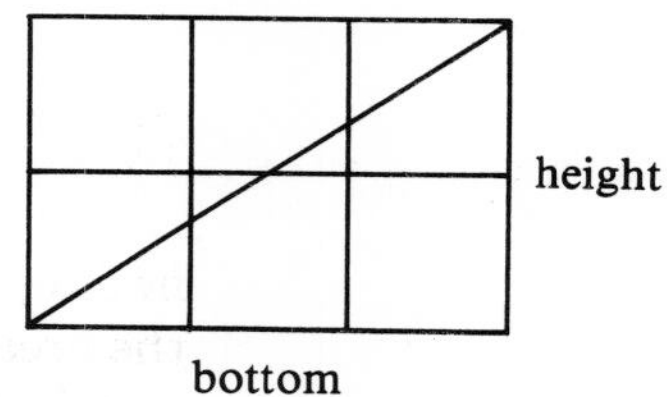

Area of □ $= b \cdot h$

$= 3 \times 2$

$= 6$ sq. units

Area of △ $= \frac{1}{2}\, b \cdot h$

$= \frac{1}{2} \times (3 \times 2)$

$= \frac{1}{2} \times 6$

$= 3$ sq. units

B. Give the child a drawing of a triangle other than a right triangle. Ask her to cover the area of the triangle with a single layer of barley. Then ask her to draw a rectangular figure onto the triangle (as shown) and see that the barley covering the original surface (√ and √ √) can be moved to just cover the new surface (√′ and √ √′). Thus, the area of a triangle is 1/2 that of a rectangle with the same base and height.

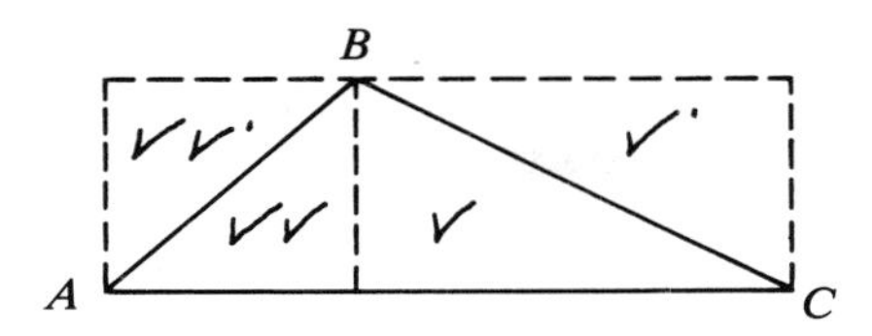

Activity 4

Area of a parallelogram

To find the area of a parallelogram region, show the child that a parallelogram may be rearranged into a rectangular region (as shown). This rearrangement involves Piaget's concept of conservation of area.

Have him identify the height and base of the newly formed rectangular region and proceed as in finding the area of a rectangular region.

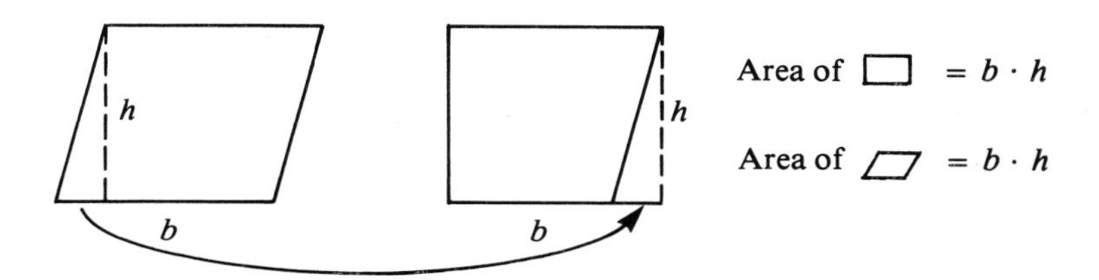

Activity 5

Area of a circular region

To show the relation of the area of a rectangular region to that of a circular region, find the area of a parallelogram region as a vehicle for finding the area of a circular region. Beginning with the principle that the area of a parallelogram region is the product of base times height, a series of substitutions may follow:

$$\text{Area } ▱ = b \cdot h \qquad \text{but } b = \tfrac{1}{2}C \text{ and } h = r$$

$$= \tfrac{1}{2}C \cdot r, \qquad \text{but } C = 2\pi r$$

$$= \tfrac{1}{2}2\pi r \cdot r,$$

$$= \tfrac{1}{2}2\pi r^2$$

$$= \left(\tfrac{1}{2} \cdot \tfrac{2}{1}\right) \pi r^2$$

$$= \overset{1}{\boxed{\tfrac{2}{2}}}\, \pi r^2$$

$$\text{Area } ○ = \pi r^2 \qquad \text{where } \pi \simeq 3.14$$

Draw a circle on a piece of heavy paper, cut the circle into equal-sized parts, and reposition the parts to construct a parallelogram (as shown).

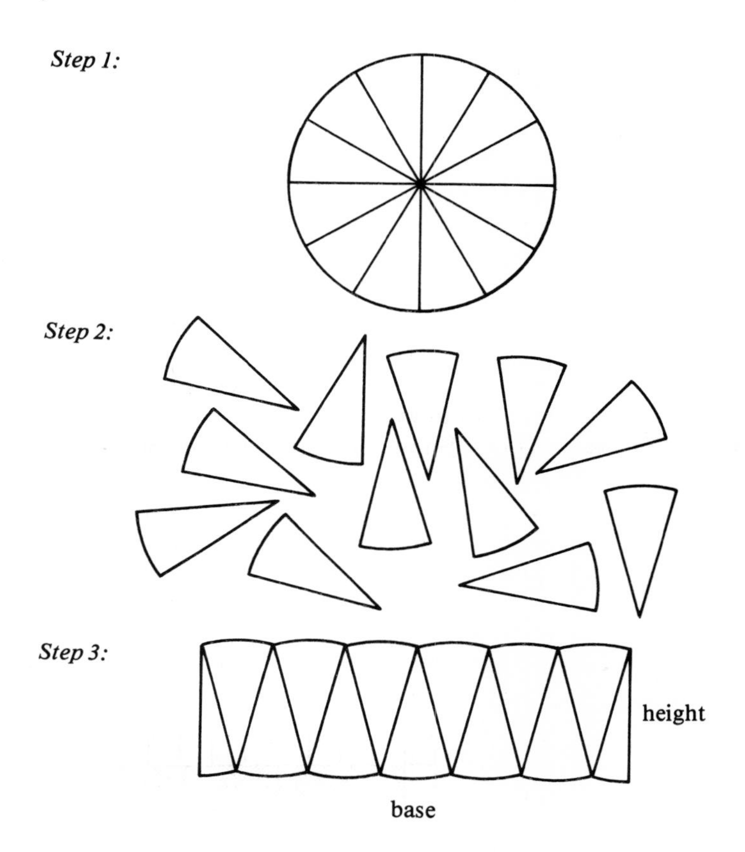

Guide the child to notice that the two longer sides of the parallelogram make up the circumference of the circle, so the base of the parallelogram is 1/2 the circumference of the circle, or $b = 1/2C$. Also, the height of the parallelogram is the radius of the circle, so $h = r$.

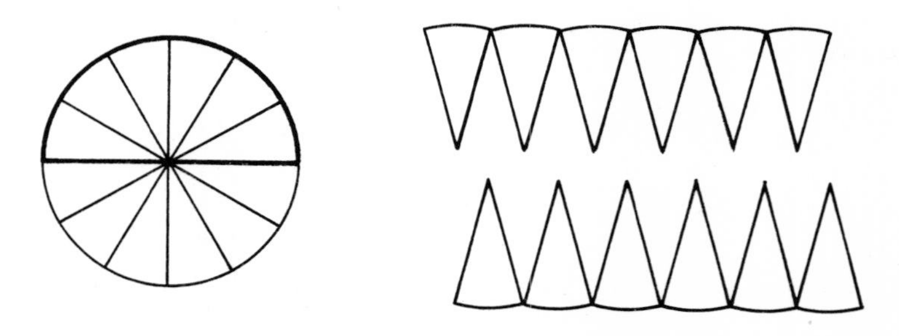

Activity 6

Area of a circle

To find the area of a circle, use is made of the π relationship; that is, the circumference is about 3.14 times the diameter. This may be expressed formally as:

$$\frac{C}{d} = \pi$$

$$\frac{{}^{1}\not{d}}{1} \bullet \frac{C}{{}_{1}\not{d}} = \pi \frac{\not{d}^{1}}{1}$$

$$C = \pi d \qquad \text{but } d = 2r$$

$$C = \pi 2r \qquad \text{(by substitution)}$$

$$C = 2\pi r \qquad \text{(usual form)}$$

Additional considerations: A Piagetian task that assesses whether or not a student states that equal-sized surfaces covered take up the same amount of space (area), regardless of their arrangements, is as follows.

Conservation of area

Show the child two green 8-by-11-inch pieces of paper. Tell the child these are two grazing pastures for neighboring cows. Show him two sets of red blocks ("these are barns") with 15 in each set. Place one barn at a time on each pasture (placing the barns simultaneously on each pasture). On one pasture, spread the barns out; on the other, keep them together. Ask the child, "Do both cows have the same amount of grass to eat?" If necessary, discuss the idea that when a barn is placed on a pasture, grazing area is lost.

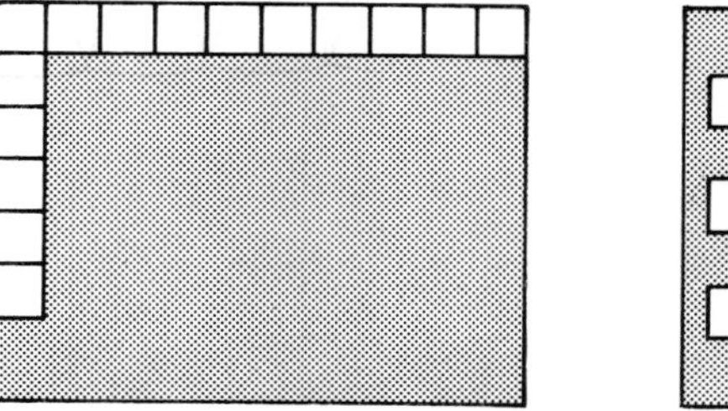

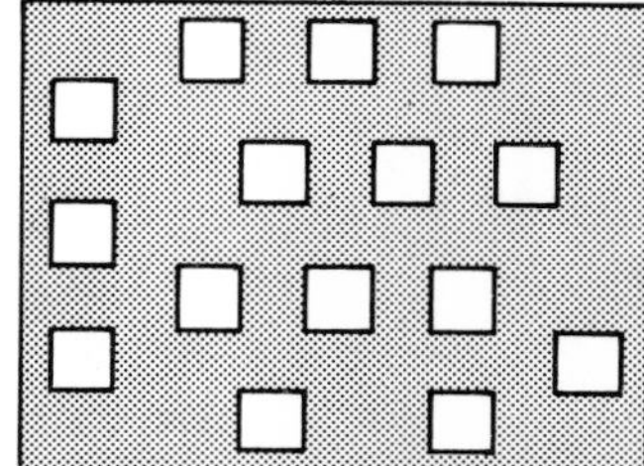

VOLUME

Volume is the amount of space taken up by a solid (three-dimensional) shape. Volume is often measured in cubic centimeters.

A procedure of creating shapes with centimeter cubes builds a foundation for measuring volume. Counting the number of cubes in a layer and then the number of layers lays a basis for finding the volumes of *prisms* (three-dimensional figures).

Mini-lesson 4.2.6 Find Volume

Vocabulary: cubic units, cube, rectangular prism, cuboid, circular prism

Possible Trouble Spots: does not conserve continuous quantity, for example, liquid; makes errors when multiplying

Requisite Objectives:

Measurement 1.1.9: Compare sizes of objects

Measurement 1.3.12: Measures linear distance using nonstandard and standard units

Graphs and Functions 3.1.14: Performs transformations involving conservation of length, time, discontinuous quantity, continuous quantity (liquid)

Multiplication 4.1.7: Multiply a 3-digit by a 2-digit number with and without renaming

Graphs and Functions 4.1.16: Recognize conservation of volume under various transformations

Geometry 4.2.5: Identify 3-dimensional geometric shapes

Activity 1

To find the volume of a rectangular prism, give the child a sheet of paper ruled with 1-cm squares and several 1-cm cubes. On the squared paper, shade 6 squares to show a 3-by-2 cm figure. Ask her to tell the

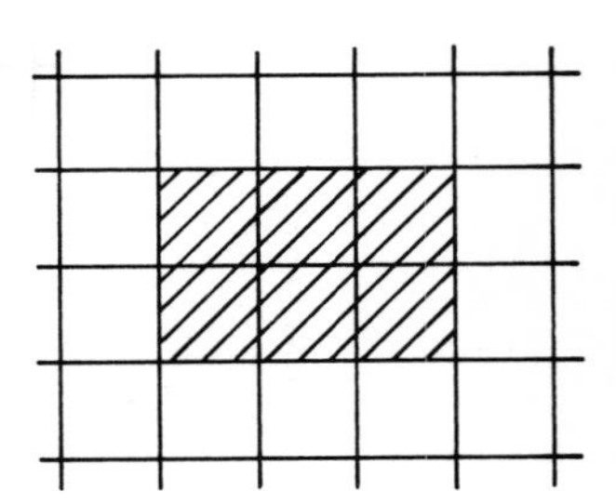

area of the shaded part. Place a cube on each square and ask her to tell the total volume of the cubes. Put a second layer of cubes atop the first set and ask the child to find the total volume of cubes. Do the same with a third and then a fourth layer. The shape built this way, based on a rectangle, is a *rectangular prism* (or *cuboid*). The volume of a rectangular prism is found, then, by multiplying the area of the bottom by the number of layers.

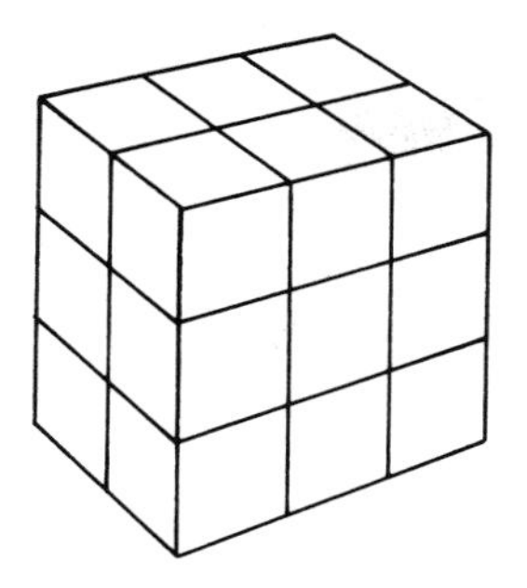

Activity 2
To find the volume of a circular prism, draw on paper ruled with 1-cm squares a circle having a 6-cm diameter. Ask the child to tell the area of the circle. Then ask him how many 1-cm cubes would be needed to cover the circle, assuming the cubes could be cut up (about 28.3-cm cubes). Tell the child that this 1-cm layer can be viewed as a 1-cm-high cylinder (circular prism). Ask him to find the volume of the cylinder. ($V = a \times h$, so $V = 28.3 \times 1 = 28.3 \text{ cm}^3$.)

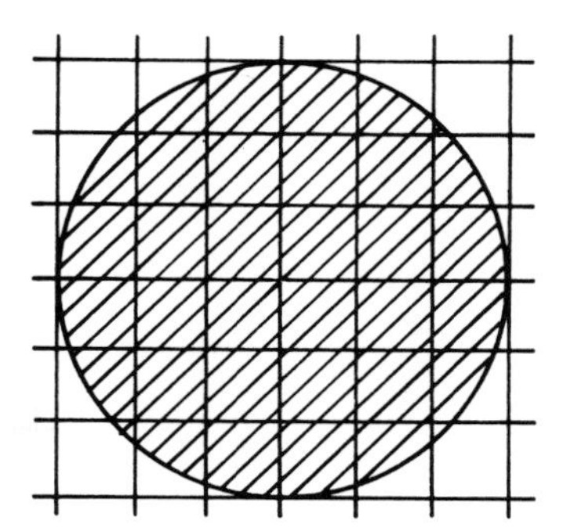

Additional considerations: It is helpful for students to pour sand from container to container to construct notions about volume. Begin with identical containers; then use different shapes but the same volume.

Evaluation of students' performance in finding volume should involve their problem-solving skills, not merely whether they obtain the correct answer or use the correct formula. Observation of the process used to find the volume of various containers should be included.

THE LITER

Volume describes the space an object occupies. Capacity describes the space inside a container. Since volume and capacity measure three-dimensional space, they are measured in cubic centimeters (cm^3), cubic meters (m^3), and so on. The *liter* is the unit for measuring liquids and represents 1000 cm^3.

Mini-lesson 4.2.7 Use the Milliliter and Liter for Measurement of Capacity

Vocabulary: liter, milliliter

Possible Trouble Spots: same as for Mini-lesson 4.2.6

Requisite Objectives:
Same as for Mini-lesson 4.2.6

Activity 1
To use the liter as a standard measure of capacity, give the child some containers—cups, jugs, flower vases, cans, bowls, buckets. Ask her to arrange them in ascending order of estimated capacity and then to tell the reasons for the choice of order. Then give the child marbles, peas, sugar cubes, sand, and water and provide experiences with these as measuring units. For many shapes, only sand or water will be suitable for measuring.

Then give her a liter cube. This may be made from thick cardboard that is painted inside and outside with oil paint to make it waterproof. The measure of the cube may be 1 dm^3. When full, the cube holds 1 liter of water. Ask her to then pour various amounts of water from the liter cube into a plastic pitcher and mark the pitcher 1 liter, 1/2 liter, and 1/4 liter.

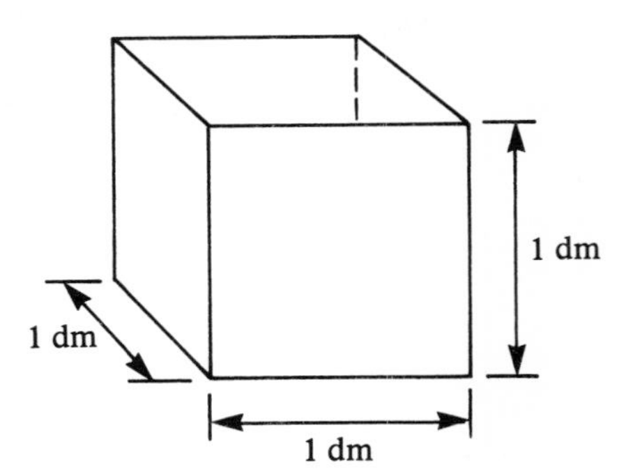

Additional considerations: Decentration is the ability to look at more than one quality of objects at a time, such as considering the width and height of a glass jar simultaneously.

Piagetian tasks that may be used to assess if a student is able to recognize that the amount of liquid remains invariant even though the shapes, sizes, and number of containers change is as follows.

Conservation of continuous quantity

Procedure A: Place two beakers or bottles of the same size and appearance before the child. Fill both with a liquid and ask the child if there is the same amount of water in both containers. If the child says yes, proceed with the activity. If she says no, tell her to make the amounts the same in both containers. Then pour the water from one of the containers into a taller, narrower bottle. Ask the child if there is the same amount of water in both containers.

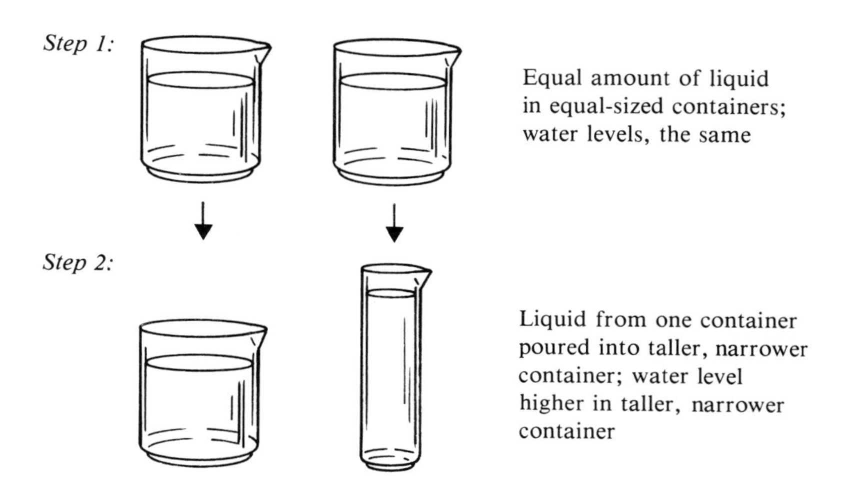

Procedure B: Same as Procedure *A*, except this time pour water into a shorter, wider container.

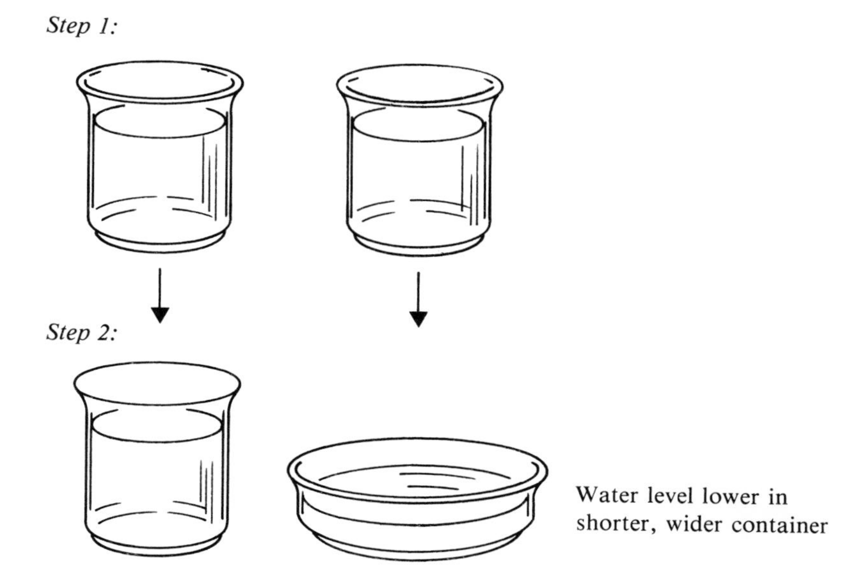

Procedure C: Same as Procedure *A*, but this time pour the liquid from one of the original containers into four small containers of equal size. Ask the child if the amount of water poured into the four little containers is the same as in the big container. The nonconserver of liquid will say there is more liquid in the four small containers. Her ability to decenter is not developed to the point of realizing that the decrease in size offsets the increase in number of containers.

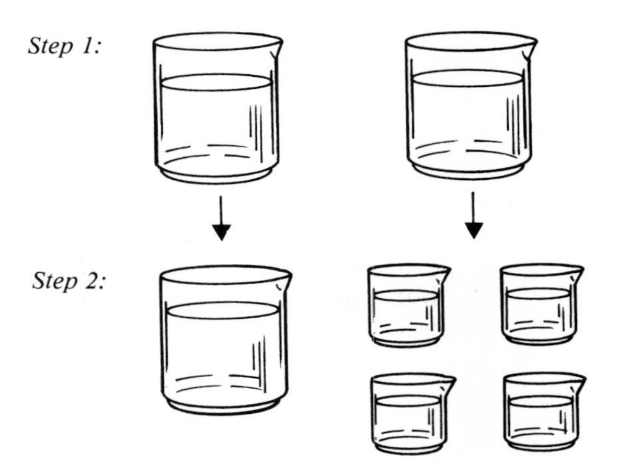

MASS

Weight

All bodies exert an attraction on one another due to the pull of the earth on an object (the object's weight). The pull of the earth on an object depends upon the distance of the object from the center of the earth. As the distance increases, the pull decreases. If the object is far enough away, the pull becomes almost nil. Thus, the weight of an object gets smaller and smaller as it is moved away from the earth. Since the earth is not a perfect sphere, the weight of an object can vary from place to place on the earth's surface. The moon's pull on objects near to it is considerable since it is also a large body. The "moon-weight" and the "earth-weight" of a person is not the same since the pulls are not the same.

Two ways of finding the weight of an object are with a spring balance and with a scale balance. An object may be attached to a spring. The spring is then stretched by the earth's pull. The amount of the extension indicates the weight of the object. In using a scale, the object to be weighed is placed in one pan of the balance and a metal weight in the other pan so that the metal weight and the object balance. The pull of the earth is the same for both; thus, they both have the same weight. Would they still balance on the moon? Yes, because the pull of the moon on both objects would be the same. Although their weights would change from earthweight, they would still be equal to each other.

The *weight* of an object is measured in units of force (called the "newton"). Weight depends on the pull or force of one body on another (usually the earth). The weight of a tennis ball, a piece of metal, a bag of feathers, or a ball of clay may all have the same measure in newtons at various distances from the center of the earth. The pull of the earth on a body is the weight of that body.

Mass

Mass is a quality of objects that involves resistance to change in motion. An object's mass does not change. It is the same whether the object is near the earth or the moon or anywhere else in space. Its weight can change but not its mass.

Mass is measured in units called *kilograms* (kg). Very small masses are measured in *grams* (g) (1000 g = 1 kg). The weight of an object depends on the pull of the earth on its mass. For example, a box of groceries is heavy because the pull of the earth makes it feel heavy, but a person in space has little weight because the pull of the earth is slight.

Mini-lesson 3.2.18 Use Grams and Kilograms for Measuring Weight

Vocabulary: gram (g), kilogram (kg), weight, mass, estimate, balance, heavier than, lighter than, scale

Possible Trouble Spots: does not conserve mass; difficulty forming generalizations regarding mass and gravity; visual perception problems that interfere with reading a scale

Requisite Objectives:
Read a scale

Use symbols for expressing notation

Measurement 1.3.12: Measures linear distance using nonstandard and standard units
Measurement 2.1.11: Use appropriate measuring device, for example, ruler or straight edge for length

Graphs and Functions 2.1.12: Select situations showing the reflexive relation

Activity
To compare masses of objects, have children engage in the following activities.

A. Have the child find a pair of objects that have the same mass (they may use a balance). Ask him to suspend these objects from equal-length pieces of elastic and to measure the lengths of the stretch. Objects with the same mass are pulled toward earth with the same force, thus extending the elastic by the same amount.

B. Have the child compare objects having the mass of 1 kg by holding them in her hands. Then ask her to use a balance to weigh different objects by balancing them with 1-kg, 1/2-kg, and 1/4-kg masses. For example, she may wish to find how many books have the same mass as 1/4 kg or 1/2 kg.

Additional considerations: It is helpful for children to relate measures to objects that are familiar. For example, a gram is the approximate weight of a paper clip or a piece of paper about the size of a dollar.

There is an interesting relationship among length, volume, and weight in the metric system of measure. One gram is the weight of 1 cubic cm of water, which in turn equals 1 ml of water. Since water temperature affects volume, the water must be at 4° Celsius. Furthermore, 1 kg is the weight of 1 cubic dm of water and in turn equals 1 l of water.

The kilogram is the standard SI unit for measuring mass. SI stands for International System of Units.

Differences in shape depend upon rearrangements of the matter's molecular structure. Differences in weight are affected by the density of the object's molecules. "Objects which are heavy for their size are thus composed of tightly packed elements; lighter ones are more loosely packed, with lots of empty spaces in between."[1]

Relation of volume, shape, and weight

The Piagetian task that assesses whether a child can recognize that equal amounts of matter will displace equal amounts of water, regardless of differences in the shape or weight of the matter (providing the weight of the object is heavier than water) is as follows.

Procedure A: Give the child two balls of clay equal in amount, shape, and size and show her two identical containers with equal amounts of water. Say, "If I put one of these in each beaker, what will happen to the water levels?" The child should answer that both water levels will rise the same distance.

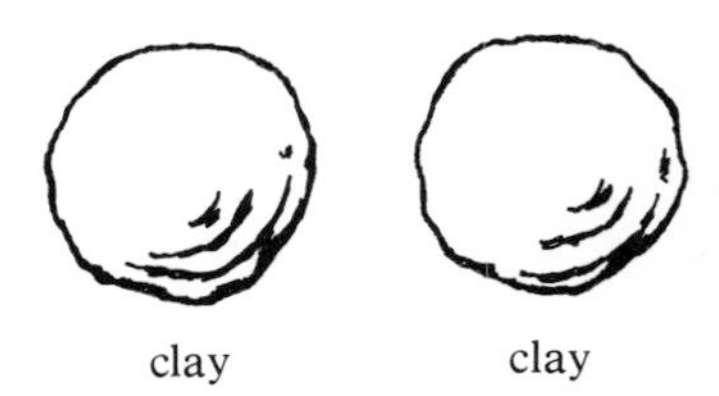

Procedure B: Give the child a ball of clay and a rock (or marble egg) equal in size and shape. The same question and answer as in Procedure *A* should follow.

1. J. H. Flavell, *The Developmental Psychology of Jean Piaget* (Princeton: Van Nostrand, 1963), p. 302.

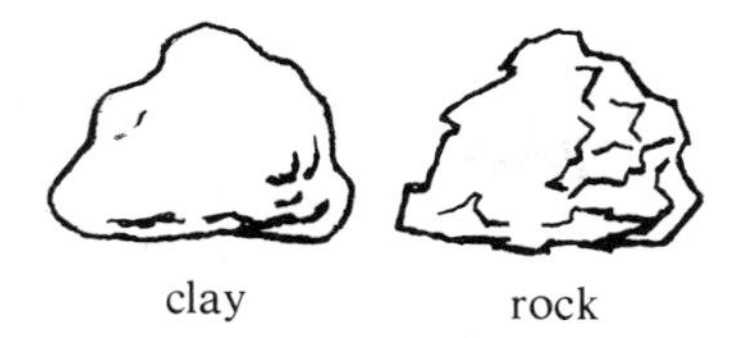

Procedure C: Show the child two balls of clay that are equal in size. Roll one into a sausage shape. The same question and answer as in Procedure *A* should follow.

Procedure D: Give the child two objects of the same size and shape but different in weight (This is the same condition as in Procedure *B*.) The same question and answer as in Procedure *A* should follow.

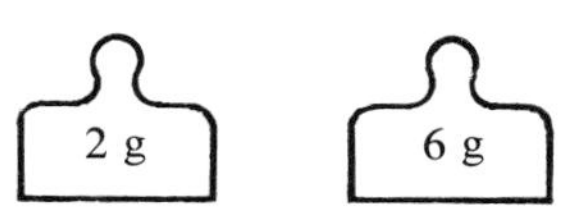

Procedure E: Give the child two objects of the same shape and weight but different in size. Ask the same question as in Procedure *A*, but, this time, the child should state that the water will be higher in the beaker with the larger-sized object. (Chart 22.2 is presented as a summary of procedures *A* to *E*.)

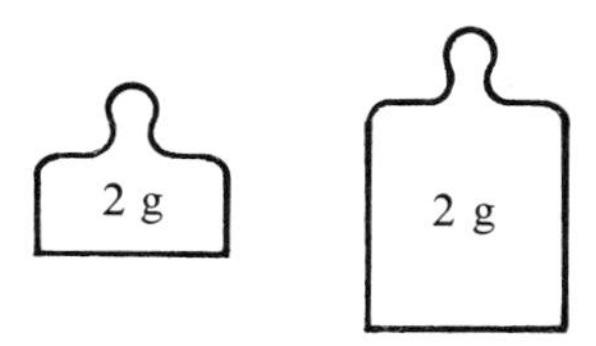

For further information on these ideas, the reader is directed to several sources.[2]

2. Jean Piaget, *The Construction of Reality in the Child* (New York: Basic Books, 1954); *Play, Dreams and Imitation in Childhood* (New York: Norton, 1962); *Six Psychological Studies* (New York: Random House, 1967); *Structuralism* (London: Routledge & Kegan Paul, 1971); Jean Piaget and Barbel Inhelder, *Mental Imagery in the Child* (New York: Basic Books, 1971); and Flavell, *Developmental Psychology of Jean Piaget.*

Chart 22.2. Summary of the Conditions and Water Displacements

Condition	Size	Weight	Shape	Water displacement
1. clay, clay	Same	Same	Same	Same
2. clay, rock	Same	Different	Same	Same
3. clay, clay	Same	Same	Different	Same
4. 2 g, 6 g	Same	Different	Same	Same
5. 2 g, 2 g	Different	Same	Same	Different

ANGLES MEASURED IN DEGREES

The *degree* is the basic unit used in measuring angles. If a complete rotation is made, the measure of the angle is 360°. A *right angle* involves a rotation of 1/4 of the way, or 90°; an *acute angle* is a rotation less than 90°; and an *obtuse angle* is a rotation greater than 90°. Perpendicular lines form right angles.

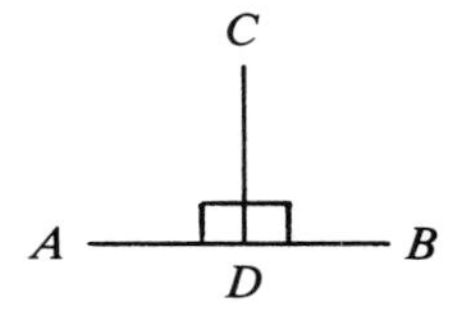

The notation to indicate that CD is perpendicular to AB is written "$CD \perp AB$."

Angles with the same measure are *congruent angles:* for the right angles α and β, $\angle\alpha \cong \angle\beta$.

If two angles share the same vertex and side and they have no common interior parts, they are said to be *adjacent angles*. The sum of the measures of adjacent angles equals the measure of the angle that they comprise, for example, $\angle ABC$ as shown below.

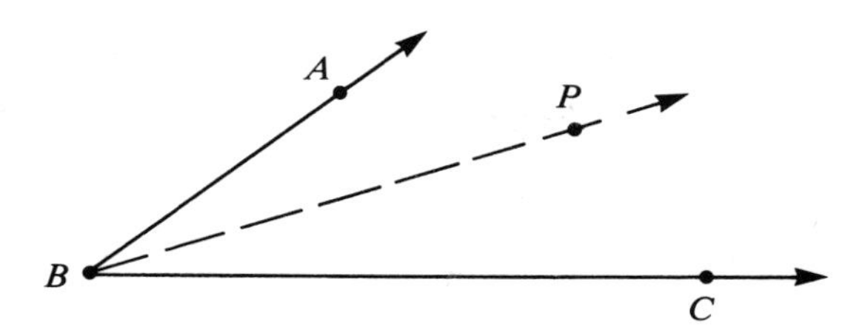

The letter m standing for "measure," $m\angle ABP + m\angle CBP = m\angle ABC$. $\overrightarrow{BP}$ is common to both $\angle ABP$ and $\angle CBP$. Point B is a common vertex.

Two angles whose measures sum to 90° are *complementary.* If two angles have measures that sum to 180°, they are *supplementary.*

Vertical angles are opposite angles formed by the intersection of two straight lines. Vertical angles of intersecting straight lines are equal.

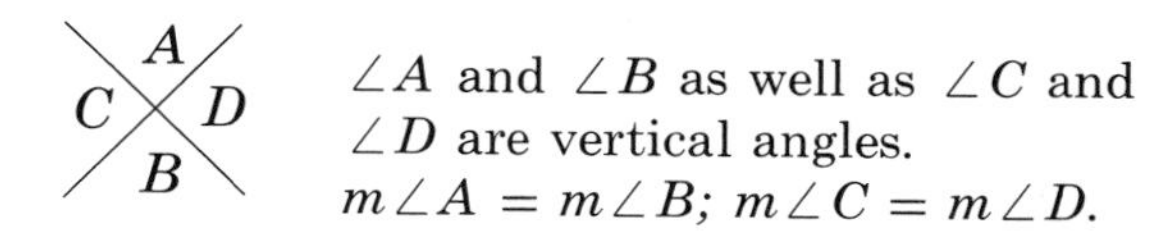

Angles may be measured by a device called a *protractor.* About 3,000 years ago, the Babylonians had used as a protractor a circle divided into 12 equal-sized parts but had found that this was not precise enough. They had then divided each of the 12 parts into 30 parts, making a circle a total of 360 parts.

An angle of 1/360 of a complete turn is considered to be 1 *degree* (1°).

Mini-lesson 6.1.6 Measure Angles of Polygons

Vocabulary: acute, obtuse, right angle, degrees, protractor

Possible Trouble Spots: visual perception difficulties interfere with reading a protractor; difficulty reading a scale; poor eye-hand coordination; poor fine motor skills

Requisite Objectives:
Measurement 1.1.9: Compare sizes of objects

Geometry 1.3.11: Identify simple plane figures

Geometry 3.2.15: Identify angles

Geometry 5.1.10: Construct geometric figures

Activity 1
Give the child a ditto copy of the half-circle protractor or an actual protractor. Show the child how to use the protractor: Place the center

of the protractor on the vertex of the angle; align the zero side with one of the rays of the angle so that the other ray intersects the protractor in some point; read the degree mark where this intersection with the second ray occurs (that number is the measure of the angle).

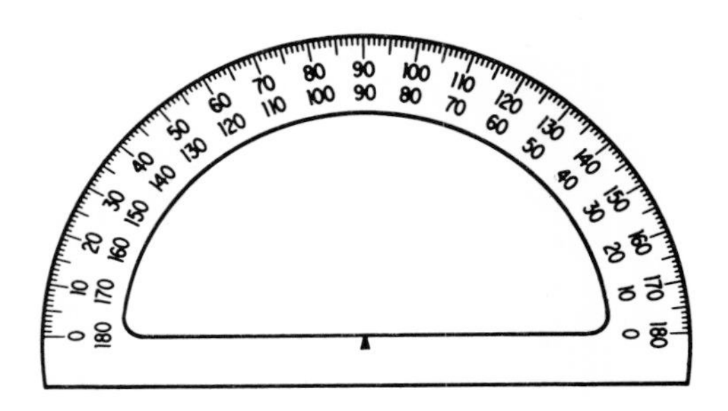

Then ask the child how many of the 180 parts are in (a) a right angle, (b) 1/3 of a complete turn, and (c) 1/6 of a complete turn. Ask the same kinds of questions in relation to 360 parts.

Activity 2
To use a protractor to find the complement and the supplement of a plane angle, ask the child to draw a straight line $\overleftrightarrow{AB}$ and then to draw a ray $\overrightarrow{AC}$ so that the measure of $\angle CAB = 10°$. Ask him to plot a point (D) to the left of $\overline{AB}$ and to find the measure of $\angle CAD$. Help him to draw an angle complementary to $\angle CAB$ and to find its measure.

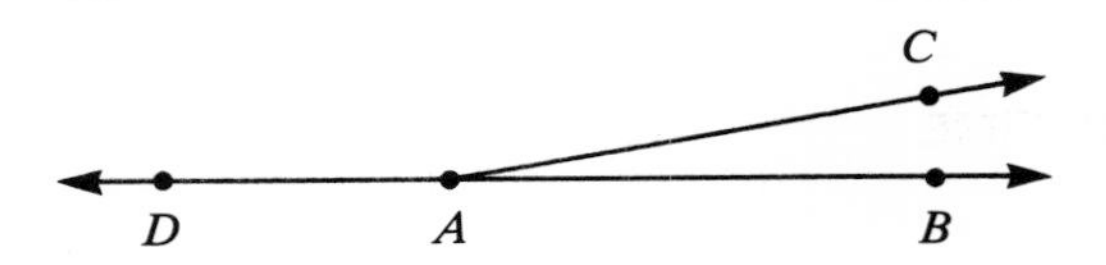

Additional considerations: This mini-lesson is a good vehicle for incorporating practice in estimating and problem solving. These skills would also serve to remove the rote drill approach that is so often involved in measuring angles of polygons.

MEASURING TEMPERATURE

The first mercury thermometer was invented by Gabriel Robert Fahrenheit in 1714. On the Fahrenheit scale, water freezes at 32° and boils at 212°, labeled 32° and 212°. This temperature scale was modified in 1742 by Anders Celsius, an astronomer from Sweden. Celsius assigned 0° to the freezing point of water and 100° to the boiling point.

He called his scale the *centigrade* scale, or 100 grades. It has become the custom to use the name Celsius instead of centigrade when using this thermometer. Figure 22.1 shows the two scales. The formulas for converting from one scale to the other are as follows:

$$\text{Celsius (C)} = \frac{5(F - 32)}{9}$$

$$\text{Fahrenheit (F)} = \frac{9C}{5} + 32$$

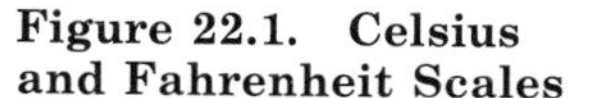

Figure 22.1. Celsius and Fahrenheit Scales

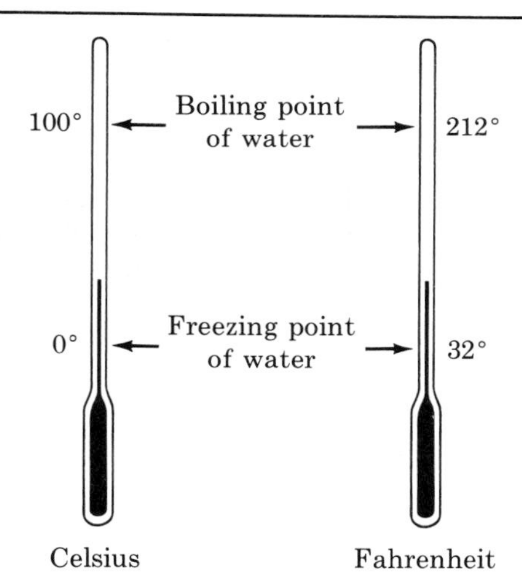

MEASURING TIME

Piaget's experiments investigating the development of time concepts show that children are not aware of the relation of differing speeds within a single time boundary until about age 8 or 9. Prior to this age, the child does not focus simultaneously on both the speeds of two objects and the distances they travel during the same time. This ability, however, is a necessary prerequisite of time measurement, since the relationship of coordinated speeds underlies how the hands on a clock face are related. The hour pointer and the minute pointer travel at different speeds and along different distances within a specified time.[3] Children about age 8 to 9 demonstrate conservation of time on the following activity.

---

3. Fredricka K. Reisman, "An Evaluative Study of Cognitive Acceleration in Mathematics in the Early School Years" (Ph.D. diss., Syracuse University, 1968) and *A Guide to the Diagnostic Teaching of Arithmetic* (Columbus, Ohio: Charles E. Merrill, 1972, 1978).

Mini-lesson 7.1.3 Compute Time

Vocabulary: —— *minutes after the hour;* 7:05, and so forth; time

Possible Trouble Spots: Cannot state that two objects starting from the same starting line at the same time (a) traverse equal distances over equal times if they go the same speed, (b) traverse unequal distances over unequal times if they go the same speed, and (c) traverse unequal distances over the same time if one goes at a faster speed

Requisite Objectives:
See Fredricka K. Reisman, *A Guide to the Diagnostic Teaching of Arithmetic* (Columbus, Ohio: Charles E. Merrill, 1978) for a task analysis on telling time to the precision of a minute.

Activity

Conservation of time

Procedure A: Tell the child that the cars will be started at the same time. Start the cars from the starting line. Make them go at the same speed. Ask the child if they stopped at the same time. They should have traversed the same distance from the starting line.

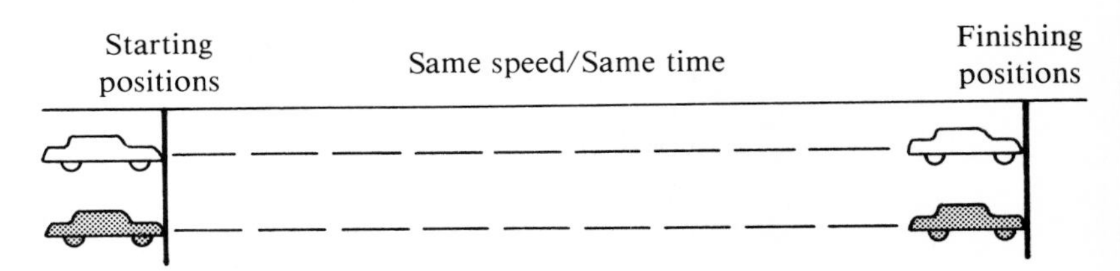

Procedure B: Tell the child that the cars will be started at the same time. Start the cars from the same starting line. Make them go the same speed but keep one going for a longer time. Ask the child if they stopped at the same time. The car going for a longer time traverses a greater distance.

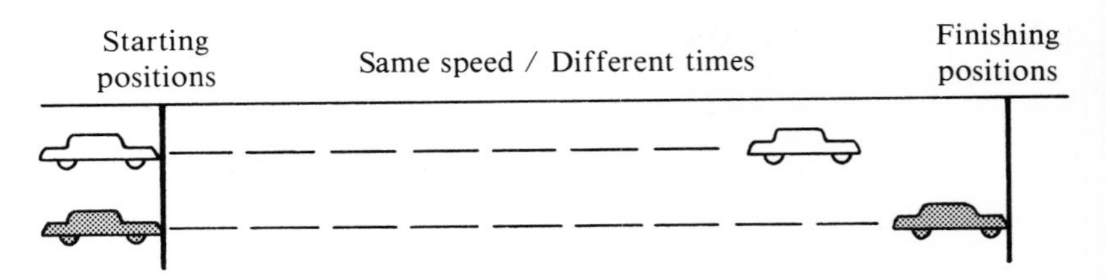

Procedure C: Tell the child that the cars will be started at the same time. Start the cars at the same time and position. Make one of them go faster but stop them at the same time. Ask the child if they stopped at the same time. The car going faster will traverse a longer distance.

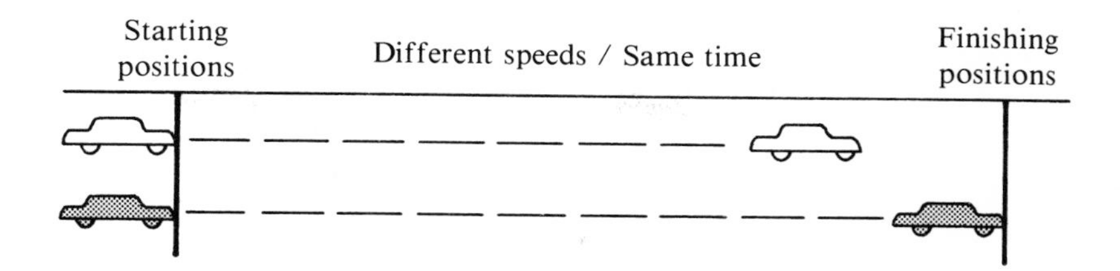

Additional considerations: Setting a clock face and reading (identifying) time on a clock face are skills, not concepts. In order for children to have a conceptual understanding on how the hands on a clock face work, they need to recognize the time-speed-distance relation. The movements of the clock-face hands are related to the movements of the cars described in Procedure *C*. From a diagnostic standpoint, if children perform Procedure *C* incorrectly, they probably do not as yet have the necessary cognitive structures for dealing with concepts of time measurement.

The skill of reading or setting a clock face at a specified time is static in nature. The measurement of the passing of time is dynamic and depends upon the time-speed-distance relations. It is for this reason that activities directing children to show the time at which they get up in the morning, go to school, eat lunch, and come home have virtually nothing to do with time measurement. These activities merely involve copying skills. The child copies a specified picture, namely, what another clock looks like at a particular moment. This has nothing to do with the idea that the hour hand and the minute hand traverse, or measure, different distances in a given span of time because they are moving at different speeds. Since many children in kindergarten to second grade have difficulty in copying from a vertical plane to a horizontal plane, these traditional first activities in time measurement appear even more inappropriate.

SCIENTIFIC NOTATION

Scientific notation allows large numbers to be expressed as a product of a mixed decimal and a power of 10. The integral part of the mixed decimal is a number from 1 to 9. Using scientific notation involves the changing of the name of the number, not its value. Study the following examples to notice the renaming pattern.

$$923345 = 9.23345 \times 10^5 \approx 9.2 \times 10^5$$
$$38611 = 3.8611 \times 10^4 \approx 3.9 \times 10^4$$
$$1806 = 1.806 \times 10^3 \approx 1.8 \times 10^3$$
$$314 = 3.14 \times 10^2 \approx 3.1 \times 10^2$$
$$29 = 2.9 \times 10^1$$

Notice that the whole number may be written as a decimal (314.0). The decimal point divides the integer by a power of 10 for each place moved to the left.

$$314 \div 10^1 = 31.4$$
$$314 \div 10^2 = 3.14$$

The number 3.14 has one digit in the units place that is between 1 and 10. This is the scientific notation form of the original number. In order to compensate for the division by 10^2, the number must also be multiplied by 10^2 so as not to change its value:

$$(314 \div 10^2) \times 10^2 = 3.14 \times 10^2$$

Notice that the movement of the decimal point in a right to left direction parallels the direction of increasing place values:

$$314.0 = 3.14 \times 10^2$$

In fact, the number of places the point is moved to the left (2) is the same as the exponent of the power of 10.

A similar notation is used for decimals, but notice the inverse direction of movement of the point.

$$.0000038 = 3.8 \times 10^{-6}$$
$$.00072 = 7.2 \times 10^{-4}$$
$$.001 = 1.0 \times 10^{-3}$$
$$.01 = 1.0 \times 10^{-2}$$
$$.314 = 3.14 \times 10^{-1}$$

In this last example, the number is multiplied and then divided:

$$(.314 \times 10^1) \div 10^1 = 3.14 \div 10^1 = \frac{3.14}{10^1}$$
$$= 3.14 \times \frac{1}{10} = 3.14 \times 10^{-1}$$

Notice that $1/10 = 10^{-1}$ in our notational system.

SIGNIFICANT DIGITS

Scientific notation is based on the idea that zeros in a numeral whose only purpose is to determine the place value of nonzero digits are *nonsignificant.* For example, the zeros are nonsignificant digits in .001 but significant digits in .100.

Significant digits may be determined as follows:

1. Any nonzero digit is significant (1, 2, 3, 4, 5, 6, 7, 8, 9).
2. Zeros used for determining the position of the decimal point are nonsignificant (as in .001 or 0.02).
3. All other zeros are significant (25.0, .390, 304).

Each of the following numbers have four significant digits:

7892   368.0   0.3681   .7500   0.0004896

Accuracy and preciseness

A numeral having a greater number of significant digits than another is said to be a more *accurate* measurement than the numeral having fewer significant digits. Thus, the measurement 239,000 miles (the approximate distance from the earth to the moon) is more accurate than 0.0000000079 inches (read "seventy-nine ten-billionths of an inch") (the estimated diameter of the average atom). Note, however, that the atom measurement is more *precise,* since inches are a more refined measurement scale than miles.

When comparing the accuracy of two numbers having the same number of significant digits, the one having the larger significant digit in the extreme left position is more accurate. The measurement 0.000387 inches is less accurate and more precise than 8.37 inches—less accurate because although both have the same number of significant digits, 8.37 has the larger significant digit in the extreme left position, and more precise because millionths is a more precise scale than hundredths. This relates to the point that although numbers such as 8 and 8.00 are equal in value, they are not the same fraction. They are equivalent fractions: $8/1 = 800/100$.

EXERCISES

I. In the figures below, square *ABCD* is congruent to square $A^1B^1C^1D^1$.

1. The measure of square region *OPQR* is $(< / > / =)$ the measure of square region *ABCD*. (Circle the correct sign.)

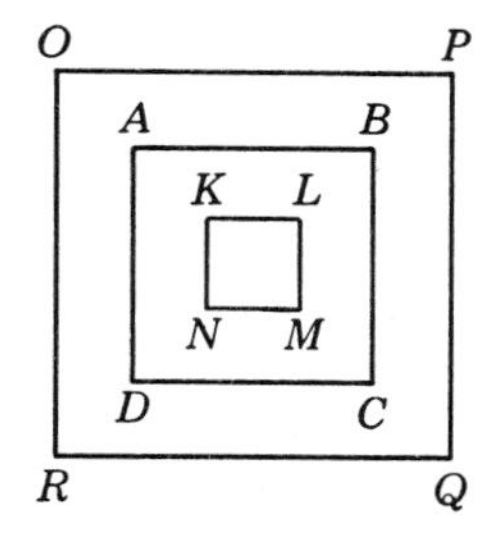

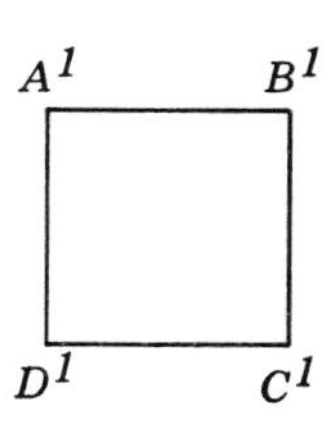

2. The measure of square region $KLMN$ is $(< / > / =)$ the measure of square region $ABCD$.
3. The measure of square region $KLMN$ is $(< / > / =)$ the measure of $OPQR$.
4. The measure of square region $ABCD$ is $(< / > / =)$ the measure of square region $A^1B^1C^1D^1$.

II. In the figure below, the measure of the region $KLMN$ is _____.

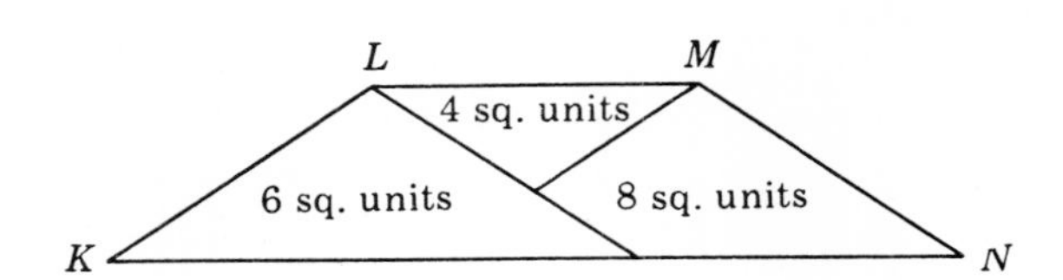

III. Match the ideas in Columns I and II.

Column I

1. volume of a cubic region
2. area of a trapezoid region
3. area of a square region
4. volume of a spherical region
5. circumference of a circle
6. area of a rectangular region
7. area of a circular region
8. area of a triangular region
9. perimeter of a square
10. perimeter of a rectangle
11. perimeter of a triangle

Column II

a. s^2
b. $\pi r^2 h$
c. $\frac{1}{3}\pi r^2 h$
d. $2h + 2w$
e. $\frac{4}{3}\pi r^3$
f. $2\pi r$
g. $4s$
h. e^3
i. bh
j. πr^2
k. $\frac{1}{2} bh$
l. $\frac{1}{2}h(b_1 + b_2)$
m. $a + b + c$

IV. Write the next three numbers for each of the following sets:

1. $\{3, 5, 7, 9, 11, \ldots\}$
2. $\{4, 12, 24, 40, 60, \ldots\}$
3. $\{5, 13, 25, 41, 61, \ldots\}$

V. 1. In number IV above, find the squares of the first elements in sets 1, 2, and 3. What pattern do you notice?

2. Find the squares of the second elements and third elements respectively. Does the pattern still hold?

VI. The three square regions below having sides 3, 4, and 5 respectively are placed so they form a right triangle. What is the area of the resulting triangular region?

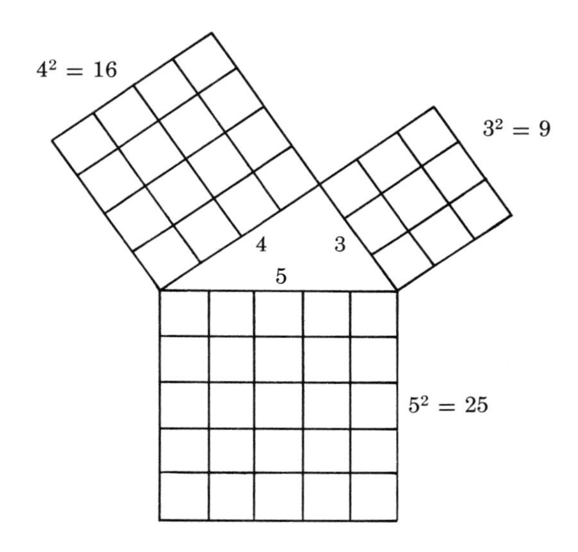

VII. Express the perimeter of a square having side x cm long.

VIII. In the illustration below, lines AB and CD intersect at 0, and $\angle BOD = 65°$. Find the measure of the following angles:

1. x
2. y
3. z

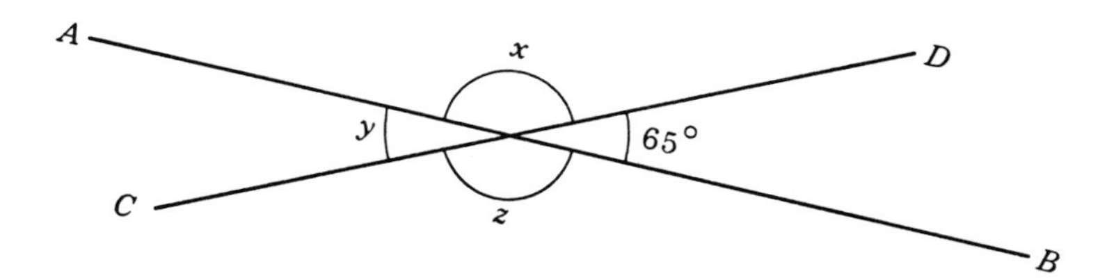

IX. 1. Make a cube by joining six congruent squares together. It is easier when some of the squares are joined as shown in a below. A shape such as a will fold to make a cube and is called a net of the cube. If each square has a side of 2 cm, what is the measure of the cube's volume?

2. Copy figures b, c, and d, cut them out, and form the resulting figures.

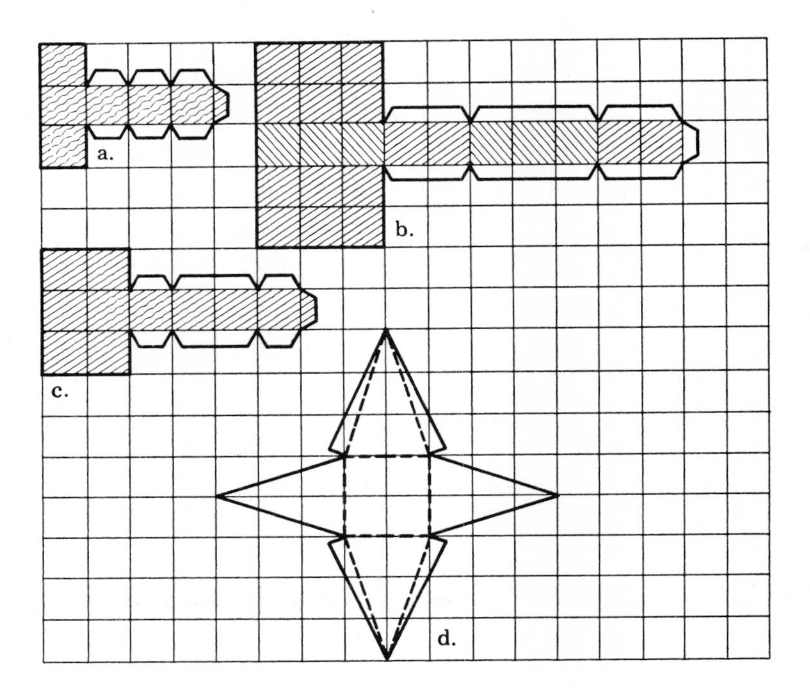

X. Find the volume of a cube having a side of:

1. 6 cm
2. 7 cm
3. 8 cm
4. 10 cm

XI. Complete the table below:

Length of side in cm	1	2	3	4	5	6	7	8	9	10
Volume of cube in cm^3	1	8								

XII. Tell the difference between the following pair of measurements:

$$3 \text{ inches and } 3\frac{0}{8} \text{ inches}$$

XIII. Tell the unit of precision in each of the measures below:

1. 6 ft. 2 in.
2. 25 min. 13 sec.
3. 4 yd. 1 ft. 10.0 in.

SUGGESTED READINGS

Bachrach, Beatrice. "Do Your First Graders Measure Up?" *The Arithmetic Teacher* 16 (November 1969): 537–38.

Dienes, Z. P., and Golding, E. W. *Exploration of Space and Practical Measurement.* New York: Herder & Herder, 1966.

Hogben, Lancelot. *The Wonderful World of Mathematics.* Garden City, N. Y.: Doubleday, 1968.

Hopkins, R. A. *The International (SI) Metric System and How It Works.* Tarzana, Calif.: Polymetric Services, 1973. Revised edition, 1974.

Hopkins, R. A., ed. *Metric in a Nutshell.* Tarzana, Calif.: American Metric Publishing Co., 1977.

Krause, Eugene F. "Elementary School Metric Geometry." *The Arithmetic Teacher* 15 (December 1968): 673–82.

Parker, Helen C. "Teaching Measurement in a Meaningful Way." *The Arithmetic Teacher* 7 (April 1960): 194–98.

Ploutz, Paul F. *The Metric System, A Programmed Approach.* Columbus, Ohio: Charles E. Merrill Co., 1972.

Sinclair, Hermine. "Number and Measurement." *Piagetian Cognitive-Develop-Research and Mathematical Education,* edited by Myron F. Rosskoph, Leslie P. Steffe, and Stanley Tabak. Washington, D.C.: National Council of Teachers of Mathematics, 1971.

Steffe, Leslie. "Thinking About Measurement." *The Arithmetic Teacher* 18 (May 1971): 332–38.

Mathematical Applications

The National Assessment of Educational Progress report completed in 1979 refers to mathematical application as follows:

> The use of the three lower cognitive levels—knowledge, skill and understanding—to solve problems. Students must use reasoning and judgment to select the appropriate procedures based on an understanding of the problem situation. They must also recall particular facts and definitions and have the skill to carry out appropriate algorithms.[1]

Application problems include consumer problems involving money and percent; using graphs and tables, probability and statistics; constructing and interpreting time schedules; and solving everyday problems by counting, measuring, estimating, judging, and inferring.

In addition to objectives in the strand entitled *Mathematical Applications* in the Appendix, objectives listed under the *Graphs and Functions* strand may be considered as mathematical applications.

---

1. National Assessment of Educational Progress, *Mathematical Applications* (Denver, Colo.: NAEP, August 1979), p. xiii.

PERCENT

Decimal fractions were described as fractions whose denominators are powers of 10. *Percents* are decimals whose denominators are hundredths. All concepts involving the decimal numeration system, principles for using fractions, and computations with decimals apply to percents.

Fractions written in hundredths are sometimes referred to as percents. For example, 30 percent means 30/100 and is often written as "30%"; 28/100 may be written as "28%" and is read "twenty-eight percent."

Mini-lesson 6.3.8 Find Percent, Rate, and Base

Vocabulary: percent, rate, %, base, percentage

Possible Trouble Spots: does not see relationship between fractions and percent; does not see relationship between decimals and percent; has difficulty with objectives in the multiplication strand; has difficulty with solving indirect open sentences, for example, "What percent of 60 is 75?"

Requisite Objectives:
Rationals/Notation 1.3.8: Identify the fractional part of a set

Rationals/Notation 4.1.3: Extend expanded notation to include fractions through thousandths

Decimals/Notation 4.2.2: Extend place value to thousandths

Decimals/Notation 4.3.2: Translate fractions to decimals

Mathematical Applications 6.3.7: Relate decimals to percent

Activity 1
To relate the concept of percent to fractions, have the children participate in the following activities.

A. Shade in 25% of a square region.

Rename 25% as .25 and .25 as 1/4.

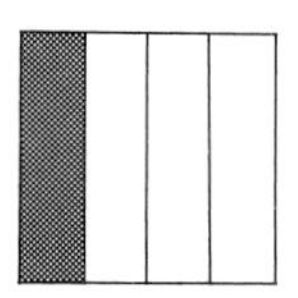

B. Shade 33% of a group. Rename $33\frac{1}{3}\%$ as $.3\overline{3}$ and $.3\overline{3}$ as 1/3. Select from groups of 3, 4, and 5 circles the group that is most appropriate to the problem.

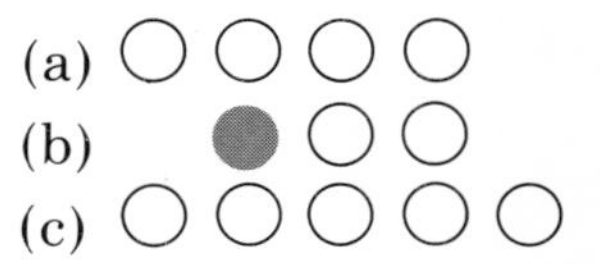

Activity 2
To compute with percents, make a percent calculator using graph paper.

A. Find 60% (the rate) of 50 (the base). Draw a line from zero on the percent axis to the base number on the base axis. The percentage is the number shown below the point at which the percent meets the diagonal line (30). Percentage is the product of the rate and the base.

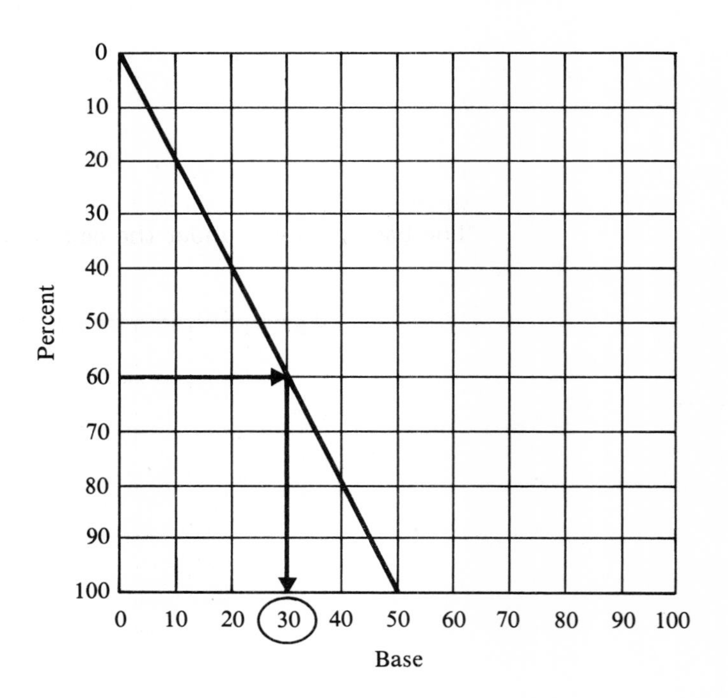

B. Find what percent 60 is of 75. Draw on the percent calculator a line from 75 on the base axis to zero on the percent axis. The percent is to the left of the point at which 60 meets the diagonal line (80%).

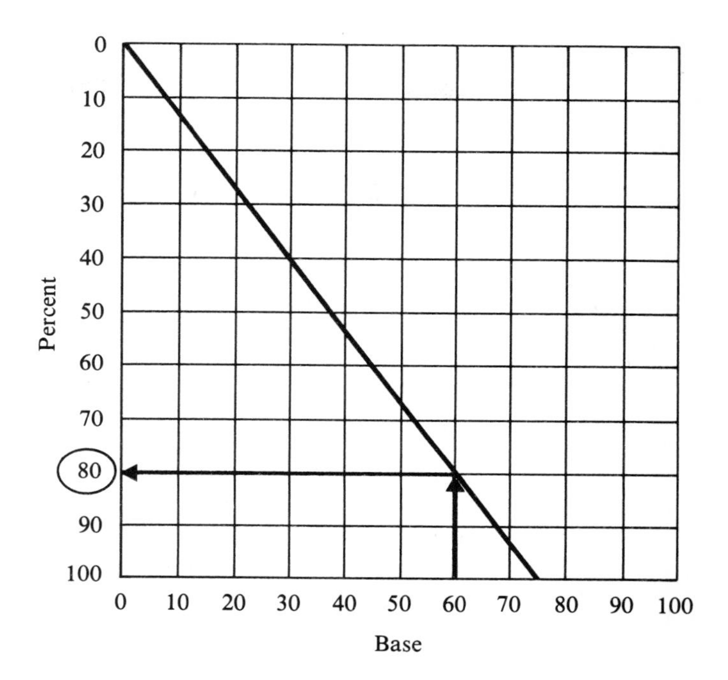

C. Find the base when the percent is 40 and the percentage is 10. Draw on the percent calculator a line from zero so that it passes through the point of intersection of the percent and of the percentage. The base is shown under the endpoint of the diagonal line (25).

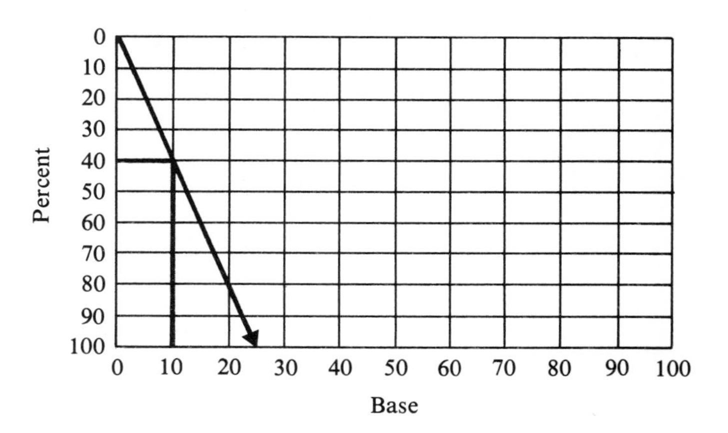

Additional considerations: In activities such as 1B, it is important to avoid cueing the students by using items that mask diagnostic data.

For example, many items that ask for identification of $\frac{1}{4}$ of a group look like this:

Fill in $\frac{1}{4}$ of the group.

○ ○ ○ ○

The students fill in as follows: ● ○ ○ ○. You have presented only one alternative, and the students may be attending to the *one* in one-fourth and may not realize that $\frac{1}{4}$ refers to one out of four parts. Then, when percents are introduced, students may have difficulty with items such as the following:

Fill in 25% of the group.

○ ○ ○ ○ ○ ○ ○ ○

Note: Mini-lesson 6.3.8 is a prerequisite for Mini-lessons 6.3.9 and 7.1.5.

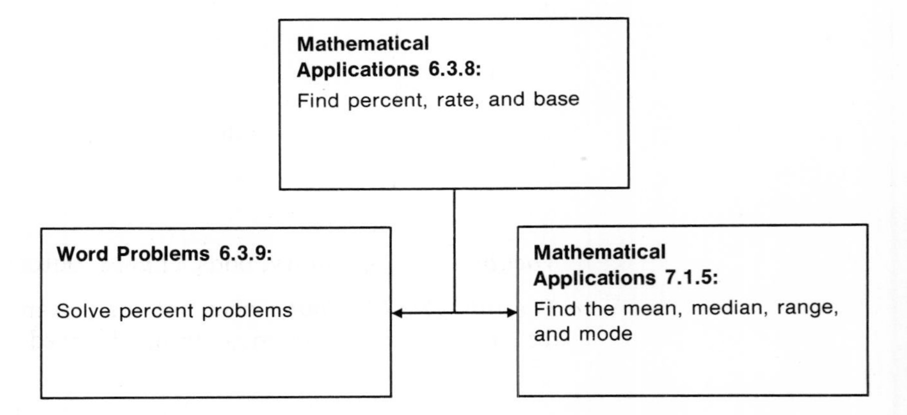

PROBABILITY

Probability deals with the idea that certain events are more likely to occur than others. Some important ideas underlying probability activities are the notions of the likelihood of events: certainty versus uncertainty as well as the predictability of happenings. When examining probabilities, the following questions should be considered.

Ideas underlying probability

1. *Are all outcomes equally certain?* When the outcomes are not equally likely, an estimated probability based upon relative frequency can be computed.

For example, suppose an insurance company finds that out of 8,000 New Yorkers born in 1919, 6,329 were still alive in 1975. This represents a ratio of 6,329/8,000 of New Yorkers born in 1919 who lived to age 56. This ratio may be expressed as .791 and is interpreted that the probability of a New Yorker born in 1919 living until age 56 is .791. Thus, when outcomes are not equally likely, probabilities that are estimates can be computed based on relative frequency of favorable outcomes compared to the total number of outcomes. As the number of events increases, the accuracy of the estimated probability becomes greater.

2. *If two or more events are involved, are they mutually exclusive?* For example, if a person draws only one card from a deck, he or she cannot draw a heart and a club since these two events cannot happen simultaneously. However, a heart and an ace can be picked at the same time (the ace of hearts is drawn).

3. *If two or more events are involved, are they independent?* Two or more events are considered independent if the occurrence of one does not affect the occurrence of the other. For example, consider the event of drawing a card from a deck and then drawing a second card from a deck. If the first card is replaced and the deck is shuffled prior to drawing the second card, the two drawings are independent.

When introducing children to probability ideas, no attempt should be made to formalize the concepts in the early years of instruction. At the primary levels, the children may discuss activities involving the notion of probability.

Mini-lesson 7.1.4 Determine Probability

Vocabulary: probability, odds, chance, outcome, ratio

Possible Trouble Spots: does not understand how to estimate; has difficulty making inferences from observed data; difficulty dealing with abstractions

Requisite Objectives:
Apply the ratio idea of fractions to probability

Distinguish between situations involving odds and probabilities

Mathematical Applications 4.1.18: Make inferences based upon judgments and estimates

Mathematical Applications 4.3.12: Identify outcomes from a given set of possible outcomes

Graphs and Functions 5.1.11: Graph probabilities

Activity 1
To explore probability ideas, tell the children the following story: "Sam placed five green beads and three yellow beads into a bag. Then he took a bead out of the bag without looking." At this point, ask the pupils what color bead they think he pulled out of the bag. Tell the

pupils that Sam found he had a yellow bead. Ask the children if they guessed correctly and then ask them to name how many beads of each color were left in the bag.

Continue the story by telling the children that Sam pulled out another bead. Ask the pupils to guess what color bead was pulled out. Tell them that Sam had picked a green bead. Ask the children if they guessed correctly.

Continue the story: "Sam's friend Kathy pulled a bead out of Sam's bag. She picked a yellow bead. She took another bead out. Sam and Kathy were surprised at the color of this last bead pulled out of the bag." Ask the children why Sam and Kathy were surprised to see the color of that bead. The pupils should respond that they were surprised because Kathy picked a yellow bead even though there were 4 green beads but only one yellow bead in the bag. A guess of taking a green bead at that point was more justified than a guess of yellow. However, it was neither impossible nor wrong to guess yellow at that time. (Encourage children to give their own reasons for saying which color of bead came next out of the bag.)

Complete the story: "Both Sam and Kathy were certain which color would come out next." Ask the children why they think Sam and Kathy were certain. (It is important to emphasize that only when the last yellow bead had been taken from the bag were the children able to be sure of the color of the next bead.)

Activity 2

To graph probability ideas, obtain a natural-colored wood cube (must be a true cube—otherwise, results will be biased) and color one of its faces black and three faces blue. Discuss with the children which color face will have the best chance of coming on top if the cube is rolled (like dice are rolled).

Tell the children to take turns rolling the cube 30 times and to record in a block graph the number of times each color comes up on top. When the graph is completed, discuss with the children the idea of frequency. (This activity involves graphing skills as well as probability concepts.)

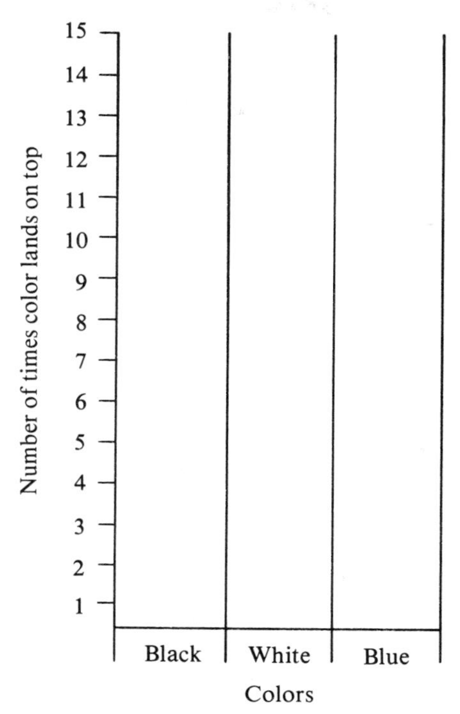

Ask the pupils to roll the cube another 30 times and to graph the results. Then ask them to add the results of the first 30 throws and the results of the second 30 throws and to make a graph showing the results for the 60 throws. Discuss with them which color now has the greatest frequency and the least frequency.

Ask the children what they think will happen if they roll the cube more times. Do they suppose the same color will still have the greatest frequency? Have the children verify their answer by throwing the cube additional times.

For additional activities and information on probability, the reader is directed to several sources.[2]

---

2. See Henry Van Engen and Douglas Grouws, "Relations, Number Sentences and Other Topics," in *Mathematics Learning in Early Childhood,* Thirty-seventh Yearbook of the National Council of Teachers of Mathematics, edited by Joseph N. Payne, Reston, Va.: National Council of Teachers of Mathematics, 1975), pp. 265–67; Rolland R. Smith, "Probability in the Elementary School," in *Enrichment Mathematics for the Grades,* Twenty-seventh Yearbook of the National Council of Teachers of Mathematics, edited by Julius H. Hlavaty, chairman; Albert A. Blank, Vincent J. Glennon, Joseph N. Payne, Richard S. Pieters, Harry D. Ruderman, and Henry W. Syer (Washington, D.C.: National Council of Teachers of Mathematics, 1963), pp. 127–33; David A. Page, "Probability," in *The Growth of Mathematical Ideas, Grades K-12,* Twenty-fourth Yearbook of the National Council of Teachers of Mathematics, edited by Phillip S. Jones, chairman; Harold P. Fawcett, Alice M. Hach, Charlotte W. Junge, Henry W. Syer, and Henry Van Engen (Washington, D.C.: National Council of Teachers of Mathematics, 1959), pp. 229–71.

Additional considerations: A fractional number may be used to state a probability. A probability cannot be less than zero or greater than 1. If the probability is zero, the event cannot happen. If the probability is 1, the event must happen. The probability of an occurrence is the ratio of the number of favorable outcomes to the number of possible outcomes. The odds of an occurrence is the ratio between the number of favorable outcomes and the number of unfavorable outcomes.

Note: Mini-lesson 7.1.4 is a prerequisite for Mini-lessons 7.1.6 and 7.2.1.

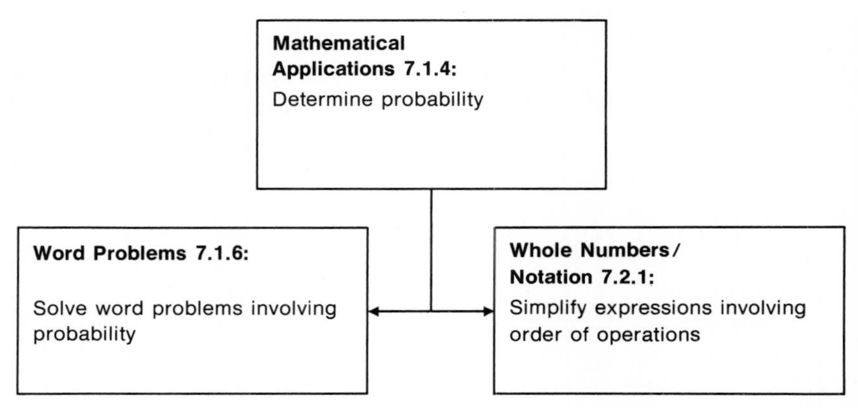

STATISTICS

John Graunt (1620–74) founded the field of statistics. His analysis of a seventeenth-century publication entitled *Bills of Mortality,* which listed births, christenings, and deaths, represented the first time social and biological phenomena were interpreted from mass data.

In probability, the chances of selecting a certain sample from a known population are determined; in *statistics,* estimates or projections about a whole population based upon a known sample are made. Probability involves working from the whole to a part; statistics deals with going from a part to the whole.

Descriptive and inferential statistics

Statistics are of two types: *descriptive* and *inferential. Descriptive statistics* summarize or describe numerical data. Young children participate in processes that underlie the concepts of descriptive statistics. Such processes, which are related to notions of probability, are "counting, collecting, measuring, classifying, comparing, recording, displaying, ordering, interpreting, and representing."[3] *Inferential statistics* estimate generalizations based on a sampling of a general population. Inferential statistics may be better suited to older children who are in Piaget's formal operational stage of cognitive development.

Graphing data

Various types of graphs are used to record numerical data. The teacher should emphasize to the children that the need for drawing a graph stems from a question that has been asked. For example: "What is the most popular girl's name in class?" "How many children in the class bring lunch?" "How many buses stop at the school's front door between seven and nine o'clock each morning?" "At what times do children brush their teeth at night?" To answer such questions, information referred to as *data* needs to be collected. This data may be shown by pictures or in table form. The completed graph picture or table indicates the answer to the original question. Subsequent questions may arise from information shown on the graphs. For example: "On what days do most children brush their teeth the latest?" A question such as this one may, in turn, give rise to additional questions, such as "How are these days different?" Perhaps the answer involves children staying up later on weekends, and a relation is seen that offers a hypothesis for the data.

Pictographs

Pictographs are a method of recording data that is within a young child's frame of reference. Pictographs and simple block graphs are also appropriate for use with the slow-learning child.

Frequency and mode

For older children working with graphs, the words *frequency* (the number of a subset) and *mode* (the most popular) should be introduced as statistical terms. Other terms appropriate for older children deal with three interpretations of the word *average:*

Interpretations of average

1. *Arithmetic mean:* the sum of all values divided by the number of values
2. *Median:* the middle value when data are arranged in order of value; the value at which half of the values are larger and half of the values are smaller
3. *Mode:* the value that occurs most frequently

Suggestions for graphs

The children may be encouraged to compare their graphs with graphs involving a greater number of children (a larger sample). For example, they may broaden a survey of names of girls in a classroom to those of girls in the entire school. The children may find that the mode for the entire school differs from the class mode.

Some other suggestions for surveys include (a) the number of boys and girls absent each day in a given week, (b) favorite television shows seen in a week, (c) types of protein eaten in a day (may be related to

3. Van Engen and Grouws, "Relations, Number Sentences and Other Topics," p. 267.

a science lesson), and (d) favorite school subjects. An analysis of how to represent the various subjects on the graph could be discussed.

When the number of elements of a particular set is large, scale representations, involving a many-to-one idea, may be used.

Mini-lesson 4.2.9 Extend Picture and Bar Graphs

Vocabulary: line, graph, pictograph, bar graph, circle graph

Possible Trouble Spots: Visual perception problems; does not understand relationship between representation with a picture or bar graph and the real world situation

Requisite Objectives:
Measurement 1.1.9: Compare sizes of objects

Whole Numbers/Notation 1.3.5: Show many-to-one and one-to-many representations of the same number

Graphs and Functions 2.2.12: Collect and record data using picture and bar graphs

Activity 1
To use pictographs, draw a pictograph showing how many cars of different sizes use a particular gas pump during a given time duration. Younger children could use toy cars to represent each vehicle counted. Point out that in the case of none of a particular type of car using the pump, the graph should contain an empty row. Suggest that the time duration for the observations be restricted in order to avoid unnecessarily high numbers of pictures in the graph.

GRAPH OF COMPACT AND REGULAR-SIZED CARS
USING GAS PUMP 8:00 TO 8:15 A.M.

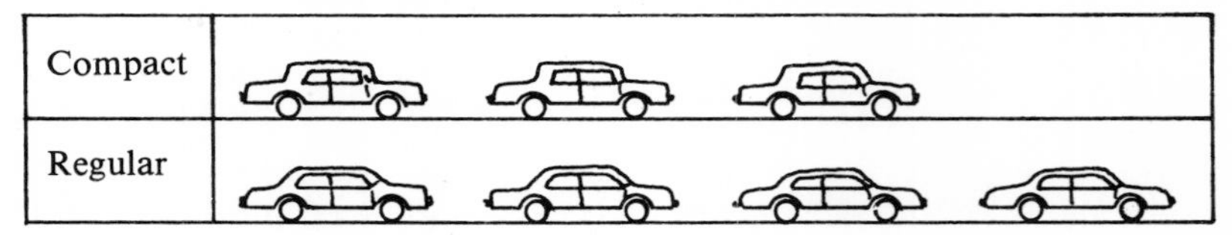

Note: Since the drawings in a pictograph (like those in Activity 1) are symbols of the objects counted, they should all be the same size; only the appearances should differ.

Activity 2
To use block graphs, construct a block graph showing the frequency of particular letters of first names of the pupils in a class.

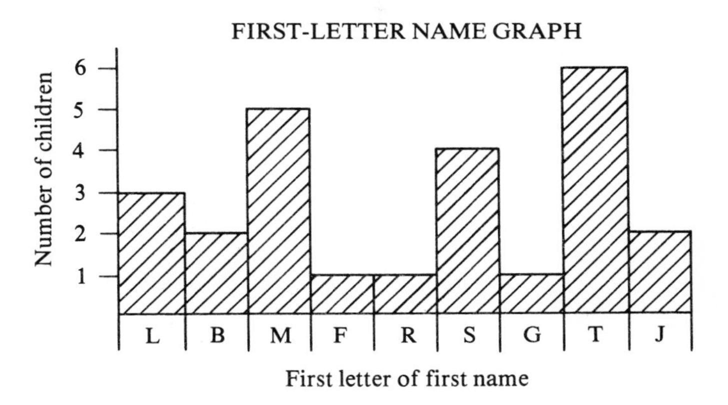

Additional considerations: The NAEP reports on mathematical applications stated that performance on items involving a circle graph "increased from age 9 to 13 to 17."[4] These problems required an awareness that fractional parts of a circle graph represented the equivalent fractional part of the total population involved. Students were found to have "difficulty in projecting to the future using past trends" shown on a graph.

Note: Mini-lesson 4.2.9 is a prerequisite for 4.2.10 and 4.2.11.

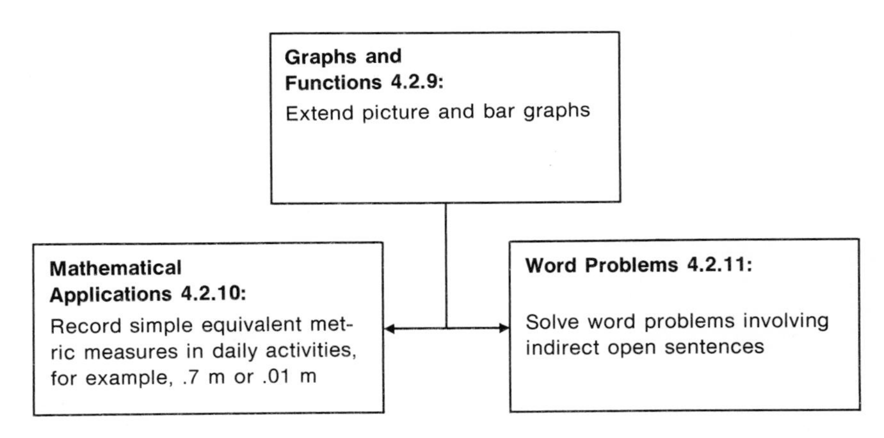

4. NAEP, *Mathematical Applications,* p. 21.

MATRIX

A *matrix* affords an orderly approach to sorting and classifying information. The idea of matrices can be introduced at the concrete level by having the child arrange objects by categories.

Mini-lesson 5.1.12 Form a Matrix Grid

Vocabulary: matrix, ordered pair

Possible Trouble Spots: directionality problems; difficulty with orientation in space; difficulty constructing concepts and, therefore, has trouble with classification tasks

Requisite Objectives:
Graphs and Functions 4.1.17: Interpret and use a grid
Complete addition grid

Activity 1
To form a matrix, set out on a table 3 pencils, 3 crayons, 3 pens, and 3 pieces of construction paper. Each of these classes of materials should be in black, red, blue. Use the labels "Colors" and "Materials"; "Black," "Red," and "Blue"; "Pencils," "Crayons," "Pens," and "Paper" to show rows and columns of the matrix. Place the black pencil in the top left cell. Place the remaining objects in the appropriate cells.

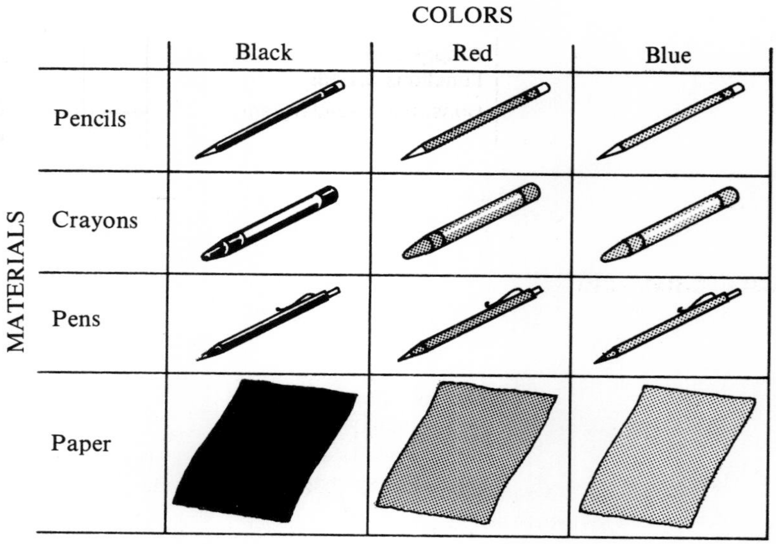

COLORS

MATERIALS	Black	Red	Blue
Pencils			
Crayons			
Pens			
Paper			

Activity 2
To form a matrix grid, place the set of numbers $\{2, 5, 12, 15, 22, 25\}$ on a grid to form a matrix as shown.

	Odd Numbers	Even Numbers
Less than 10	5	2
From 11 to 20	15	12
From 21 to 30	25	22

Additional considerations: The ability to interpret matrices (plural of matrix) is helpful in solving cross-product multiplication problems. The cross-product model for multiplication is discussed in Chapter 6.

Note: Mini-lesson 5.1.12 is a prerequisite for Mini-lessons 5.1.13 and 4.2.4.

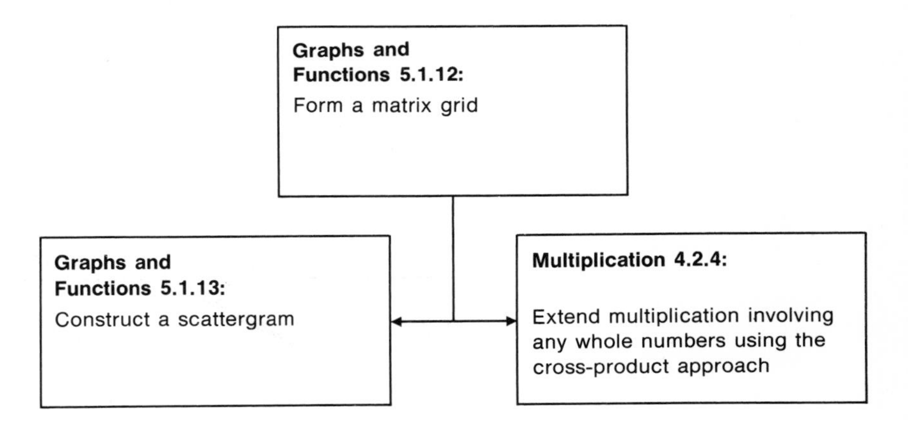

SCATTERGRAMS

The frequency of information shown on a matrix may be displayed on a graph called a *scattergram.*

Mini-lesson 5.1.13 Construct a Scattergram

Vocabulary: scattergram, matrix, graph

Possible Trouble Spots: difficulty inferring from observations; difficulty with spatial perception

Requisite Objective:
Graphs and Functions 5.1.12: Form a matrix grid

Activity 1
To construct a scattergram, have children participate in the following activities.

A. Compare scattergrams for odd and even numbers, given the set of whole numbers greater than 2 and less than 30 $\{3, \ldots, 29\}$. The scattergram shows, for example, that there are four odd numbers from 0 to 9 $\{3, 5, 7, 9\}$ and three even numbers $\{4, 6, 8\}$.

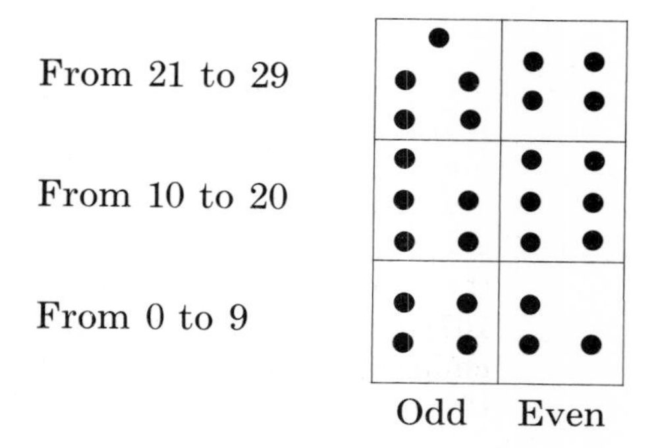

B. Ask a child how many children there are in his family including himself. Ask him to tell how many are boys and how many are girls. If, for example, the child responds that there are 5 children in his family—3 boys and 2 girls—ask him to place a dot in the appropriate square on a graph. The dot may be placed anywhere in the square. Ask each child in the classroom to place a dot reflecting the number of children in his or her family. Do not ask two pupils in the same family to record information.

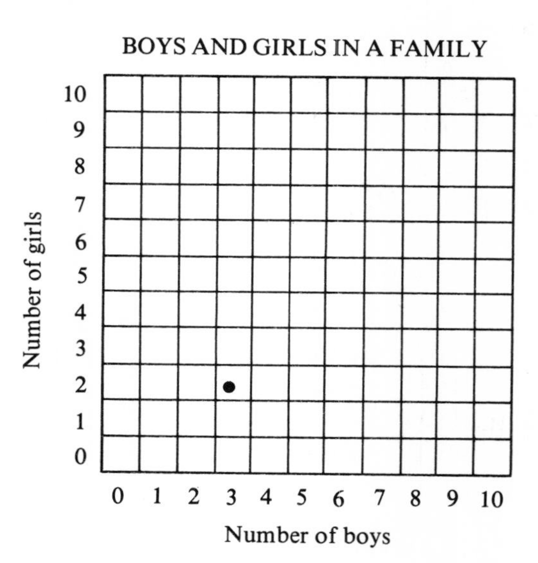

Count the number of dots in each square on the scattergram. Record the number for each square in the corresponding square on a duplicate grid. The information recorded in this manner is in a matrix form.

Ask the children the following questions about the data:

1. How many families of 3 children are there?
2. How many families of 2 children?
3. What is the most common-sized family?

C. Discuss with children such generalizations as the tallest people wear the largest shoes or the larger the car engine the smaller the number of miles per gallon (or kilometers per liter).

Have the class collect data from either 100 men or 100 women including fourth graders on up. They should ask each person's height (height categories can be listed in groups of 3 cm, for example, 164–167) and shoe size and then enter the information on a graph. Tell the children that half sizes should be rounded as the next higher shoe size (for example, size 7 1/2 is to be recorded as size 8).

Activity 2
To interpret data from scattergrams, present the three scattergrams shown and ask the children to match the patterns with the following labels:

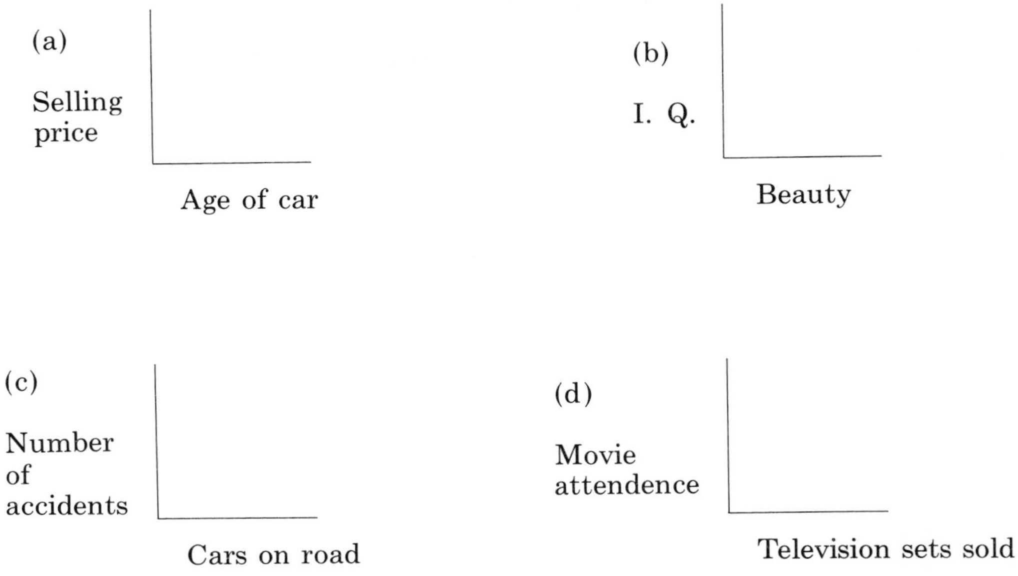

(a)

Selling price

Age of car

(b)

I. Q.

Beauty

(c)

Number of accidents

Cars on road

(d)

Movie attendence

Television sets sold

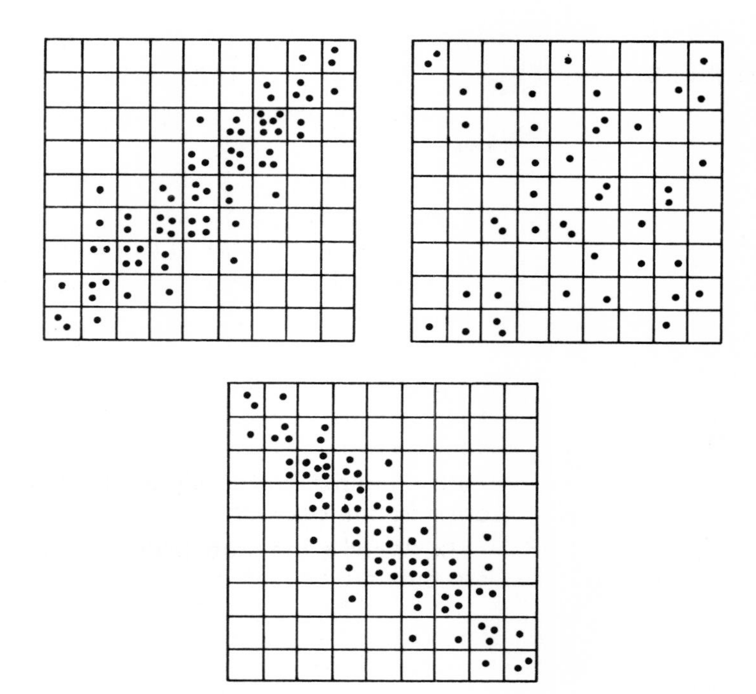

Additional considerations: Scattergrams involve ability to estimate relationships that are portrayed spatially. This type of graph is less precise than bar graphs or pictographs and may be more difficult to interpret because of the lack of specific number values. However, a cursory look at data such as are available in a scattergram can illustrate trends without involving unnecessary calculation.

Note: Mini-lesson 5.1.13 is a prerequisite for Mini-lessons 5.2.3 and 5.3.14.

Graphs and Functions 5.1.13:
Construct a scattergram

Multiplication 5.2.3:
Use factor trees to find prime factorization of a composite number

Graphs and Functions 5.3.14:
Interpret line graphs

EXERCISES

I. Find the probability (P) of getting heads when flipping a dime.

II. Find the probability of drawing an ace from a shuffled deck of cards.

III. Find the probability of rolling a die and obtaining an even number.

IV. Find the probability of rolling a die and obtaining a 14.

V. Find the probability of rolling a die and obtaining a number less than 14.

VI. Find the odds against drawing a club from a deck of cards.

VII. A candy manufacturer finds that of 25,000 candy wrappers, 1,300 came off in packaging. What is the probability that a piece of candy selected at random will be unwrapped?

VIII. Find the probability of drawing a club or an ace on a single draw from a deck of cards.

IX. What is the probability of drawing two jacks when two cards are drawn from a deck and

1. the first card is replaced and the deck is shuffled before the second draw?
2. the first card is not replaced?

X. Find the probability of drawing a heart from a deck of cards and then rolling an even number on a pair of dice.

XI. When throwing a die, what is the probability of

1. scoring 6?
2. not scoring 6?

XII.
1. How many scores above 3 are there on a die?
2. What is the probability of throwing a number greater than 3?
3. What is the probability of throwing a number less than 3?

XIII. Ten cards marked 0, 1, 2, 3, 4, 5, 6, 7, 8, 9 are placed face down. Tell the probability of selecting a card marked

1. 6
2. 9
3. 0
4. $n > 7$
5. $n < 5$
6. $3 < n < 7$

XIV. A box contains 4 socks, 1 red, 1 blue, 1 yellow, and 1 green. Find the probability of selecting one that is

1. blue
2. not blue

XV. In a box are 3 black socks and 7 white socks. Find the probability of selecting a sock that is

1. black
2. white

XVI. 40 percent of 80 is what number?

XVII. What percent of 40 is 5?

XVIII. 60 percent of what number is 120?

XIX. A lady was 60 pounds overweight. She lost 40 percent of this excess weight. What percent did she lose?

SUGGESTED READINGS

Fey, J. T. "Probability, Integers, and Pi." *The Mathematics Teacher* 64 (April 1971): 329–32.

National Assessment of Educational Progress. *Mathematical Applications.* Denver, Colo.: National Assessment of Educational Progress, August 1979.

Rudd, D. "A Problem in Probability." *The Mathematics Teacher* 67 (February 1974): 180–81.

Schmalz, R. "The Teaching of Percent." *The Mathematics Teacher* 70 (April 1977): 340–43.

Sharron, Sidney, and Reys, Robert E. *Applications in School Mathematics.* Reston, Va.: National Council of Teachers of Mathematics, 1980.

Starr, N. "A Paradox in Probability Theory." *The Mathematics Teacher* 66 (February 1973): 166–68.

Solving Word Problems

Riedesel and Burns summarized specific techniques reported in a number of studies for improving problem solving:[1]

1. Using drawings and diagrams
2. Following and discussing a model problem
3. Having pupils write their own problems and solve each others' problems
4. Using problems without numbers
5. Using orally presented problems
6. Emphasizing vocabulary
7. Writing mathematical sentences
8. Using problems of proper difficulty level
9. Helping pupils correct problems
10. Praising pupil progress
11. Sequencing problem sets from easy to hard

1. C. A. Riedesel and P. C. Burns, "Research on the Teaching of Elementary-School Mathematics." In *Second Handbook of Research on Teaching,* edited by Robert M. W. Travers (Chicago: Rand McNally, 1973), pp. 1160–61.

They also reported other studies that found the following:

1. Students made higher scores when the numbers were in a sequence necessary to solve the problem.[2]
Example: "If I have 5 apples and give 2 away, how many do I still have?" $5 - 2 = 3$

2. Problems with a single classification were easier for first-graders.[3]
Example: "I have 3 *cookies* and 2 *cookies*. How many *cookies* do I have?" rather than "I have 3 *kittens* and 2 *goldfish*. How many *pets* do I have?"

3. Problems with the question asked first were easier.[4]
Example: "How many cookies do I have? I have 3 cookies and 2 cookies."

4. Direct teaching of reading skills and vocabulary is directly related to improving problem-solving achievement.[5]

METACOGNITION APPLIED TO SOLVING WORD PROBLEMS

In the future, research on solving word problems will attempt to expand the work on metacognition beyond memory, attention, communication, and comprehension to include any situation that calls for problem solving. Brown suggests the following:

> Checking the results of an operation against certain criteria of effectiveness, economy and common sense reality is a metacognitive skill applicable whether the task under consideration is solving a math problem, memorizing a prose passage, following a recipe, or assembling an automobile or piece of furniture. Self-interrogation concerning the current state of one's knowledge during problem solving is an essential skill in a wide variety of situations, those of the laboratory, the school, or everyday life.[6]

---

2. P. C. Burns and J. L. Yonally, "Does the Order of Presentation of Numerical Data in Multi-Step Arithmetic Problems Affect Their Difficulty?" *School Science and Mathematics* 64 (1964): 267–70.

3. L. R. Steffe, "The Effects of Two Variables on the Problem-Solving Abilities of First Grade Children," *Teaching Report #21* (Madison, Wis.: University of Wisconsin, 1967).

4. M. H. Williams and R. W. McCreight, "Shall We Move the Question?" *The Arithmetic Teacher* 12 (1965): 418–21.

5. L. F. Vanderlinde, "Does the Study of Quantitative Vocabulary Improve Problem-Solving?" *Elementary School Journal* 65 (1964): 143–52; J. P. Treacy, "The Relationship of Reading Skills to the Ability To Solve Arithmetic Problems," *Journal of Educational Research* 38 (1944): 89-96.

6. A. L. Brown, "Knowing When, Where, and How To Remember: A Problem of Metacognition." In *Advances in Instructional Psychology,* edited by R. Glaser. (Hillsdale, N. J.: Lawrence Erlbaum Associates, in press).

As discussed in Chapter 1, *metacognition* (also called "cognitive monitoring") *is thinking about your thinking.* Metacognition is rooted in Mead's definition of reflective thought: "*Reflective thought is the ability to dissect* (one's) *action in terms of the foreseen consequences of alternative courses of action*" (italics mine).[7]

Brainerd discussed the related issue of self-awareness as follows:

> The child must discover for himself the inconsistencies between his beliefs and the results of his behavior. Whether or not he makes use of such information is, of course, quite another matter.[8]

Flavell suggested that we should not make sharp distinctions among cognition, memory, communication, and comprehension processes. His discussion that follows serves as a basis for applying metacognition to solving word problems.

> Psychologists tend to think that there is one cognitive phenomenon called "memory" and another, wholly different and unrelated one, called "communication." . . . The reality is that there are deep commonalities . . . and explicating these commonalities may give us a more integrated view of cognitive development.[9]

A first step for seeing these commonalities, as suggested by Flavell, is to examine the basic meaning of the concept of *audience.* "The function of an audience is to receive and interpret information."[10]

In addition to being a listener, one may also be an audience (receive information) by reading. This may be referred to as the *message.* A message could be a word problem. It appears that difficulty in solving word problems may be developmental; that is, aside from reading skills such as word recognition and inflectional endings, reasons for poor problem solving may be different for younger students than for those in the middle grades. For example, younger students were found to do the following:

> Younger students will indicate comprehension of a message even when the message is ambiguous or lacks critical information.[11]

---

7. G. H. Mead, *Mind, Self and Society* (Chicago: University of Chicago Press, 1934), p. xxvi.

8. C. J. Brainerd, "Learning Research and Piagetian Theory." In *Alternatives to Piaget,* edited by Linda S. Siegel and Charles J. Brainerd (New York: Academic Press, 1978), p. 73.

9. J. H. Flavell, "Metacognitive Development." In *Structural/Process Models of Complex Human Behavior,* edited by Joseph M. Sandura and Charles J. Brainerd (The Netherlands: Sijthoff & Noordhoff, 1978), pp. 232–33.

10. Flavell, "Metacognitive Development," p. 233.

11. J. D. Karabenick and S. A. Miller, "The Effects of Age, Sex, and Listener Feedback on Grade-School Children's Referential Communication." In *Child Development,* in press.

Younger students are less likely than older to probe for more information.[12]

Gurova studied fifth and sixth grader's ability to solve a problem and the student's ability to analyze his or her mental processes in solving it. He found that awareness of one's own mental operations was a necessary factor in solving relatively complex problems requiring logical reasoning.[13]

Students well into middle grades education have trouble distinguishing salient story elements.[14]

Pilot data described by Flavell suggest that 6- and 8-year-olds show detection of comprehension failure by looking puzzled, slowly repeating problematic directions to themselves, questioning, and replaying taped directions *which would be analagous to rereading trouble spots in word problems* (italics mine).[15]

Baron[16] and Brown[17] proposed that metacognitive strategies monitor and direct other strategies. For example, a metacognitive strategy might involve categorizing situations according to the strategy needed. This may be applied to solving word problems when students identify a particular problem as one-step, two-step, and so forth; if they involve combining operations, separating operations, or a combination of the two; or if a problem written in the present tense, past tense, future tense, conditional (if . . . , then . . .), or mixed tenses is to be translated to an equation.[18]

ONE-STEP WORD PROBLEMS

This class of problems requires that students use only one of the basic operations (add, subtract, multiply, divide). The NAEP tested

12. K. B. Alvy, "Relationship of Age to Children's Egocentric and Cooperative Communication," *Journal of Genetic Psychology* 112 (1968): 275–86; J. M. Cosgrove and C. J. Patterson, "Plans and the Development of Listener Skills." In *Developmental Psychology,* in press.

13. L. L. Gurova, "Schoolchildren's Awareness of Their Own Mental Operations in Solving Arithmetic Problems, Volume III, Problem Solving in Arithmetic and Algebra." In *Soviet Studies in the Psychology of Learning and Teaching Mathematics* (Chicago: University of Chicago Press, 1969), pp. 97–102.

14. A. L. Brown and S. S. Smiley, "Rating the Importance of Structural Units of Prose Passages: A Problem of Metacognitive Development," *Child Development* 48 (1977): 1–8; F. W. Danner, "Children's Understanding of Intersentence Organization in the Recall of Short Descriptive Passages," *Journal of Educational Psychology* 68 (1976): 174–83.

15. Flavell, "Metacognitive Development," p. 236.

16. J. Baron, "Intelligence and General Strategies." In *Strategies in Information Processing,* edited by G. Underwood (London: Academic Press, in press).

17. A. L. Brown, "The Role of Strategic Behavior in Retardate Memory." In *International Review of Research in Mental Retardation,* edited by N. R. Ellis (New York: Academic Press, 1974).

18. Fredricka K. Reisman and S. H. Kauffman, *Teaching Mathematics to Children with Special Needs* (Columbus, Ohio: Charles E. Merrill, 1980).

students' performance on computation and related word problems by using the same numbers in both a word problem and a computation problem as follows.[19]

Word Problem	*Related Computation Task*
Paul has 21 stamps in his collection.	21
He buys 54 more from a stamp dealer.	+54
How many does he have after he buys them?	

Eighty-two percent of the 9-year-olds correctly responded to the word problem, whereas 90% of this same group computed the vertical algorithm correctly. The report stated that the 13-year-olds did well on both the computation and word problems.

The report implied that the operation sign provided a cue in helping children to solve the computation task, whereas "9-year-olds may have had difficulty deciding which operation to use when solving the word problem."[20] This may be the case, but I believe the reason why more children were able to solve the computation task is that the word problem is a more complex setting. It would be of interest to vary conditions of a problem to ascertain empirically if indeed students could not decide on which operation to use or if the more complex nature of the task were the trouble spot. For example, use the same problem but *change the tense*. Observe if more or fewer students solve the problem correctly as the tense changes. Note that equations are read in the present tense.

Example (past tense):
Paul had 21 stamps in his collection. He bought 54 more from a stamp dealer. How many did he have after he bought them?

Example (future tense):
Paul is going to buy 21 stamps for his collection. He will then buy 54 more from a stamp dealer. How many will he have after he buys them?

Example (mixed):
If Paul has 21 stamps in his collection and he wants to buy 54 more from a stamp dealer, how many would he have after he bought them?

It was reported that students had more difficulty with problems containing extraneous numerical information (only 47 percent of 9-year-olds and 56 percent of 13-year-olds obtained the correct answer). The following item was reported: "One rabbit eats 2 pounds of food

19. National Assessment of Education Progress, *Mathematical Applications* (Denver, Colo.: National Assessment of Education Progress, August 1979), p. 3.

20. NAEP, *Mathematical Applications,* p. 3.

each week. There are 52 weeks in a year. How much food will 5 rabbits eat in one week?"[21]

Results for one-step word problems involving fractions showed that students did better when pictures of fractional quantities of liquids in two measuring cups to be added were presented.[22] Children found particular difficulty with percent problems.

MULTI-STEP WORD PROBLEMS

These problems involve more than one operation. In general, it was found that the more exposure and practice students had with multi-step problems, the better they performed.[23]

The following excerpt from Menchinskaya and Moro sums up a variety of ways that aid solving word problems:

1. Do not start calculating until you have carefully studied the condition of the problem as a whole:
   a) in reading through the whole problem, pay particular attention to the question;
   b) return to the problem's condition and select related data; it often helps to write out the condition briefly;
   c) if problems of a familiar sort can easily be distinguished in a given complex problem, solve them; then the problem is less complex and will be easier to solve.

2. In solving a hard problem, use different methods:
   a) first, try to imagine exactly what the problem is talking about. To do this, it is helpful to modify the problem: Replace large numbers with small ones, invent a similar problem from your own life or, the reverse, ask yourself what you know from the condition and what you need to find out in the problem, and try to reproduce its contents in mathematical language;
   b) a drawing or diagram made in class can help greatly in solving a difficult problem. The drawing or diagram should express the correlations between the given and the unknown (the question). When making a drawing or diagram, check yourself continually and construct the drawing from what is said in the text of the problem. Watch for mistakes in your drawing. If you find a mistake, correct it at once;
   c) once you start to solve, performing every operation with numbers, keep asking yourself what you have learned by this operation and

---

21. NAEP, *Mathematical Applications,* p. 3.

22. NAEP, *Mathematical Applications,* p. 7.

23. NAEP, *Mathematical Applications,* p. 13.

whether it was necessary to perform this operation in terms of the question asked in the problem.[24]

EXERCISES

I. Analyze word problems contained in a basal textbook series. Look for the following conditions across the grade levels:

1. Do the problems that occur early in each of the texts use present tense?
2. Are complete sentences or telescopic language used?
3. Are the problems categorized by arithmetic operation or is the student allowed to make that judgment?
4. Do the problems relate to student experiences or are they removed from reality?

II. Make the same comparisons as in number I across three series but at the same grade level.

SUGGESTED READINGS

Alvy, K. B. "Relationship of Age to Children's Egocentric and Cooperative Communication." *Journal of Genetic Psychology* 112 (1968): 275–86.

Baron, J. "Intelligence and General Strategies." In *Strategies in Information Processing,* edited by G. Underwood. London: Academic Press, in press.

Brainerd, C. J. "Learning Research and Piagetian Theory." In *Alternatives to Piaget,* edited by Linda S. Siegel and Charles J. Brainerd. New York: Academic Press, 1978.

Brown, A. L. "Knowing When, Where, and How To Remember: A Problem of Metacognition." In *Advances in Instructional Psychology,* edited by R. Glaser. Hillsdale, N. J.: Lawrence Erlbaum Associates, in press.

Brown, A. L. "The Role of Strategic Behavior in Retardate Memory." In *International Review of Research in Mental Retardation.* Vol. 7, edited by N. R. Ellis. New York: Academic Press, 1974.

24. N. A. Menchinskaya and M. I. Moro, "Teaching Problem Solving," Volume XIV. Edited by Joseph R. Hooten. In *Soviet Studies in the Psychology of Learning and Teaching Mathematics,* edited by Jeremy Kilpatrick, Izaak Wirszup, Edward G. Begle, and James W. Wilson (Stanford: School Mathematics Study Group, 1975), p. 105.

Burns, P. C., and Yonally, J. L. "Does the Order of Presentation of Numerical Data in Multi-Step Arithmetic Problems Affect Their Difficulty?" *School Science and Mathematics* 64, (1964): 267–70.

Cosgrove, J. M., and Patterson, C. J. "Plans and the Development of Listener Skills." In *Developmental Psychology,* in press.

Danner, F. W. "Children's Understanding of Intersentence Organization in the Recall of Short Descriptive Passages." *Journal of Educational Psycology* 68 (1976): 174–83.

Flavell, J. H. "Metacognitive Development." In *Structural/Process Models of Complex Human Behavior,* edited by Joseph M. Scandura and Charles J. Brainerd. The Netherlands: Sijthoff & Noordhoff, 1978. Pp. 213–45.

Karabenick, J. D., and Miller, S. A. "The Effects of Age, Sex, and Listener Feedback on Grade-School Children's Referential Communication." *Child Development,* in press.

Mead, G. H. *Mind, Self and Society.* Chicago: University of Chicago Press, 1934.

Menchinskaya, N. A., and Moro, M. I. "Teaching Problem Solving." Vol XIV. Edited by Joseph R. Hooten. In *Soviet Studies in the Psychology of Learning and Teaching Mathematics,* edited by Jeremy Kilpatrick, Izaak Wirszup, Edward G. Begle, and James W. Wilson. Stanford: School Mathematics Study Group, 1975. Pp. 89–143.

National Assessment of Education Progress. *Mathematical Applications.* Denver, Colo.: National Assessment of Education Progress, August 1979.

Riedesel, C. A., and Burns, P. C. "Research on the Teaching of Elementary-School Mathematics." In *Second Handbook of Research on Teaching,* edited by Robert M. W. Travers. Chicago: Rand McNally, 1973. Pp. 1149–76.

Scandura, J. M., and Brainerd, C. J. *Structural/Process Models of Complex Human Behavior.* The Netherlands: Sijthoff & Noordhoff, 1978.

Simon, H. A. *Models of Thought.* New Haven, Conn.: Yale University Press, 1979.

Steffe, L. R. "The Effects of Two Variables on the Problem-Solving Abilities of First Grade Children." *Teaching Report #21.* Madison, Wis.: Wisconsin Research and Development Center for Cognitive Learning. University of Wisconsin, 1967.

Treacy, J. P. "The Relationship of Reading Skills to the Ability To Solve Arithmetic Problems." *Journal of Educational Research* 38 (1944): 89–96.

Vanderlinde, L. F. "Does the Study of Quantitative Vocabulary Improve Problem-Solving?" *Elementary School Journal* 65 (1964): 143–52.

Williams, M. H., and McCreight, R. W. "Shall We Move the Question?" *The Arithmetic Teacher* 12 (1965): 418–21.

Appendix A/Answers to Exercises

CHAPTER 2

a. 1 **b.** 2 **c.** 6 **d.** 13 **e.** 2 **f.** 24 **g.** 21 **h.** 1 **i.** 6 **j.** 2
k. 1 **l.** 17 **m.** 24 **n.** 24 **o.** 19 **p.** 13 **q.** 29 **r.** 15 **s.** 15
t. 24 **u.** 13

CHAPTER 3

I. 1. Originated with the Hindus; brought to Western Europe by the Arabs.
2. Because it's based on base ten.
3. 300 times as great (300:1).
(*Hint:* In the numeral ③33, how many times as great is the value of the encircled 3 in relation to the value of the underlined 3? [300:3 = 100:1].)
4. Its value is divided by 10; it is one-tenth its original value.

II. 1. $1{,}435 = (1 \times 1000) + (4 \times 100) + (3 \times 10) + (5 \times 1)$
$= (1 \times 10^3) + (4 \times 10^2) + (3 \times 10^1) + (5 \times 10^0)$
2. $796 = (7 \times 100) + (9 \times 10) + (6 \times 1) = (7 \times 10^2) + (9 \times 10^1) + (6 \times 10^0)$
3. $1{,}000{,}431 = (1 \times 1{,}000{,}000) + (4 \times 100) + (3 \times 10) + (1 \times 1)$
4. $3{,}060 = (3 \times 1000) + (0 \times 100) + (6 \times 10) + (0 \times 1)$

III. 1. The count-on model represents a change in position after a count of 9; the exchange model represents the ten-for-ten number equivalence that is shown by bundling 10 objects to stand for the number 10, which is referred to as a ten-for-one exchange due to the exchange of 10 objects for 1 object of an equivalent value.

CHAPTER 4

I. 1. Additive identity
2. Associative Property for Addition
3. Closure for addition of whole numbers
4. Commutative Property for Addition

II. 1. False
2. False
3. True
4. False

CHAPTER 5

I. Both involve renaming of partial sums; both may be added from right to left or left to right.

II. 1. Expanded form
 2. Associative Property for Addition
 3. Commutative Property for Addition
 4. Associative Property for Addition
 5. Addition facts
 6. Table for addition of whole numbers

III. 1. Error in additive identity. The child thinks "$n + 0 = 0$."
 2. Lack of using Associative Property for Addition. The child is adding only two addends.
 3. The child subtracted units. He or she either ignored addition sign for units or does not know how to add with renaming.
 4. Renamed hundreds instead of tens, probably because of the zero in the top addend. Ask the child to solve the problem

$$\begin{array}{r} 326 \\ +\ 19 \\ \hline \end{array}$$

 to see if he or she does not understand that the renaming occurs immediately in the next place value, or if indeed the zero in the addend confused the child.

IV. Associative Property for Addition

CHAPTER 6

I. Generating place values; telling time on a clock face

II. Union of equivalent disjoint sets; cross product

III. Makes no distinction between number of groups and size of a group

IV. When two nonequivalent sets are involved

VI. : : : : 2 rows, 4 columns

VII. See answer for numeral VIII.

VIII. 1. Commutative Property for Multiplication
 2. Associative Property for Multiplication
 3. Closure for multiplication of whole numbers
 4. Closure for multiplication of even whole numbers
 5. Distributive Property for Multiplication over Addition
 6. Multiplicative identity

IX. Basic multiplication facts

X. $(4 \times 1) + (4 \times 3)$; $(4 \times 2) + (4 \times 2)$

CHAPTER 7

I. Renaming; Distributive Property of Multiplication over Addition; table for multiplication; table for addition

II. Renaming 12 as 3×4; Associative Property for Multiplication; closure for multiplication; closure for addition; Distributive Property for Multiplication over Addition

III. **1.** $3a + 3b$

2. $3ax + 3bx$

3. $4x^2 + 32x$

4. $64x + 24$

5.

$$\begin{array}{r} 8x + 3 \\ \underline{7x + 1} \\ 8x + 3 \\ \underline{56x^2 + 21x \phantom{{}+3}} \\ 56x^2 + 29x + 3 \end{array}$$

6.

$$\begin{array}{r} 2y + 4x \\ \underline{y + 3x} \leftarrow \text{(Commutative Property for Addition)} \\ 6xy + 12x^2 \\ \underline{2y^2 + 4xy \phantom{{}+12x^2}} \\ 2y^2 + 10xy + 12x^2 \end{array}$$

IV. Numbers c and e; possibly number d; no evidence given to judge numbers f, g, and h, which shows the importance of gathering background information on the child to better understand possible causes for trouble spots in order to build appropriate teaching activities; perhaps number 11 (there may be a language communication aspect that hinders understanding of the teacher's instruction).

V. 3, 4, 5, 6, 8, 9

VI. **1.** 7, 14, 21, 28, 35, 42, 49, 56, 63

2. 70, 140, 210, 280, 350, 420, 490, 560, 630

VII.

$$\begin{array}{r} 90 \times 7 = 630 \\ \underline{3 \times 7 = 21} \\ 93 \times 7 = 651 \end{array}$$

VIII. **1.** m can have the values 1 to 9.

2. n can have the values 0 to 9.

3. $100l + 10m + n$

IX. Commutative

X.

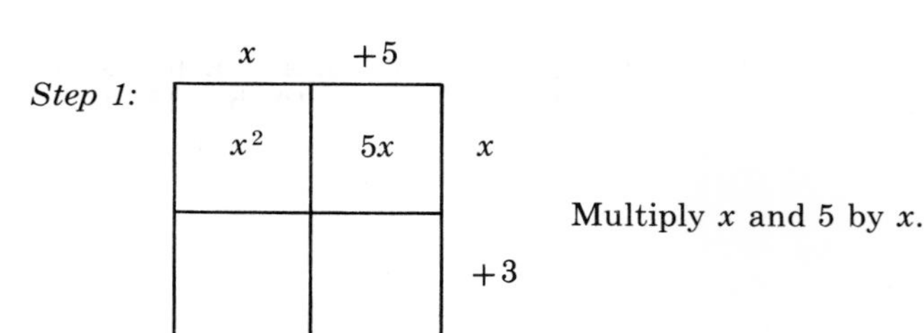

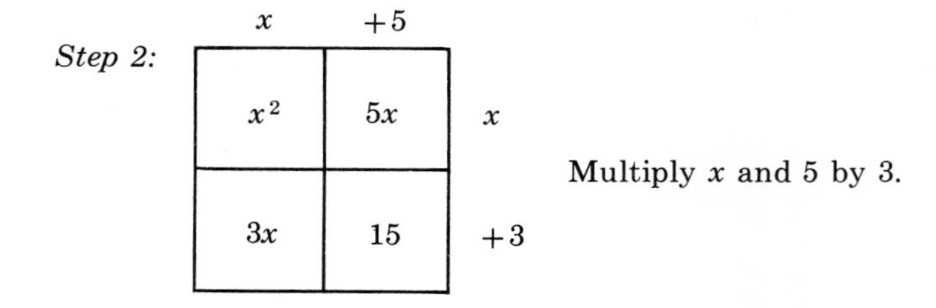

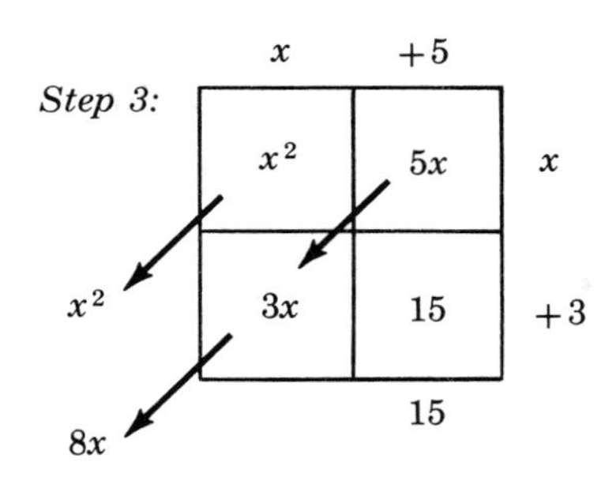

Add the partial products to obtain $x^2 + 8x + 15$.

CHAPTER 8

1. 8 **2.** 21 **3.** $^-5$ **4.** 20 **5.** 4

6. a, b, e, g

Note: Addition of integers involves the same properties as for addition and multiplication of whole numbers in addition to closure for subtraction.

7. Confuses negative numbers and place-value notation; students must count jumps, not points, when computing on the number line

CHAPTER 9

I. **1.** 5 **2.** 9 **3.** 33 **4.** 41

II. $a = b + c$

III. is not

IV. cannot

V. $\neq$

VI. cannot

VII. $0 + 6$

VIII. n

IX. **1.** a **2.** a **3.** a **4.** b **5.** b **6.** b **7.** c **8.** c **9.** c **10.** a **11.** a **12.** b **13.** c

X. Each operation cancels the effect of the other, $(a + b) - b = a$; $(a - b) + b = a$.

CHAPTER 10

I. **1.**

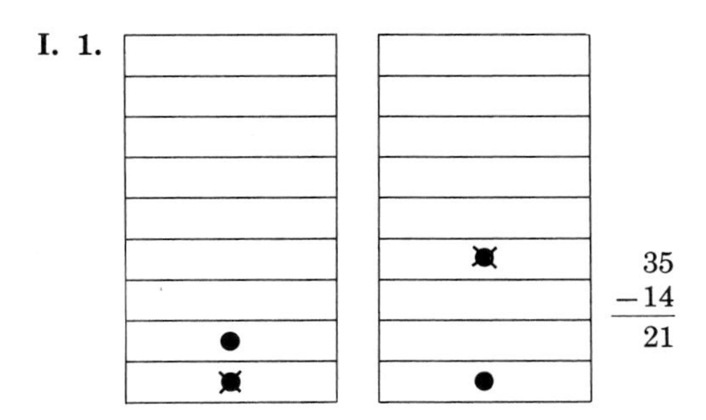

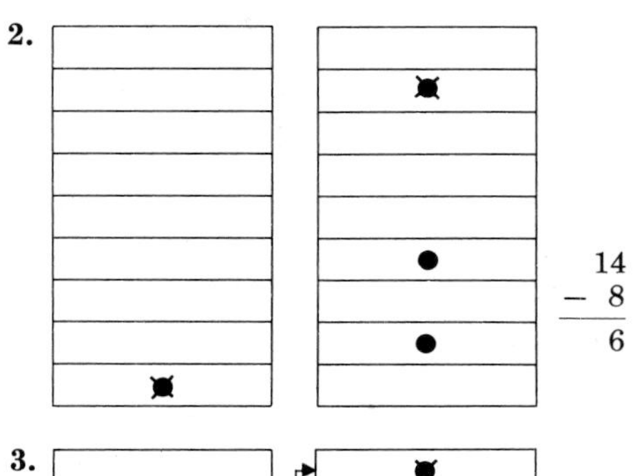

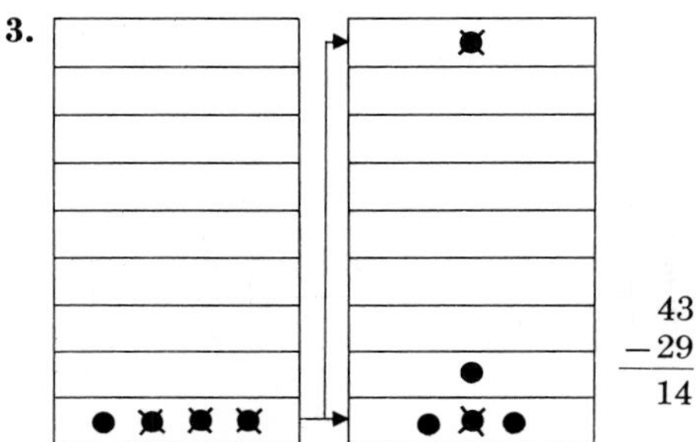

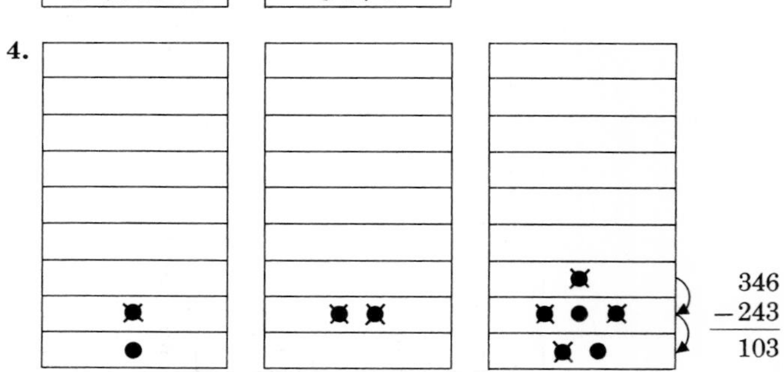

II. **1.** Error in basic subtraction fact $8 - 3 = 5$
2. Does not complete algorithm
3. Does not apply additive identity concept to subtraction
4. Using 1 as multiplicative identity
5. Subtracts ones number from all others
6. No renaming in tens column
7. Subtracts smaller number from larger number to avoid renaming

III. **1.**
$$\begin{array}{rcr} 385 & \rightarrow & 7 \\ -168 & \rightarrow & -6 \\ \hline 217 & \rightarrow & 1 \end{array}$$

2.
$$\begin{array}{rcr} 8736 & \rightarrow & 6 \\ -2578 & \rightarrow & -4 \\ \hline 6158 & \rightarrow & 2 \end{array}$$

CHAPTER 11 I.

	Closure				*Associativity*				*Commutativity*				*Identity element*				*Inverse element*				*Distributive property*	
	+	×	−	÷	+	×	−	÷	+	×	−	÷	+	×	−*	÷*	+	×	−	÷	×	÷†
Natural numbers	yes	yes	no	no	yes	yes	no	no	yes	yes	no	no	no	yes	no	no	no	no	no	no	yes, for multiplication over + and −	right only
Whole numbers	yes	yes	no	no	yes	yes	no	no	yes	yes	no	no	yes	yes	no	no	no	no	no	no	yes, for multiplication over + and −	right only
Integers	yes	yes	yes	no	yes	yes	no	no	yes	yes	no	no	yes	yes	no	no	yes	no	yes	no	yes, for multiplication over + and −	right only

* Right-hand identity element only for subtraction (0); right hand identity element only for division (1).

† Right-hand distribution only when the operation being distributed is on the right side of the main operation sign.

For example: $12 \div 4 = (8+4) \div 4$
$= (8 \div 4) + (4 \div 4)$
$= 2 + 1$
$3 = 3$

but

$12 \div 4 \neq 12 \div (3+1)$
$\neq (12 \div 3) + (12 \div 1)$
$\neq 4 + 12$
$\neq 16$

II. $15 \div 5 = 3$

III. **1.** partitive **2.** measurement

IV.

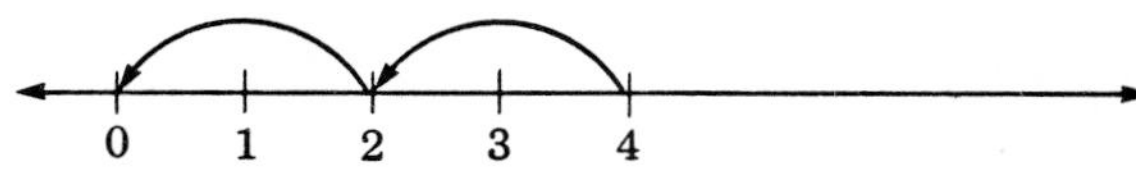

V. undefined

VI.

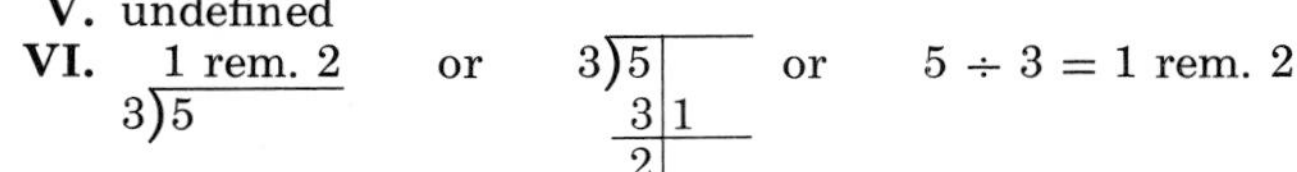

$3\overline{)5}$ = 1 rem. 2 or $3\overline{)5}$, 3 | 1, 2 or $5 \div 3 = 1$ rem. 2

Note: $3\overline{)5}$ = 1 rem. 2/3 does not reflect this statement.

VII. They are inverse relations.

VIII. Subtract the divisor from the dividend the number of times indicated by the quotient. The final remainder should then be the same as the remainder obtained in the division.

CHAPTER 12

I. **1.** Error in basic division fact
2. Fails to complete the algorithm
3. Leaves the remainder greater than the divisor
4. Incomplete subtraction
5. Incorrect subtraction
6. Omits zero in the quotient

II. $168 \div 8\ (160 + 8) \div 8$
$= (160 \div 8) + (8 \div 8)$
$= 20 + 1$
$= 21$

III. Closed

IV. n, 1

V. $=$

VI. Is not

CHAPTER 13

I. **1.** A prime number has only itself and 1 as factors, thus the intersection showing factors of prime numbers would contain only the number 1, since 1 is a factor of every number.
2. 1 (They are relatively prime.)
3. $3\overline{)21}$ = 7, $3\overline{)63}$, $2\overline{)126}$; $2\overline{)14}$ = 7, $2\overline{)28}$; $7\overline{)49}$ = 7

$126 = 2 \cdot \boxed{7} \cdot 3 \cdot 3$ GCF = 7
$28 = 2 \cdot \boxed{7} \cdot 2$
$49 = \boxed{7} \cdot 7$

4. 1

II. **1.** 2, 4, 6, 8, 10, 12, 14, 16, 18, . . . 28, (30)
3, 6, 9, 12, 15, 18, 21, 24, 27, (30)
5, 10, 15, 20, 25, (30)

2. 2, 4, 6, 8, 10, 14, 16, 18, 20
4, 8, 12, 16, 20
5, 10, 15, 20

III. $6 = \{2, 3\}$ $\quad$ $\text{LCM} = 2 \cdot 3 \cdot 3 = 18$
$9 = \{3, 3\}$ $\quad$ $\text{GCF} = 3$

IV. **1.** true
2. true

V. 6, 12, 18, 24
8, 16, 24

VI. Prime

VII. Composite

VIII. $\{1, 2, 3, 6, 9, 18\}$

IX. Yes, 1×18, 2×9, 3×6

X. Yes, 1×24, 2×12, 3×8, 4×6 $\quad$ $\{1, 2, 3, 4, 6, 8, 12, 24\}$

XI. Itself

XII. **1.** no $\{1, 5, 25\}$
2. no $\{1, 2, 3, 4, 6, 9, 12, 18, 36\}$
3. no $\{1, 7, 49\}$
4. no $\{1, 2, 4, 8, 16, 32, 64\}$
5. no $\{1, 3, 9, 27, 81\}$
6. All are square numbers.

XIII. Odd number of elements

XIV. The Fundamental Theorem of Arithmetic

XV. $372 = 2 \cdot 2 \cdot 3 \cdot 31$ or $2^2 \cdot 3 \cdot 31$

$$\begin{array}{r|l} & 31 \\ 3 & 93 \\ 2 & 186 \\ 2 & 372 \end{array}$$

XVI. **1.** True
Example: $5^2 - 1 = 25 - 1 = 24$. 24 is not prime.
2. True
Example: $24 = 11 + 13$
3. True
Example: If $n = 4$, $2n = 8$; 5 and 7 are primes that lie between 4 and 8.
4. True
Examples: 29 is prime, 30 is a multiple of 6; 67 is prime, 66 is a multiple of 6.
5. True
Examples: 3, 5; 11, 13; 59, 61; 107, 109
6. True

XVII. The greatest common divisor, the greatest common factor

XVIII. 12

CHAPTER 14

I. $\frac{3}{4}$; .75, 75%, 3:4

$\frac{2}{5}$, .40, 40%, 2:5

II. **1.** True **2.** True **3.** True **4.** True

III. **1.** $3\frac{1}{2}$ **2.** $2\frac{3}{8}$ **3.** $6\frac{1}{4}$ **4.** $2\frac{3}{5}$

IV. **1.** $\frac{5}{3}$ **2.** $\frac{31}{5}$ **3.** $\frac{19}{8}$ **4.** $\frac{15}{8}$

CHAPTER 15

I. Commutative

II. $\left(\frac{x}{y} \cdot \frac{z}{w}\right) \cdot \frac{u}{v}$

III. $\frac{0}{b}$, $b \neq 0$

IV. $a \cdot d = b \cdot c$

V. Multiplicative inverse

VI. $\frac{40}{64} = \frac{8 \times 5}{8 \times 8} = \frac{5}{8}$; $\frac{14}{42} = \frac{7 \times 2}{7 \times 6} = \frac{1}{3}$; $\frac{12}{24} = \frac{1}{2}$

Obtain a common denominator: 8, 16, 24
3, 6, 9, 12, 15, 18, 21, 24
2, 4, 6, 8, 10, . . . , 24

Find equivalent fractions:

$$\frac{5}{8} = \frac{15}{24};\ \frac{1}{3} = \frac{8}{24};\ \frac{1}{2} = \frac{12}{24}$$

Solution: $\frac{15}{24} > \frac{12}{24} > \frac{8}{24}$ so $\frac{40}{64} > \frac{12}{24} > \frac{14}{42}$

VII. Multiplicative identity

VIII. Multiplicative inverse

IX. Closed

X. 4.

XI. 4.

XII. 4.

XIII. **1.** $\frac{1}{4}$ **2.** $\frac{1}{2}$ **3.** $\frac{1}{2}$ **4.** $\frac{1}{3}$

XIV. **1.** $\frac{4}{6}$ **2.** $\frac{6}{8}$ **3.** $\frac{1}{2}$ **4.** $\frac{2}{3}$

XV. $\frac{-2}{1} = {}^{-}2 \div 1 = {}^{-}2$; $\frac{2}{{}^{-}1} = 2 \div {}^{-}1 = {}^{-}2$

$\frac{-2}{1} = {}^{-}1 \times \frac{2}{1}$; $\frac{2}{{}^{-}1} = \frac{1}{{}^{-}1} \times \frac{2}{1}$; $\frac{2}{{}^{-}1} = \frac{1}{{}^{-}1} \times (2 \div 1) = \frac{1 \times 2}{{}^{-}1} = {}^{-}2$;

$\frac{{}^{-}2}{1} = {}^{-}1 \times (2 \div 1) = {}^{-}1 \times 2 = -2$

CHAPTER 16

I. **1.** $1\frac{1}{4}$ **2.** $2\frac{2}{3}$ **3.** $\frac{3}{4}$ **4.** $\frac{1}{2}$ **5.** $\frac{1}{4}$ **6.** $\frac{1}{4}$ **7.** $\frac{1}{6}$

II. **1.** $\frac{12}{18}$ or $\frac{2}{3}$ **2.** $\frac{9}{16}$ **3.** $\frac{8}{10}$ or $\frac{4}{5}$

III. **1.** $\frac{7}{4}$ **2.** $\frac{11}{5}$ **3.** $\frac{14}{3}$ **4.** $\frac{45}{16}$

IV. **1.** $4\frac{1}{5}$ **2.** 10 **3.** 15 **4.** $1\frac{1}{2}$ **5.** $1\frac{1}{5}$ **6.** $2\frac{1}{2}$

V. =

VI. Nonzero whole numbers
VII. Commutative
VIII. Associative
IX. $\frac{0}{k}$
X. **1.** Associative Property for Addition
2. Commutative Property for Addition
3. Additive identity
4. Closure for addition of fractions
XI. 1.

CHAPTER 17

I. Expanded form:
1. $.3 = 3 \times 1/10$ **2.** $.15 = (1 \times 1/10) + (5 \times 1/100)$
3. $.0004 = 4 \times \frac{1}{1000}$
4. $908.6 = (9 \times 100) + (0 \times 10) + (8 \times 1) + (6 \times 1/10)$
Exponential notation:
1. $3 \times 1/10^1$ or 3×10^{-1} **2.** $(1 \times 1/10^1) + (5 \times 1/10^2)$ or $(1 \times 10^{-1}) + (5 \times 10^{-2})$
3. $4 \times 1/10^4$ or 4×10^{-4} **4.** $(9 \times 10^2) + (0 \times 10^1) + (8 \times 10^0) + (6 \times 10^{-1})$

II. **1.** $\frac{316}{1000}$ **2.** $\frac{61}{100}$ **3.** $\frac{1}{100}$ **4.** $11\frac{93}{1000}$ **5.** $8\frac{8}{1000}$ or $\frac{8008}{1000}$
III. **1.** .34 **2.** .683 **3.** 68.3 **4.** .000035 **5.** 1.8 **6.** 3.0 **7.** 1.4 **8.** −.31
IV. $\frac{13}{15} = .8\overline{6}$ Factoring the denominator (3.5) indicates a repeating decimal.
V. $\frac{5}{10}$
VI. **1.** c **2.** a **3.** b
VII. **1.** $.1818\overline{18}$
2. $.\overline{052631578947368421}$
3. $6.\overline{6}$
4. $-.7\overline{3}$
VIII. No
IX. Rational numbers
X. Irrational numbers
XI. $\sqrt{2} = 1.414214\ldots$
$\sqrt{3} = 1.732051\ldots$
$\pi = 3.141592653589793\ldots$
$\frac{1}{\sqrt{2}} = .707107\ldots$
$\frac{1}{\pi} = .318309886\ldots$
$\cdot^{2} = 9.8696044036667637\ldots$

XII.

	100	10	1	$\frac{1}{10}$	$\frac{1}{100}$
1.		6	6	3	4
2.		1	9	0	4
3.	9	0	0	0	1

XIII. 1.

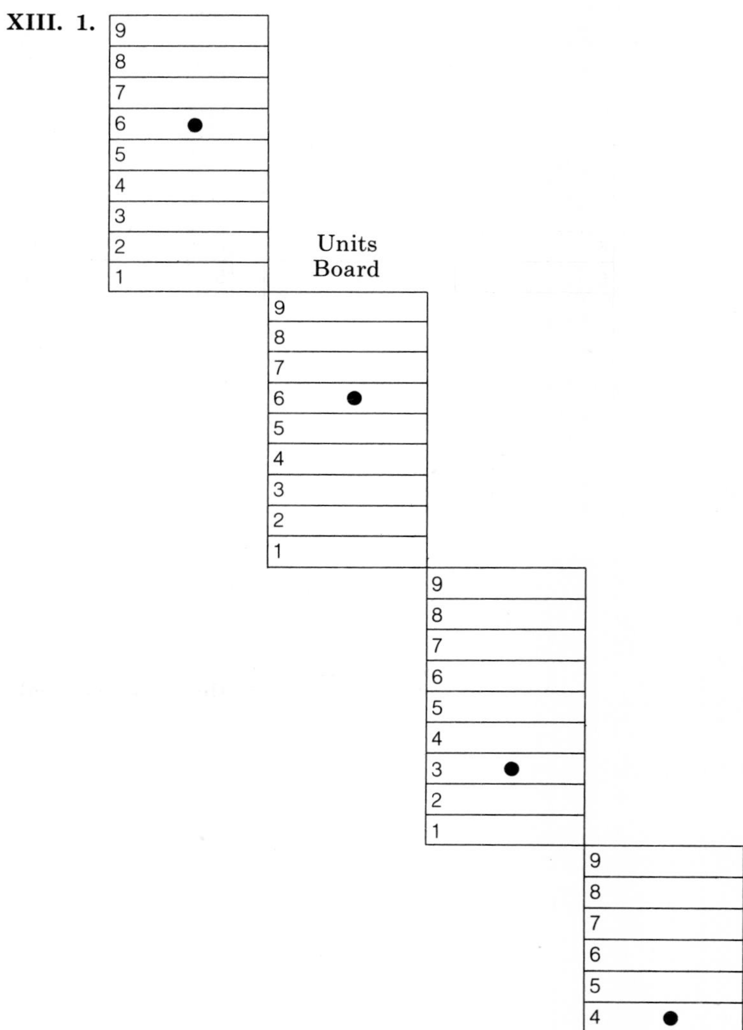

Units Board

2.

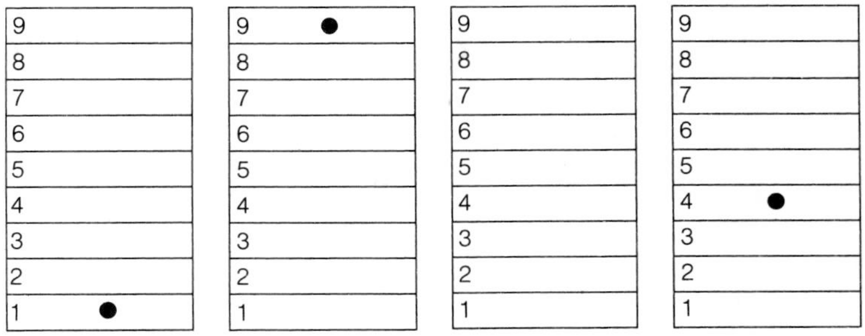

9	9 ●	9	9
8	8	8	8
7	7	7	7
6	6	6	6
5	5	5	5
4	4	4	4 ●
3	3	3	3
2	2	2	2
1 ●	1	1	1

Units Board

3.

9 ●	9	9	9	9
8	8	8	8	8
7	7	7	7	7
6	6	6	6	6
5	5	5	5	5
4	4	4	4	4
3	3	3	3	3
2	2	2	2	2
1	1	1	1	1 ●
9	0	0	0	1

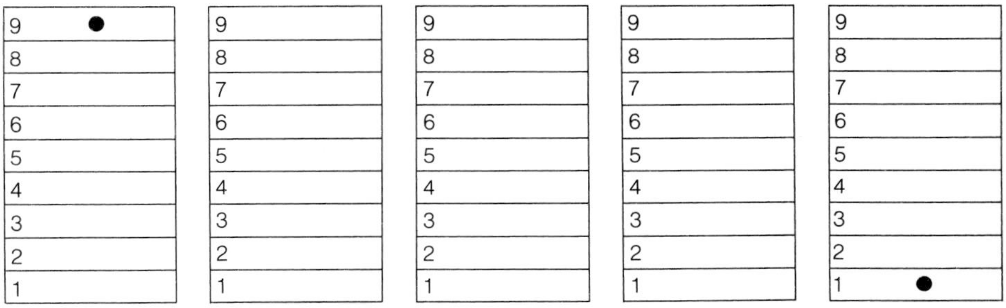

XIV. **1.** .13 **2.** .4 **3.** .024 **4.** .115 **5.** .02

XV. **1.** .17 **2.** .87 **3.** 1.21 **4.** 47.33
.2 .9 1.2 47.3
0 1 1 47

XVI. **1.** 4.8, 4.9, 5.0, . . . , 5.2, 5.3, 5.4
2. 63.5, 63.6, 63.7, 63.8, 63.9, 64.0

XVII. **1.** .5, .556, .56
2. .03, .070, .2
3. .300, .40, .7
(*Hint:* Rename as the same place value.)

CHAPTER 18

I. 1.

units	tenths
9	9
8	8
7	7
6 ●	6
5	5
4	4 ●
3	3 ●
2 ●	2
1	1
8	.7

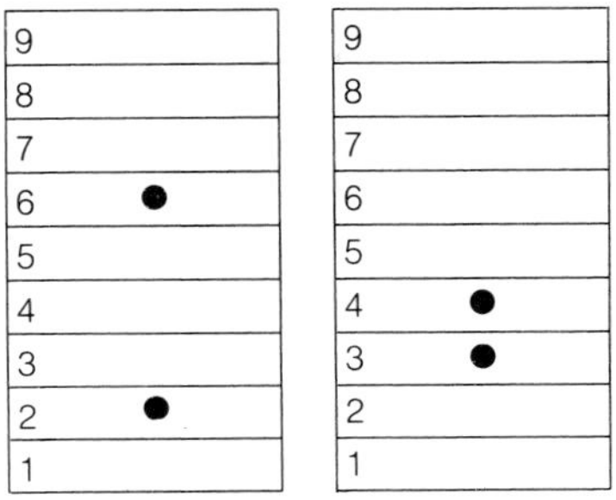

2.

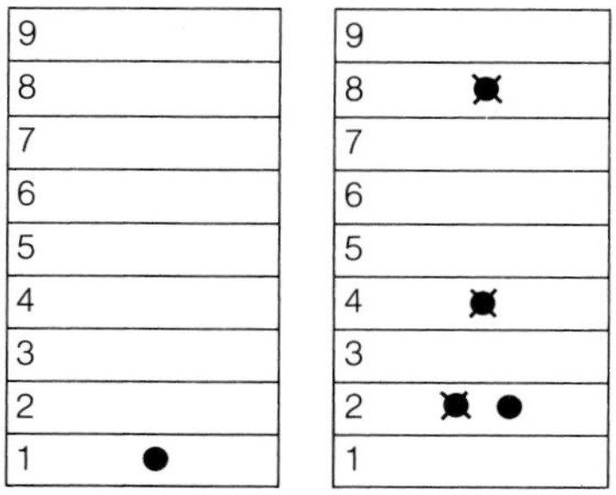

1 .2

Change 4-space token to two 2-space.

Make a forward move to the 1-space on the units board by exchanging the 2-space and 8-space tokens on the tenths board.

3.

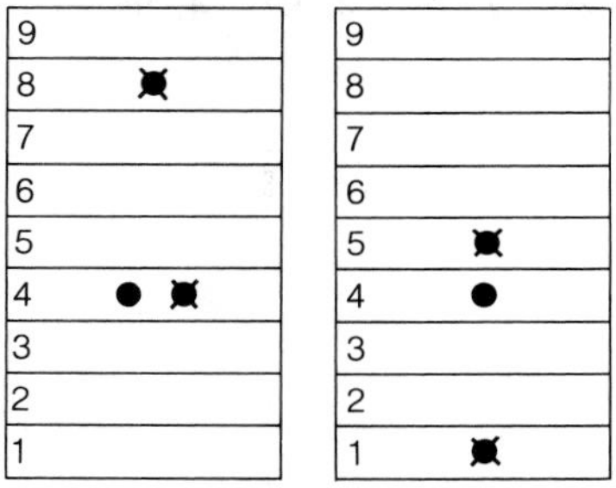

8	.5
−4	.1
4	.4

4.

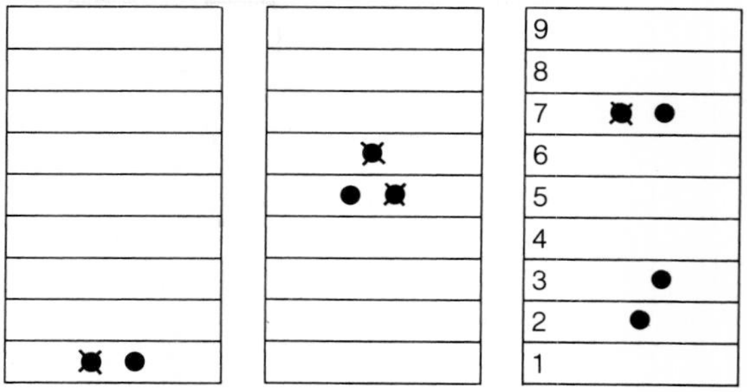

1	6	.2
−1	5	.7
		.5

This procedure shows null matches that represent adding inverses to obtain zero. The boards have the final value of 5 tenths.

II. Upper left: 1.0
Right: .9
Bottom: 1.1

III. **1.** 98.487
2. 7.546
3. 269.39

IV. 4, 5, 2, 1, 3

V. $\begin{array}{r} 1\,80. \\ .02\overline{)3.60} \end{array}$

$\begin{array}{r} 180 \\ \underline{.02} \\ 3.60 \end{array}$

CHAPTER 19

I. **1.** Three to the fourth power.
2. Five cubed.

II. **1.** $9^3 = 729$
2. $4^4 = 256$

III. **1.** 5 **2.** 4 **3.** 3 **4.** 2

IV. **1.** 2^4 **2.** 3^3 **3.** 6^6 **4.** 9^2

V. **1.** 10^3 **2.** 10^5 **3.** 10^2 **4.** 10^6

VI. **1.** 8×10^4 **2.** 2×10^5

VII. **1.** 3^5 **2.** 7^4 **3.** 11^3 **4.** 2^6 **5.** 5^5 **6.** 13^4

VIII. **1.** 400 **2.** 7,690 **3.** 10,000 **4.** 210,600

CHAPTER 20

Set A

I. For example, point out square tiles on the floor and ceiling; rectangular window frames; circles associated with clocks, bottles, plant jars; triangles of various shapes, including classmates' arms and legs bent at the elbows and knees.

II. **1.**

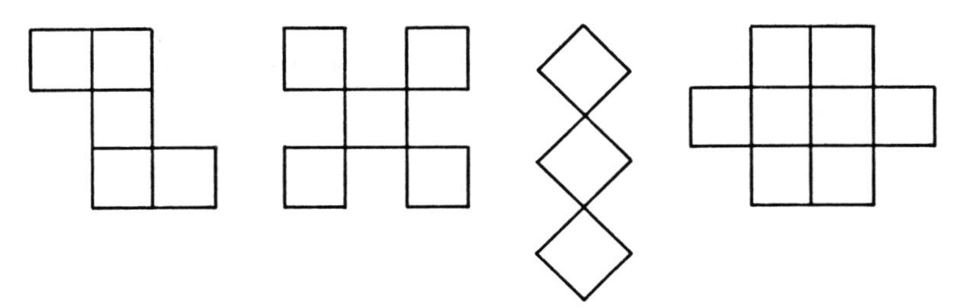

2.

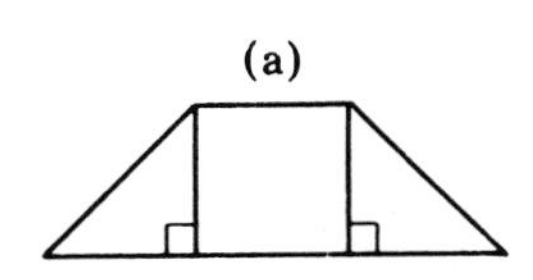

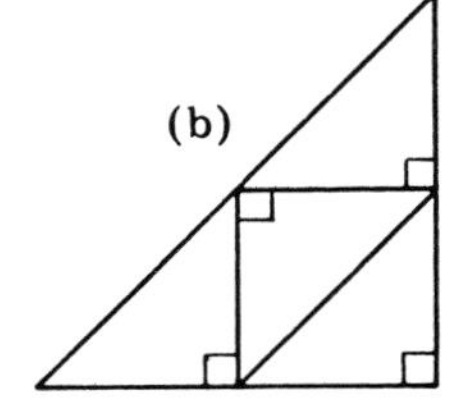

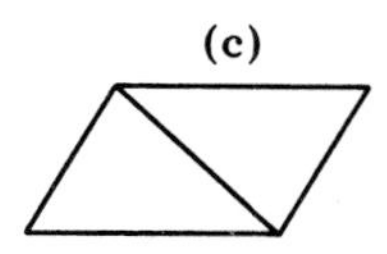

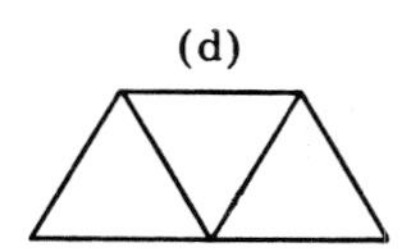

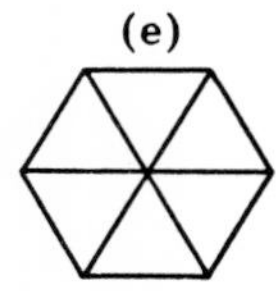

III. 1.

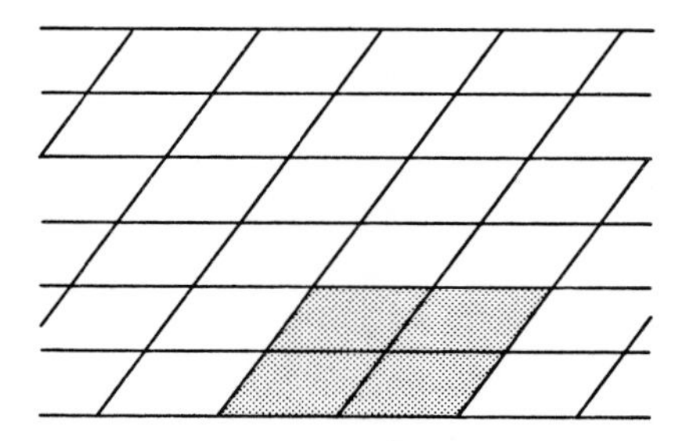

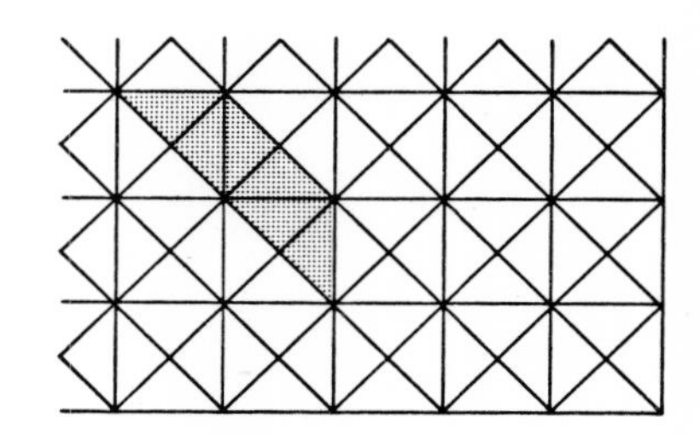

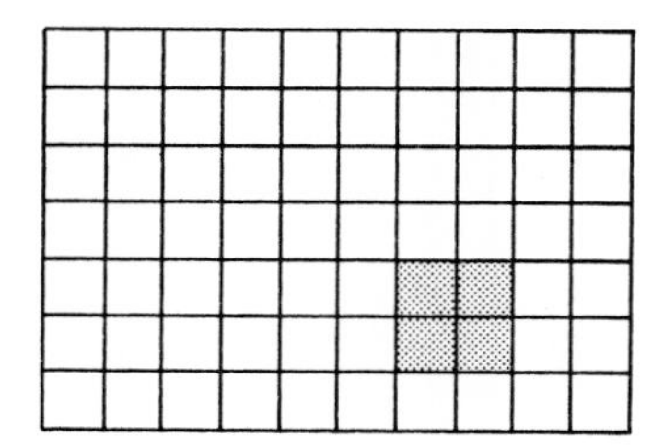

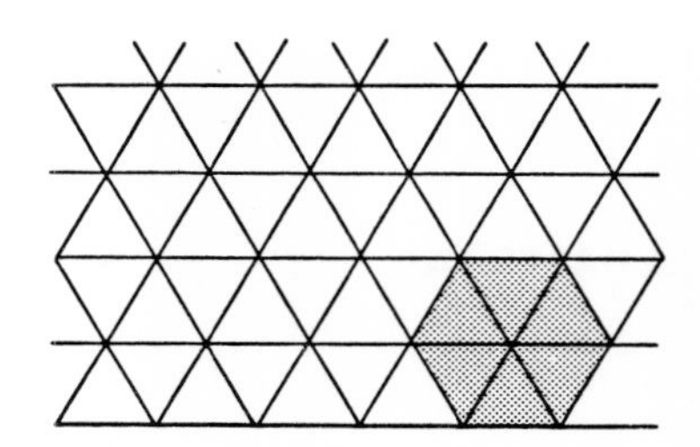

2.

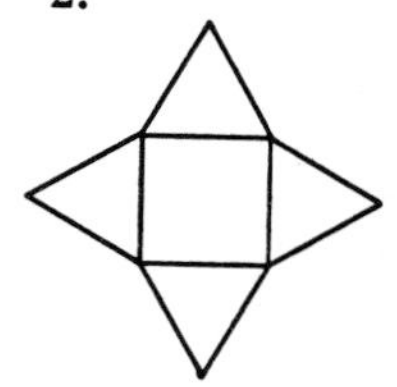

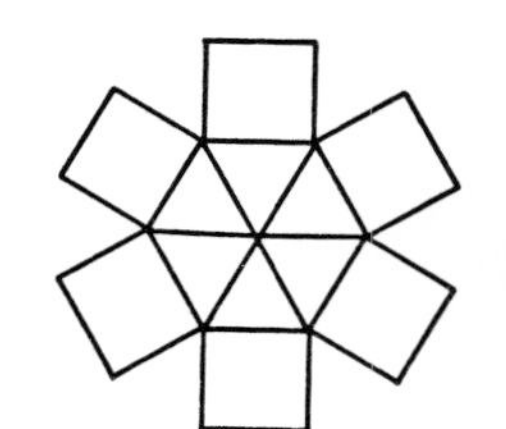

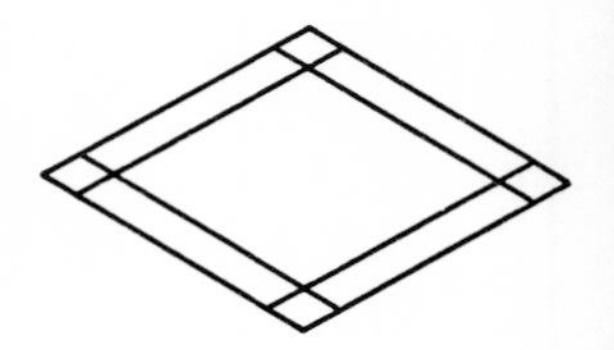

3.

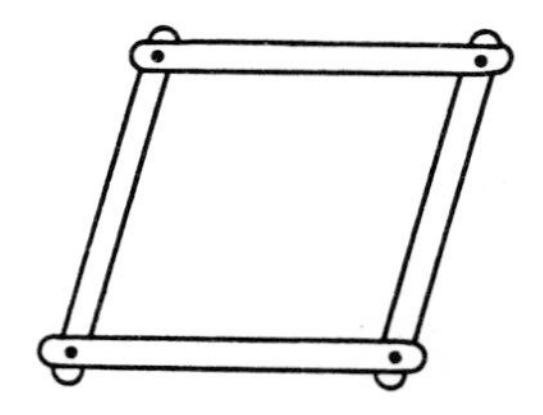

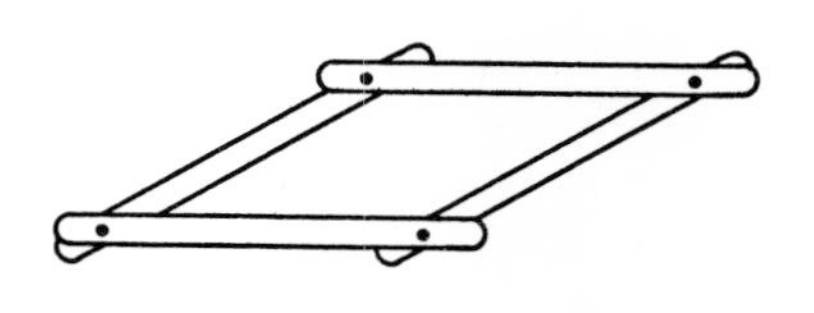

IV. line (3) line segment (1) ray (2)
V. line
VI. **1.** yes **2.** no **3.** triangle

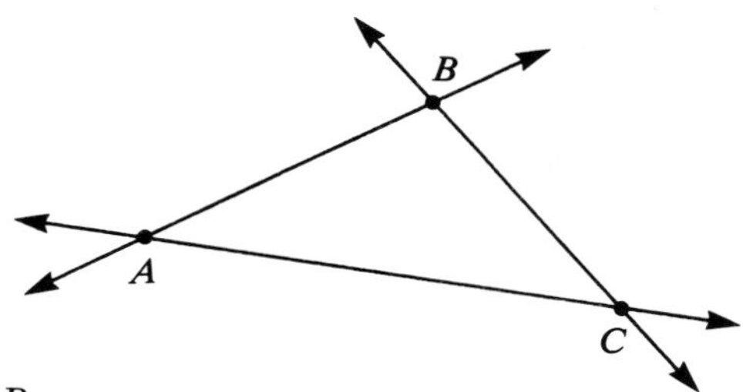

Set B

I. **1.**

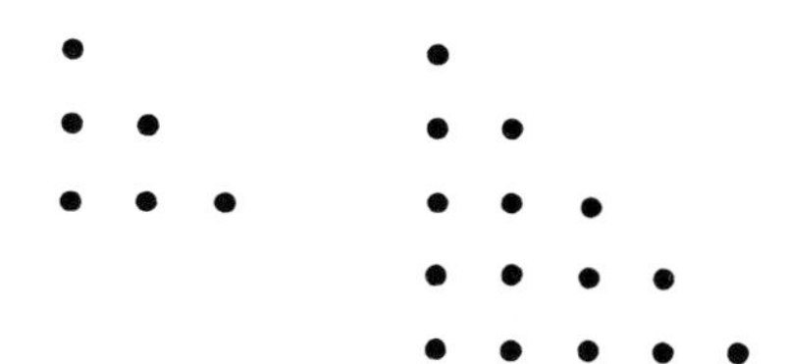

2. Triangular number

II. **1.**

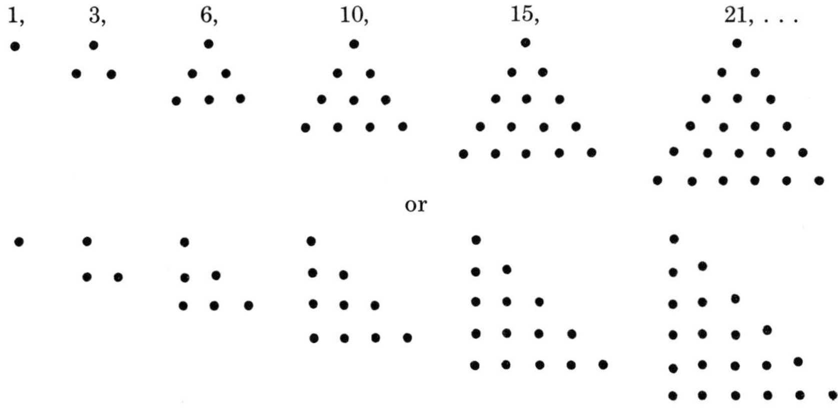

2. Each dot pattern of a triangular number is obtained from the previous one by adding one more row with an extra dot in it.
3. The tenth triangular number has 10 dots in its longest row and has 55 dots in all. $(1 + 2 + 3 + 4 + 5 + 6 + 7 + 8 + 9 + 10 = 55)$

III. **1.** 1, 3, 6, 10, 15
2. They are triangular numbers.
3. Yes, since it is included in the series of triangular numbers.

IV. $1 + 3 = 4 = 2^2$; $3 + 6 = 9 = 3^2$; $6 + 10 = 16 = 4^2$; $10 + 15 = 25 = 5^2$

V.

Position of triangular number	1	2	3	4	5	6
Triangular number	1	3	6	10	15	21
(Triangular number) × 2	2	6	12	20	30	42

(position) × (position + 1) = (triangular number) × 2

Examples: $4 \times 5 = 20$ $5 \times 6 = 30$ $6 \times 7 = 42$

Note: The triangular number is generated by $\frac{n(n+1)}{2}$, where n is the position.

VI.

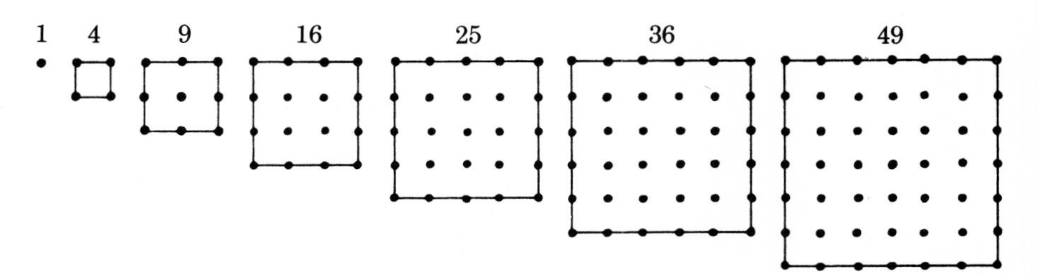

VII. **1.** $1 + 3 = 4 = 2^2$
2. $1 + 3 + 5 = 9 = 3^2$
3. $1 + 3 + 5 + 7 = 16 = 4^2$

VIII.

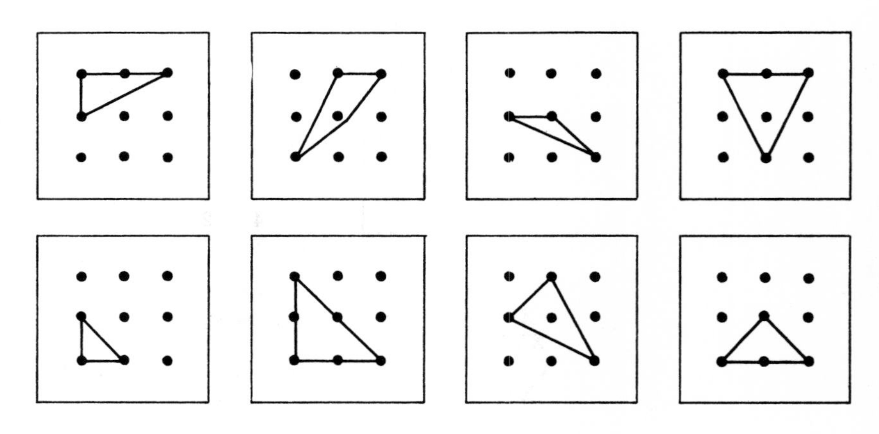

1. 4, right triangles
2. 5, isosceles triangles

IX. Scalene triangle

X. 15 From each of the 6 points, you can draw 5 lines, one to each of the other points. This totals (6 × 5), or 30, lines. However, this procedure counts each line twice. The line from A to B is the same line as from B to A. So there are $\frac{6 \times 5}{2}$, or 15, lines. Use this method for number XI.

XI. **1.** 6 **2.** 21 **3.** 36 **4.** 190 **5.** 4950

XII. **1.** **e.** 10 **f.** 15 **g.** 21 **h.** 28 **i.** 36 **j.** 45
2. They are triangular numbers.

XIII. Triangular patterns from 6 on can be changed to rectangular patterns.

XIV. The number of lines in the mystic rose is $\frac{24 \times 23}{2} = 276$.

XV.

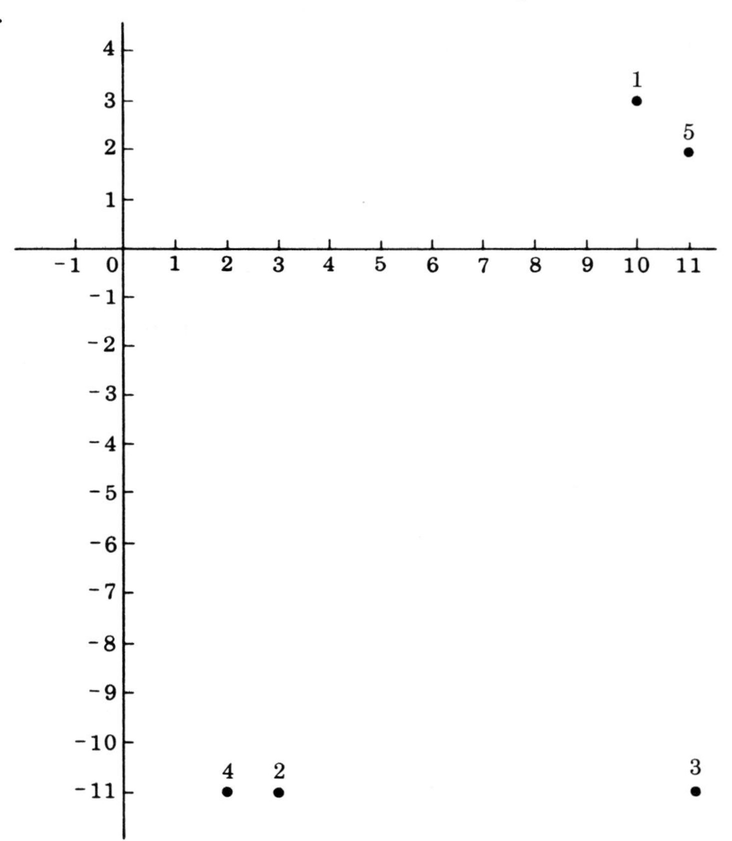

XVI. Vertical, or vertically opposite, angles

XVII. Square

XVIII. (2, 1), (22, 2), (22, 11), (2, 11) rectangle

XIX.

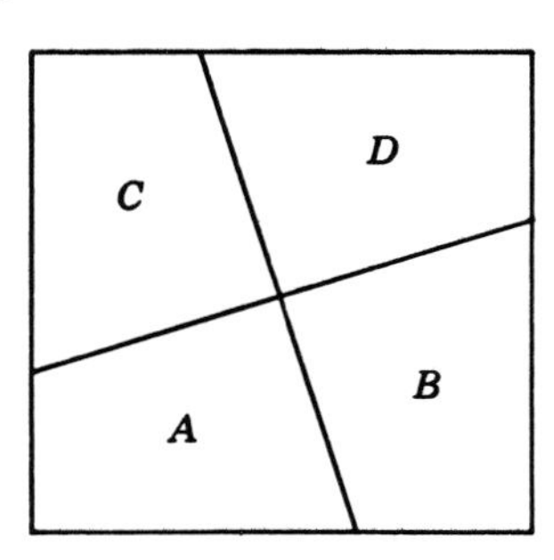

XX.

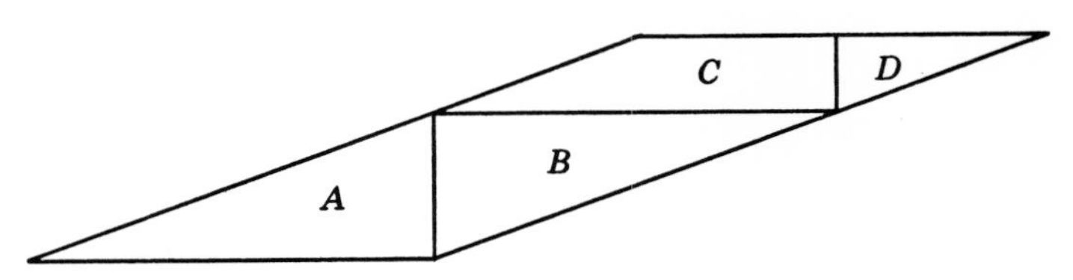

Set C

I. **1.**

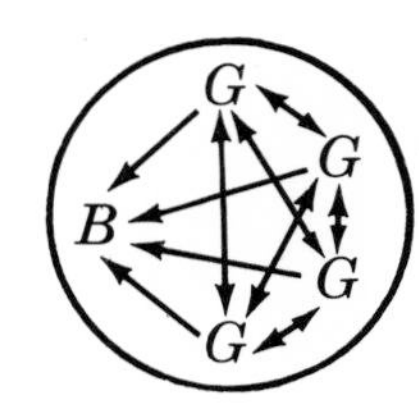

2.

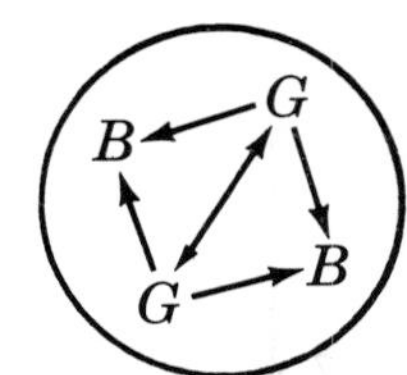

or

B G B G

II.

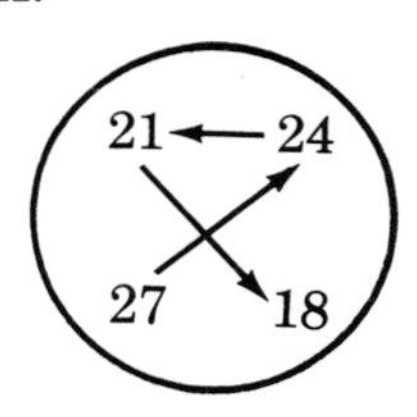

III. "Is a third of"

IV.

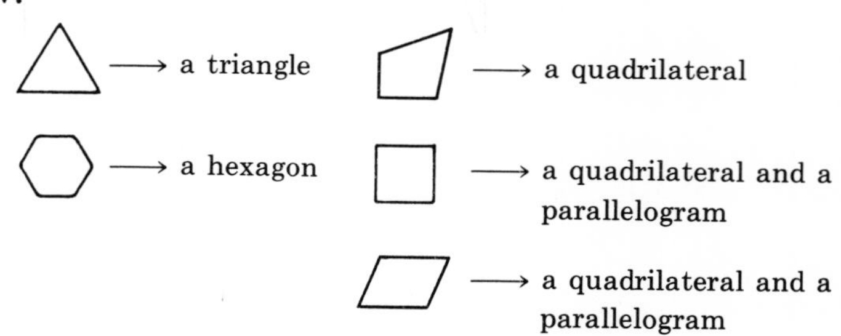

V. **1.** Is one less than
2. Is $\frac{1}{2}$ of
3. Is three times

VI. **1.** $\{(5, 3), (7, 5), (9, 7), (11, 9)\}$
2. $\{(30, 10), (60, 20), (90, 30), (120, 40)\}$
3. $\{(3, 30), (4, 40), (5, 50), (6, 60), (7, 70)\}$

VII.

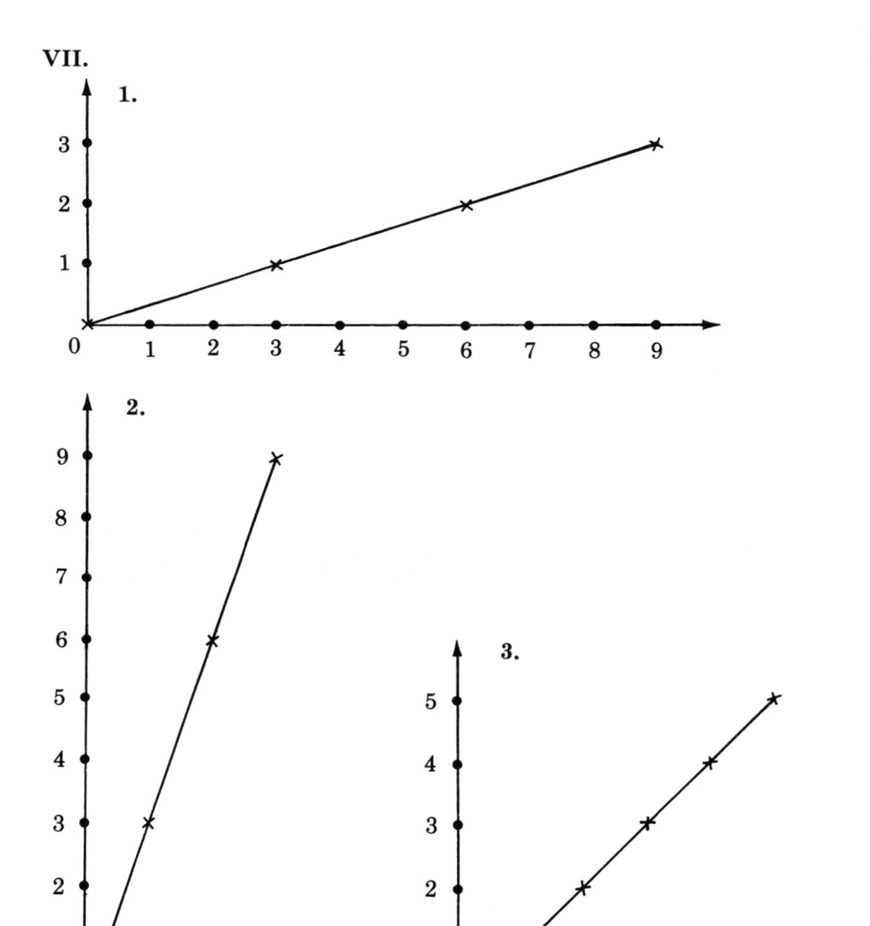

1. Domain is $\{0, 3, 6, 9\}$.
Range is $\{0, 1, 2, 3\}$.
2. Domain is $\{0, 1, 2, 3\}$.
Range is $\{0, 3, 6, 9\}$.
3. Domain is $\{1, 2, 3, 4, 5\}$.
Range is $\{1, 2, 3, 4, 5\}$.

VIII. **1.** $(4, 8)$ **2.** $(6, 10)$ **3.** $(16, 20)$ **4.** $(\Delta, \Delta + 4)$ **5.** yes

IX.

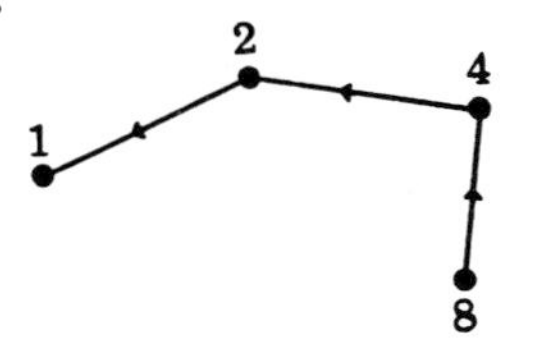

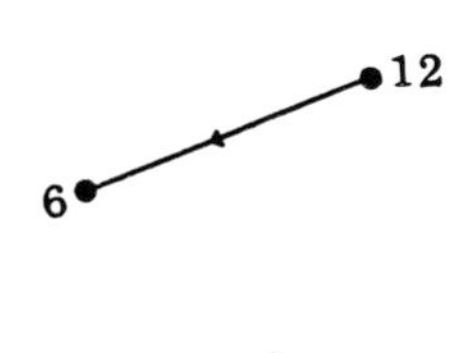

X. **1.** "Is 3 less than"
2. "Is 3 more than"

XI.

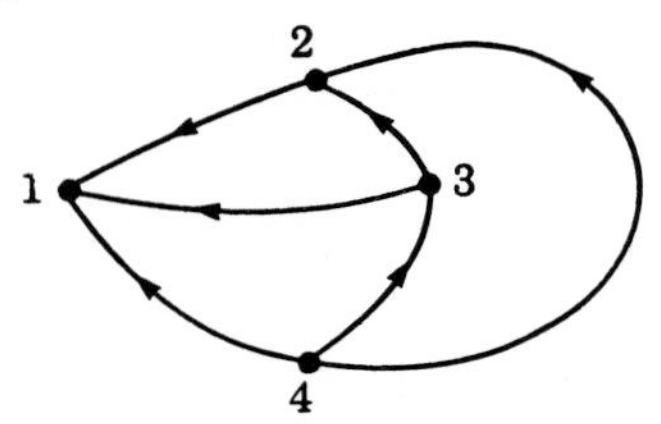

XII. **1.** Is half of
2. Is younger than
3. Is the same as
4. Is the child of
5. Lives in the same house as

XIII. 1, 3, 5, and 6 are symmetric relations.

XIV.

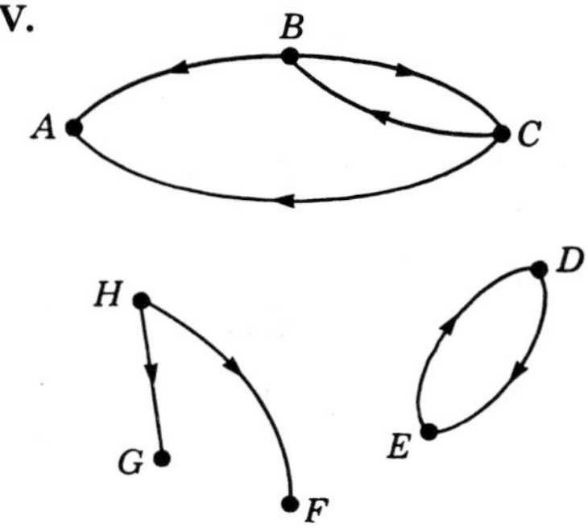

XV. **1.** C is the brother of A and C; G and H are brothers; F is the brother of D and E.
2. C, F, G, H are boys; A, B, D, E are girls.

XVI.

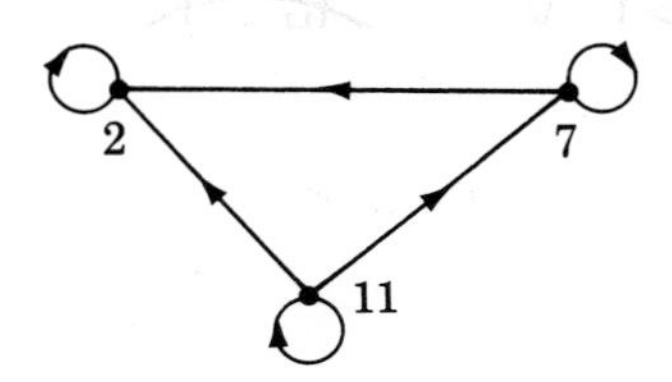

XVII. *Example:* $\{1, 2, 3, 6\}$

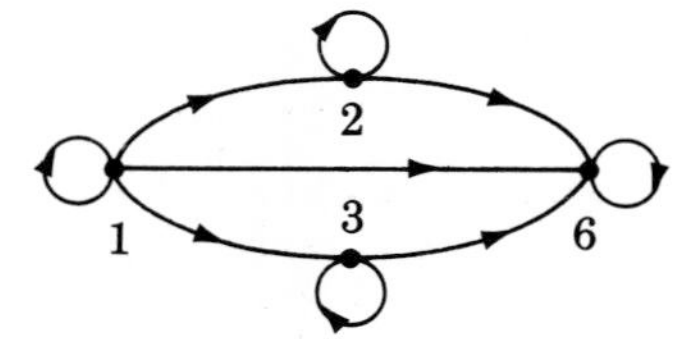

Each number is a factor of itself. The relation is reflexive but not symmetric.

XVIII.

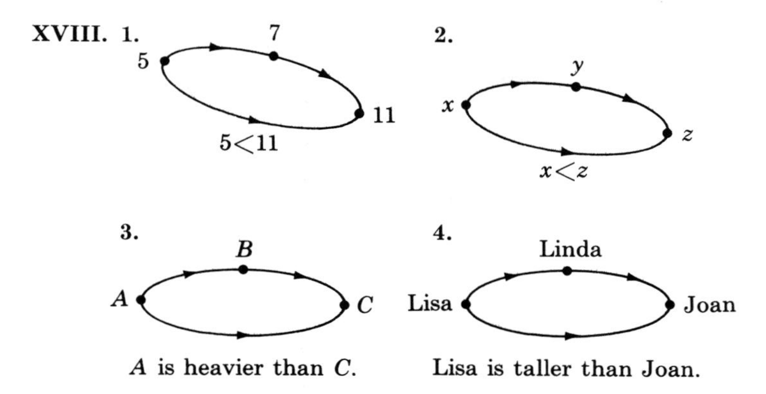

XIX. 1, 3, and 5 are equivalence relations.

CHAPTER 21

I. 1, 2, and 6 have vertical lines; 3 has a diagonal line; 4 and 5 have horizontal lines through their centers.

II.

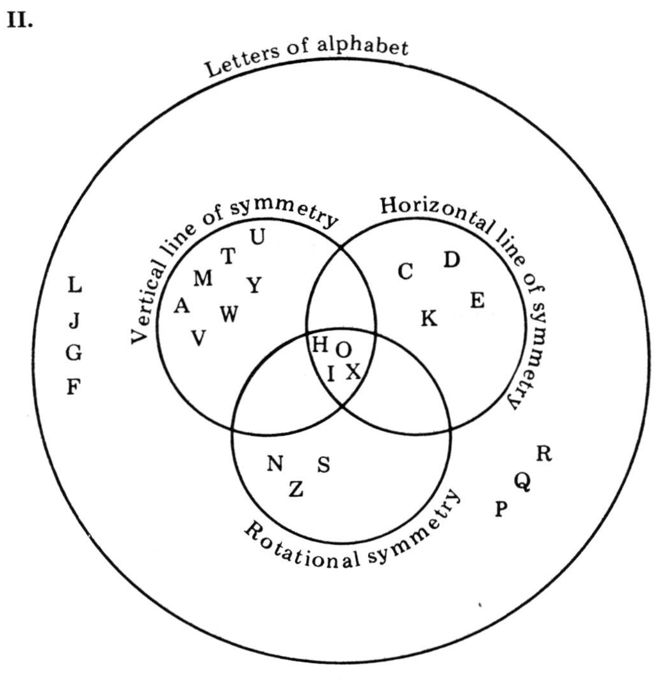

III. 1. Traversable (2 odd vertices)
2. Nontraversable (4 odd vertices)
3. Nontraversable (6 odd vertices)
4. Traversable (No odd vertices)

IV.

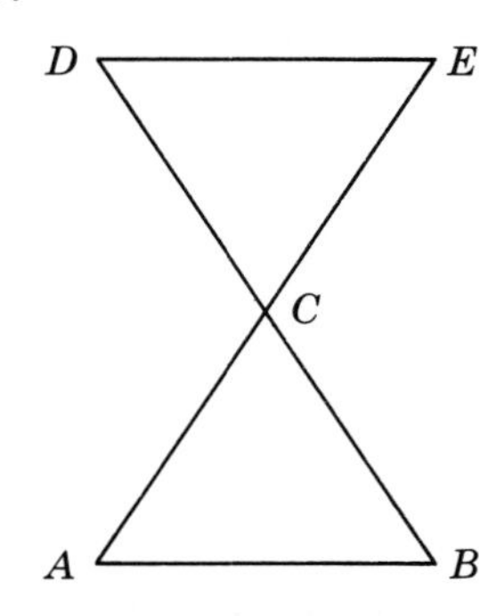

traversable

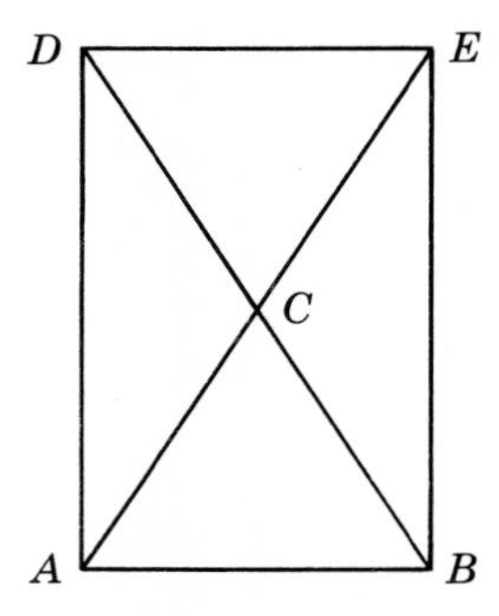

nontraversable

V. slide: 2 turn: 4, 5 flip: 1, 3

CHAPTER 22

I. **1.** > **2.** < **3.** < **4.** =

II. 6 + 4 + 8 = 18 square units

III. **1.** h **2.** l **3.** a **4.** e **5.** f **6.** i **7.** j **8.** k **9.** g **10.** d **11.** m

IV. **1.** 13, 15, 17
2. 84, 112, 144
3. 85, 113, 145

V. **1.** $3^2 = 9$, $4^2 = 16$, $5^2 = 25$; pattern, $9 + 16 = 25$
2. $5^2 = 25$, $12^2 = 144$, $13^2 = 169$; yes, $25 + 144 = 169$
$7^2 = 49$, $24^2 = 576$, $25^2 = 625$; yes, $49 + 576 = 625$
(The pattern is $a^2 + b^2 = c^2$)

VI. $4^2 + 3^2 = 5^2$ or $16 + 9 = 25$ sq. units

VII. $4x$ cm

VIII. **1.** 115° (180° minus 65°)
2. 65°
3. 115°

IX. 8 sq. cm

b., c.

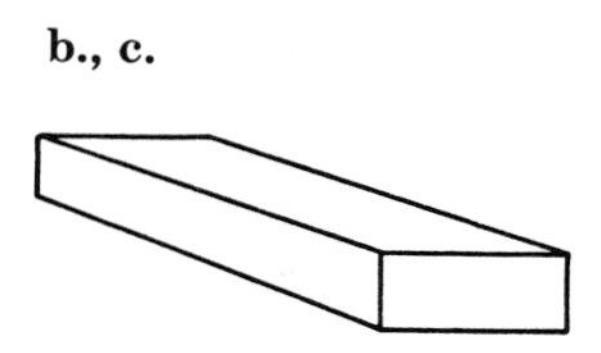

d.

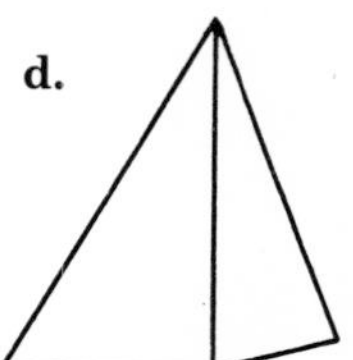

X. **1.** 216 cm^3
2. 343 cm^3
3. 512 cm^3
4. 1000 cm^3

XI.

L	1	2	3	4	5	6	7	8	9	10
V	1	8	27	64	125	216	343	512	729	1000

XII. litre

XIII. 3 in. means correct to the nearest inch, or $3 \pm \frac{1}{2}$ in.; $3\frac{0}{8}$ in. means correct to the nearest $\frac{1}{8}$ in., or 3 in. $\pm \frac{1}{16}$ in.

XIV. **1.** 1 in.
2. 1 sec.
3. 0.1 in.

CHAPTER 23

I. There are two possible outcomes—heads or tails, so $P(\text{heads}) = \frac{1}{2}$.

II. There are 4 aces (ace of clubs, ace of diamonds, ace of hearts, ace of spades) and 52 different possible outcomes. $P(\text{ace}) = \frac{4}{52} = \frac{1}{13}$.

III. The numbers 1, 2, 3, 4, 5, 6 are represented on the six different faces of the die by dots. Of these six outcomes, three are even numbers: 2, 4, 6. $P(\text{even}) = \frac{3}{6} = \frac{1}{2}$.

IV. Of the six possible outcomes, 1, 2, 3, 4, 5, 6, none correspond to a 14. $P(14) = \frac{0}{6} = 0$. (The P of any impossible event is zero.)

V. Of the six possible outcomes, 1, 2, 3, 4, 5, 6, there are six successes, since every outcome is less than 14. P(number less than 14) $= \frac{6}{6} = 1$. (The probability of any certain event is 1.)

VI. $P(\text{club}) = \frac{13}{52} = \frac{1}{4}$

$P(\text{no club}) = \frac{39}{52} = \frac{3}{4}$

Odds against drawing a club are $\frac{3/4}{1/4} = \frac{3}{1} = 3{:}1$.

VII. $P(\text{unwrapped}) = \frac{\text{number found to be unwrapped}}{\text{total number}} = \frac{1300}{25000} = .052$.

VIII. There are 52 possible outcomes, 13 clubs, 4 aces. Count the ace of clubs as an outcome only once. $P(\text{club or ace}) = (13 + 3)/52 = \frac{16}{52} = .307$.

IX. **1.** There are $52 \cdot 52$, or 2,704, possible outcomes, with 16 possible favorable outcomes. P(drawing two jacks) $= \frac{16}{2704} = \frac{1}{169}$.

2. There are $52 \cdot 51$, or 2,652, possible outcomes, with 12 possible favorable outcomes. P(drawing two jacks) $= \frac{12}{2652} = \frac{1}{221}$.

X. The probability that two independent events both occur is the product of their probabilities.

P (drawing a heart) $= \frac{13}{52} = \frac{1}{4}$

P (rolling an even number) $= \frac{18}{36} = \frac{1}{2}$

P (drawing a heart and then rolling an even number) $=$ $\frac{1}{4} \times \frac{1}{2} = \frac{1}{8}$

XI. **1.** $\frac{1}{6}$ **2.** $\frac{5}{6}$

XII. **1.** 3 **2.** $\frac{1}{2}$ **3.** $\frac{1}{2}$

XIII. **1.** $\frac{1}{10}$ **2.** $\frac{1}{10}$ **3.** $\frac{1}{10}$ **4.** $\frac{1}{5}$ **5.** $\frac{1}{2}$ **6.** $\frac{3}{10}$

XIV. **1.** $\frac{1}{4}$ **2.** $\frac{3}{4}$

XV. **1.** $\frac{3}{10}$ **2.** $\frac{7}{10}$

XVI. 32

XVII. 12.5

XVIII. 200

XIX. 24

Appendix B/Scope and Sequence Chart

	1.1	1.2	1.3
Sets	1. Denote elements of a set.* 2. State the attributes common to two or more sets. 3. Recognize equivalent and nonequivalent sets. 4. Compare nonequivalent sets.	1. Combine disjoint sets.	1. Separate sets into subsets. 2. Remove a subset from a given set.
Whole Numbers / Notation	5. Indicate the cardinal number property of sets. 6. Recognize the ordinal number property of sets. 7. Write the digits 0 to 9.	2. Count and write numbers from 0 to 9 in sequence.	3. Use count-on place-value model. 4. Count and write numbers through 19 in sequence. 5. Show many-to-one and one-to-many representations of the same number. 6. Count by 2s, 5s, and 10s.
Integers / Notation			
Rationals / Notation			7. Identify the fractional part of a whole. 8. Identify the fractional part of a set.
Decimals / Notation			
Addition		3. Add numbers with sums 0 to 9. 4. Write the operation sign for addition.	9. Add two 1-digit numbers with sums to 18.
Subtraction			10. Perform take-away subtraction with minuend to 18.

* Also appropriate for kindergarten.

	2.1	2.2	2.3
Sets	1. Represent set relationships and operations with Venn diagrams at the concrete level.		
Whole Numbers / Notation	2. Apply many-to-one generalization to exchange model of place value. 3. Extend number notation to 99. 4. Recognize odd and even numbers.	1. Use exchange model for place value. 2. Extend place value to hundreds place.	
Integers / Notation			
Rationals / Notation	5. Select concrete representation of equivalent fractions, e.g., halves and fourths, thirds and sixths.		1. Select fractional equivalents to the number 1 at concrete and picture levels.
Decimals / Notation			
Addition	6. Add two numbers, with no renaming, sums to 99 and check.	3. Use properties as addition. 4. Add three 1-digit numbers with sums less than 20. 5. Add three numbers, sums to 99, with no renaming. 6. Add numbers with sums to 999, with no renaming.	2. Add two numbers with sums to 999, with one renaming. 3. Add with sums to 999, with two renamings. 4. Add like fractions, with sums up to 1.
Subtraction	7. Subtract numbers with minuend less than 100, with no renaming.	7. Investigate the addition properties as they apply to subtraction. 8. Subtract numbers with minuend less than 100, with no renaming.	5. Subtract with minuends less than 100, with renaming. 6. Subtract with minuends less than 1,000, with one renaming. 7. Subtract like fractions, with no renaming.

	1.1	1.2	1.3
Multiplication		5. Become aware of beginning multiplication ideas.	
Division			
Geometry	8. Show topological relationships: proximity, enclosure, separateness, order.*	6. Recognize topologically equivalent shapes.	11. Identify simple plane figures.
Measurement	9. Compare sizes of objects.*	7. Arrange geometric figures by number of sides and size.	12. Measures linear distance using nonstandard and standard units. 13. Identify time on a clockface to the hour and half-hour. 14. Identify time on a calendar to the year, month, day.
Graphs and Functions			
Mathematical Applications		8. Recognize symmetry in nature and in own body.	15. Make simple binary judgments. 16. Compare the value of penny, nickel and dime.
Word Problems	10. Orally describe simple topological relationships.* 11. Solve simple word problems involving proximity, enclosure, separateness, order.	9. Orally describe simple addition situations. 10. Translate simple addition word problems to corresponding basic addition fact and solve.	17. Orally describe simple take-away subtraction situations. 18. Translate simple take-away word problems to corresponding basic subtraction fact and solve.

	2.1	2.2	2.3
Multiplication	8. Show multiplication situations related to many-to-one relationships.		
Division			
Geometry	9. Show rigid transformation topological relationships, e.g., slides, turns, flips. 10. Select situations showing reflexive relations, e.g., "the same as."		8. Identify line, line segment, parallel lines.
Measurement	11. Perform transformations involving conservation of numbers.	9. Use centimeter for linear measure. 10. Show so many minutes after an hour on a clockface.	9. Estimate measures including meter, decimeter, and centimeter.
Graphs and Functions	12. Perform transformations involving conservation of number and mass.	11. Read picture and bar graphs. 12. Collect and record data using picture and bar graphs.	
Mathematical Applications	13. Match month with holidays, e.g., Thanksgiving in November, Christmas in December. 14. Perform money-related tasks in everyday life situations.	13. Compare the values of a quarter and half-dollar.	10. Compute value of coins, with sums less than one dollar.
Word Problems	15. Translate simple addition word problems to corresponding algorithm and solve. 16. Solve simple word problems involving present tense, inflectional endings, and prepositions.	14. Translate simple additive take-away subtraction problems to number sentence and solve. 15. Solve comparative subtraction word problems, present tense.	11. Solve word problems involving addition and subtraction of fractions with like denominators, sums and minuends equal to or less than 1.

	3.1	3.2	3.3
Sets	1. Combine disjoint equivalent sets.	1. Remove all disjoint equivalent sets from a given set.	
Whole Numbers / Notation	2. Extend place value through thousands. 3. Write numerals in expanded notation.	2. Use the inequality sign in number relationships.	
Integers / Notation			1. Use inverse relationship between addition and subtraction. 2. Identify the additive inverse of a number. 3. Add inverses to obtain the additive identity (zero).
Rationals / Notation		3. Identify and compare equivalent fractions. 4. Extend identification of any fractional number.	4. Use inverse relationship between multiplication and division. 5. Identify multiplicative inverse as the reciprocal of a number. 6. Multiply reciprocals to obtain the identity element for multiplication.
Decimals / Notation			
Addition	4. Add with renaming.		
Subtraction	5. Subtract with zero in minuend, with renaming.		7. Perform comparison subtraction. 8. Perform additive subtraction.

	4.1	4.2	4.3
Sets			
Whole Numbers/ Notation	1. Extend place value through millions. 2. Round whole numbers.		
Integers/Notation			
Rationals/ Notation	3. Extend expanded notation to include fractions through thousandths.	1. Write fractional numbers equivalent to 1, from graphic representation.	1. Investigate graphic representation of multiplication of simple fractions using rectangular regions.
Decimals/Notation		2. Extend place value to thousandths.	2. Translate fractions to decimals.
Addition	4. Estimate the sum.		3. Add decimal fractions, with no renaming.
Subtraction	5. Estimate the difference.		4. Subtract decimal fractions, with no renaming.

	3.1	3.2	3.3
Multiplication	6. Show multiplication as the union of several equivalent disjoint sets. 7. Use multiplication properties. 8. To multiply with 1, 2, 3, 4, 5, 6, 7, 8, 9 as factors. 9. Compute basic multiplication facts.	5. Use the Distributive Property of Multiplication over Addition as related to the multiplication algorithm. 6. Multiply two-digit by one-digit numbers, with no renaming. 7. Extend multiplication of whole numbers, with no renaming. 8. Multiply two-digit by one-digit numbers, with renaming. 9. Extend multiplication of whole numbers, without and with renaming to two factors times two factors.	
Division		10. Become aware of the division operation. 11. Identify measurement and partitive division situations. 12. Compute basic division facts. 13. Divide two-digit by one-digit numbers, without remainder and with remainder, with no renaming in dividend. 14. Use division algorithm with one-digit divisor, renaming in dividend, with and without remainder.	
Geometry		15. Identify angles. 16. Extend identification of geometric shapes.	
Measurement	10. Extend use of the calendar. 11. Interpret written notation of time, e.g., 3:01, 3:45, etc. 12. Write time shown on a clockface.	17. Record measures of temperature. 18. Use grams and kilograms for measuring weight.	
Graphs and Functions	13. Identify inequality relations. 14. Performs transformations involving conservation of length, time, discontinuous quantity, continuous quantity (liquid).	19. Record weight measures and temperature on a graph. 20. Use a number line.	9. Extend use of a number line to set of integers.
Mathematical Applications	15. Compute value of money, with sums greater than one dollar. 16. Subtract monetary values. 17. State approximate money value of objects, e.g., crayons, movie ticket, postage stamp.		
Word Problems	18. Solve simple word problems involving multiplication. 19. Solve simple addition, subtraction, multiplication word problems involving past tense. 20. Solve addition and subtraction problems involving money.	21. Solve simple division problems involving present and past tense. 22. Select appropriate operation to solve a word problem.	

	4.1	4.2	4.3
Multiplication	6. Multiply a three-digit by a one-digit number, with and without renaming. 7. Multiply a three-digit by a two-digit number, with and without renaming. 8. Estimate the products.	3. Multiply three-digit by three-digit factors. 4. Extend multiplication involving any whole numbers using cross product approach.	5. Show use of indirect open sentence in finding equivalent fractions by multiplying. 6. Multiply a whole number times a fraction based on picture representation.
Division	9. Investigate properties of multiplication in regard to division. 10. Divide any whole number by a one-digit number.		7. Divide a three-digit number by a two-digit number with remainder. 8. Estimate quotient. 9. Show use of indirect open sentence in finding equivalent fractions by dividing.
Geometry		5. Identify three-dimensional geometric shapes.	10. Use notation for line and line segments.
Measurement	11. Use kilometer and millimeter as linear measures. 12. Find perimeter of polygons. 13. Find area of polygon regions.	6. Find volume. 7. Use the milliliter and liter for measurement of capacity.	11. Record temperature shown on a celsius thermometer.
Graphs and Functions	14. Select situations showing symmetric relations, e.g., "is as tall as." 15. Select situations showing transitive relations, e.g., "If John is taller than Bill, and Bill is taller than Glenn, then John is taller than Glenn." 16. Recognize conservation of volume under various transformations. 17. Interpret and use a grid.	8. Read circle graphs. 9. Extend picture and bar graphs.	
Mathematical Applications	18. Make inferences based upon judgments and estimates.	10. Record simple equivalent metric measures in daily activities, e.g., .7 m or .01 m.	12. Identify outcomes from a given set of possible outcomes.
Word Problems	19. Solve word problems involving future tense. 20. Solve word problems involving measures of polygons.	11. Solve word problems involving indirect open sentences. 12. Solve cross product type word problems.	13. Solve problems involving mixed tenses and conditional situations for whole numbers.

	5.1	5.2	5.3
Sets			
Whole Numbers / Notation			
Integers / Notation			
Rationals / Notation		1. Compare the least common multiple to the least common denominator. 2. Simplify a fraction before multiplying. 3. Find the least common denominator using least common multiple as the denominator, and the multiplicative identity property.	1. Find the simplest name for a fraction using the greatest common factor as the fractional multiplicative identity. 2. Find equivalent fractions, using the multiplicative identity property. 3. Find least common denominator using prime factors. 4. Rename improper fractions as mixed numbers. 5. Translate mixed number to equivalent improper fraction.
Decimals / Notation	1. Extend expanded notation to thousandths.		
Addition			6. Add fractions with unlike denominators, with sums equal to or less than one. 7. Add fractions, with renaming to mixed numbers. 8. Add mixed numbers.
Subtraction			9. Subtract fractions with unlike denominators, minuends equal to or less than one. 10. Subtract fractions from mixed numbers, with and without renaming in minuend. 11. Subtract mixed numbers with and without renaming. 12. Subtract a fraction from a whole number. 13. Subtract a mixed number from a whole number.

	6.1	6.2	6.3
Sets			
Whole Numbers / Notation			
Integers / Notation			
Rationals / Notation	1. Apply number properties to operations on fractions.		
Decimals / Notation	2. Round decimal numbers. 3. Order decimal numbers.	1. Identify patterns of decimal place value using mappings. 2. Complete mathematical sentences showing patterns related to decimal notation.	
Addition	4. Add decimals with renaming.	3. Add columns of decimal numbers with renaming.	1. Add a mixed decimal to a mixed decimal.
Subtraction	5. Subtract decimals with renaming.		2. Subtract a mixed decimal from a mixed decimal. 3. Subtract a decimal fraction from a whole number. 4. Subtract a mixed decimal fraction number from a whole number.

	5.1	5.2	5.3
Multiplication	2. Find the least common multiple. 3. Compare prime and composite numbers. 4. Identify prime numbers. 5. Multiply two common fractions, products less than 1.	3. Use factor trees to find prime factorization of a composite number. 4. Identify numbers that are relatively prime. 5. Find the greatest common factor.	
Division	6. Divide fractions using horizontal algorithm, no inversion of divisor. 7. Divide any whole number by a two-digit number.	6. Find the prime factors of a composite number using the division algorithm. 7. Use divisibility rules for finding prime factors.	
Geometry	8. Perform simple projective transformations. 9. Construct and investigate the Möbius strip. 10. Construct geometric figures.		
Measurement			
Graphs and Functions	1. Graph probabilities. 2. Form a matrix grid. 3. Construct a scattergram.		14. Interpret line graphs. 15. Interpret coordinates of a point on a grid.
Mathematical Applications			
Word Problems			16. Solve word problems involving fractions.

	6.1	6.2	6.3
Multiplication		4. Multiply decimal numbers.	5. Multiply a mixed decimal by a mixed decimal.
Division		5. Divide decimal by whole number. 6. Divide decimal by decimal.	6. Divide a mixed decimal by a mixed decimal.
Geometry		7. Identify types of triangles.	
Measurement	6. Measure angles of polygons.	8. Find area of triangular region. 9. Find area of parallelogram region. 10. Find area of circular region.	
Graphs and Functions			
Mathematical Applications	7. Construct time schedules, e.g., TV, homework. 8. Interpret public transportation schedules. 9. Explain daylight savings time.	11. Explain meaning of time and a half and double time. 12. Compute time across time zones. 13. Interpret sizes of clothing. 14. Relate legend on map to estimate distances.	7. Relate decimals to percent. 8. Find percent, rate, and base.
Word Problems	10. Solve word problems using decimal numbers.		9. Solve percent problems.

	7.1	7.2	7.3
Sets	1. Use set notation.		
Whole Numbers/ Notation		1. Simplify expressions involving order of operations. 2. Use nonnegative exponential notation.	
Integers/Notation			
Rationals/ Notation			
Decimals/Notation			
Addition			
Subtraction			

	8.1	8.2	8.3
Sets	1. Recognize difference between equal and equivalent sets.		
Whole Numbers / Notation			
Integers / Notation		1. Apply whole number properties to integers. 2. Simplify expressions of integers (order of operations).	
Rationals / Notation			
Decimals / Notation			
Addition	2. Add integers.		
Subtraction	3. Subtract integers.		

	7.1	7.2	7.3
Multiplication		3. Multiply using nonnegative exponents.	
Division			
Geometry	2. Use set notation with nonmetric geometry.		1. Tell if a network is traversable using Euler's procedure. 2. Find relationships between the number of spaces, vertices, and arcs in any network.
Measurement	3. Compute time.	4. Find equivalent standard measures. 5. Use formula to compute perimeter, circumference, area, volume, and distance.	
Graphs and Functions			3. Supply the missing element of ordered pairs. 4. Find a formula to describe a set of ordered pairs.
Mathematical Applications	4. Determine probability. 5. Find the mean, median, range, and mode.	6. Complete patterns of series. 7. Compute sales tax. 8. Interpret bank statement.	5. Solve simple one variable equations. 6. Solve simple geometric problems.
Word Problems	6. Solve word problems involving probability.	9. Solve problems using formulas.	7. Select a word problem appropriate to a given mathematical expression.

	8.1	8.2	8.3
Multiplication	4. Multiply integers.		
Division	5. Divide integers.		
Geometry			
Measurement	6. Compute temperature.		
Graphs and Functions	7. Graph ordered pairs.		
Mathematical Applications		3. Select appropriate units of measure for real life situations. 4. Solve word problems involving simple analogies.	1. Display understanding of various taxes, e.g., state and federal, social security, gas. 2. State pros and cons of installment buying. 3. Compute simple and compound interest. 4. Balance a checkbook.
Word Problems			5. Solve word problems involving consumer mathematics.

Index